Springer Series in
OPTICAL SCIENCES 95

Founded by H.K.V. Lotsch

Springer
New York
Berlin
Heidelberg
Hong Kong
London
Milan
Paris
Tokyo

Springer Series in
OPTICAL SCIENCES

New editions of volumes prior to volume 70

1 **Solid-State Laser Engineering**
By W. Koechner, 5th revised and updated ed. 1999, 472 figs., 55 tabs., XII, 746 pages
14 **Laser Crystals**
Their Physics and Properties
By A. A. Kaminskii, 2nd ed. 1990, 89 figs., 56 tabs., XVI, 456 pages
15 **X-Ray Spectroscopy**
An Introduction
By B. K. Agarwal, 2nd ed. 1991, 239 figs., XV, 419 pages
36 **Transmission Electron Microscopy**
Physics of Image Formation and Microanalysis
By L. Reimer, 4th ed. 1997, 273 figs. XVI, 584 pages
45 **Scanning Electron Microscopy**
Physics of Image Formation and Microanalysis
By L. Reimer, 2nd completely revised and updated ed. 1998, 260 figs., XIV, 527 pages

Published titles since volume 70

70 **Electron Holography**
By A. Tonomura, 2nd, enlarged ed. 1999, 127 figs., XII, 162 pages
71 **Energy-Filtering Transmission Electron Microscopy**
By L. Reimer (Ed.), 1995, 199 figs., XIV, 424 pages
72 **Nonlinear Optical Effects and Materials**
By P. Günter (Ed.), 2000, 174 figs., 43 tabs., XIV, 540 pages
73 **Evanescent Waves**
From Newtonian Optics to Atomic Optics
By F. de Fornel, 2001, 277 figs., XVIII, 268 pages
74 **International Trends in Optics and Photonics**
ICO IV
By T. Asakura (Ed.), 1999, 190 figs., 14 tabs., XX, 426 pages
75 **Advanced Optical Imaging Theory**
By M. Gu, 2000, 93 figs., XII, 214 pages
76 **Holographic Data Storage**
By H.J. Coufal, D. Psaltis, G.T. Sincerbox (Eds.), 2000 228 figs., 64 in color, 12 tabs., XXVI, 486 pages
77 **Solid-State Lasers for Materials Processing**
Fundamental Relations and Technical Realizations
By R. Iffländer, 2001, 230 figs., 73 tabs., XVIII, 350 pages
78 **Holography**
The First 50 Years
By J.-M. Fournier (Ed.), 2001, 266 figs., XII, 460 pages
79 **Mathematical Methods of Quantum Optics**
By R.R. Puri, 2001, 13 figs., XIV, 285 pages
80 **Optical Properties of Photonic Crystals**
By K. Sakoda, 2001, 95 figs., 28 tabs., XII, 223 pages
81 **Photonic Analog-to-Digital Conversion**
By B.L. Shoop, 2001, 259 figs., 11 tabs., XIV, 330 pages
82 **Spatial Solitons**
By S. Trillo, W.E. Torruellas (Eds), 2001, 194 figs., 7 tabs., XX, 454 pages
83 **Nonimaging Fresnel Lenses**
Design and Performance of Solar Concentrators
By R. Leutz, A. Suzuki, 2001, 139 figs., 44 tabs., XII, 272 pages
84 **Nano-Optics**
By S. Kawata, M. Ohtsu, M. Irie (Eds.), 2002, 258 figs., 2 tabs., XVI, 321 pages
85 **Sensing with Terahertz Radiation**
By D. Mittleman (Ed.), 2003, 207 figs., 14 tabs., XVI, 337 pages

Ferenc Krausz Georg Korn Paul Corkum
Ian A. Walmsley
Editors

Ultrafast Optics IV

Selected Contributions to the 4th International Conference on Ultrafast Optics, Vienna, Austria

With 281 Figures

Springer

Ferenc Krausz
Photonics Institute
Vienna Institute of Technology
Gusshausstr. 27/387
1040 Vienna
Austria
ferenc.krausz@tuwien.ac.at

Georg Korn
Katana Technologies GmbH
Medizintechnik
Albert-Einstein-Ring 7
14532 Kleinmachnow
Germany
g.korn@katanalaser.com

Paul Corkum
Steacie Institute for Molecular Sciences
National Research Council of Canada
100 Sussex Drive, Room 3005
Ottawa, Ontario K1A 0R6
Canada
Paul.Corkum@nrc.ca

Ian A. Walmsley
The Institute of Optics
University of Rochester
Rochester, NY 14627
USA
walmsley@optics.rochester.edu

Library of Congress Cataloging-in-Publication Data
Ultrafast optics IV : selected contributions to the 4th International Conference on Ultrafast Optics, Vienna, Austria / Ferenc Krausz ... [et al.].
p. cm. — (Springer series in optical sciences ; v. 95)
Includes bibliographical references and index.
ISBN 0-387-40091-5 (acid-free paper)
1. Laser pulses, Ultrashort—Congresses. 2. Picosecond pulses—Industrial applications—Congresses. I. Krausz, Ferenc. II. International Conference on Ultrafast Optics (4th : 2003 : Vienna, Austria) III. Series.

QC689.5.L37U466 2003
621.36′6—dc22 2003070770

ISBN 0-387-40091-5 Printed on acid-free paper.

Printed in the United States of America. (SBA)

9 8 7 6 5 4 3 2 1 SPIN 10928882

Springer-Verlag is a part of *Springer Science+Business Media*

springeronline.com

Preface

The papers in this volume cover the major areas of research activity in the field of ultrafast optics at the present time, and they have been selected to provide an overview of the current state of the art. The purview of the field is the methods for the generation, amplification, and characterization of electromagnetic pulses with durations from the pico- to the attosecond range, as well as the technical issues surrounding the application of these pulses in physics, chemistry, and biology. The contributions were solicited from the participants in the Ultrafast Optics IV Conference, held in Vienna, Austria, in June 2003. The purpose of the conference is similar to that of this book: to provide a forum for the latest advances in ultrafast optical technology.

Ultrafast light sources provide a means to observe and manipulate events on the scale of atomic and molecular dynamics. This is possible either through appropriate shaping of the time-dependent electric field, or through the application of fields whose strength is comparable to the binding forces of the electrons in atoms and molecules. Recent advances discussed here include the generation of pulses shorter than two optical cycles, and the ability to measure and to shape them in all degrees of freedom with unprecedented precision, and to amplify them to the Zettawatt/cm^2 (10^{21} W/cm^2) range. This, in turn, allows the extension into new wavelength ranges, and thus makes possible the generation of attosecond pulses in the XUV spectral region, as well as their application in atomic physics and surface science. Other technologies are also arising, such as ultrafast electron diffraction and X-ray pulse generation, with applications in chemistry and molecular biology.

We are greatly indebted to Evgeni Sorokin, of the Vienna University of Technology, and Margaret Mitchell of Springer-Verlag, whose sterling work in organizing and editing the contributions to this book have made it an authentic record of the current state of ultrafast optics. And, of course, thanks to all authors for allowing this book to cover the most exciting activities at the frontiers of ultrafast science.

Vienna, Austria — *Ferenc Krausz*
Berlin, Germany — *Georg Korn*
Ottawa, Canada — *Paul Corkum*
Oxford, UK — *Ian Walmsley*

September 2003

Contents

Part III Measurement and Control of the Carrier-Envelope Phase

Part IV Extreme UV Radiation and Sub-Femtosecond Pulses

Part VI Ultrafast Nonlinear Optics

Part VII Novel Applications of Ultrashort Light Pulses

Part I

Generation of Femtosecond Laser Pulses

Compact Femtosecond Oscillators

W. Sibbett, B. Agate, C. T. A. Brown, A. A. Lagatsky, C. G. Leburn, B. Stormont, and E. U. Rafailov

School of Physics & Astronomy, University of St Andrews, North Haugh, St Andrews, Fife, KY16 9SS, Scotland

Summary. Configurations are described for compact femtosecond Cr:LiSAF laser oscillators that operate with low pump-power thresholds. By way of illustration of resonator simplification, a progression of low-loss cavity designs from dual-prism through single-prism to prismless schemes is reviewed. A complete femtosecond laser system is described including the laser diode pump system and drivers that has a footprint as small as 22 cm $\times$ 28 cm, and a total electrical drive power of less than 1W! Examples of efficient and rugged femtosecond Yb:tungstate and Cr:YAG lasers are also described together with a brief outline of a demonstration in optical datacomms at rates above 1Tb/s for a scheme incorporating the latter laser.

1 Introduction

The development of compact femtosecond lasers that are compatible with applications in photobiology/medicine, microscopy and imaging has been an aspiration of researchers for some time. Portable, robust and low-cost system options are sought especially for diagnostic measurements that could be made under field conditions. Our recent work at the University of St Andrews has enabled us to demonstrate a range of diode-pumped, modelocked Cr:LiSAF laser oscillators that, in our opinion, offer at least some of the desired characteristics of compactness with efficiency and practicality. Moreover, with this as our archetypal system we have proceeded to show that other material systems such as Yb-doped tungstates and Cr-doped YAG crystals also lend themselves to meeting attractive specifications on efficiency and resonator simplification. A review of key achievements within this recent programme of work constitutes the content of this invited paper.

The modelocking of a Cr:LiSAF laser was first reported by Miller and co-workers in 1992 [1], but this laser was not diode-pumped. However, in the following year, French *et al.* [2] described the modelocking of a diode-pumped Cr:LiSAF laser. Both of these lasers were configured to facilitate Kerr-lens modelocking (KLM). The first demonstration of semiconductor saturable absorber mirror (SESAM) mediated modelocking [3-5], that affords greater tolerance through increased flexibility in the resonator design came in 1994 through the research of Kopf *et al.* [6]. Indeed, more recent demonstrations of directly diode-pumped Cr:LiSAF and Cr:LiSGaF lasers, that produce output average powers in the range of 125mW [7-9] to 500mW [10], and pulse durations as short as 10fs [11] (KLM) and 45fs [12] ("soliton modelocking"), represent a significant advance in the progression towards more practical and robust 'real world' ultrashort-pulse lasers. We describe here how a significant reduction in size and complexity in low power, femtosecond laser oscillators can be achieved through the implementation of a range of novel resonator designs.

The conditions required to achieve a low pump threshold and access to optical nonlinearity have been identified by Alfrey [13] and Piché [14] respectively. In addition to the obvious requirement for low losses, the primary objective is that of coincident tight foci of pump and laser modes on the front face of the gain (Kerr) medium, which ensures the required strong nonlinear lensing for successful low-power modelocking. The cw threshold is further reduced if a gain medium is used which exhibits a high pump absorption coefficient and $\sigma\tau$ product. Piché's work [14] also confirmed that the exploitation of optical nonlinearity favours small beam spot sizes. Fortunately, advances in the development of single transverse mode semiconductor diode lasers have made available an attractive range of pump diodes that are fully consistent with these exacting requirements.

With resonator compactness in mind for the femtosecond oscillators, there are a number of approaches that can be used to reduce the footprint of the overall laser configuration. The first and most obvious method is to decrease the overall dimensions of the laser cavity by using simplified cavity designs that involve fewer optical components. For example, 3-mirror Ti:sapphire resonators have been developed by Ramaswamy-Paye, Bouma and Fujimoto [15, 16] and subsequently demonstrated in both Cr^{4+}:YAG [17, 18] and Cr:LiSAF lasers [19]. Such laser cavities have inherent size and mechanical stability advantages and permit simpler alignment procedures for Kerr-lens modelocking [15, 20-22]. The use of asymmetric cavity arrangements for the optimisation of ultrashort-pulse lasers for direct diode-pumping was suggested [23] and demonstrated later [24]. These lasers retain the superior alignment stability properties of the standard 4-mirror, 7-element cavities, while focusing the laser mode more tightly in the gain medium and reducing the overall cavity length. Further simplification of the pump arrangement was pioneered in our research group where the lower power, diffraction-limited, narrow-stripe, single-spatial-mode AlGaInP laser diodes [19, 25-27] could be used.

A major limitation to reducing cavity lengths has been the minimum separation of the prism pair that is required to provide compensation for the group velocity dispersion. Alternative dispersion-control schemes that offer superior features of compactness for femtosecond laser configurations have been demonstrated [15-17, 28, 29]. Recently, a prismless femtosecond Ti:sapphire laser was reported [30] with a repetition frequency of 2GHz, using a ring cavity with a perimeter of less than 150mm.

In this review we describe a number of very compact laser oscillators that are pumped using narrow-stripe laser diodes. First, a femtosecond Cr:LiSAF oscillator having a single dispersion-compensating prism in each arm of its resonator is described. Second, on the basis of experimental observations and related dispersion analyses, the design and demonstration of an ultra-compact and simplified femtosecond laser is reported. This laser has only a single prism for dispersion compensation and its footprint, including the pump source and drivers, is only 22cm×28cm. This is then followed by a description of a prismless cavity option that involves a mirror design that offers integral group delay dispersion compensation [27]. Extracavity frequency-doubling of the output of a broadly tunable version of our compact Cr:LiSAF laser is then shown to be a relatively simple and efficient means of producing femtosecond pulses in the blue spectral region. As emphasised in the text, it is interesting to follow the improving trend in efficiency (both optical-to-optical and electrical-to-optical) for these simplified Cr:LiSAF lasers where some can be operated at total drive powers of less than 1W!

From this foundation we proceed to describe a very efficient femtosecond Yb:KYW laser operating around 1050 nm and a robust Kerr-lens mode-locked Cr:YAG laser that produces a sequence of femtosecond pulses in the principal telecomms window around 1500nm. Results are also included for a datacomms application involving spectral slicing to demonstrate that this KLM-based Cr:YAG laser is an appropriate digital optical source for data rates above 1Tb/s. Some commentary is also included as an introduction to the possibilities that arise from the prospects of using quantum-dot semiconductors as broad-bandwidth gain media and further miniaturisation through photonic bandgap structures in semiconductor devices.

2 Low-threshold Femtosecond Lasers

The basic conditions for low pump thresholds were readily satisfied with the favourable emission-cross-section- upper-state-lifetime product ($\sigma\tau$) in the Cr^{3+}:LiSrAlF_6 (Cr:LiSAF) laser crystal and the use of narrow-stripe pump AlGaInP laser diodes having near-diffraction-limited output beams [19, 25, 27]. This facilitated a low cw laser threshold [13] and figure 1 shows the basic cavity configuration reported earlier [25]. In this case the Cr:LiSAF crystal was a 7mm Brewster/Brewster cut crystal with 1.5% Cr-doping (pump absorption coefficient, $\alpha_p \sim 9\text{cm}^{-1}$). The high degree of asymmetry in folding angles around the Brewster-angle cut gain medium is a necessary condition for the lowest possible cw threshold [31]. The large angle between the long cavity arm and the folding section entirely compensates for the astigmatism introduced by the laser crystal. By arranging the angle between the folding section and the short cavity arm to be small, the cavity mode may then be focused tightly in both the sagittal and tangential planes and made to coincide with the focused pump beam at the crystal facet. The asymmetry in cavity arm lengths permits the focused spot size on the modelocking element, a SESAM in this case [32, 33]) to be varied easily by adjusting the length of the short cavity arm [34]. This aspect of the design is especially important in lasers operating in a low-power regime because the saturation of the SESAM must be optimised for a range of modest intracavity powers. The SESAM used with this Cr:LiSAF laser was a low-loss antiresonant Fabry-Perot saturable absorber that is compatible with low intracavity powers [10, 12]. The incorporation of the SESAM assists significantly in the initiation and stabilisation of the mode locking process at our low intracavity powers.

Initially, two AlGaInP laser diodes were polarisation coupled to provide up to 90mW of incident pump power. The beams from these diodes were collimated using standard ×20 anti-reflection-coated microscope objectives but no further reshaping of the 3:1 elliptical mode was employed. When a small battery pack containing six 1.5V AA-type penlight batteries and a low dropout voltage regulator circuit powered the diodes, the laser operated for more than 14 hours.

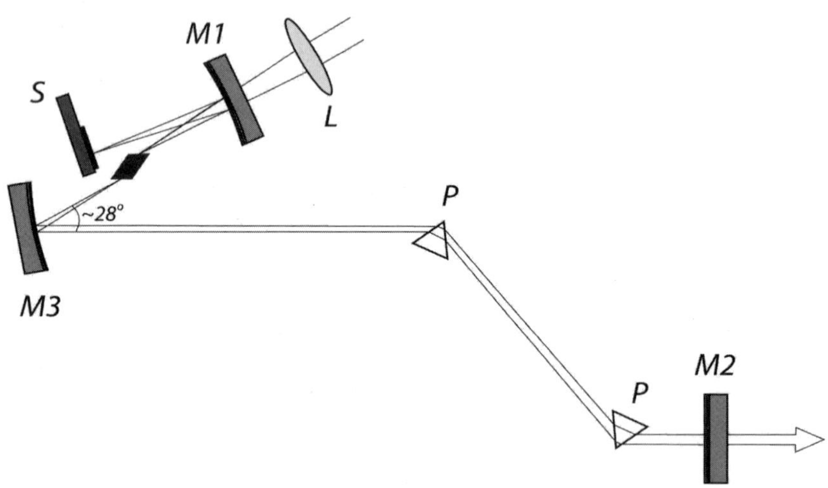

Fig. 1. Schematic of the archetypal resonator configuration for a femtosecond laser. S: SESAM; M2: plane mirror; M1: 50mm ROC mirror; M3: 75mm ROC mirror; P: LaK31 prisms; L: 50mm pump focusing lens.

With a 1.5% output coupling mirror the cw threshold was less than 20mW and the threshold power for modelocking was 22mW. This laser produced 57fs pulses at an average output power of 6.5mW for an incident pump power of 72mW. The pulses were bandwidth-limited, having a spectral width of 13.5nm centred on 855nm. It is also notable that this laser exhibited very low noise characteristics, viz, just 73fs of rms timing jitter in the 500-5000Hz range [25].

3 Compact Cavities with a Pair of Prisms for Dispersion Compensation

In the laser configuration of figure 1 the necessary separation of the dispersion compensating prisms constitutes a large proportion of the cavity "real estate". Although much more compact dispersion compensation schemes involving elements such as chirped mirrors [28] and dispersive SESAMs [29] exist for higher power lasers, their insertion losses are still too high for our low-threshold conditions.

Aoshima and co-workers [35] showed that a standard prism pair could be incorporated in a more compact, symmetric, self-mode locked Z-fold cavity by placing one prism in each of the cavity arms. A schematic of this cavity is shown in figure 2a where it can been seen that the necessary distance between the folding mirrors for tight focusing, and the required inter-prism path lengths, could be combined very conveniently. This permits a considerable reduction in the footprint of an ultrashort-pulse laser.

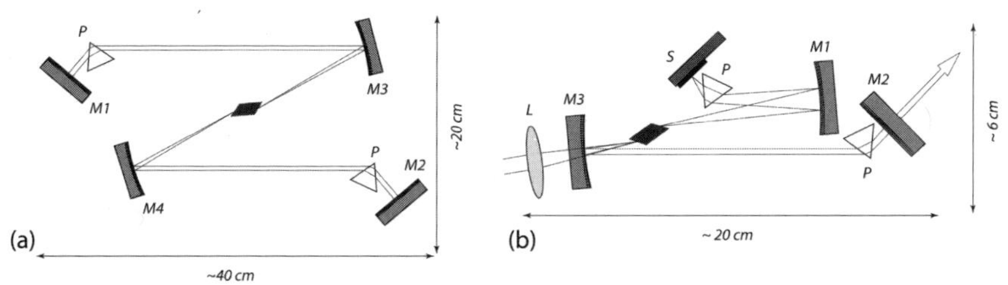

Fig. 2. (a) Schematic of the 4-mirror symmetric "Aoshima-type" Cr:LiSAF cavity described in [35]. (b) Adapted compact resonator design with compatibility to low-threshold operation. S: SESAM.

The separation of the prisms was determined on the simplified assumption that they would act in the same way as a prism-pair in a single arm of the cavity. There was a significant observed increase in pulse duration from 30fs to 90fs and this was attributed to a spatially dispersed beam in the Cr:LiSAF crystal which imposed a bandwidth limiting effect [35]. Despite the limitations in this reported performance we decided to adapt this type of arrangement for our low-threshold laser arrangement and figure 2b shows a schematic of the highly asymmetric four-mirror Z-cavity with a prism in each arm. With this design, the entire cavity optics were accommodated within a 20cm×6cm footprint. The cw threshold of the laser was just 20mW with the p-polarised diode laser and with both diodes supplying a total of 76mW at the pump lens, the Cr:LiSAF laser produced an average power of 1.5mW with a highly reflecting plane end-mirror. This laser generated stable, bandwidth-limited 115fs pulses at a repetition frequency of 407MHz.

Initially it was assumed that the prisms should be oriented with their apices oriented in the same direction, because the standard description of the operation of a prism pair would imply a reversal of the ray direction through the cavity focus. Experimentally, however, we found that this simple ray picture did not describe accurately the behaviour of the resonator. The prism in the longer arm was then 'flipped' so that its apex pointed in the opposite direction to the prism in the short arm. In this configuration, the laser produced 100fs pulses at a repetition frequency of 353MHz where this longer cavity length was necessary to achieve femtosecond pulse operation. The time-bandwidth product of the pulses was 0.43 confirming a frequency-chirped output. The average output power was 1.1mW for a pump power of 76mW and the modelocking threshold was 61mW. This behaviour was not predicted by a simple ray picture but a more detailed analysis of the Aoshima-type cavities provided a better explanation of the role of the prisms and their contributions towards dispersion in this design of resonator [26].

4 A Highly Compact Femtosecond Laser with a Single Prism for Dispersion Compensation

A conclusion from our detailed modelling of the highly asymmetric cavities [26] was that the prism in the short arm did not contribute significantly to the intracavity negative dispersion because of its short distance from the folding mirror. This implied that this prism may not be necessary to achieve an acceptable amount of negative dispersion and so we predicted that a single prism in the longer arm might suffice for femtosecond-pulse operation. The removal of the prism from the tightly focused short arm simplified considerably the laser cavity and removed the alignment and optimisation requirements associated with this prism.

The laser cavity was set up as in figure 3, where the LaK31 prism used previously in the short arm was removed and the fold angle of the 75mm ROC mirror set to 11°. The distance from this folding mirror to the Cr:LiSAF crystal was 95mm and the length of the short arm was 60mm. The cw threshold was 26mW with a 0.75% output-coupling mirror centred at 856nm. When mode locked this laser produced 160fs pulses with an average output power of 6.1mW at a pulse repetition frequency of 435MHz.

To reconfigure the cavity within a yet smaller footprint, the LaK31 prism was replaced with an SF10 glass prism that exhibited a larger dispersion. This prism was placed 100mm from the 50mm ROC pump folding mirror with a space after the prism of 75mm. In this configuration the calculated GVD was -368fs^2, but the laser produced 330fs pulses at a repetition frequency of 400MHz. The footprint of the laser cavity including a compact version of the laser diode pump system and a battery pack was reduced to 15cm×6cm.

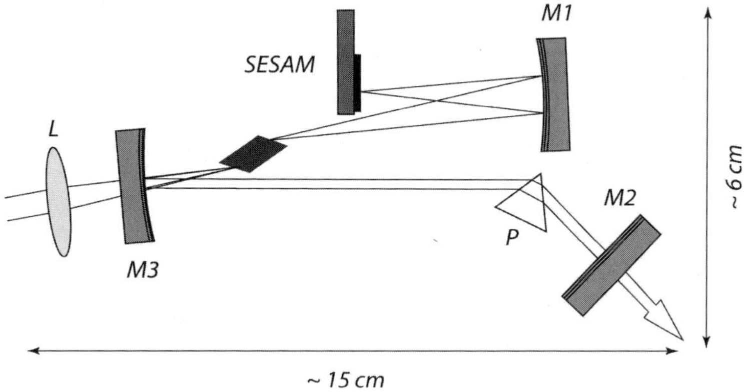

Fig. 3. Femtosecond Cr:LiSAF laser with single intracavity prism. L: 50mm pump focus lens; M1: 75mm ROC folding mirror; M3: 50mm ROC folding mirror; M2: 0.75% output-coupling mirror.

5 Performance Enhancement with an Improved Pump Module

The major residual factor that limited the performance of our compact, ultrafast Cr:LiSAF lasers was the available pump power. The high beam quality of narrow-stripe diodes plays a vital part in achieving low thresholds and high efficiencies, such that switching to the more expensive, higher power, broad area diodes was not an option. Using single spatial mode diodes, which are at present less than $30 each, reduces substantially the cost of this type of laser system.

Currently, the output power of commercially available narrow-stripe diodes in the red spectral region is limited to 50-60mW. Thus the losses in coupling the pump beam into the laser crystal must be minimised and a high degree of overlap between the pump and laser modes must be provided. The high ellipticity of most laser diode sources exacerbates the latter problem. In the case of high power pump diodes, this problem has been elegantly ameliorated using an elliptical resonator mode [10]. However, the excellent beam quality of our low power diode lasers means that the simpler solution of circularising the pump mode can be utilised and it is this approach that we have used in our work. Standard commercial diodes were custom microlensed using a single cylindrical lens inside the existing diode can.

To increase the available pump power, the lasers described in previous sections used two polarisation-coupled pump diodes but this is a rather inefficient method of beam combining. With the optics available to us, we achieved a coupling efficiency of 91% of the power emitted by the two diodes. Inevitably, the resulting beam is of mixed p and s polarisations. This has two inherent disadvantages. First, the need to couple into a Brewster-cut gain crystal means there is a loss of approximately 10% on the s-polarised component of the pump beam. Second, the s-component experiences a lower absorption coefficient in the Cr:LiSAF crystal, thereby reducing the slope efficiency and raising the threshold of the laser [13].

The broad absorption bandwidth of Cr:LiSAF in the red region of the spectrum means that these problems can be avoided if wavelength combining [36] is used. We used a dielectric mirror to combine the beam from a 660nm, 50mW diode laser with that from a 690nm, 60mW diode laser. In this way we were able to achieve a coupling efficiency of 98% and 96% absorption of the launched pump power in a 3mm, 5.5 atm% Cr:LiSAF crystal. By contrast, with polarisation coupling, only ~86% of the launched power is absorbed in the same crystal.

This pump module was used with a laser cavity similar to that described in figure 3, but with the 7mm 1.5 atm% Cr:LiSAF crystal replaced by a 3mm 5.5 atm% Cr:LiSAF crystal. A mirror set with marginally higher pump transmission was also used. With this arrangement modelocked output powers were generated at greater than 20 mW in transform-limited 150fs pulses representing an efficiency of 19% with respect to launched pump power and the electrical-to-optical efficiency was raised to almost 5%. At slightly lower average powers (15mW) transform-limited pulses as short as 113fs have been generated [26].

6 Low-threshold Prismless Femtosecond Cr:LiSAF Lasers

In our prismless configuration [27], the resonator included only the 3mm (5.5 atm% doped) Cr:LiSAF crystal, two dichroic high-reflectors, a SESAM and an output coupler as illustrated in figure 4. When pumped with two wavelength-combined AlGaInP diodes (660nm and 685nm), this laser produced transform-limited 136fs pulses centred at 859nm at a repetition-rate of 470MHz. Average output powers of up to 20mW were achieved for 95mW of incident pump power, giving an optical-to-optical efficiency of 21%. With the spectral bandwidth of 6nm this implied a time-bandwidth product of 0.33.

Interestingly, in this case the intracavity dispersion compensation was provided by the dichroic folding mirrors ('HR' in figure 4), which were found to exhibit negative group delay dispersion (GDD) in the spectral region of laser operation. Although not a specified property of the dielectric coating, this identified feature facilitated a more integrated and compact methodology for dispersion compensation and this has some similarity with the SESAM-based method reported by Kopf *et al.* [29]. For our particular mirrors there was adequate negative GDD across sufficient bandwidth to enable transform-limited femtosecond pulses to be established. It was therefore not necessary

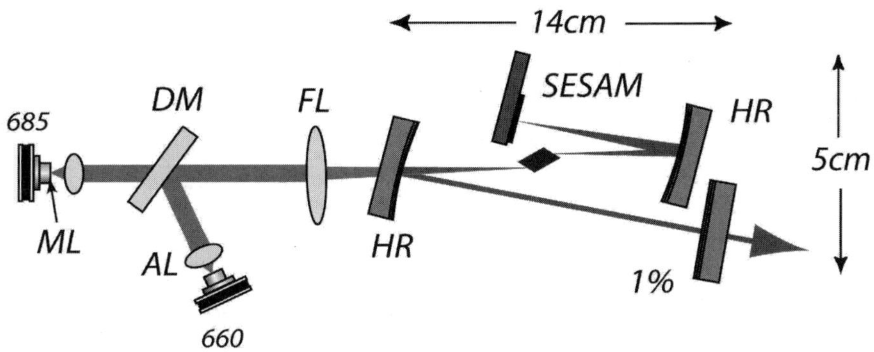

Fig. 4. A diode-pumped, compact femtosecond laser with integrated dispersion compensation [27]. FL: 38mm lens; ML: circularising microlens; DM: dichroic mirror; AL: aspheric collimating lens.

to tune the laser to a specific wavelength to exploit this inherent negative GDD characteristic

We performed a dispersion measurement on our HR mirrors using white-light interferometry [37], and found that the key feature of the optical coating was a periodic drop in group delay (GD). Fortunately, this extended over a broad spectral band that had good overlap with the gain spectrum of the Cr:LiSAF crystal. From the GDD-versus-wavelength characteristic, reproduced in figure 5, it can be seen that the dispersion curve of the HR mirrors is similar to that of a GTI type of mirror structure [38] which affords suitable features for dispersion compensation. Our investigations to date suggest that the observed group-delay profile is consistent with a series of high-finesse GTI-type structures.

To take full advantage of this cavity simplification and associated reduction in its footprint we constructed a version of the laser using small-scale optomechanical components. This miniature femtosecond laser, shown in figure 6, is powered by just six penlight (AA) batteries (the electrical drive power is less than 1W) and is therefore portable. The battery pack enabled modelocked operation to be achieved for more than 12 hours. The entire laser system, including the pump stage, power source and drive electronics, fits on a 22cm×28cm breadboard, and operates with an electrical-to-optical efficiency of approximately 4%. Initial results have demonstrated pulses of 130fs, at an average output power of 14mW and a time-bandwidth product of 0.42.

We have also demonstrated the flexibility of this compact, low-threshold laser system by pumping it with two pairs of diode lasers (total pump power ~180mW) to achieve kilowatt pulse peak powers [27] as well as gigahertz repetition-rates [39]. Specifically, we have generated 122fs pulses at average powers of 35mW – an optical-to-optical efficiency of 20%. This corresponded

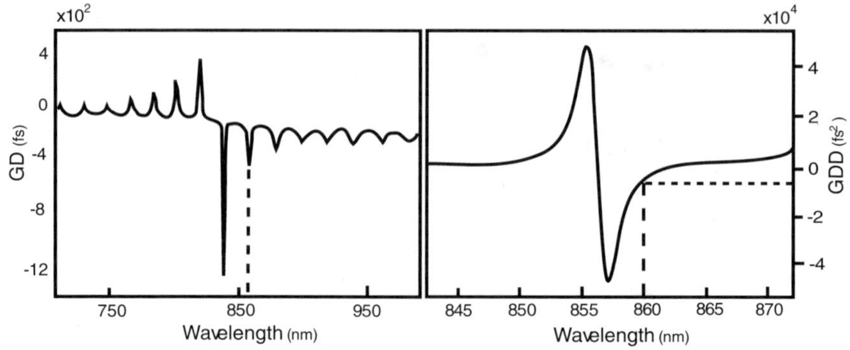

Fig. 5. Group delay (GD) and group delay dispersion (GDD) versus wavelength for the HR mirrors. The location of the central lasing wavelength is shown by a dashed line.

to peak powers of 1.2kW and a time-bandwidth product of 0.3. The 1.002GHz pulse-repetition-rate system, which is to our knowledge the highest reported for a directly diode-pumped, non-harmonically modelocked bulk femtosecond laser, produced transform-limited 146fs pulses at an average output power of 3mW. Both the kW and GHz lasers utilise very simple cavity geometries and again demonstrate the effectiveness of a compact and integrated approach to dispersion compensation. Indeed, because of the short cavity length (150mm) and low round-trip losses, the GHz cavity would be very difficult to implement using less integrated methods of dispersion compensation.

7 Portable Ultrafast Blue-light Sources

By incorporating a single prism into the long cavity arm of the kW-laser configuration [27], we have been able to demonstrate broad tunability (827nm – 877nm) from such a femtosecond Cr:LiSAF laser that maintains the low operational thresholds, high efficiency and overall design simplicity. With such a tunable near-infrared source we have utilised a straightforward extracavity frequency-conversion scheme to efficiently produce femtosecond pulses in the blue spectral region [40-43]. Both birefringent and quasi-phasematching techniques at room temperature have been utilised in bulk and waveguide doubling crystals respectively, to achieve second harmonic generation (SHG) in attractively practical, single-pass arrangements (figure 7).

500fs pulses of blue light, with an average power of 12mW have been generated from a 3mm bulk $KNbO_3$ (potassium niobate) crystal, with only 45mW of fundamental and a corresponding optical-to-optical SHG efficiency as high as 30%. Because the tunable fundamental source laser was pumped with single narrow-stripe laser diodes, the required electrical drive power was

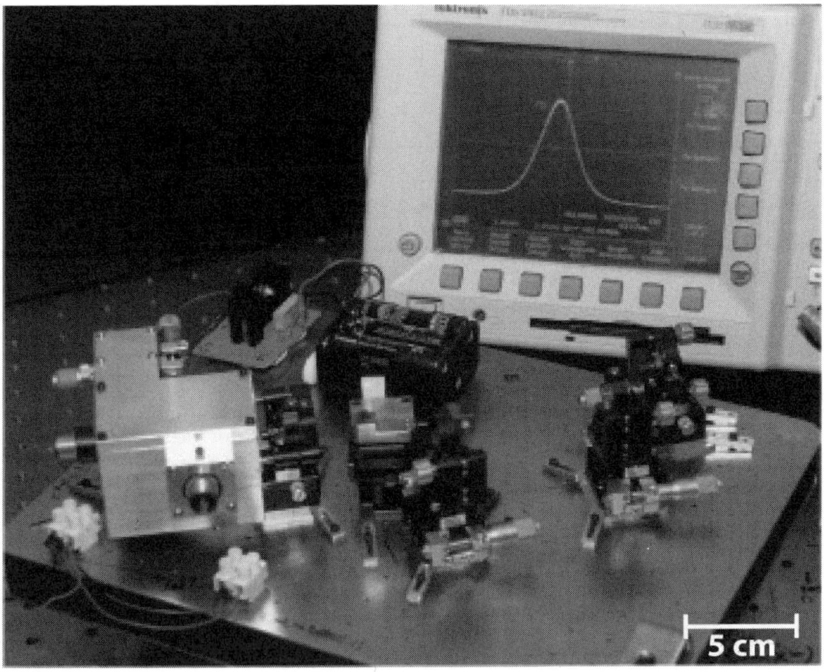

Fig. 6. Photograph of a highly efficient, self-contained, compact femtosecond laser [27].

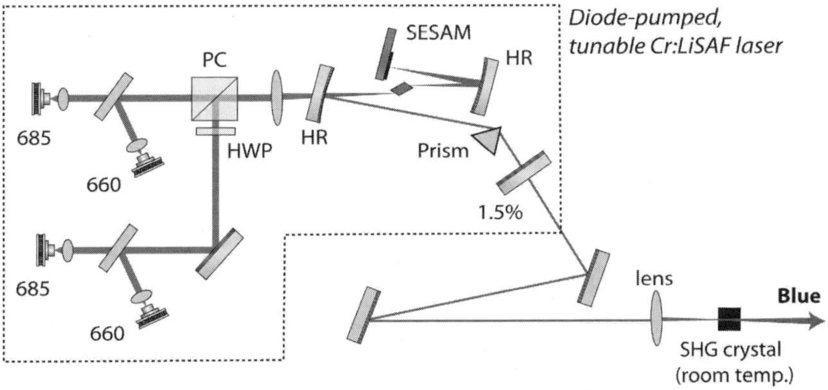

Fig. 7. Extracavity frequency-doubling of a compact femtosecond Cr:LiSAF laser [40].

less than 1.2W, resulting in an overall electrical-to-blue conversion efficiency of 1% – the highest reported efficiency of any femtosecond blue source to date [40]. The high beam quality of the blue light was observed in an M^2 measurement of 1.8. Complementary results using a periodically-poled and waveguided KTP crystal provided 5.6mW of blue light from only 27mW of fundamental – with an internal SHG conversion efficiency of up to 37%, and an overall electrical-to-blue conversion of 0.5% [41-43]. This miniaturised, portable Cr:LiSAF laser of figure 6, combined with the simple and efficient SHG scheme described here, thus represents a practical and effective route to designing a portable femtosecond blue light source that is rugged, compact and battery-powered.

We are currently investigating the means by which a related frequency conversion scheme could be used to access the ultraviolet spectral region via the third harmonic of a Cr:LiSAF laser. A suitable periodically-poled structure of $LiTaO_3$ has been designed for this purpose. This portable femtosecond source operating in the UVB spectral region is being developed for planned photo-biological studies within our research group.

8 High-efficiency Femtosecond Yb:KYW Laser

Yb-doped materials have attracted considerable interest during past few years in the development of efficient, simple and robust diode-pumped femtosecond lasers in the 1μm spectral region [44-48]. It is convenient that the strong absorption bands of these laser materials can be pumped by commercially available 980nm InGaAs laser diodes and the small Stokes shift (~600cm^{-1}) between absorption and emission reduces the thermal loading and increases the inherent laser efficiency.

The Yb-doped double tungstates of $Yb^{3+}:KY(WO_4)_2$ (Yb:KYW) and $Yb^{3+}:KGd(WO_4)_2$ (Yb:KGdW) possess similar spectroscopic properties [49, 50] and have the best lasing characteristics among the Yb-doped materials [51, 52]. A remarkable feature of both crystals is that they exhibit relatively broad emission bandwidths, which could support sub-100fs pulses. A particular advantage of the KYW host reported here is that it offers a facility for Yb-doping to very high concentrations [53] that permits the use of relatively short crystals for better overlap of pump and laser modes. In the first reported femtosecond Yb:KGW laser [46], 176fs pulses were generated at an average power of 1.1W with SESAM-stabilised mode locking and pulses of 71fs duration were obtained from a Kerr-lens mode-locked Yb:KYW laser. The disadvantage of Yb^{3+} is that the final laser level is thermally populated (in a quasi-three-level energy manifold) and so high efficiency operation requires good pump and laser mode overlap and high pump intensities when the laser operates at room temperature. These requirements can be satisfied readily by using a diffraction-limited pump beam which meets the additional requirement of a small beam waist inside the laser medium to reduce the saturation energy and to facilitate stable modelocking [54].

Klopp and co-workers described a passively mode-locked Yb:KYW laser pumped by a 2W high beam quality tapered diode laser [55] but its conversion efficiency with respect to the absorbed pump power(≈14%) did not exceed the efficiency of the system reported in reference 46, where broad stripe diodes were used as the pump source. As mentioned already, the development of moderately high power narrow-stripe lasers, which produce diffraction-limited output beams, offers the potential for the design of efficient, compact femtosecond laser sources [26, 27, 56].

We report here the development of a low-threshold and highly efficient femtosecond Yb:KYW laser that is pumped by a narrow-stripe InGaAs laser diode and incorporates a low-loss SESAM for self-starting the modelocking process. Remarkably, 123fs pulses were generated with 107mW of output power for only 308mW of pump power. This corresponds to a quite exceptional optical-to-optical efficiency of 35% which is more than a factor of two higher than anything reported previously [55, 56]. The corresponding electrical-to-optical efficiency with respect to the input electrical power to the diode laser exceeded 14%, which represents an improvement of more than three times in comparison with the best results reported previously from the femtosecond Cr:LiSAF laser described in section 6 [27]. We believe this to represent the highest reported efficiency for any femtosecond laser to date.

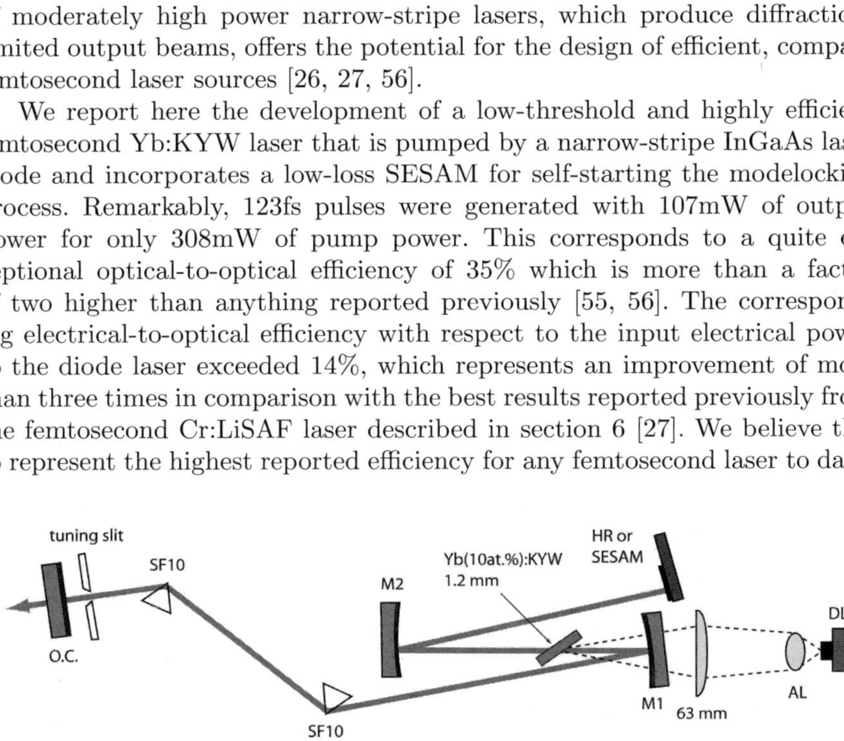

Fig. 8. Schematic of the femtosecond Yb:KYW laser. DL: narrow-stripe InGaAs laser diode; AL: 6.2mm aspherical lens; M1 and M2: 100mm ROC folding mirrors [57].

For our laser experiments a 1.2-mm-long 10at.% Yb^{3+}-doped $KY(WO_4)_2$ crystal at Brewster incidence was employed. The pump source was a single, narrow-stripe InGaAs diode laser (p-polarised) rated at maximum output power of 350mW at 981nm with a spectral width of 0.5nm. A highly asymmetric, astigmatically compensated Z-fold cavity was constructed and this arrangement served to facilitate efficient operation in the cw and modelocked regimes. With such a cavity, the laser mode radius inside the gain medium was calculated to be $40\mu m \times 20\mu m$ and a beam waist was $41\mu m$ on the SESAM. In a cw configuration and with an incident pump power of 308mW on the crystal, we obtained a maximum output power of 163mW at 1046nm, which corresponds to an optical-to-optical efficiency of 53%. The slope efficiency was 73% with respect to the absorbed pump power. By inserting a prism into the cavity, tuning was possible from 1012nm to 1079nm.

Using the set-up illustrated in figure 8 [57] a modelocking regime was realised with a stable pulse sequence at a 114MHz repetition frequency and a maximum average output power of 107mW for an output coupling of 3%. Two SF10 prisms separated by 39cm provided suitable group-velocity dispersion compensation. The pulse duration was measured to be 123fs, and the corresponding spectral width of 9.2nm near 1047nm implied a duration bandwidth product of 0.31. The shortest pulse duration of 114fs was obtained with 2% output coupling at a centre wavelength of 1049nm with 79mW of average output power. The threshold for self-starting mode-locked operation was as low as 120mW. With a vertical slit located between the second prism and the output coupler this Yb:KYW laser was tunable from 1040nm to 1070nm with a maximum output power of 65mW near 1050nm for an output coupling of 1%.

9 A Compact Femtosecond Cr:YAG Laser

High-repetition-rate Cr^{4+}:YAG femtosecond lasers operating around 1500nm are of particular relevance to the study of optical datacomms where the attractive feature of GHz pulse repetition frequencies range has already been demonstrated. Collings and co-workers [58] reported 200fs pulses at 0.9GHz while Tomaru *et al.* [59, 60] have demonstrated 3-element and 2-element cavities operating at 1.2GHz and 2.6GHz respectively. In our work we have developed a 3-element Kerr-lens modelocked femtosecond Cr^{4+}:YAG laser with a pulse repetition frequency just above 4GHz.

Figure 9 shows the configuration of the 3-element cavity that we used. The 11.6mm long crystal had one plane facet with a dielectric coating that provided broadband high reflectivity centred at 1550nm and high transmissivity at the pump wavelength whereas the second was Brewster-angled. The folding mirror had a radius of curvature of -15mm and the Littrow prism completed the cavity configuration. Typically, 5 - 6mm of fused silica was inserted to provide the necessary intracavity dispersion compensation. The back face of this element was coated to give a reflectivity of 99.75 % and thereby served as the output coupler.

The pump source was a Yb:fibre laser (up to 10W) operating at 1064nm, which is close to the peak absorption in Cr^{4+}:YAG. Its output beam was passed through a simple telescope system and focusing lens onto the plane surface of the crystal rod. An important requirement for KLM operation is compensation for astigmatism and this was achieved by providing a folding angle between 72° and 80° at mirror M2.

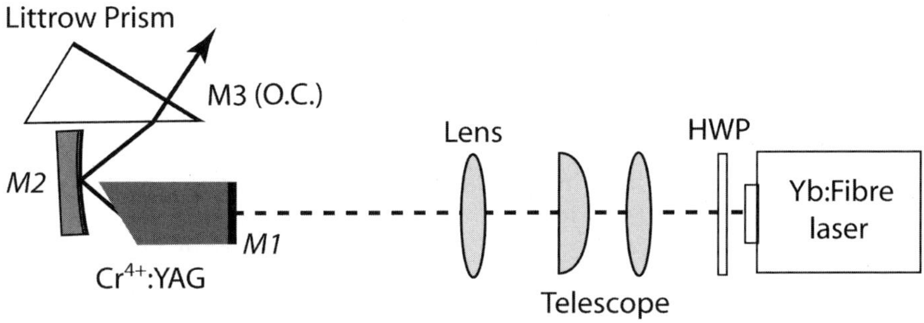

Fig. 9. Schematic diagram of a compact KLM Cr^{4+}:YAG laser [61].

Typical pulse durations were measured to be 70fs and when taken with the corresponding spectral width of 47nm, the deduced duration bandwidth product was 0.39, indicating that the pulses were near-transform limited. For these data, the pulse repetition frequency was 3.42GHz and an output power of 63mW was obtained [61]. A further reduction in the cavity size also allowed a repetition-rate of 4.05GHz to be achieved. In this more compact configuration, the pulse duration was 107fs with a spectral width of 24nm and an output power of 20mW. Further investigations are also underway to determine whether Kerr-lens modelocking can be retained to achieve pulse repetition frequencies beyond 5GHz.

A novel OTDM/WDM datacomms demonstration based on the spectral slicing of this Cr^{4+}:YAG laser has also been carried out [62]. Total capacities up to 1.36Tb/s with spectral efficiencies of 0.2b/s/Hz and 0.4b/s/Hz have been achieved. This demonstrates for the first time the compatibility of an ultrahigh capacity spectral slicing scheme with a femtosecond Cr^{4+}:YAG laser.

10 A Future Perspective on Low-threshold Femtosecond Oscillators

In applications such as those in ultrafast optical processing there are some fascinating perspectives based on future developments that can already be envisaged. In particular, the exploitation of photonics-based semiconductor devices in design-specific modules that promise functionality with integration begins to open up some intriguing possibilities. Within this research area, for example, we are especially interested in quantum-dot semiconductor materials because these can have the potential to offer bandwidths that are sufficiently broad to support the generation and subsequent amplification of femtosecond optical pulses. Indeed, some preliminary work carried out in our labs has shown that an amplification factor of 18dB can be achieved for 200fs pulses propagating in a structure comprising InGaAs dots in GaAs/GaAlAs [63]. Additionally, our vision extends to the possible use of photonic bandgap structures and devices as options for the design of future chip-based components such as pulse compressors and routers. Our initial assessments have indicated, for example, that temporal compression in the picosecond regime is feasible in semiconductors incorporating micro-structured and waveguided photonic bandgap structures for total device lengths around 10μm. In a more speculative direction, work that is being undertaken on organic semiconductors in the University of St Andrews has also opened up the prospect of exploiting thin film waveguided polymer structures as gain media for laser oscillators and amplifiers. If the levels of wall-plug efficiency that have been reported here for the vibronic crystals could be reproduced in these light-emitting polymers then it becomes likely that practical femtosecond organic-semiconductor oscillators and amplifiers could become available. Moreover, with already demonstrated basic micro-structuring using soft lithography techniques, it is not inconceivable that cheap and disposable plastic femtosecond lasers may yet enter the volume commodity market place. This is a most exciting prospect as we look towards our horizons for future developments in compact femtosecond sources!

Acknowledgement. Grateful thanks are due to several former group members who made many significant contributions to this work. They are Drs Burns, Hopkins, Valentine and Kemp who are now at the Institute of Photonics in Strathclyde University, Glasgow. The provision of SESAM devices from Professor Keller (ETH Zurich) for the Cr:LiSAF laser and from Professor Okhotnikov (Tampere University of Technology, Finland) for the Yb:KYW laser is acknowledged with thanks. We also acknowledge the funding of this work by the UK Engineering and Physical Sciences Research Council (EPSRC) and the EPSRC-supported Ultrafast Photonics Collaboration.

References

1. A. Miller, P. LiKamWa, B. H. T. Chai, and E. W. Van Stryland, *Optics Letters*, vol. 17, pp. 195-197, 1992.
2. P. M. W. French, R. Mellish, J. R. Taylor, P. J. Delfyett, and L. T. Florez, *Optics Letters*, vol. 18, pp. 1934-1936, 1993.
3. F. X. Kartner and U. Keller, *Optics Letters*, vol. 20, pp. 16-18, 1995.
4. I. D. Jung, F. X. Kartner, L. R. Brovelli, M. Kamp, and U. Keller, *Optics Letters*, vol. 20, pp. 1892-1894, 1995.
5. F.X. Kartner, I. D. Jung, and U. Keller, *IEEE Journal of Selected Topics in Quantum Electronics*, vol. 2, pp. 540-556, 1996.
6. D. Kopf, K. J. Weingarten, L. R. Brovelli, M. Kamp, and U. Keller, *Optics Letters*, vol. 19, 1994.
7. M. J. P. Dymott and A. I. Ferguson, *Applied Physics B-Lasers and Optics*, vol. 65, pp. 227-234, 1997.

8. V. P. Yanovsky, A. Korytin, F. W. Wise, A. Cassanho, and H. P. Jenssen, *IEEE Journal of Selected Topics in Quantum Electronics*, vol. 2, pp. 465-472, 1996.
9. R. Mellish, S. C. W. Hyde, N. P. Barry, R. Jones, P. M. W. French, J. R. Taylor, C. J. vanderPoel, and A. Valster, *Applied Physics B-Lasers and Optics*, vol. 65, pp. 221-226, 1997.
10. D. Kopf, K. J. Weingarten, G. Zhang, M. Moser, M. A. Emanuel, R. J. Beach, J. A. Skidmore, and U. Keller, ' *Applied Physics B-Lasers and Optics*, vol. 65, pp. 235-243, 1997.
11. S. Uemura and K. Torizuka, *Japanese Journal of Applied Physics*, vol. 39, pp. 3472-3473, 2000.
12. D. Kopf, A. Prasad, G. Zhang, M. Moser, and U. Keller, *Optics Letters*, vol. 22, pp. 621-623, 1997.
13. A. J. Alfrey, *IEEE Journal of Quantum Electronics*, vol. 25, pp. 760-766, 1989.
14. M. Piché, *Optics Communications*, vol. 86, pp. 156-160, 1991.
15. B. E. Bouma and J. G. Fujimoto, *Optics Letters*, vol. 21, pp. 134-136, 1996.
16. M. Ramaswamy-Paye and J. G. Fujimoto, *Optics Letters*, vol. 19, pp. 1756-1758, 1994.
17. R. Mellish, S. V. Chernikov, P. M. W. French, and J. R. Taylor, *Electronics Letters*, vol. 34, pp. 552-553, 1998.
18. G. J. Valentine and W. Sibbett, Paper CTuM47 presented at Conference on Lasers and Electro-Optics (CLEO), San Fransisco, 1998.
19. G. J. Valentine, J.-M. Hopkins, P. Loza-Alvarez, G. T. Kennedy, W. Sibbett, D. Burns, and A. Valster, *Optics Letters*, vol. 22, pp. 1639-1641, 1997.
20. X. G. Huang, F. R. Huang, W. K. Lee, and M. R. Wang, *Optics Communications*, vol. 142, pp. 249-252, 1997.
21. V. L. Kalashnikov, V. P. Kalosha, I. G. Poloyko, and V. P. Mikhailov, *Journal of the Optical Society of America B-Optical Physics*, vol. 14, pp. 964-969, 1997.
22. X. G. Huang and M. R. Wang, *Optics Communications*, vol. 158, pp. 322-330, 1998.
23. G. J. Valentine, M. P. Critten, G. T. Kennedy, W. Sibbett, D. Burns, and A. Valster, Paper CMI1 presented at Conference on Lasers and Electro-Optics (CLEO), Baltimore, 1997.
24. S. Uemura and K. Miyazaki, *Optics Communications*, vol. 138, pp. 330-332, 1997.
25. J.-M. Hopkins, G. J. Valentine, W. Sibbett, J. A. der Au, F. Morier-Genoud, U. Keller, and A. Valster, *Optics Communications*, vol. 154, pp. 54-58, 1998.
26. J.-M. Hopkins, G. J. Valentine, B. Agate, A. J. Kemp, U. Keller, and W. Sibbett, *IEEE Journal of Quantum Electronics*, vol. 38, pp. 360-368, 2002.
27. B. Agate, B. Stormont, A. J. Kemp, C. T. A. Brown, U. Keller, and W. Sibbett, *Optics Communications*, vol. 205, pp. 207-213, 2002.
28. K. Gabel, P. Russbuldt, R. Lebert, and A. Valster, *Optics Communications*, vol. 157, pp. 327-334, 1998.
29. D. Kopf, G. Zhang, R. Fluck, M. Moser, and U. Keller, *Optics Letters*, vol. 21, pp. 486-488, 1996.
30. A. Bartels, T. Dekorsy, and H. Kurz, *Optics Letters*, vol. 24, 1999.
31. D. Burns, M. P. Critten, and W. Sibbett, *Optics Letters*, vol. 21, pp. 477-479, 1996.
32. U. Keller, D. A. B. Miller, G. D. Boyd, D. G. Chiu, J. F. Ferguson, and M. T. Asom, *Optics Letters*, vol. 17, pp. 505, 1992.

33. U. Keller, K. J. Weingarten, F. X. Kartner, D. Kopf, B. Braun, I. D. Jung, R. Fluck, C. Honninger, N. Matuschek, and J. A. derAu, *IEEE Journal of Selected Topics in Quantum Electronics*, vol. 2, pp. 435-453, 1996.
34. H. Kogelnik, *The Bell Systems Technical Journal*, vol. 45, pp. 455-494, 1965.
35. S. Aoshima, H. Itoh, and Y. Tsuchiya, *IEEE Journal of Selected Topics in Quantum Electronics*, vol. 3, pp. 95-99, 1997
36. I. Sorokina, T, S. Naumov, E. Sorokin, E. Wintner, and A. V. Shestakov, *Optics Letters*, vol. 24, pp. 1578-1580, 1999.
37. K. Naganuma, K. Mogi, and H. Yamada, *Optics Letters*, vol. 15, pp. 393-395, 1990.
38. I. Gires and P. Tournois, *C. R. Acad. Sci.*, vol. 258, pp. 6112, 1964.
39. A. J. Kemp, B. Stormont, B. Agate, C. T. A. Brown, U. Keller, and W. Sibbett, *Electronics Letters*, vol. 37, pp. 1457-1458, 2001.
40. B. Agate, A. J. Kemp, C. T. A. Brown, and W. Sibbett, *Optics Express*, vol. 10, pp. 824-831, 2002.
41. B. Agate, E. U. Rafailov, W. Sibbett, S. M. Saltiel, P. Battle, T. Fry, and D. Mohatt, Paper CD-10596 presented at CLEO Europe, Munich, Germany, 2003.
42. B. Agate, E. U. Rafailov, W. Sibbett, S. M. Saltiel, P. Battle, T. Fry, and E. Noonan, *Optics Letters (accepted)*, 2003.
43. B. Agate, A. J. Kemp, D. J. L. B. Birkin, E. U. Rafailov, C. T. A. Brown, W. Sibbett, P. Battle, T. Fry, and D. Mohatt, Paper WJ2 presented at LEOS Annual Meeting, Glasgow, Scotland, 2002.
44. C. Honninger, R. Paschotta, M. Graf, F. Morier-Genoud, G. Zhang, M. Moser, S. Biswal, J. Nees, A. Braun, G. A. Mourou, I. Johannsen, A. Giesen, W. Seeber, and U. Keller, *Applied Physics B-Lasers and Optics*, vol. 69, pp. 3-17, 1999.
45. F. Druon, F. Balembois, P. Georges, A. Brun, A. Courjaud, C. Honninger, F. Salin, A. Aron, F. Mougel, G. Aka, and D. Vivien, *Optics Letters*, vol. 25, pp. 423-425, 2000.
46. F. Brunner, G. J. Spuhler, J. A. der Au, L. Krainer, F. Morier-Genoud, R. Paschotta, N. Lichtenstein, S. Weiss, C. Harder, A. A. Lagatsky, A. Abdolvand, N. V. Kuleshov, and U. Keller, *Optics Letters*, vol. 25, pp. 1119-1121, 2000.
47. F. Druon, S. Chenais, P. Raybaut, F. Balembois, P. Georges, R. Gaume, G. Aka, B. Viana, S. Mohr, and D. Kopf, *Optics Letters*, vol. 27, pp. 197-199, 2002.
48. M. J. Lederer, M. Hildebrandt, V. Z. Kolev, B. Luther-Davies, B. Taylor, J. Dawes, P. Dekker, J. Piper, H. H. Tan, and C. Jagadish, *Optics Letters*, vol. 27, pp. 436-438, 2002.
49. N. V. Kuleshov, A. A. Lagatsky, V. G. Shcherbitsky, V. P. Mikhailov, E. Heumann, T. Jensen, A. Diening, and G. Huber, *Applied Physics B-Lasers and Optics*, vol. 64, pp. 409-413, 1997.
50. N. V. Kuleshov, A. A. Lagatsky, A. V. Podlipensky, V. P. Mikhailov, E. Heumann, A. Diening, and G. Huber, in *OSA Trends in Optics and Photonics Vol.10, Advanced Solid State Lasers*, C. R. P. a. W. R. Bosenberg, Ed. Washington, DC: OSA, 1997, pp. 415-419.
51. N. V. Kuleshov, A. A. Lagatsky, A. V. Podlipensky, V. P. Mikhailov, and G. Huber, ' *Optics Letters*, vol. 22, pp. 1317-1319, 1997.
52. A. Brenier, *Journal of Luminescence*, vol. 92, pp. 199-204, 2001.

53. M. C. Pujol, M. A. Bursukova, F. Guell, X. Mateos, R. Sole, J. Gavalda, M. Aguilo, J. Massons, F. Diaz, P. Klopp, U. Griebner, and V. Petrov, *Physical Review B*, vol. 65, pp. art. no.-165121, 2002.
54. C. Honninger, R. Paschotta, F. Morier-Genoud, M. Moser, and U. Keller, *Journal of the Optical Society of America B-Optical Physics*, vol. 16, pp. 46-56, 1999.
55. P. Klopp, V. Petrov, U. Griebner, and G. Erbert, *Optics Express*, vol. 10, pp. 108-113, 2002.
56. G. J. Valentine, A. J. Kemp, D. J. L. B. Birkin, D. Burns, F. Balembois, P. Georges, H. Bernas, A. Aron, G. Aka, W. Sibbett, A. Brun, M. D. Dawson, and E. Bente, *Electronics Letters*, vol. 36, pp. 1621-1623, 2000.
57. A. A. Lagatsky, E. U. Rafailov, C. G. Leburn, C. T. A. Brown, N. Xiang, O. G. Okhotnikov, and W. Sibbett, *Electronics Letters (accepted)*, 2003.
58. B. C. Collings, J. B. Stark, S. Tsuda, W. H. Knox, J. E. Cunningham, W. Y. Jan, R. Pathak, and K. Bergman, *Optics Letters*, vol. 21, pp. 1171-1173, 1996.
59. T. Tomaru and H. Petek, *Optics Letters*, vol. 25, pp. 584-586, 2000.
60. T. Tomaru, *Optics Letters*, vol. 26, pp. 1439-1441, 2001.
61. C. G. Leburn, A. A. Lagatsky, C. T. A. Brown, and W. Sibbett, Paper CThF6 presented at Conference on Lasers and Electro-Optics (CLEO), Baltimore, MD, 2003.
62. Y. J. Chai, C. G. Leburn, A. A. Lagatsky, C. T. A. Brown, R. V. Penty, I. H. White, and W. Sibbett, to be presented at Lasers and Electro-Optic Society (LEOS) Annual Meeting, Tuscon, Arizona, 2003.
63. E. U. Rafailov, P. Loza-Alvarez, W. Sibbett, G. S. Sokolovskii, D. A. Livshits, A. E. Zhukov, and V. M. Ustinov, *IEEE Photonics Technology Letters (submitted)*, 2003.

Novel 10- to 40-GHz Telecom Pulse Generating Sources with High Average Power

G. J. Spühler[1,2], L. Krainer[1,2], S. Lecomte[1], S. C. Zeller[1], K. J. Weingarten[2], R. Paschotta[1], and U. Keller[1]

[1] Ultrafast Laser Physics, Institute of Quantum Electronics, Swiss Federal Institute of Technology, ETH Hönggerberg HPT, 8093 Zürich, Switzerland
spuehler@phys.ethz.ch

[2] GigaTera, Inc., Technoparkstrasse 1, CH-8005 Zurich, Switzerland

1 Introduction

As data transmission rates continue to increase, pulsed lasers are becoming increasingly important for telecom applications. Novel transmission systems at 10 Gb/s and higher often use return-to-zero (RZ) formats [1, 2] and soliton dispersion management techniques [3, 4], which both rely on clean optical pulses. Therefore, these approaches benefit greatly from the availability of simple, compact, and efficient optical pulse generators. There are many compelling reasons to use a pulsed laser directly as a source in the transmitter of optical telecommunication systems, rather than an externally modulated continuous-wave (cw) source. First, they eliminate the need for a high-end modulator to create the pulses and thereby simplify system architecture, increase efficiency, and reduce cost. Secondly, the contrast ratio of pulsed lasers is typically much higher than for modulated cw sources. This improves system signal-to-noise and allows further scaling to higher repetition rates through optical time-division multiplexing (OTDM). Apart from the transmitter side, there are interesting applications of pulsed lasers also in the receivers of transmission systems, e.g. for demultiplexing and clock recovery [5–7]. Furthermore, applications outside the telecom area, in the fields of optical clocking [8, 9], high-speed electro-optic sampling [10, 11], frequency metrology (similar to the work presented in Ref. [12]) or generation of polarized electron beams for particle accelerators [13] become increasingly important.

Besides a repetition rate that is matched to the standards for 10, 40 Gb/s or higher transmission rates [14], the requirements for an ideal pulse source for telecom applications contain the following features: low timing jitter, high contrast ratio, wavelength tunability in the regime of interest (mostly in the C-band from 1530–1565 nm) and a high efficiency. The requirement for the pulse duration depends somewhat on the coding format. Usually it is roughly a third of the bit period. Also simplicity, reliability and robustness are important features.

Multi-GHz pulse sources have so far almost always involved either an edge-emitting semiconductor laser [15], which is usually actively or hybrid mode-locked, or a harmonically mode-locked fiber ring laser [16]. Edge-emitting semiconductor lasers appear very attractive due to their very compact and stable optical setup, but expensive electronics are required for active mode locking, and the structures required for a semiconductor laser with this performance are complicated and difficult to produce. Multi-GHz fiber lasers can easily generate high-quality pulses, but they have very long laser cavities and thus require sophisticated means to obtain stable harmonic mode locking with a large number of precisely equidistant pulses in the cavity. Additionally, individual pulses generated by harmonic mode locking do not necessarily exhibit a fixed phase relation. This excludes certain promising coding formats such as return-to-zero differential phase shift keying [17, 18] which carry the data in the phase of the pulses, rather than in the amplitude.

Here, we report on novel alternative pulse sources for telecom applications, based on passively fundamental mode-locked solid-state lasers and synchronously pumped optical parametric oscillators.

Up to a few years ago, the repetition rate of passively mode-locked solid-state lasers has been limited to a few GHz. Especially in the telecom wavelength regime, where only few solid-state gain media are available multi-GHz pulse repetition rates have not been possible [19, 20]. In recent years, the consequent exploitation of the flexibility of semiconductor saturable absorber mirrors (SESAMs) [21, 22] in combination with increased understanding of the limitations [23, 24] allowed us to develop passively mode-locked lasers with multi-GHz pulse repetition rates, very good pulse quality, comparatively high output powers and wavelengths in the regime of interest. Passive mode locking implies that the pulses are achieved without using any multi-GHz electronics. In addition, the pulses originate from fundamental mode locking. Thus, every output pulse is a copy of the same single pulse which bounces back and forth in the cavity. Therefore, pulse-to-pulse variations are minimized and the phase of the pulses is constant.

Synchronously pumped optical parametric oscillators (OPOs) are well known for their potential to provide very large wavelength tuning ranges – significantly larger than possible with most laser media. This makes them interesting for telecom pulse sources. However, the repetition rate was limited until recently to values of a few GHz [25] because higher repetition rates make it difficult to exceed the threshold value of the pump peak power of the OPO. By pushing the performance of passively mode-locked lasers at 1 μm in terms of power and repetition rate we could develop widely tunable OPOs with 10-GHz pulse repetition rate and high average output powers.

2 Challenges

The main challenge one has to face when pushing the repetition rate of a passively mode-locked laser to the multi-GHz regime, is the suppression of Q-switched mode locking (QML) [23]. Q-switched mode locking is a state of operation where the relaxation oscillations are destabilized; the damping effect for increased intracavity powers due to gain saturation is not strong enough to suppress the growth due to increased saturation of the absorber. A high repetition rate naturally leads to relatively low intracavity pulse energies which limit the saturation of the gain and thus increase the tendency for QML. In addition, the choice of gain media in the wavelength regime of interest is rather limited. Mainly Er:Yb:glass and Cr:YAG are used for solid state lasers in the 1.5-μm regime. The latter has very strict requirements on the pump source due to its weak pump absorption and it exhibits relatively low efficiencies, due to limited crystal quality and relatively high induced losses [26, 27]. Therefore, the only valuable candidate is Er:Yb:glass, which however has very small cross sections and thus a high saturation energy which again increases the tendency towards QML. The way we overcome these limitations is explained in Section 3. In Section 4, we describe an alternative approach to get pulses with high quality, high repetition rate and an even wider tuning range in the 1.5-μm wavelength regime than obtained with Er:Yb:glass lasers: A synchronously pumped optical parametric oscillator pumped by a passively mode-locked Nd:YVO_4 laser with high repetition rate and high average power.

3 Er:Yb:glass laser

Generally, Er:Yb:glass is well suited for telecom applications. Its gain bandwidth covers essentially the whole C-band, it can be pumped with standard 980-nm laser diodes also used in erbium-doped fiber amplifiers (EDFAs), it can be produced in excellent quality and large amounts, and it is resistant and cheap. As mentioned in Section 2, its drawback is a relatively strong tendency towards QML. In order to suppress QML in our Er:Yb:glass lasers we rely on the following concepts: The SESAM allows to custom design the main absorber parameters within a certain range, mainly by adopting the structure of the device [22, 28]. We exploit that range for minimized saturation fluence and modulation depth which both helps to suppress QML [23]. For fundamental mode locking, the repetition rate dictated by the application determines the cavity length. With the constraint of the length of the cavity and the SESAM being a flat end mirror of the laser, we designed the cavity in such a way that it has minimized mode sizes on the SESAM and in the gain medium, leading to a three-mirror-cavity as shown in Fig. 1. This minimizes the saturation energy of the gain and the absorber, also helping to suppress QML. The laser glass is inserted into the cavity under Brewster angle. A third important point is the use of a single-mode 980 nm pump diode. It enables optimum mode matching of pump and laser mode even though the laser mode size in the gain is very small. This allows for good efficiency and mode quality. Figure 1 shows the cavity design for a 10-GHz and 40-GHz laser of which the total cavity lengths are 14 mm and 3.5 mm, respectively.

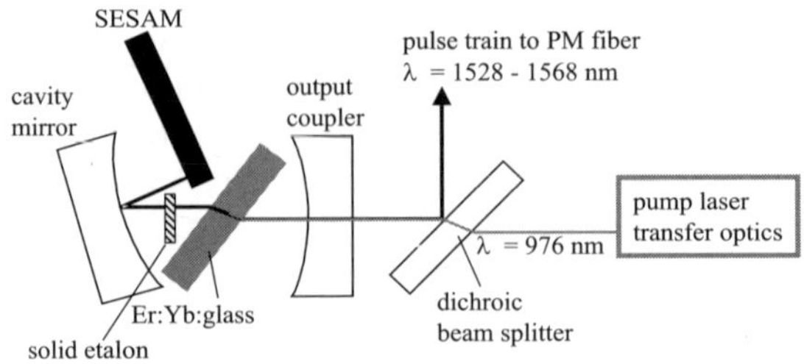

Fig. 1. Setup of the 10-GHz Er:Yb:glass laser. Cavity length ≈14 mm. 40-GHz version with cavity length ≈3.5 mm and without etalon.

Varying the SESAM parameters and the intracavity filter effect from the etalon in the 10-GHz laser, pulse durations from 1 to 19 ps could be obtained [29]. With the etalon the laser wavelength could be adjusted within 1528 – 1568 nm (full C-band) by tuning the tilt angle (Fig. 2a). Only week variations of the main laser parameters, such as pulse duration, spectral bandwidth and power were observed [30]. With a single set of optics, the repetition rate could be varied from 9.2 to 12.7 GHz by changing the cavity length. We typically couple the output of the laser into a polarization-maintaining fiber where we get 10 to 20 mW of average power for a pump power of about 220 mW.

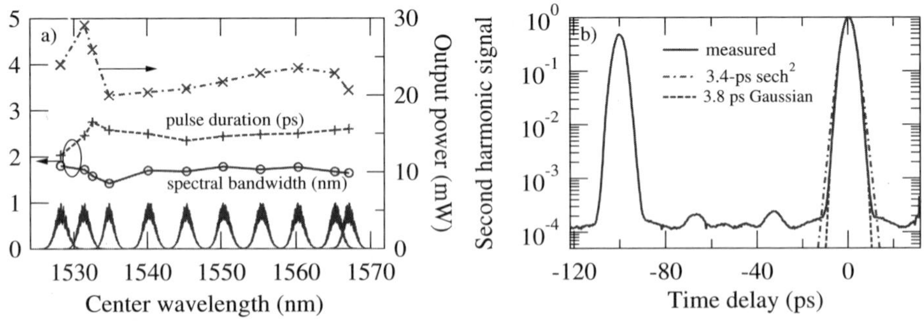

Fig. 2. a) Pulse duration (dashed), spectral bandwidth (solid) and output power (dashed-dotted) as a function of center wavelength for a 10-GHz Er:Yb:glass laser. b) Slow scan autocorrelation. The autocorrelation scan range exceeds the pulse separation. Maximum signal level between the pulses is −36 dBc. The Gaussian fits the measured pulse better than a sech^2 pulse.

The pulse train is signalized by a very high contrast ratio of typically > 30 dB and a very stable optical spectrum (Fig. 2). Fig. 2b shows a slow scan autocorrelation with a scan range exceeding the pulse separation. This reveals the first cross correlation. The highest signal between the pulses is 36 dB below the carrier showing the excellent contrast ratio. The measurement is shown together with a Gaussian and a sech^2 fit, indicating that the measured pulses are fitted better by a Gaussian pulse. In order to show the suitability of the Er:Yb:glass laser for high bit rates, we show in Fig. 3a an autocorrelation of a pulse train which has been obtained by optical time division multiplexing (OTDM) of a 10-GHz laser up to 160 GHz with commercially available multiplexers (PriTel Inc., OCM-4). First of all, we can see that the pulses are short enough (in this case 1 ps) for such a high bit rate. Secondly, the contrast ratio is high enough even for 160 Gb/s transmission to avoid a significant noise level between the multiplexed pulses. To illustrate a pulse train containing data, Fig. 3b shows an eye diagram for a 40 Gb/s back-to-back measurement (i.e. the data has not passed any transmission span). Data has been modulated onto the pulse train using a $LiNbO_3$ data modulator. The time resolution is limited by a 40-GHz detector on the sampling oscilloscope.

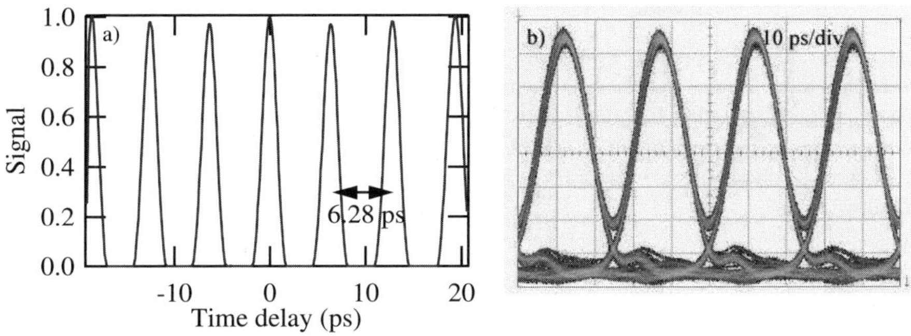

Fig. 3. a) Autocorrelation trace of a pulse train multiplexed from 10 GHz to 160 GHz and b) 40-Gb/s eye diagram from back-to-back measurement taken with a 40 GHz detector and a 50 GHz sampling head. labelf3

We have also developed an Er:Yb:glass laser directly generating a pulse train with 40-GHz repetition rate [31]. The laser is based on the same principles and the same cavity setup as shown in Fig. 1. Also the pulse parameters are similar to the 10-GHz laser (18-mW average output power, 4.3-ps pulses). Due to space constraints in the extremely short cavity, we did not prove tunability yet.

For telecom applications it is crucial that the pulses are synchronous to a clock signal, which is supplied by the system. This synchronization is naturally given for actively or hybrid mode-locked lasers. However, a passively mode-locked laser needs to be actively synchronized to the clock signal. This can be done by a simple phase-locked-loop (PLL), which locks the cavity length to the clock signal [32,33]. In order to do so, we mounted the SESAM on a piezo element. With a relatively low loop bandwidth of less than 50 kHz, we obtained rms timing jitter values below 80 fs, when integrating the phase noise power spectral density from 10 Hz to 1 MHz (Fig. 4b). The simple PLL approach allows the clock signal to vary within a certain range while the laser remains phase locked. This flexibility is required for applications in the receiver end of a transmission system, as during transmission the clock signal can be deteriorated significantly [7].

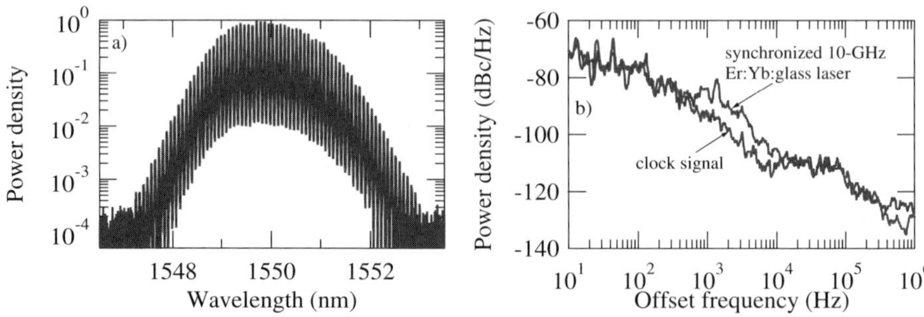

Fig. 4. a) Typical optical spectrum of a 10-GHz Er:Yb:glass laser taken with a resolution of 0.01 nm resolving the individual longitudinal modes. b) Phase noise power spectral density for clock signal and synchronized Er:Yb:glass laser. Apart from a narrow window around 1 kHz the laser noise is limited by the noise of the clock signal. The integrated (10 Hz – 1 MHz) phase noise yields an rms timing jitter of below 80 fs.

So far, we have only made use of the output of the mode-locked laser in the time domain. One can also think of applications in the wavelength domain. In the wavelength domain, fundamental-mode-locked lasers generate a stable comb-shaped optical spectrum (Fig. 4a), where the spacing of the longitudinal modes exactly equals the pulse repetition rate and shows a high optical signal-to-noise ratio (OSNR). Therefore, pulse-generating lasers are intrinsically wavelength-stable multi-channel sources which can be used as comb-generating lasers in dense wavelength-division-multiplexing (DWDM) [34,35]. Varying the cavity length and comparing/locking a single longitudinal mode of the laser to an ITU grid line or any other reference, the entire optical comb is automatically locked to the ITU grid. The distinct channels of the laser can be separated with appropriate means (e.g. an arrayed wave guide grating [36]), individually modulated and recombined to launch into the network [37]. Compared to the classic DWDM approach with one fixed-wavelength-laser per channel, the comb-generating laser approach brings many benefits to the system, such as decreased number of components (i.e. only one set of driver electronics and one wavelength locker are required), smaller inventory, space and power requirements, etc.

As current channel spacing standards are 25 GHz, 50 GHz, 100 GHz, or higher (International Telecommunication Union, ITU), we developed a 25 GHz version of the Er:Yb:glass laser [38]. Fig. 1 still holds for the cavity design. The overall cavity length is then about 6 mm. Also the output parameters are very similar to the 10-GHz and 40-GHz lasers: 25-mW average power, 1.9-ps pulse duration, 30-dB contrast ratio, and a spectral width of 3.1 nm FWHM. We could tune the laser in the range of 1528 - 1561 nm. It is important to state that tilting the etalon moves the envelope of the spectrum across the C-band, whereas changing the cavity length moves the individual longitudinal modes underneath the envelope. In this way, the comb spectrum can be tuned to any desired wavelength in the C-band, i.e. to the ITU grid, essentially without changing the grid spacing. With a 25-GHz channel spacing, the maximum frequency offset between a longitudinal mode and a required reference line is 12.5 GHz. To remove this offset, the cavity length has to be changed by a quarter wavelength leading to a change in repetition rate (and thus channel spacing) of only 1.6 MHz, which is well below the tolerance of the ITU channels and typical bandwidths of DFB lasers. The reason for this small change in repetition rate is that the carrier frequency is several orders of magnitude higher than the repetition rate. Therefore, a sub-wavelength change in cavity length has only a minimum impact on the channel spacing.

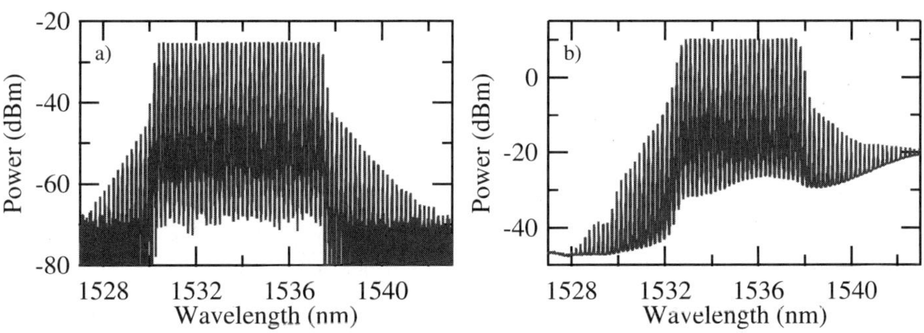

Fig. 5. Optical spectrum from a) 25-GHz laser flattened with DGE and b) 25-GHz laser flattened with DGE and amplified with an EDFA.

For a DWDM application each channel (longitudinal mode) should have the same power rather than the Gaussian like spectrum of a typical mode-locked laser. Therefore, we combined the 25-GHz laser with a dynamic gain equalizer (DGE, Silicon Light Machines, Model:SLM2200). The DGE allows setting the attenuation of each channel individually and in real time, controlled by a computer. Fig. 5a shows the optical spectrum of the laser (without etalon) in conjunction with the DGE. A fraction of the output is detected and checked for flatness. This gives an error signal which is fed back to the DGE. Using the maximum dynamic range of the DGE (17 dB), we could get up to 32 channels with a flatness of better than 0.4 dB. In order to increase the power per channel, we added an EDFA to the system (after the DGE). The gain narrowing reduced the number of channels to 25, however each carrying 10 dBm average power, and an OSNR exceeding 35 dB (Fig.

5b). No additional filters have been used to suppress the amplified spontaneous emission of the EDFA. The system is very stable, as the original laser spectrum is stable and adaptive flattening is applied in addition. One single wavelength locker with an appropriate feedback circuit controlling the cavity length allows locking all these channels to a reference.

Keeping in mind the simplicity of this approach, its potential for DWDM system applications becomes clear. An additional stage for super-continuum generation can even increase the number of obtained channels substantially [37] and would benefit from the spectral purity and stability of the presented laser. In first super-continuum generation experiments carried out at 10 GHz we obtained a stable smooth spectrum ranging from 1450 nm to 1650 nm. For that the pulse train from the laser was amplified in an EDFA and launched into a highly nonlinear fiber, generating a huge number of additional longitudinal modes, still exactly spaced by the repetition rate of the seed laser.

For future development, repetition rates beyond 50 GHz seem to be appealing for larger channel spacings and possibly higher bit rates. Possible limiting factors are the extremely small cavity dimensions and Q-switching instabilities. Due to the wide gain bandwidth of Er:Yb:glass, pulses in the range of several hundred fs could be possible. Also, somewhat higher output powers should be feasible.

4 Synchronously pumped optical parametric oscillator

Compared to Er:Yb:glass, Nd:YVO_4 is much better adapted for passively mode-locked lasers with very high repetition rates, as the emission cross-section of $114 \cdot 10^{-20}$ cm^2 [39] is ≈ 140 times higher than the one of Er:Yb:glass [40] resulting in significantly stronger suppression of Q-switching instabilities. Additionally, due to its crystalline host, it can withstand much higher intensities than Er:Yb:glass. Indeed, we have presented passively mode-locked Nd:YVO_4 lasers at 10 GHz with 2.1-W average power and pushed the repetition rates up to 157 GHz [41]. However, the emission wavelength of 1064 nm does not match the telecom requirements.

We exploited the strong suppression of Q-switching instabilities of Nd:YVO_4 to develop a high power, high repetition rate laser [41], which yields high enough pulse energies to synchronously pump an optical parametric oscillator (OPO). In this way, we gain access to a very broad tuning range. The wavelength range accessible with an OPO pumped at 1064 nm extends to more than the S, C and L-band.

The cavity setup of the 10-GHz high-power Nd:YVO_4 laser is shown on the left in Fig. 6. Here, the pump source was a beam-shaped laser diode (from unique-m.o.d.e.) with 5.6-W output power and a symmetric 20-times diffraction limited beam. With this pump power, up to 2.1 W of mode-locked average output power were achieved at a repetition rate of 10 GHz [41]. Because of the high thermal load, a relatively low (0.5%) doping level of the Nd:YVO_4 crystal had to be used, requiring a 1-mm crystal length for efficient pump absorption. Higher doping levels might lead to a thermal roll-over or even fracture of the crystal. Pump beam quality and crystal length then limited the pump spot radius to ≈50 μm in air, and for stable single transverse mode operation we had to design the cavity to have a even slightly larger laser mode. This limited the modulation depth of the SESAM that we could work with without Q-switching instabilities [23] and thus relatively long pulses of 14-ps duration were obtained.

It becomes apparent that the need to suppress Q-switching instabilities leads to a compromise between highest output power, highest repetition rate, and shortest pulse durations. The pump beam quality plays an important role in that interplay, as it dictates the obtainable mode size in the gain element.

In spite of its high repetition rate, this laser had enough peak power to synchronously pump an OPO [42]. The OPO is shown on the right of Fig. 6 as part of the whole system with the pump laser. It contains an AR-coated 50-mm long multigrating periodically poled $LiNbO_3$ (PPLN) crystal which is mounted in an oven. Limiting factors for the length of the PPLN are group velocity mismatch and manufacturing limits. OPO spot size in the PPLN and pump focusing are then designed according to the general guidelines for an OPO [43]. A simple cavity with appropriate spot size contains two curved cavity mirrors. The free spectral range of the OPO-cavity is only 1.1 GHz (corresponding to a physical cavity length of 74.4 mm), so that with a pump repetition rate of 10 GHz we have nine pulses simultaneously circulating in the cavity. Although this is reminiscent of harmonic mode locking, it does not introduce problems with timing jitter, as the timing of all circulating pulses is determined by the pump pulses. The advantage of the relatively long cavity is that it allows us to use a longer nonlinear crystal with correspondingly higher gain and lower threshold. The output coupler has 97% transmission at 1064 nm and 0.22% transmission at 1550 nm. The second mirror has a high reflectivity for the signal as well as for the pump to obtain a double pass of the pump. Both mirror coatings have $>$ 85% idler transmission to suppress significant idler feedback. A piezo below the high reflector allows to fine adjust the cavity length.

The OPO threshold was around 300 mW with 14-ps pump pulses. Using different PPLN gratings with periods between 28.9 μm and 29.9 μm and varying the crystal temperature between 150°C and 180°C, we could generate any signal wavelength in the range 1497.7 nm to 1605.9 nm. Due to imperfect quality of the different poled regions, the signal output power varied between 4 mW and 23 mW. We used up to 460 mW of pump power incident on the OPO for the tuning experiment.

The tuning range exceeding 100 nm indicates the potential of the OPO approach. The tuning range was mainly limited by the bandwidth of the coatings. But not only in terms of tuning range, also in terms of power the OPO approach is superior to alternative methods for generating multi-GHz

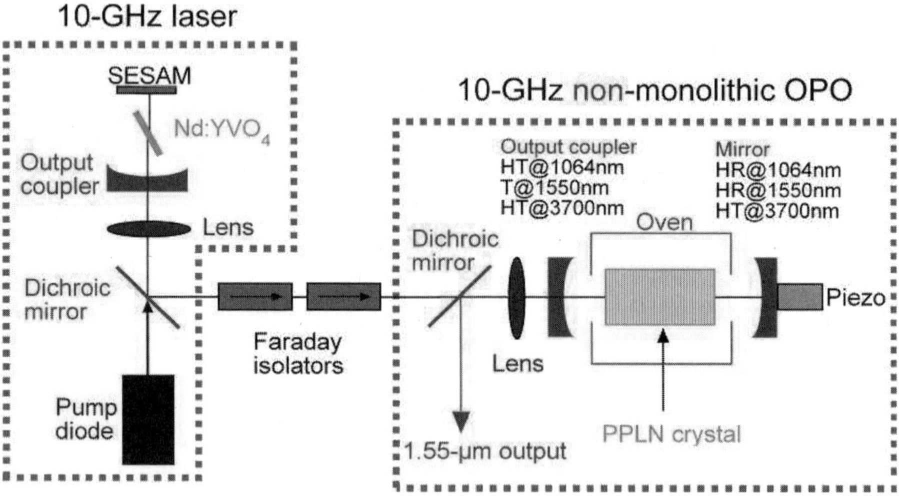

Fig. 6. Setup of synchronously pumped optical parametric oscillator with the 10-GHz $Nd:YVO_4$ pump laser and the non-monolithic OPO.

pulse streams at 1.5 μm: With 1.3 W of pump power in 13-ps pulses incident on the OPO, we measured 120 mW of average output power in 10-ps pulses with a 0.26-nm bandwidth around 1555 nm (time-bandwidth product of 0.32) (Fig. 7).

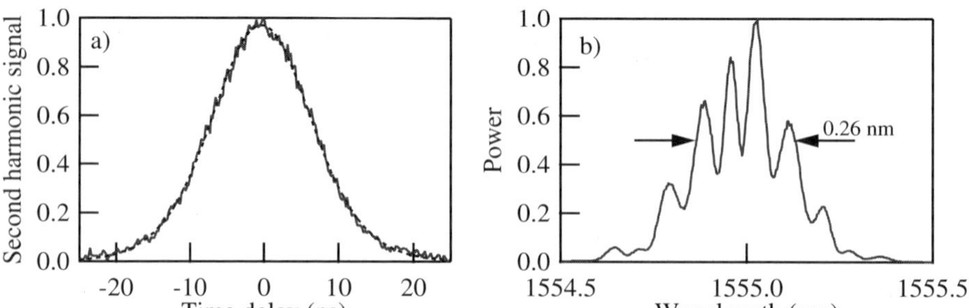

Fig. 7. Autocorrelation a) and optical spectrum (taken with a resolution of 0.08 nm) b)of the 10-GHz OPO.

There seems to be still room to increase the pulse repetition rate of synchronously pumped OPOs, by increasing the gain, by a longer PPLN, decreasing the output coupling transmission or by making use of a non-integer ratio of pump repetition rate and OPO round-trip frequency with a lowest common multiple in the multi-GHz range.

5 Conclusions

With the motivation of telecom applications, we have introduced various novel multi-GHz sources, all based on diode-pumped, passively mode-locked solid-state lasers. The application of a semiconductor saturable absorber mirror in combination with optimized cavity designs allowed progression in performance regimes where these lasers become interesting for telecom applications. They have outstanding pulse properties and due to the fact that they are pulsed all passively, they present a simple and cheap way to produce a pulse train which can be synchronized to a clock signal.

A simple approach appears to be the diode-pumped fundamental mode-locked Er:Yb:glass laser generating high-quality picosecond pulses at multi-GHz repetition rates with pulse parameters variable in a wide range. Pulse durations between 1 ps and 19 ps and pulse repetition rates of 10, 25, and 40 GHz enable OTDM and DWDM applications. C-band tunability of the 10 and 25 GHz lasers have been demonstrated. Higher repetition rates, higher powers, and shorter pulses for higher bit rates appear to be feasible.

For an even wider tuning range including S-, C-, and L-band, an optical parametric oscillator, synchronously pumped by a high-power high-repetition-rate Nd:YVO_4 laser can be applied. It also yields significantly higher average powers. So far, up to 120 mW average output power and up to 108 nm tuning range in the 1.5-μm region have been achieved with 10-GHz OPOs, but even higher power levels, higher pulse repetition rates and wider tuning ranges seem to be possible.

References

1. H. Taga, M. Suzuki, and Y. Namihira, Polarization mode dispersion tolerance of 10Gbit/s NRZ and RZ optical signals, *Electron. Lett.*, vol. 34, pp. 2098-100, 1998.
2. L.-P. Chen and C.-X. Shi, Experimental Comparison of Transmission System Penalty of 10 Gbit/s RZ and NRZ Optical Signals Due to Polarization-Mode Dispersion, *Microwave and Optical Technology Letters*, vol. 25, pp. 297-300, 2000.
3. L. F. Mollenauer and P. V. Mamyshev, Massive wavelength-division multiplexing with solitons, *IEEE J. Quantum Electron.*, vol. 34, pp. 2089-2102, 1998.
4. M. Nakazawa, H. Kubota, K. Suzuki, E. Yamada, and A. Sahara, Ultrahigh-Speed Long-Distance TDM and WDM Soliton Transmission Technologies, *IEEE J. Sel. Topics in Quantum Electron.*, vol. 6, pp. 363-396, 2000.
5. S. Bigo, O. Leclerc, and E. Desurvire, All-Optical Fiber Signal Processing and Regenration for Soliton Communications, *IEEE J. Sel. Topics in Quantum Electron.*, vol. 3, pp. 1208-23, 1997.
6. T. Yamamoto, L. K. Oxenløwe, C. Schmitt, C. Schubert, E. Hilliger, U. Feiste, J. Berger, R. Ludwig, and H. G. Weber, Clock recovery from 160 Gbit/s data signals using phase-locked loop with interferometric optical switch based on semiconductor optical amplifier, *Electron. Lett.*, vol. 37, pp. 509-10, 2001.

7. L. Oxenløwe, A. Siahlo, K. Berg, A. T. Clausen, B. Sørensen, K. Yvind, P. Jeppesen, K. Hansen, K. Hoppe, and J. Hanberg, A novel 160 Gb/s receiver configuration including a glass crystal pulsed laser, photonic crystal fiber and a simple dynamic clock, presented at 29th European Conference on Optical Communication, Rimini, Italy, 2003.
8. D. A. B. Miller, Optics for low-energy communication inside digital processors: quantum detectors, sources, and modulators as efficient impedance converters, *Opt. Lett.*, vol. 14, pp. 146-148, 1989.
9. A. Bhatnagar, C. Debaes, R. Chen, N. C. Helman, G. A. Keeler, D. Agarwal, H. Thienpont, and D. A. B. Miller, Receiver-less clocking of a CMOS digital circuit using short optical pulses, presented at 2002 IEEE/LEOS Annual Meeting; Conference Proceedings, 15th Annual Meeting of the IEEE Lasers & Electro-Optics Society, Glasgow, Scotland, 2002.
10. J. A. Valdmanis and G. A. Mourou, Subpicosecond electrooptic sampling: principles and applications, *IEEE J. Quantum Electron.*, vol. 22, pp. 69-78, 1986.
11. K. J. Weingarten, M. J. W. Rodwell, and D. M. Bloom, Picosecond optical sampling of GaAs integrated circuits, *IEEE J. Quantum Electron.*, vol. 24, pp. 198-220, 1988.
12. H. S. Margolis, C. S. Edwards, G. P. Barwood, and P. Gill, A passive optical frequency comb for frequency metrology in the 1.5 micron optical communications band, presented at 2002 IEEE/LEOS Annual Meeting; Conference Proceedings, 15th Annual Meeting of the IEEE Lasers & Electro-Optics Society, Glasgow, Scotland, 2002.
13. A. Hatziefremidis, D. N. Papadopoulos, D. Fraser, and H. Avramopoulos, Laser sources for polarized electron beams in cw and pulsed accelerators, *Nucl. Instrum. Meth. A*, vol. 431, pp. 46-52, 1999.
14. TechFest: `http://www.techfest.com/networking/wan/sonet.htm`, 2003.
15. K. Sato, Semiconductor light sources for 40-Gb/s transmission systems, *IEEE J. Lightwave Techn.*, vol. 20, pp. 2035-2043, 2002.
16. E. Yoshida, N. Shimizu, and M. Nakazawa, A 40-GHz 0.9-ps regeneratively mode-locked fiber laser with a tuning range of 1530-1560 nm, *IEEE Phot. Techn. Lett.*, vol. 11, pp. 1587-1589, 1999.
17. J. Leibrich, C. Wree, and W. Rosenkranz, CF-RZ-DPSK for Suppression of XPM on Dispersion-Managed Long-Haul Optical WDM Transmission on Standard Single-Mode Fiber, *IEEE Photon. Tech. Lett.*, vol. 14, pp. 155-7, 2002.
18. C. Xu, X. Liu, L. F. Mollenauer, and X. Wei, Comparison of Return-to-Zero Differential Phase-Shift Keying and On-Off Keying in Long-Haul Dispersion Managed Transmission, *IEEE Photon. Tech. Lett.*, vol. 15, pp. 617-9, 2003.
19. B. C. Collings, K. Bergman, and W. H. Knox, True fundamental solitons in a passively mode-locked short cavity Cr^{4+}:YAG laser, *Opt. Lett.*, vol. 22, pp. 1098-1100, 1997.
20. P. Laporta, S. Longhi, M. Marchesi, S. Taccheo, and O. Svelto, 2.5 GHz and 5 GHz Harmonic Mode-Locking of a Diode-Pumped Bulk Erbium-Ytterbium Glass Laser at 1.5 Microns, *IEEE Photon. Technol. Lett.*, vol. 7, pp. 155-157, 1995.
21. U. Keller, D. A. B. Miller, G. D. Boyd, T. H. Chiu, J. F. Ferguson, and M. T. Asom, Solid-state low-loss intracavity saturable absorber for Nd:YLF lasers: an antiresonant semiconductor Fabry-Perot saturable absorber, *Opt. Lett.*, vol. 17, pp. 505-507, 1992.

22. U. Keller, K. J. Weingarten, F. X. Kärtner, D. Kopf, B. Braun, I. D. Jung, R. Fluck, C. Hönninger, N. Matuschek, and J. Aus der Au, Semiconductor saturable absorber mirrors (SESAMs) for femtosecond to nanosecond pulse generation in solid-state lasers, *IEEE J. Sel. Top. Quantum Electron.*, vol. 2, pp. 435-453, 1996.
23. C. Hönninger, R. Paschotta, F. Morier-Genoud, M. Moser, and U. Keller, Q-switching stability limits of continuous-wave passive mode locking, *J. Opt. Soc. Am. B*, vol. 16, pp. 46-56, 1999.
24. R. Paschotta and U. Keller, Passive mode locking with slow saturable absorbers, *Appl. Phys. B*, vol. 73, pp. 653-662, 2001.
25. A. Robertson, M. E. Klein, M. A. Tremont, K.-J. Boller, and R. Wallenstein, 2.5-GHz repetition rate singly resonant optical parametric oscillator synchronously pumped by a mode-locked diode oscillator amplifier system, *Opt. Lett.*, vol. 25, pp. 657-659, 2000.
26. A. Sennaroglu, Analysis and optimization of lifetime thermal loading in continuous-wave Cr^{4+}-doped solid-state lasers, *J. Opt. Soc. Am B*, vol. 18, pp. 1578-86, 2001.
27. A. Sennaroglu, Optimization of power performance in room-temperature contunuous-wave Cr^{4+}:YAG lasers, *Opt. Comm.*, vol. 192, 83–89, 2001.
28. U. Keller, Semiconductor nonlinearities for solid-state laser modelocking and Q-switching, in *Nonlinear Optics in Semiconductors*, vol. 59, *Semiconductors and Semimetals*, E. Garmire and A. Kost, Eds. Boston: Academic Press, Inc., 1999, pp. 211-286.
29. L. Krainer, R. Paschotta, G. J. Spühler, I. Klimov, C. Y. Teisset, K. J. Weingarten, and U. Keller, Tunable picosecond pulse-generating laser with a repetition rate exceeding 10 GHz, *Electron. Lett.*, vol. 38, pp. 225-227, 2002.
30. G. J. Spühler, L. Krainer, I. Kilburn, M. Dymott, K. J. Weingarten, R. Paschotta, and U. Keller, Picosecond pulse-generating laser with repetition rate of 10 GHz and rms timing jitter <100 fs fully tunable over C-band, presented at Optical Fiber Communication Conference and Exposition, Atlanta, U.S.A., 2003.
31. S. C. Zeller, L. Krainer, G. J. Spühler, K. J. Weingarten, R. Paschotta, and U. Keller, Passively modelocked 40-GHz Er:Yb:glass laser, *Appl. Phys. B*, vol. 76, pp. 1181-1182, 2003.
32. H. Tsuchida, Pulse timing stabilization of a mode-locked Cr:LiSAF laser, *Opt. Lett.*, vol. 24, pp. 1641-1643, 1999.
33. G. J. Spühler, M. Dymott, I. Klimov, G. Luntz, L. Baraldi, I. Kilburn, P. Crosby, S. Thomas, O. Zehnder, C. Y. Teisset, M. Brownell, K. J. Weingarten, R. Dangel, B. J. Offrein, G. L. Bona, O. Boccafusca, Y. Kaneko, L. Krainer, R. Paschotta, and U. Keller, 40 GHz pulse generating source with less than 350 fs timing jitter, *Electron. Lett.*, vol. 38, pp. 1031-1033, 2002.
34. E. A. D. Souza, M. C. Nuss, W. H. Knox, and D. A. B. Miller, Wavelength-division multiplexing with femtosecond pulses, *Opt. Lett.*, vol. 20, pp. 1166-1168, 1995.
35. L. Boivin, M. Wegmueller, M. C. Nuss, and W. H. Knox, 110 Channels x 2.35 Gb/s from a single femtosecond laser, *IEEE Phot. Techn. Lett.*, vol. 11, pp. 466-468, 1999.
36. H. Takahashi, I. Nishi, and Y. Hibino, 10 GHz spacing optical frequency-division multiplexer based on arrayed waveguide grating, *Electron. Lett.*, vol. 28, pp. 380-382, 1992.

37. H. Takara, E. Yamada, T. Ohara, K. Sato, K. Jinguji, Y. Inoue, T. Shibata, and T. Morioka, 106 x 10Gb/s, 25 GHz spaced, 640 km DWDM transmission employing a single super-continuum multi-carrier source, presented at Conference on Lasers and Electro-Optics CLEO '01, Baltimore, 2001.
38. G. J. Spühler, P. S. Golding, L. Krainer, I. J. Kilburn, P. A. Crosby, M. Brownell, K. J. Weingarten, R. Paschotta, M. Haiml, R. Grange, and U. Keller, Novel multi-wavelength source with 25-GHz channel spacing tunable over the C-band, *Electron. Lett.*, vol. 39, pp. 778-80, 2003.
39. R. D. Peterson, H. P. Jenssen, and A. Cassanho, Investigation of the spectroscopic properties of Nd:YVO_4, presented at Advanced Solid–State Lasers (ASSL 2002), Quebec, Canada, 2002.
40. Kigre Inc., QX Laser Glasses, Data Sheet, 1996.
41. L. Krainer, R. Paschotta, S. Lecomte, M. Moser, K. J. Weingarten, and U. Keller, Compact Nd:YVO_4 lasers with pulse repetition rates up to 160 GHz, *IEEE J. Quantum Electron.*, vol. 38, pp. 1331-1338, 2002.
42. S. Lecomte, L. Krainer, R. Paschotta, M. J. P. Dymott, K. J. Weingarten, and U. Keller, Optical parametric oscillator with 10 GHz repetition rate and 100 mW average output power in the 1.5-μm spectral region, *Opt. Lett.*, vol. 27, pp. 1714-1717, 2002.
43. R. L. Sutherland, *Handbook of Nonlinear Optics*. New York: Marcel Dekker, Inc., 1996.

Single-Cycle Optical Pulses Produced by Coherent Molecular Oscillations

A. V. Sokolov

Department of Physics and Institute for Quantum Studies, Texas A&M University, College Station, TX 77843-4242, U. S. A.
E-mail: sokol@physics.tamu.edu

Summary. We have demonstrated that coherent molecular motion can result in a collinear generation of equidistant mutually-coherent spectral sidebands, extending in frequency from infrared to far ultraviolet. Our technique is based on adiabatic preparation of a highly coherent molecular superposition-state, which is achieved by using two narrow-linewidth lasers slightly detuned from a Raman resonance. The phases of the resultant Stokes and anti-Stokes sidebands are then adjusted in order to synthesize desired single-cycle waveforms at the target [A.V. Sokolov and S. E. Harris, J. Opt. B **5**, R1 (2003)]. By the very nature of the generation process, this light source produces trains of pulses, which are perfectly synchronized with the molecular motion in the given molecular system and provide a unique tool for studying molecular and electronic dynamics. We envision producing a coherent molecular oscillation, applying a tightly focused train of perfectly timed pulses, adjusting the delay, and studying electronic properties as functions of molecular coordinates. In the future, this Raman source may also produce sub-cycle optical pulses, and allow synthesis of waveforms where the electric field is a predetermined function of time, not limited to quasi-sinusoidal oscillations. As a result, a direct and precise control of electron trajectories in photoionization and high-order harmonic generation will become possible.

1 Introduction

According to the Heisenberg uncertainty relationship, access to the ultra-short time scale requires a high energy per particle. The fastest processes in nature occur in high-energy collisions and nuclear reactions. The techniques of high-energy and nuclear physics allow only a theoretical reconstruction of the time-domain evolution of individual events. On the contrary, ultrashort laser pulses allow direct observation and precise control of ultrafast fundamental processes in atoms and molecules.

For over a decade and until very recently, the shortest optical pulses were obtained by expanding the spectrum of a mode-locked laser by self-phase modulation in an optical fiber, and then compensating for group velocity dispersion (GVD) by diffraction grating and prism pairs [1]. With this technology it became possible to see how atoms in a molecule move during a chemical reaction [2]. In combination with the spectral modification technique [3], this technology allowed coherent control of molecular dynamics on a time scale approaching few femtoseconds [4]. Electronic motion in atoms occurs on a shorter, subfemtosecond time scale. However the production of shorter laser pulses requires new approaches. When produced, subfemtosecond pulses may be used to observe and control rapid motion of inner-shell electrons, or could provide information about fast ionization processes [5].

In this paper, after a brief overview of several emerging subfemtosecond laser techniques, I will describe the adiabatic Raman technique recently proposed in Steve Harris's group [6]. This technique holds great promise for generation of short pulses with a spectrum centered around the visible region, where subfemtosecond pulse duration implies the possibility for a sub-cycle field-shape control. I will review the unique features of this technique, such as the possibility to synthesize pulse trains that are perfectly synchronized with molecular oscillations, and discuss perspectives for this new light source in ultrafast studies of atomic and molecular motion.

Short pulse generation requires a wide phase-locked spectrum. A recent development in this direction is the use of photonic crystal fibers to produce a super-continuum of light [7]. A unique feature of these photonic band-gap structures is a low dispersion over a wide frequency range, which has allowed a significant expansion of the mode-locked laser spectrum and lead to construction of an optical clock [8]. When such a wide coherent bandwidth is produced, spectral phase adjustment can result in ultrashort pulse compression [7]. Other authors have discussed generation of subfemtosecond pulses by phase-locking a spectrum of equidistant frequency components [9]. Workers in the field of high-order harmonic generation (HHG) have predicted the formation of attosecond pulse trains [10], and have proposed methods for a single pulse selection [11]. Very recently workers in the field of HHG have measured subfemtosecond pulses in the X-ray spectral region [12], and used these pulses to trace the relaxation dynamics of core-excited atoms with subfemtosecond resolution [13]. In a related work, it was suggested that the atomic behavior could also be studied by an ultrafast electron wavepacket scattering, which is similar to the process leading to HHG [14]. HHG is a unique source of X-ray pulses, but by their very nature these pulses are difficult to control because of intrinsic problems of X-ray optics; besides, the conversion efficiency into these pulses is very low (typically 10^{-5}) [10, 11].

Another promising source of a wide bandwidth radiation is coherent Raman generation. Imasaka and colleagues have demonstrated the generation of a broad rotational Raman spectrum in molecular hydrogen (H_2) and discussed possibilities for phase-locking this spectrum [15]; Kaplan and Shkolnikov have predicted the existence of 2π Raman solitons with a phase-locked spectrum that Fourier-transforms into a train of subfemtosecond pulses [16]. Other authors have studied the collinear multicolor Raman generation [17]. There has also been significant work on dynamics of an intense femtosecond laser pulse in a Raman-active medium, including predictions of 2π pulse formation and attosecond pulse generation [18]. Recent experiments have shown that a weak probe pulse can be compressed by molecular oscillations which are excited impulsively by a strong pump pulse [19]. It was also demonstrated that molecular wavepacket revivals can produce frequency chirp, which in turn would allow femtosecond pulse compression by normal dispersion in a thin output window [20].

2 The technique of molecular modulation

We have investigated a broadband Raman light source, which is based on collinear generation of a wide spectrum of equidistant, mutually-coherent Raman sidebands [21]. The essence of this technique is the use of a Raman transition with a sufficiently large coherence that the generation length and the phase-slip length are of the same order. The technique utilizes ideas of electromagnetically induced transparency (EIT) [22] and relies on adiabatic preparation of a coherent superposition-state. The coherence is established by driving the molecular transition with two single-mode laser fields, slightly detuned from the Raman resonance so as to excite a single molecular eigenstate (Fig. 1). Molecular motion, in turn, modulates the driving laser frequencies, causing the collinear generation of a very broad spectrum. This process is considerably different from traditional stimulated Raman scattering, and is termed "molecular modulation" [6].

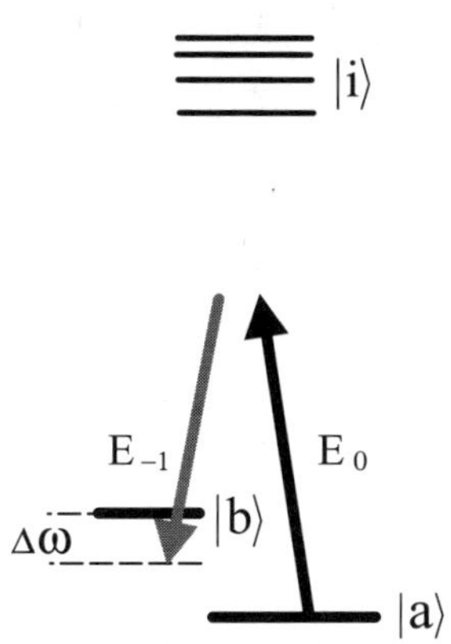

Fig. 1. Energy level schematic for establishing coherence ρ_{ab} in a molecular system. $|a\rangle$ is the ground state and $|b\rangle$ is the excited molecular state. E_0 and E_{-1} are the applied laser fields. The Raman detuning $\Delta\omega$ is set by the driving laser frequencies. The detuning from the electronic states $|\imath\rangle$ is comparable to the applied laser frequencies.

In our experiments we used two synchronized transform-limited laser pulses (typically 80 mJ / 15 ns at 10 Hz repetition rate) at wavelengths of 1.06 μm and 807 nm, such that the (tunable) laser-frequency difference was approximately equal to the fundamental vibrational frequency in D_2 (2994 cm^{-1}). The laser beams were combined on a dichroic beamsplitter and loosely focused into a D_2 cell. When we tuned the driving infrared lasers to within 1 GHz from the Raman resonance, we saw a bright beam of white light at the output of the D_2 cell. We dispersed the spectrum with a prism and projected onto a white scintillating screen. We observed up to thirteen anti-Stokes sidebands and two Stokes sidebands, in addition to the two driving frequencies, with energy above 1 mJ per 10 ns pulse for each of the nine central sidebands [23]. The generated sidebands are spaced by 2994 cm^{-1} and range from 2.94 μm to 195 nm in wavelength. The next Stokes sideband has a wavelength of 24 μm and is difficult to detect. The next (fourteenth) anti-Stokes sideband has a wavelength of 184 nm and is absorbed by air. Figure 2 shows a typical picture of the spectrum taken with a digital color camera at a single-shot exposure. The smooth near-Gaussian beam profiles for nearly all sidebands demonstrate collinear anti-Stokes generation in a regime of high molecular coherence. Very recently the same technique has allowed collinear generation of rotational sidebands in H_2 with a 100% conversion efficiency [25].

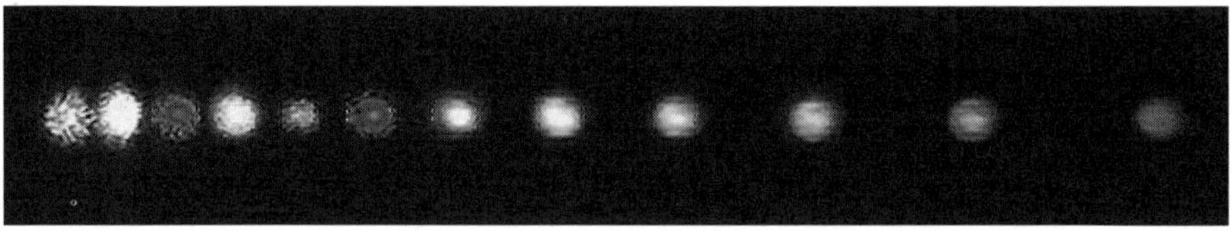

Fig. 2. Typical spectrum produced by molecular modulation in D_2. More detailed data can be found in Ref. [23]. Starting from the left, the first two sidebands are the driving frequencies (infrared). The next four anti-Stokes sidebands correspond to red, green, blue, and violet light. Beginning at the fifth anti-Stokes, the sidebands are in the ultraviolet and only fluorescence is visible. Sideband generation is collinear and highly efficient.

3 Ionization by a superposition of coherent sidebands

In further experiments we have shown good mutual coherence of the Raman sidebands produced by the molecular modulation [24, 26]. The mutual coherence across sidebands spatial and temporal profiles allows to recombine them and use spectral modification techniques [3] to synthesize specified femtosecond time structures in a target cell (Fig. 3). We used this light source to demonstrate coherent control of multiphoton ionization with shaped few-femtosecond pulses, in a regime not accessible by other sources [26]. Furthermore, we used the multiphoton ionization process to characterize our light source. We performed cross-correlation measurements, in a collinear configuration. We used five Raman sidebands (at 1.06 μm, 807 nm, 650 nm, 544 nm, and 468 nm). We varied the delay of the 544 nm and 468 nm beams with respect to the 650 nm, 807 nm and 1.06 μm beams (for details see Ref. [26]). The result of this measurement is shown in Fig. 4. The peaks in this Figure can be interpreted as optical field oscillations. The experiment agrees well with theory [26].

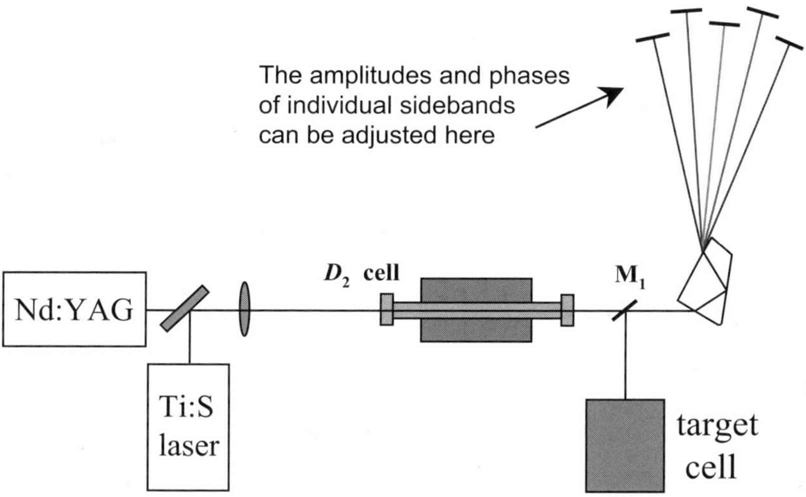

Fig. 3. Experimental setup for coherent molecular excitation, collinear Raman generation, and spectral phase adjustment. We varied sideband phases independently with tilted glass plates (not shown) inserted into individual beam paths. The mirror M1, which was displaced in the vertical plane, picked off the slightly offset retroreflected beam. The target cell was used to observe photo-ionization of low-pressure xenon gas.

It is known that a phase-locked spectrum, with a bandwidth equal to one octave, corresponds to single-cycle pulses. In the past, single-cycle (and subcycle) pulses were available only in the THz spectral region, with a duration of about 0.5 ps [27]. Our experimental results (in particular the high contrast of the ion signal in Fig. 4, and the theory/experiment agreement), establish evidence for a good mutual coherence among frequency components, which extend over an octave of optical bandwidth, and prove our ability to phase-lock these components. Thus we infer that in this experiment we are able to synthesize trains of single-cycle pulses (with a repetition period of 11 fs and a pulse duration of 2 fs). Photo-ionization with single-cycle pulses is a first step toward studying atomic dynamics on the time scale of electronic motion.

Several time scales are important for an experiment of this type: (a) the shortest accessible time interval is below 10^{-15} s; (b) the period of the syn-

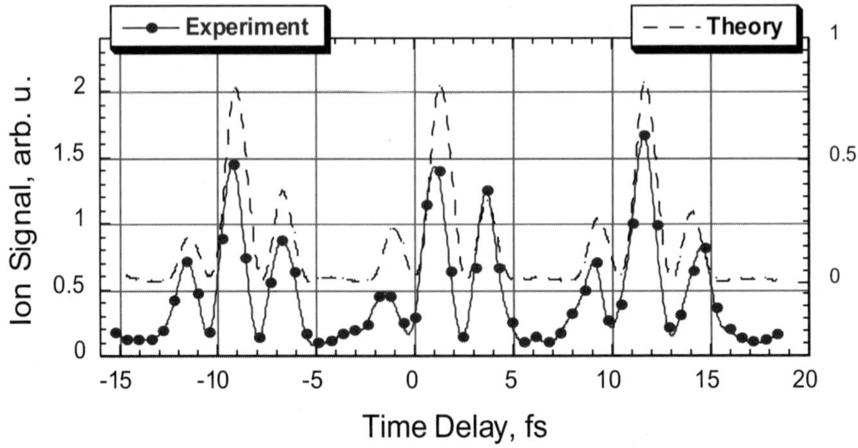

Fig. 4. Correlation of waveforms, synthesized by subsets of Raman sidebands: experiment (solid line) and theory (dashed line) (for details see Ref. [26], from where the data is adopted).

thesized waveform is equal to the molecular period (around 10^{-12}-10^{-14} s); (c) the duration of the ultra-short pulse train is approximately equal to the driving pulse duration (10^{-8} s); (d) these trains are produced at 10 Hz laser firing rate (10^{-1} s scale); (e) mechanical drifts in the spectral control setup have to be smaller than the laser wavelength on a time scale of 10^3 s, in order to allow controllable phase adjustments for the generated sidebands.

4 Synchronization of the pulse train with respect to the molecular oscillations

As discussed above, our technique will produce a train of ultrashort pulses with a very high repetition rate. While there are several possibilities for single pulse production, it should be emphasized that trains of pulses have their own advantages. There are two possibilities. When studying processes with a relaxation rate faster than the pulse repetition rate, the system will respond to each individual pulse independently; in this case the signals produced by different pulses will simply add up. The other possibility is to synchronize the pulse train with the process of interest [28].

By the very nature of the generation process, the pulses produced by molecular modulation are perfectly synchronized with the molecular motion in the given molecular system, and provide a unique tool for studying molecular and electronic dynamics in that system. The key idea is to use the same molecular species for pulse generation and for the target. We envision producing a coherent molecular oscillation in the target cell, applying a tightly focused train of perfectly timed pulses, and studying electronic properties as function of molecular coordinates [28] (Fig. 5). Possible extensions of this general technique range from studying complicated multi-mode motion of complex molecules, to probing ultrafast electronic dynamics in atoms.

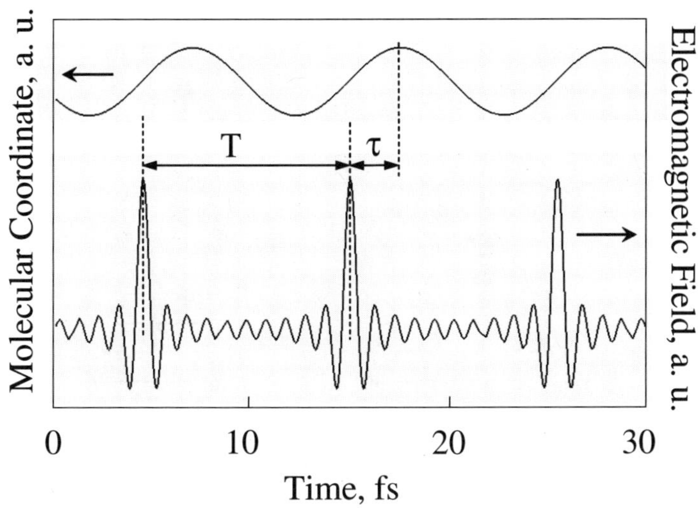

Fig. 5. Synchronization of the pulse train with respect to the molecular oscillation, as shown in the time domain.

In the frequency domain, molecular motion corresponds to a coherent superposition of molecular states. The fact that the pulse train can be synchronized with the molecular motion means that there is a possibility for different multiphoton paths, starting from different molecular levels, to interfere constructively or destructively, as shown in Fig. 6 (a). Figure 6 (b) alludes to the possibility that multiple molecular transitions can be excited. There is an analogy of this technique to EIT [22]. EIT relies on negation of refraction and absorption by quantum interference and allows unperturbed propagation of resonant laser beams through optically thick media. In the frequency-domain representation (Fig. 6) the analogy to EIT is particularly clear: with appropriate sideband phasing (or rather "anti-phasing") we can expect an EIT-like molecular stabilization against multiphoton ionization [29].

5 Perspectives for sub-cycle field shaping

Femtosecond pulses, which are now readily available in many laboratories around the world, consist of few optical cycles and allow synthesis of shaped waveforms with controlled amplitude and frequency. These waveforms are limited to quasi-sinusoidal pulse shapes. Subfemtosecond pulse duration is shorter than a single optical cycle, for the visible spectral region. When used

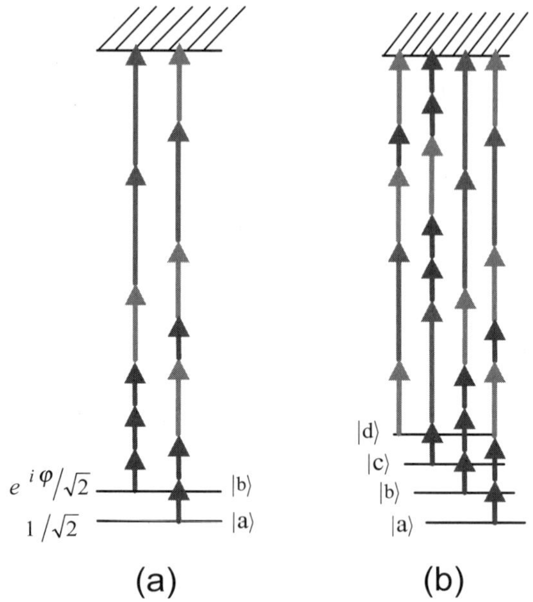

Fig. 6. Synchronization of the pulse train with respect to the molecular oscillation, as shown in the frequency domain. Part (a) shows a molecular transition at maximal coherence. Part (b) corresponds to a coherent excitation of several molecular transitions (wave-packet motion).

for pulse shaping, sub-cycle pulses will provide an unprecedented opportunity to control not just the pulse envelope (amplitude and frequency as functions of time), but also the shape of the electric field itself [28].

In our earlier experiments one of the driving laser frequencies was fixed, and the generated sideband frequencies were incommensurate. In the future experiments we propose to use two tunable driving lasers, adjust the driving frequencies to satisfy the detuning requirements for the Raman resonance, and at the same time make these frequencies equal to the multiple of their frequency difference. As the generated broadband spectrum is then used for the sub-cycle pulse synthesis, the carrier-envelope phase of the individual pulses (as defined in Refs. [30, 31], and otherwise known as the absolute phase) will be fixed.

If no special measures are taken, the carrier-envelope phase will change from shot to shot (even though it will be the same for the pulses within the same train (for a single shot). In order to achieve sub-cycle shape control, one can apply a second harmonic together with the fundamental driving lasers. If laser frequencies are equal to a multiple of their frequency-difference, then one of the Raman sidebands will coincide in frequency with the second harmonic. One can then select laser shots based on the interference result (constructive or destructive) of this second harmonic and the corresponding

Raman sideband. It can be shown [30] that the interference provides sufficient information for determining the carrier-envelope phase. As it may not be practical (with current technologies) to stabilize the relative phase of two laser oscillators on the time scale of minutes, we will not attempt to stabilize the carrier-envelope phase. Rather, we will sort shots by the carrier-envelope phase. This sorting will be sufficient for studying absolute phase phenomena, and for sub-cycle pulse shaping.

As a first application of this light source, we can look at ion and electron yields in low-pressure gases as functions of the carrier-envelope phase of the synthesized pulse trains. We can study correlations between the ion yield and the type of interference (constructive or destructive) of the second harmonic field with a high-order Raman sideband. Another possibility would be to look at the spatial distribution of photoelectrons as a function of the sub-cycle pulse shape [31]. In our earlier experiment [26] we have shown that molecular modulation provides sufficient laser intensities to produce ionization of xenon. In that experiment we used incommensurate laser frequencies; as a result, different multiphoton-order processes lead to different ionized electron energy states, and could not interfere with each other. When the sideband frequencies are made equal to a multiple of their frequency difference, different order multiphoton path will interfere, and the absolute phase effects will come into play. This is the situation shown in Fig. 6 for molecules, but even for simple atomic systems the observation of such phenomena will be a significant first step. At increased intensities, electron re-collision processes will become important, and a complex atomic behavior (revealed by the ion and electron yields) as a function of the sub-cycle pulse shape, will be expected [32].

The possibility to produce non-sinusoidal waveforms where the electric field is an arbitrary predetermined function of time will also open new opportunities for a direct and precise control of electronic motion in HHG. In the three-step model for HHG [33], the atomic potential is strongly perturbed by an ultra-strong laser field. A free electron is produced by tunneling ionization and then moves under the action of the sinusoidal laser field. The electron is driven away from its parent ion, and is then accelerated to a high energy as it moves back. The recollision of the electron with the ion results in a production of an attosecond X-ray burst, which in the frequency domain gives the well-known plateau spectrum for high-order harmonics with a cut-off frequency proportional to the electrons pondermotive energy. HHG is an example of a process with a few-femtosecond "relaxation time", which is a result of a fast spatial spread of the wavefunction for the free electron. This kind of a process is ideal for studying with high-repetition-rate pulse trains.

If in an HHG experiment one is able to adjust the sub-cycle pulse shape, one will have a precise control over the electron trajectory, the recollision time, and the recollision energy; that will open a range of possibilities for studying fundamental processes in atoms. Subfemtosecond control of electron trajectories was recently demonstrated by using shaped mode-locked laser pulses [34], and also by using femtosecond pulses with controlled absolute phase [35]. It was also shown that a light wave can accelerate or decelerate an electron wave packet when a subfemtosecond ionizing X-ray pulse is timed to the phase of the light wave [36]. A particularly exciting possibility is to use sub-cycle pulses synchronized with molecular oscillations to study HHG in aligned molecules [37]. There one will be able to observe structural changes (probed by HHG) as a molecule undergoes a large-amplitude motion.

The idea of producing sub-cycle pulses synchronized with molecular motion can be combined with a recent proposal to merge the adiabatic Raman technique with the impulsive technique. According to this method, a short probe pulse can be compressed into a singlet, doublet or triplet of subfemtosecond pulses by beating with an adiabatically prepared Raman coherence [38]. This method has the advantage of the adiabatic pumping in producing a large Raman coherence and the advantage of the impulsive excitation in reducing the number of pulses per train. The proposal relies on a thorough understanding of the mechanisms of molecular modulation [39]. Additional improvements may come from putting the molecular medium, which provides the wide-band generation, into a hollow-core photonic crystal fiber [40]. The ability to engineer the dispersion characteristics of microstructure fibers may allow control over the intensity distribution in the generated spectrum, by setting modified phase-matching conditions for the generated frequencies [41].

In conclusion, among the techniques that show promise for subfemtosecond pulse generation, the adiabatic Raman technique is unique in providing the possibility to produce sub-cycle pulses and non-sinusoidal fields synchronized with molecular oscillations.

The author thanks Andrea Burzo, Deniz Yavuz, David Walker, G. Y. Yin, and Steve Harris for valuable contributions, and acknowledges support from Texas Advanced Research Program and from The Welch Foundation. This work was also supported in part by the U. S. Army Research Office, the U. S. Air Force Office of Scientific Research, and the U. S. Office of Naval Research.

References

1. R. L. Fork, C. H. Brito-Cruz, P. C. Becker, C. V. Shank, Opt. Lett. **12**, 483 (1987); A. Baltuska, Z. Wei, M. S. Pshenichnikov, D. A. Wiersman, Opt. Lett. **22**, 102 (1997); M. Nisoli, S. DeSilvestri, O. Svelto, R. Szipocs, K. Ferencz, Ch. Spielmann, S. Sartania, F. Krausz, Opt. Lett. **22**, 522 (1997).
2. A. Zewail, J. Phys. Chem. A, **104**, 5660 (2000).
3. C. W. Hillegas, J. X. Tull, D. Goswami, D. Strickland, W. S. Warren, Opt. Lett. **19**, 737 (1994); A. M. Weiner, Prog. Quant. Electr. **19**, 161 (1995); T. Baumert, T. Brixner, V. Seyfried, M. Strehle, G. Gerber, Appl. Phys. B, **65**, 779 (1997).

4. S. A. Rice and M. Zhao, *Optical Control of Molecular Dynamics* (Wiley, New York, 2000); M. Shapiro and P. Brumer, Adv. At. Mol. Opt. Phys. **42**, 287 (2000); R. J. Levis and H. A. Rabitz, J. Phys. Chem. A, **106**, 6427 (2002).
5. F. Krausz, Phys. World **14**, 41, (2001).
6. A. V. Sokolov and S. E. Harris, J. Opt. B, **5**, R1-R26 (2003).
7. J. K. Ranka, R. S. Windeler, A. J. Stentz, Opt. Lett. **25**, 25 (2000); A. M. Zheltikov, Phys. Usp. **45**, 687 (2002); P. St. J. Russell, Science **299**, 358 (2003).
8. T. Udem, R. Holzwarth, T. W. Hansch, Nature **416**, 233 (2002).
9. T. W. Hansch, Opt. Commun. **80**, 71, (1990).
10. Gy. Farcas and Cs. Toth, Phys. Lett. A **168**, 447, (1992); S. E. Harris, J. J. Macklin, T. W. Hansch, Opt. Commun. **100**, 487 (1993); P. Antoine, A. L.'Huiller, M. Lewenstein, Phys. Rev. Lett. **77**, 1234 (1996); K. J. Schafer and K. C. Kulander, Phys. Rev. Lett. **78**, 638 (1997).
11. P. B. Corkum, N. H. Burnett, M. Y. Ivanov, Opt. Lett. **19**, 1870, (1994); I. P. Christov, M. M. Murnane, H. C. Kapteyn, Phys. Rev. Lett. **78**, 1251 (1997).
12. P. M. Paul, E. S. Toma, P. Breger, G. Mullot, F. Auge, Ph. Balcou, H. G. Muller, P. Agostini, Science **292**, 1689, (2001); M. Hentschel, R. Kienberger, Ch. Spielmann, G. A. Reider, N. Milosevic, T. Brabec, P. Corkum, U. Heinzmann, M. Drescher, F. Krausz, Nature **414**, 509, (2001).
13. M. Drescher, M. Hentschel, R. Kienberger, M. Uiberacker, V. Yakovlev, A. Scrinzi, Th. Westerwalbesloh, U. Kleineberg, U. Heinzmann, F. Krausz, Nature **419**, 803 (2002).
14. H. Niikura, F.Legare, R. Hasbani, A. D. Bandrauk, M. Yu. Ivanov, D. M. Villeneuve, P. B. Corkum, Nature **417**, 917 (2002).
15. S. Yoshikawa and T. Imasaka, Opt. Commun. **96**, 94 (1993); H. Kawano, Y. Hirakawa, T. Imasaka, IEEE J. Quant. Electr. **34**, 260, (1998).
16. A. E. Kaplan, Phys. Rev. Lett. **73**, 1243 (1994); A. E. Kaplan and P. L. Shkolnikov, J. Opt. Soc. Am. B **13**, 347 (1996).
17. V. Schulz-von der Gathen, T. Bornemann, V. Kornas, H. F. Dobele, IEEE J. Quant. Electr. **26**, 739, (1990); L. L. Losev and A. P. Lutsenko, Quantum Electron. **23**, 919, (1993); G. S. McDonald, G. H. C. New, L. L. Losev, A. P. Lutsenko, M. Shaw, Opt. Lett. **19**, 1400 (1994); J. Z. Li, M. Katsuragawa, M. Suzuki, K. Hakuta, Phys. Rev. A **58**, R58 (1998).
18. E. M. Belenov, A. V. Nazarkin, I. P. Prokopovich, Pis'ma Zh. Eksp. Teor. Fiz. **55**, 223 (1992); E. M. Belenov, P. G. Kryukov, A. V. Nazarkin, I. P. Prokopovich, Zh. Eksp. Teor. Fiz. **105**, 28 (1994); I. P. Prokopovich and A. A. Khrushckinskii, Laser Phys. **7**, 305 (1997).
19. M. Wittmann, A. Nazarkin, G. Korn, Phys. Rev. Lett. **84**, 5508 (2000); Opt. Lett. **26**, 298 (2001); N. Zhavoronkov and G. Korn, Phys. Rev. Lett. **88**, 203901 (2002).
20. V. Kalosha, M. Spanner, J. Herrmann, M. Ivanov, Phys. Rev. Lett. **88**, 103901 (2002); R. A. Bartels, T. C. Weinacht, N. Wagner, M. Baertschy, Chris H. Greene, M. M. Murnane, H. C. Kapteyn, Phys. Rev. Lett. **88**, 013903 (2002).
21. S. E. Harris and A. V. Sokolov, Phys. Rev. A **55**, R4019 (1997); S. E. Harris and A. V. Sokolov, Phys. Rev. Lett. **81**, 2894 (1998); A. V. Sokolov, D. D. Yavuz, S. E. Harris, Opt. Lett. **24**, 557 (1999); F. L. Kien, J. Q. Liang, M. Katsuragawa, K. Ohtsuki, K. Hakuta, A. V. Sokolov, Phys. Rev. A **60**, 1562 (1999); A. V. Sokolov, Opt. Lett. **24**, 1248 (1999).
22. For review see S. E. Harris, Physics Today **50**, 36 (1997); M. O. Scully and M. S. Zubairy, *Quantum Optics* (Cambridge University Press, Cambridge, 1997).

23. A. V. Sokolov, D. R. Walker, D. D. Yavuz, G. Y. Yin, S. E. Harris, Phys. Rev. Lett. **85**, 562 (2000).
24. A. V. Sokolov, D. D. Yavuz, D. R. Walker, G. Y. Yin, S. E. Harris, Phys. Rev. A **63**, 051801(R) (2001).
25. D. D. Yavuz, D. R. Walker, G. Y. Yin, S. E. Harris, Opt. Lett. **27**, 769 (2002).
26. A. V. Sokolov, D. R. Walker, D. D. Yavuz, G. Y. Yin, S. E. Harris, Phys. Rev. Lett. **87**, 033402 (2001).
27. R. R. Jones, D. You, P. H. Bucksbaum, Phys. Rev. Lett. **70**, 1236 (1993).
28. A. V. Sokolov, Appl. Phys. B **77**, 343 (2003).
29. A. Burzo and A. V. Sokolov, to be published.
30. H.R. Telle, G. Steinmeyer, A.E. Dunlop, J. Stenger, D.H. Sutter, U. Keller, Appl. Phys. B **69**, 327-332 (1999); D. J. Jones, S. A. Diddams, J. K. Ranka, A. Stentz, R. S. Windeler, J. L. Hall, S. T. Cundiff, Science **288**, 635 (2000); A. Apolonski, A. Poppe, G. Tempea, Ch. Spielmann, Th. Udem, R. Holzwarth, T. W. Hansch, F. Krausz, Phys. Rev. Lett. **85**, 740 (2000).
31. G. G. Paulus, F. Grasbon, H. Walther, P. Villoresi, M. Nisoli, S. Stagira, E. Priori, S. DeSilvestri, Nature (London) **414**, 182 (2001).
32. This idea is related to the work on phase control of multiphoton excitations using a second or a third harmonic together with the fundamental laser frequency [for references, see R. A. Blank and M. Shapiro, Phys. Rev. A **52**, 4278 (1995); M. Protopapas, C. H. Keitel , P. L. Knight , Rep. Prog. Phys. **60**, 454 (1997)].
33. P. B. Corkum, Phys. Rev. Lett. **71**, 1994 (1993).
34. R. Bartels, S. Backus, E. Zeek, L. Misoguti, G. Vdovin, I. P. Christov, M. M. Murnane, H. C. Kapteyn, Nature **406**, 164 (2000); I. P. Christov, R. Bartels, H. C. Kapteyn, M. M. Murnane, Phys. Rev. Lett. **86**, 5458 (2001).
35. A. Baltuska, Th. Udem, M. Uiberacker, M. Hentschel, E. Goulielmakis, Ch. Gohle, R. Holzwarth, V. S. Yakovlev, A. Scrinzi, T. W. Hansch, F. Krausz, Nature **421**, 611 (2003).
36. R. Kienberger, M. Hentschel, M. Uiberacker, Ch. Spielmann, M. Kitzler, A. Scrinzi, M. Wieland, Th. Westerwalbesloh, U. Kleineberg, U. Heinzmann, M. Drescher, F. Krausz, Science **297**, 1144 (2002).
37. R. Velotta, N. Hay, M. B. Mason, M. Castillejo, J. P. Marangos, Phys. Rev. Lett. **87**, 183901 (2001); M. Lein, N. Hay, R. Velotta, J. P. Marangos, P. L. Knight, Phys. Rev. Lett. **88**, 183903 (2002).
38. Fam Le Kien, Nguyen Hong Shon, K. Hakuta, Phys. Rev. A **64**, 051803(R) (2001).
39. Fam Le Kien, K. Hakuta, A. V. Sokolov, Phys. Rev. A **66**, 023813 (2002); Phys. Rev. A **66**, 055802 (2002).
40. F. Benabid, J. C. Knight, G. Antonopoulos, P. St. J. Russell, Science **298** 399 (2002).
41. Nguyen Hong Shon, Fam Le Kien, K. Hakuta, A. V. Sokolov, Phys. Rev. A **65**, 033809 (2002).

A Robust New Concept for Octave-Spanning Chirped Mirrors with Minimal Dispersion Ripple

Günter Steinmeyer

Max-Born-Institute for Nonlinear Optics and Short-Pulse Spectroscopy,
Max-Born-Straße 2a, D-12489 Berlin, Germany steinmey@mbi-berlin.de

1 Introduction

The generation of ultrafast optical pulses has always relied strongly on the capabilities of dispersion compensation schemes. Some recent advances into the sub-5-fs regime have only been possible by improved dispersion compensation, be it by using chirped mirrors [1–3], liquid-crystal phase modulators in a zero dispersion delay line [4–6], or Raman sideband generation [7, 8]. However, neither method has been demonstrated with spectra exceeding a bandwidth of one optical octave. Here we address a novel route towards compression of ultrabroadband continua, based on a new generation of chirped mirrors.

In the case of chirped mirrors, the fundamental barrier to further extension of its bandwidth can be tracked down to parasitic Gires-Tournois interferometer (GTI, [9]) effects, mainly arising at the interface between the chirped mirror stack and the ambient medium. Together with a reflection inside the mirror stack, the parasitic top reflection forms a variable-path GTI, which causes a modulation of the group delay dispersion vs. wavelength. This dispersion ripple overlays the desired dispersion characteristics of a chirped mirror and creates satellite pulses, which merge into a broad temporal pedestal upon multiple bounces off chirped mirrors [10]. Generally, a tradeoff between bandwidth and suppression of dispersion ripple can be seen and limits the usefulness of chirped mirrors to much less than an optical octave. Several ways to reduce the dispersion ripple of chirped mirrors have been developed. The most straightforward idea is suppression of the disturbing top reflection, either by an antireflection (AR) coating [11, 12] or by passage through a substrate [13, 14]. The former approach impedance-matches from air as the ambient medium to one of the coating layer materials, whereas the latter uses an ambient medium with an index identical to one of the coating layer materials. Neither approach works well without additional impedance matching measures inside the layer stack [13]. For this purpose, double-chirping inside the layer stack is used [11]. Generally, these ideas work well for chirped mirrors of moderate bandwidth. The AR coating approach is limited to about 0.6 optical octaves by bandwidth issues of the AR coating. Passage through the substrate can extend the useful coating bandwidth to an optical octave. However, this approach is mechanically cumbersome because extremely thin substrates must be used for a net negative dispersion of the mirror.

A third straightforward approach to suppress unwanted reflections, namely Brewster-angle orientation, has not been exploited yet. This method avoids the bandwidth issues of AR coatings and also does not require a complicated optical construction [15]. We will explore the capabilities of the Brewster-angled chirped mirrors in the following.

2 A Coating Design Example for Brewster-Angled Chirped Mirrors

A schematic drawing of a Brewster-angled chirped mirror and one representative optical path in this structure are depicted in Fig. 1. The design of such a mirror structure is widely identical to normal-incidence mirrors [13] . For calculation of the optical path, however, one has to account for the internal angle ϑ_i inside the respective layer, i.e. physical layer thicknesses t_i relate to the optical path length l_i via $l_i = n_i t_i \cos\vartheta_i$, where n_i is the refractive index of the ith layer. A second adjustment from normal to Brewster incidence is dictated by the reduced reflectivity at an interface between high-index and low-index material. For the typically used material combination TiO_2 and SiO_2, e.g., one calculates a 35% decrease of the amplitude reflectivity per interface in the stack. This decrease has to be compensated for by additional layers. As Brewster-angled mirrors do not require the matching sections of conventional chirped mirrors, however, one can also save up to 20 layers. Therefore, given the same constraints of reflectivity and bandwidth, one finds that Brewster-angled mirrors require about the same number of layers as conventional chirped mirrors. Only for wider bandwidth the number

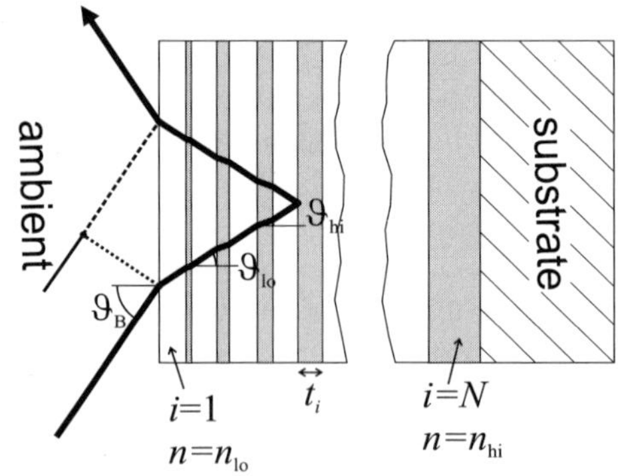

Fig. 1. Beam path inside a chirped mirror structure oriented at Brewster's angle relative to the incident beam.

of layers has to be increased. As a coarse estimate, one has to allocate about 100 layers per octave bandwidth for the Brewster-angle design approach.

In the following, we will discuss a design that is based on 120 layers and covers 1.2 octaves bandwidth. The 10 μm thick design compensates for 40 fs^2 of group delay dispersion (GDD). This is equivalent to the dispersion of 1-mm path length inside a light optical glas. The initial design and an improved computer-optimized version are shown in Fig. 2. As the top layer material,

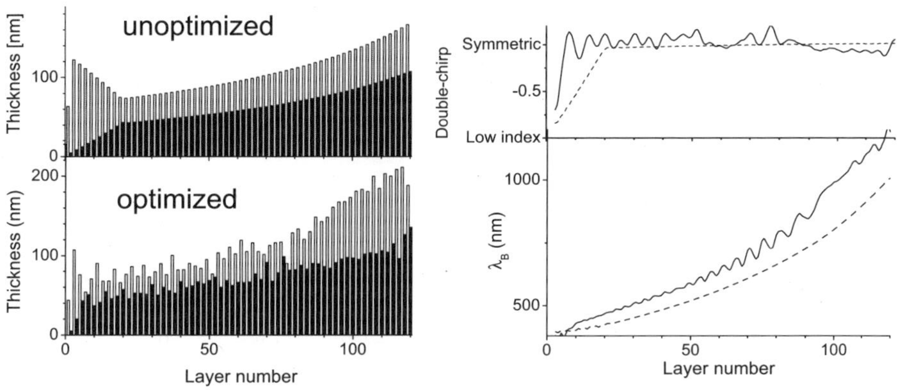

Fig. 2. Coating designs for Brewster incidence. Shown are mechanical thicknesses of the layers. Unshaded bars refer to low-index layers, shaded layers to high-index layers. The initial design is shown in the top left, the bottom left design is computer-optimized. Dashed lines on the right refer to the unoptimized design, solid lines to the optimized one. On the right the double-chirp (i.e. the layer pair symmetry) and the Bragg wavelength, as extracted from the coating designs, are also shown.

SiO_2 with an index of refraction of 1.49 has been assumed. This material results in a Brewster angle of $\vartheta_B = 56.2°$. The spectral characteristics of the coating at Brewster incidence are depicted in Fig. 3. Even without further computer-optimization the coating already exhibits a high reflectivity over more than one octave together with a reasonably smooth GDD. The GDD ripple amounts to about 50 fs^2 (rms) for the octave from 400 – 800 nm. The GD exhibits similar oscillations with an rms value of 2.5 fs. Tuning the incident angle away from Brewster's angle dramatically increases these values, as shown in Fig. 4. The bandwidth of the coating can be extended and the dispersion ripple can be further suppressed by computer optimization, using a local gradient optimization strategy. Comparing the optimized and the initial design, one can see that the improved coating sequence now covers the range from 430 to 1000 nm, both with smooth dispersion and reasonably high reflectivity. The optimized coating exhibits an average reflectivity of 99.5 %, a GD ripple of 0.5 fs (rms), and a GDD ripple of only 8 fs^2 in an 1.2 octave

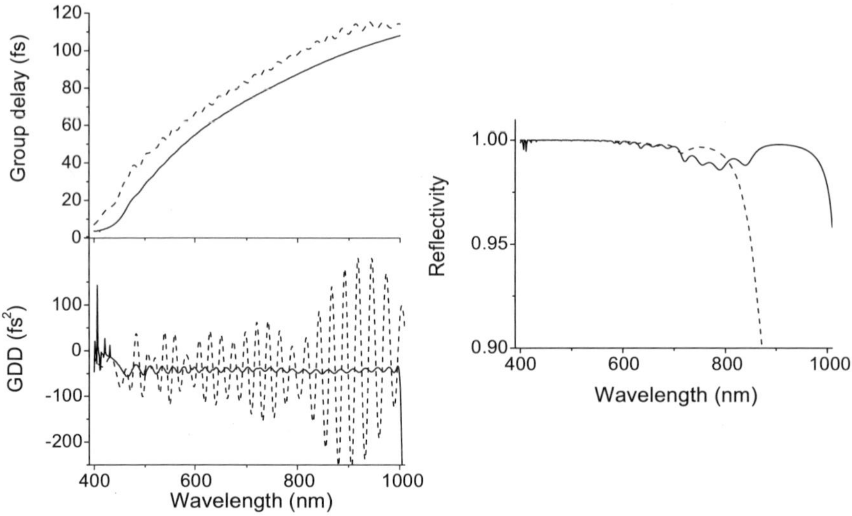

Fig. 3. Reflective amplitude and phase properties of the design of Fig. 2 for Brewster incidence. Dashed lines refer to the unoptimized design, solid lines to the optimized one. On the left the group delay and its dispersion are shown. On the right the power reflectivity of the designs is depicted.

bandwidth. This coating therefore covers nearly double the bandwidth of the most broadband standard DCMs demonstrated to date.

Most remarkably, the basic structure of the layer sequence remains intact during the optimization process. Close to the interface to air, the main effect in the optimization process seems to be a modification of the double-chirp section. The simple linear chirp in the duty-cycle is replaced by a more complex function with a strong oscillatory component in the first 40 layers close to the interface to air. This corrugated double-chirping serves to further suppress the dispersion oscillations. Additionally, changes at the bottom of the layer structure greatly increase the reflectivity of the coating beyond 800 nm.

For a further test of the usefulness of Brewster-angled chirped mirrors, the recompression of a pulse with a 140-THz bandwidth was simulated. It was assumed that the pulse carries a chirp from 400 fs^2 material dispersion. Perfect recompression by 10 bounces off the chirped mirror should then yield the bandwidth-limited duration of 3.2 fs. Figure 4 shows that only the optimized design provides nearly perfect recompression to the sub-4-fs regime, whereas only 7 to 8 fs pulse duration are achieved with the unoptimized design with its stronger dispersion ripple. A good criterion for the deterioration of the pulse shape by residual dispersion ripple is the rms width [10]. The angular dependence of the rms width and the average dispersion ripple amplitude are also shown in Figure 4. One can conclude that for the 10-bounce recompression a severe degradation of pulse quality sets in at about 5 de-

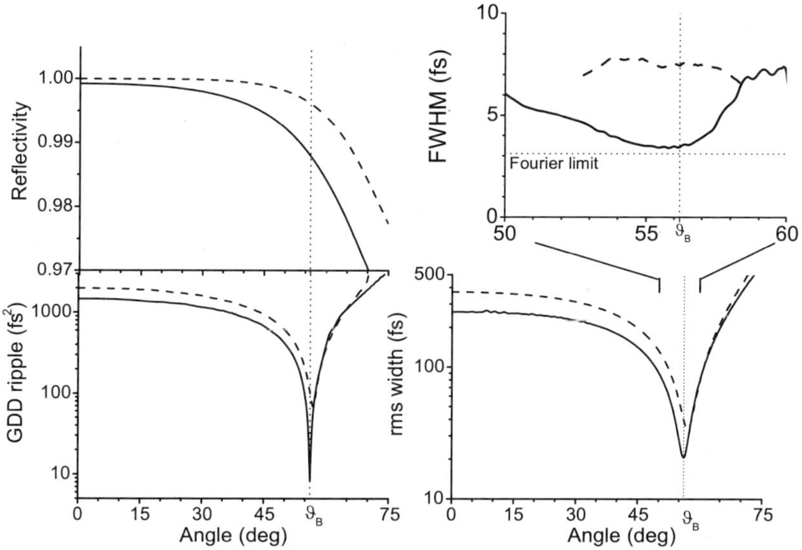

Fig. 4. Coating properties vs. incident angle. Top left: Reflectivity. Bottom left: rms value of the GDD dispersion ripple. Right: Effects of the varying dispersion ripple on pulse shape parameters after 10 reflections off the chirped mirror. Top: Full width at half maximum (FWHM) pulse width. Ideal recompression of the pulse would yield a 3.2 fs pulse duration, as indicated by the solid black line. Note that only the range of a meaningful FWHM-duration is shown. Bottom: rms width. Values for the unoptimized design are shown as a dashed line, values for the computer-optimized design as a solid line. The dotted vertical line indicates Brewster's angle, the dotted horizontal line the Fourier limit of the test pulse.

grees deviation from the nominal Brewster's angle. At below 52° and above 58° incidence, the pulse is compressed to 5 fs rather than the optimum 3.5 fs at Brewster incidence. This is a surprisingly large window for the operation of Brewster-angled mirrors, which makes it clear that slight errors in the angular alignment on the order of one degree can easily be tolerated in this design approach.

3 Conclusion

A new approach to chirped mirror design is introduced. Use of chirped mirrors at Brewster incidence can strongly reduce dispersion ripple and avoid impedance mismatch problems at the interface to the ambient medium. We show by example that it is possible to compensate for group delay dispersion over a bandwidth of about 1.2 octaves whilst maintaining a high reflectivity. This is significantly more bandwidth than any previous mirror design technique allowed so far. It also exceeds the bandwidth of any other dispersion compensation technology, including prism or grating sequences and $4f$-shapers, which are limited to one octave by overlapping grating orders.

The Brewster-angle approach may not yet be the universal solution to intracavity dispersion compensation, but it appears excellently suited for extracavity compression of white-light continua. This compression is currently the most demanding application in terms of bandwidth. As no complicated matching sections are required the layer sequences in the Brewster-angle design approach are structurally relatively simple. This simplicity results in a relatively high tolerance of the approach towards growth errors. Given current growth monitoring capabilities, reliable manufacturing of such mirrors therefore appears promising. Chirped mirror structures at non-normal angles of incidence open up interesting new possibilities for dispersion compensation of white-light continua and other ultra-wideband sources.

References

1. A. Baltuška, Z. Wei, M. S. Pshenichnikov, D. A. Wiersma, and R. Szipöcs, *Appl. Phys. B* **65**, 175–88 (1997).
2. D. H. Sutter, G. Steinmeyer, L. Gallmann, N. Matuschek, F. Morier-Genoud, U. Keller, V. Scheuer, G. Angelow, and T. Tschudi, *Opt. Lett.* **24**, 631–3 (1999).
3. F. X. Kärtner, U. Morgner, R. Ell, T. Schibli, J. G. Fujimoto, E. P. Ippen, V. Scheuer, G. Angelow, and T. Tschudi, *J. Opt. Soc. Am. B* **18**, 882–5 (2001).
4. L. Xu, L. Liming, N. Nakagawa, R. Morita, and M. Yamashita, *IEEE Phot. Technol. Lett.* **12**, 1540–2 (2000).
5. L. Xu, N. Nakagawa, R. Morita, H. Shigekawa, and M. Yamashita, *IEEE J. Quantum Electron.* **36**, 893–9 (2000).
6. B. Schenkel, J. Biegert, U. Keller, C. Vozzi, M. Nisoli, G. Sansone, S. Stagira, S. De Silvestri, and O. Svelto, *Opt. Lett.* **28**, 1987-9 (2003).
7. N. Zhavoronkov and G. Korn, *Phys. Rev. Lett.* **88**, 203901 (2002).
8. V. Kalosha and J. Herrmann, *Phys. Rev. A* **67**, 031801 (2003).
9. F. Gires and P. Tournois, *C. R. Acad. Sci. Paris* **258**, 6112 (1964).
10. G. Steinmeyer, *IEEE J. Quantum Electron.* **39**, 1027-34 (2003).
11. F. X. Kärtner, N. Matuschek, T. Schibli, U. Keller, H. A. Haus, C. Heine, R. Morf, V. Scheuer, M. Tilsch, and T. Tschudi, *Opt. Lett.* **22**, 831–3 (1997).
12. N. Matuschek, F. X. Kärtner, and U. Keller, *IEEE J. Quantum Electron.* **35**, 129–37 (1999).
13. N. Matuschek, L. Gallmann, D. H. Sutter, G. Steinmeyer, and U. Keller, *Appl. Phys. B* **71**, 509–22, (2000).
14. G. Tempea, V. Yakovlev, B. Bacovic, F. Krausz, and K. Ferencz, *J. Opt. Soc. Am. B* **18**, 1747–50, (2001).
15. G. Steinmeyer, *Opt. Express*, **11**, 2385–96, (2003).

10-fs, Diode-Pumped Cr:LiCAF Laser

Philipp Wagenblast[1], Richard Ell[1], Uwe Morgner[1,2], Felix Grawert[3], and Franz Kärtner[3]

[1] High-Frequency and Quantum Electronics Laboratory, University of Karlsruhe, Germany, wagenblast@ihq.uni-karlsruhe.de
[2] Max-Planck-Institut für Kernphysik, Heidelberg, Germany
[3] Research Laboratory of Electronics, Massachusetts Institute of Technology, Cambridge, MA

1 Abstract

10-fs pulses from a diode-pumped, soft-aperture Kerr-lens mode-locked Cr^{3+}: LiCAF laser with a spectral bandwidth of 150 nm and 40 mW of output power are demonstrated at a repetition rate of 110 MHz. For dispersion compensation, double-chirped mirrors and prisms are used. The pulses are characterized using spectral shearing interferometry (SPIDER).

2 Introduction

Cr^{3+}:LiCAF and other colquiriite crystals enable sub-10 fs pulse generation from a directly diode-pumped and, therefore, low cost system. 10-fs pulses with mode-locked output powers of 2 mW have been demonstrated [1] from a prism dispersion compensated hard-aperture Kerr-lens mode-locked (KLM) Cr^{3+}:LiSAF laser by optimum choice of prism material. Furthermore, a diode-pumped KLM Cr^{3+}:LiCAF laser has been reported [2] generating pulses as short as 20 fs by using chirped dielectric mirrors for dispersion compensation. Recently, we showed that the bandwidth of Cr^{3+}:LiCAF is sufficient to generate 9-fs pulses [3], when pumped with a Ti:Sapphire laser. Diode-pumped, 10-fs colquiriite lasers can be a cost-effective replacement for Ti:sapphire systems for many applications like time-resolved spectroscopy, metrology, terahertz wave generation and detection, and Optical Coherence Tomography. Among all colquiriites, LiCAF shows the lowest quantum defect and the best thermal properties [4] [5], allowing for high doping and best power extraction. In this paper, we present a reliable, ultrafast 10-fs diode-pumped Cr^{3+}:LiCAF laser with 40 mW of average output power, which is suitable for many of the above-mentioned applications. Broadband dispersion compensation is achieved by the use of broadband, second- and third-order dispersion compensating double-chirped mirrors [3] [6].

Mode-locked, sub-10 fs lasers usually exploit the Kerr nonlinearity via self-focusing of the laser beam through a hard or soft aperture using a real or gain aperture, respectively. For soft aperturing, a good pump beam quality is crucial, because the pump beam, and thereby the inverted volume inside the laser crystal, has to be smaller than the cw laser mode volume. High brightness laser diodes generally emit multimode beams which are often more than ten times beyond the diffraction limit. Such beams cannot be focused into a small volume. Diode pumped lasers have therefore been mode-locked by the use of intracavity hard apertures [1] [2]. One possible way to lower the drawbacks of multimode pumping in KLM lasers is to reduce the inverted volume by use of highly doped crystals with short absorption length. We present in this paper for the first time a diode-pumped, soft-aperture KLM laser with a short, 2 mm long, highly Cr-doped LiCAF crystal.

The standard theory for KLM with hard or soft aperture [7] [8] [9] generally assumes that the nonlinear change of the resonator eigenmode size is responsible for self amplitude modulation, which stabilizes the KLM process. In the case of a hard aperture, this is attributed to reduced diffraction losses, because the beam diameter at the aperture decreases with increasing peak power, whereas for the soft aperture, a better mode-matching geometry in the gain medium provides larger gain at higher intensities, which results in positive dynamic gain necessary for stable pulsed operation [10]. In [8] and [9] the role of thermal lensing and gain guiding for stable KLM has been pointed out. Thermal lensing modifies the resonator stability regions and gain guiding stabilizes otherwise unstable resonators [11] [12]. By mapping experimentally the stability of the resonator and comparing the result with theory, we found, that in fact in our laser thermal lensing and gain guiding significantly influence the resonator stability regions. Strong gain guiding leads to a non-vanishing phase front curvature at the plane end mirrors [11], and is clearly observed experimentally in cw mode of operation. It is of major importance for KLM action in our laser, because it significantly extends the stability regions of the laser into otherwise unstable regions, where the nonlinear self-amplitude modulation is sufficiently high for mode-locking without aperture.

3 Experimental

The set-up, shown in Fig.1, is a standard z-folded resonator at 110 MHz including a prism sequence. The radius of curvature of the folding mirrors is 75 mm, resulting in resonator mode radii in the sagittal and tangential plane of 20 μm and 30 μm in the 2 mm long, Brewster-cut laser crystal. The absorption length of the 10% Cr-doped crystal at the pump wavelength of 675 nm is 0.7 mm. Our pump sources are 500 mW laser diodes with an emitting area of 1×100 μm (Coherent), where the large divergence of the fast axis is collimated by a fiber lens. The diodes are transversely multimode with a beam quality factor M^2 around 15 for the slow axis, and in the order of one for the fast axis. The beams of two diodes are polarization-combined and scaled down by a f_{100}/f_{50} telescope, resulting in a beam of 20×70 μm in diameter in the crystal, with focal depth of 0.8 mm, which roughly equals the absorption length and therefore satisfies the optimum mode-matching condition. The

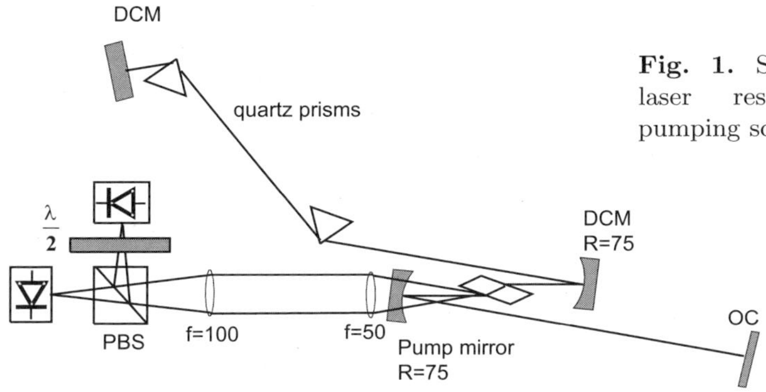

Fig. 1. Set-up of the laser resonator and pumping scheme.

total absorbed pump power amounts to 800 mW at maximum, and in cw-mode of operation, the laser emits up to 150 mW at a wavelength of 780 nm, and operates eight times above threshold. We use three different kinds of laser optics. First, double-chirped mirrors (DCMs) providing negative second-order dispersion (SOD) and the appropriate third-order dispersion to compensate for the crystal and the prisms [3]. Second, the pump light is coupled through a low dispersive dichroic mirror with high pump transmission at 675 nm into the laser cavity. And third, a low dispersive output coupler with 0.8% transmission is used. The bandwidth of this cavity is about 200 nm around 800 nm. A pair of fused quartz prisms with a separation of 30 cm and an insertion of 8 mm provides dispersion tunability around zero SOD.

Mode-locking is achieved by soft-aperture KLM. The appropriate resonator configuration is set by the separation z of the curved mirrors and the distance x between the crystal and the mirror close to the pump. Neglecting gain guiding effects, a stability gap would show up in the asymmetric resonator design with respect to the mirror separation z. The fact that this gap is

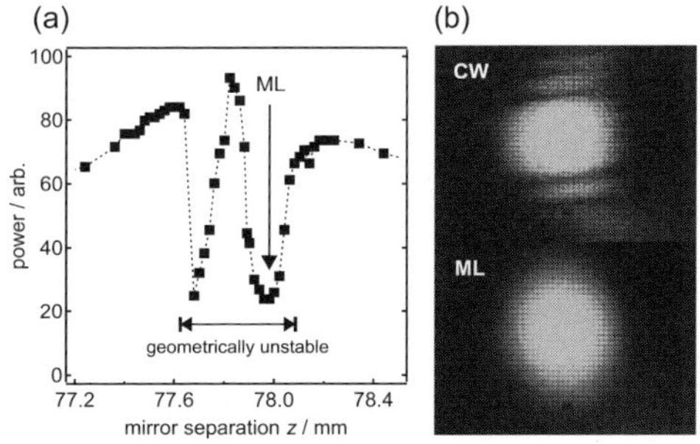

Fig. 2. a) cw output power versus mirror separation. The laser runs in the geometrically unstable regime because of gain guiding effects. The proper mirror distance for mode-locking is indicated by the arrow (ML). b) photograph of cw (top) and pulsed (bottom) output beam on a rectangular grid.

not observed in the experiment (see Fig. 2a) together with the non-vanishing phase front curvature emphasizes the importance of gain guiding. To start mode-locking, the cavity is aligned into this stability gap, close to the edge of geometric stability, where the cw output beam shape shows strong aberrations as can be seen in the upper part of Fig. 2b. By shaking an end mirror, soft-aperture Kerr-lens mode-locking is initiated, and the output beam transforms into a circular, near-diffraction limited TEM_{00} beam with a measured M^2 factor of 1.17. In the mode-locked state of operation, the laser emits up to 40 mW of average power, which is a 30% increase from cw operation. With an intracavity pulse energy of 25 to 45 nJ we calculate a nonlinear phase shift of 0.2 to 0.4 rad per round trip. Although the nonlinear phase shift is smaller than in typical KLM lasers, and although we expect the self-amplitude modulation by the soft-aperture to be much smaller than in the case of diffraction limited pumping, once initiated, mode-locking is maintained for hours.

The mode-locked spectrum is shown in Fig. 3a. It extends from 750 to 900 nm and shows an additional peak at 980 nm, which is off the main high reflectivity range of the pump mirror (Fig. 4) but located at the next maximum of the pump mirror reflectivity. The modulation in the main part

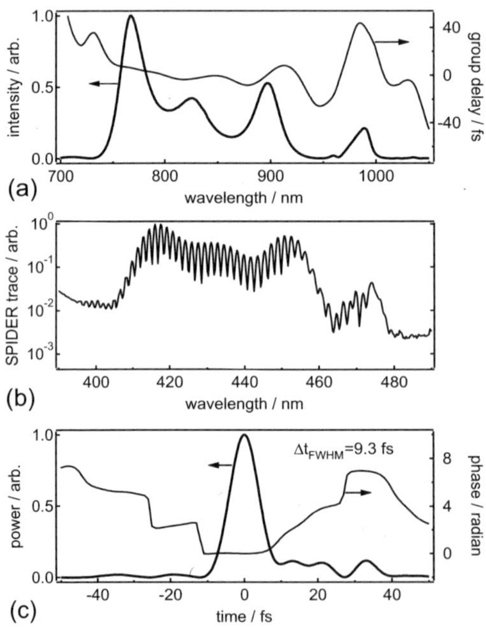

Fig. 3. Results of the SPIDER characterization of the mode-locked pulses. a) mode-locked spectrum and group delay, b) sheared interferogram, c) Fourier-transformed pulse and temporal phase.

of the spectrum results from the dispersion oscillations of the double-chirped mirrors [3] [13]. Assuming a flat phase over the spectrum, we calculate a pulse duration of 8.4 fs. The actual pulse shape has been characterized by spectral shearing interferometry (SPIDER) [14], a method which directly reconstructs the spectral phase and group delay, which is also shown in Fig. 3a, and by Fourier transform the pulse envelope and phase, which are plotted in Fig. 3c. The duration of this pulse is 9.3 fs (FWHM), and pre-pulses are suppressed by about one order of magnitude. The interferogram of the upconverted, spectrally sheared pulses is shown in Fig. 3b on a logarithmic scale. The post-pulses visible in Fig. 3c can be partly assigned to the long-wavelength peak at 980 nm, where the group delay delivered by SPIDER is largest. This spectral peak occurs at the position of the next maximum of the pump mirror reflectivity, as can be seen from Fig. 4, where the measured reflectivity and group delay of the pump mirror are plotted. The associated

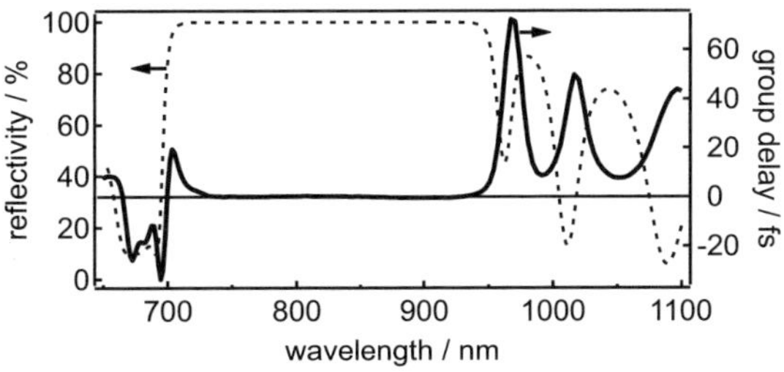

Fig. 4. Group delay and reflectivity characteristics of the pump mirror.

group delay resulting from the mirror at 980 nm is positive, retarding these spectral components with respect to the main part of the pulse. An additional post-pulse arriving 33 fs after the main part of the pulse is a result of spurious reflections on the air-to-mirror impedance matching structure on the DCMs [3] [13].

4 Conclusion

We demonstrate sub-10-fs pulses from a diode-pumped Cr^{3+}:LiCAF laser at an average output power of 40 mW. To our knowledge these are the first sub-10-fs pulses generated with this laser material. The reported output power marks a twentyfold improvement in comparison to previously published results from a diode-pumped 10-fs colquiriite laser, at the same pump power level. These results were achieved by the use of broad-band second and third order dispersion compensating mirrors, and soft-aperture Kerr-lens mode-locking in a thin crystal with short pump absorption length.

This work has been supported by the Deutsche Forschungsgemeinschaft within the Center for Functional Nanostructures (CFN) at University of Karlsruhe (TH) and within the contract Mo850/2-1. The work of F. X. Kärtner is supported by NSF grant ECS-0119452.

References

1. S. Uemura and K. Torizuka. *Development of a Diode-Pumped Kerr-Lens Mode-Locked Cr:LiSAF Laser.* IEEE Journ. of Quantum Electron., **39**, 68 (2003).

2. K. M. Gäbel, P. Rußbüldt, R. Lebert, and A. Valster. *Diode pumped* Cr^{3+}*:LiCAF fs-Laser.* Opt. Commun., **157**, 327 (1998).
3. P. Wagenblast, U. Morgner, F. Grawert, T. R. Schibli, F. X. Kärtner, V. Scheuer, G. Angelow, and M. J. Lederer. *Generation of sub-10-fs pulses from a Kerr-lens mode-locked* Cr^{3+}*:LiCAF laser oscillator by use of third-order dispersion-compensating double-chirped mirrors.* Opt. Lett., **27**, 1726 (2002).
4. L. K. Smith, S. A. Payne, W. L. Kway, and B. H. T. Chai. *Investigation of the Laser Properties of* Cr^{3+}*:*$LiSrGaF_6$. IEEE Journ. of Quantum Electron., **28**, 2612 (1992).
5. M. Stalder, M. Bass, and B. H. T. Chai. *Thermal quenching of fluorescence in chromium-doped fluoride laser crystals.* J. Opt. Soc. Am. B, **9**, 2271 (1992).
6. F. X. Kärtner, N. Matuschek, T. Schibli, U. Keller, H. A. Haus, C. Heine, R. Morf, V. Scheuer, M. Tilsch, and T. Tschudi. *Design and fabrication of double-chirped mirrors.* Opt. Lett., **22**, 831 (1997).
7. V. Magni, G. Cerullo, and S. De Silvestri. *ABCD matrix analysis of propagation of gaussian beams through Kerr media.* Opt. Commun., **96**, 348 (1993).
8. M. Mehendale, T. R. Nelson, F. G. Omenetto, and W. A. Schroeder. *Thermal effects in laser pumped Kerr-lens modelocked Ti:sapphire lasers.* Opt. Commun., **136**, 150 (1997).
9. A. Ritsataki, P. French, and G. New. *A numerical model of Kerr-lens mode-locking.* Opt. Commun., **142**, 315 (1997).
10. S. Namiki, E. Ippen, H. Haus, and C. Yu. *Energy rate equations for mode-locked lasers.* J. Opt. Soc. Am. B, **14**, 2099 (1997).
11. F. Salin and J. Squier. *Gain guiding in solid-state lasers.* Opt. Lett., **17**, 1352 (1992).
12. A. E. Siegman. *Lasers.* University Science Books, Mill Valley (1986).
13. F. X. Kärtner, U. Morgner, R. Ell, T. Schibli, J. G. Fujimoto, E. P. Ippen, V. Scheuer, G. Angelow, and T. Tschudi. *Ultrabroadband double-chirped mirror pairs for generation of octave spectra.* J. Opt. Soc. Am. B, **18**, 882 (2001).
14. C. Iaconis and I. A. Walmsley. *Self-Referencing Spectral Interferometry for Measuring Ultrashort Optical Pulses.* IEEE Journ. of Quantum Electron., **35**, 501 (1999).

Broadband Phase-Coherent Optical Frequency Synthesis With Actively Linked Ti:Sapphire and Cr:Forsterite Femtosecond Lasers

A. Bartels, N. R. Newbury, I. Thomann, L. Hollberg, and S. A. Diddams

National Institute of Standards and Technology, 325 Broadway, Boulder, CO 80305, USA albrecht@boulder.nist.gov

1 Introduction

Extremely broadband femtosecond pulse sources with well defined carrier-envelope phase evolution have been a field of intense research over the past years [1–3]. The time-domain motivation for the push towards increased bandwidth is the generation of optical pulses that approach the single-cycle regime and can be used for phase-sensitive nonlinear experiments [4]. The frequency-domain motivation is the interest in extending the bandwidth over which optical frequencies can be precisely synthesized and subsequently used for absolute optical frequency measurements or comparisons [1, 5, 6]. Nonlinear frequency conversion is one route to overcome the bandwidth limitations of current systems based on *one* femtosecond laser. A second attractive route is linking *two* femtosecond lasers with overlapping emission spectra. Phase-locking the repetition rates of two femtosecond lasers with different gain media has recently been demonstrated. Here, we use a Ti:sapphire laser and a Cr:forsterite laser to extend this approach and additionally equalize the rate at which the carrier-waves move underneath their pulse envelopes. In other words, we link the absolute position of the two frequency combs, thereby establishing a true phase-coherence between all modes of their combined spectrum. The laser repetition rates are phase-locked by using a nonlinear cross-correlation technique. The difference $\Delta f_0 = f_{0,Ti} - f_{0,Cr}$ between the carrier-envelope offset frequency of the Ti:sapphire laser, $f_{0,Ti}$, and that of the Cr:forsterite laser $f_{0,Cr}$ is measured by beating the portions of the their spectra around 1.1 μm wavelength against each other. The phase-lock between $f_{0,Ti}$ and $f_{0,Cr}$ is established by controlling the pump power of the Cr:forsterite laser. As a final step, the carrier-envelope offset frequency of the combined comb $F_0 = |2f_{0,Cr} - f_{0,Ti}|$ is measured, which is necessary to perform a phase-sensitive measurement or an absolute frequency measurement with the combined comb .

2 Phase-coherently linked Cr:forsterite and Ti:sapphire femtosecond lasers

The broadband Ti:sapphire laser and the Cr:Forsterite laser we employ have been described elsewhere [7,8]. Their output spectra are shown in Fig. 1. The combined coverage extends from 0.57 μm to 1.45 μm at a power level of 1 nW per frequency mode. It is important to point out that the spectra overlap without the necessity for external spectral broadening. The ring oscillators have a repetition rate of 433 MHz. Their output powers are both approximately 500 mW. Both lasers have one mirror mounted onto a piezoelectric transducer (PZT) by which we control their repetition rates. An acousto-optic modulator (AOM) in the pump laser beams allows us to independently control the carrier-envelope-offset frequencies.

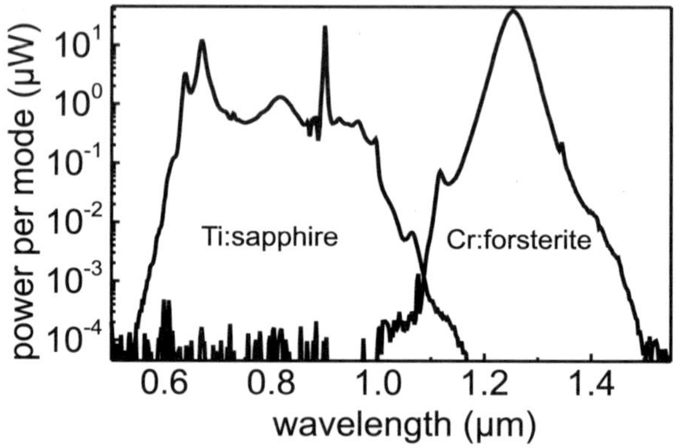

Fig. 1. Spectra of the Ti:sapphire and the Cr:forsterite laser.

To create a combined phase-coherent frequency comb from the two lasers we employ the scheme sketched in Fig. 2. The mode spacings of the combs are equalized by phase-locking the repetition rate of the Ti:sapphire laser, $f_{R,Ti}$, to that of the Cr:forsterite laser $f_{R,Cr}$. The phase-lock is based on a nonlinear cross-correlation signal between the laser outputs that is generated in a type-I phase matched BBO crystal in a non-collinear configuration and serves as error signal in our feedback-loop. It is offset from zero, filtered, amplified and drives the piezoelectric transducer behind a cavity mirror of the Ti:sapphire laser. Thus, the cross-correlation signal is locked to a non-zero value and the timing between the two laser pulse trains is precisely stabilized. The advantage of this method over conventional methods employing photodetection of the pulse trains and phase-locking electronic microwave signals at the repetition rate or one of its harmonics is that it has an inherently higher phase-sensitivity at the lock point and therefore allows for tighter phase-locking. A timing-noise spectrum of the two lasers as measured

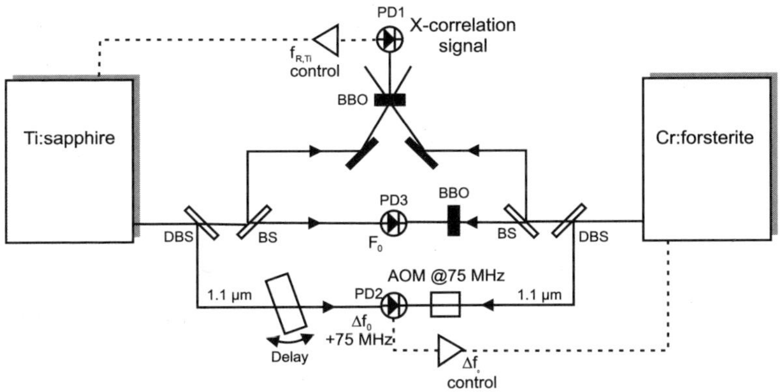

Fig. 2. Setup to phase-lock the Ti:sapphire and Cr:forsterite lasers. Most of the laser spectra transmits the dichroic beamsplitters DBS. 50% are then reflected by the beamsplitters BS and generate a cross-correlation signal between the two lasers in a BBO crystal that is detected with photodiode PD1. This signal fed back to the PZT in the Ti:sapphire laser. The portions of the spectra around 1.1 μm that are reflected by DBS are combined on photodiode PD2 to generate a heterodyne beat signal at Δf_0. A thick quartz plate is used to adjust the timing of the pulses. The light from the Cr:forsterite that passes BS is frequency doubled and generates a heterodyne beat with the fundamental Ti:sapphire laser output on photodiode PD3.

with an out-of-loop nonlinear cross-correlator (not shown in Fig. 2) is shown in Fig. 3a). It exhibits a pronounced contribution around 30 kHz that cannot be cancelled by our feedback loop due to the limited bandwidth of the employed PZT. It is important to note, however, that compared to passively mode-locked Ti:sapphire lasers this pronounced timing noise at frequencies of tens of kHz is quite unusual (see e.g. reference [9] for comparison). We believe that it is due to excess power noise present on the pump laser[1] for the Cr:forsterite laser in the same frequency range. This writes itself onto the Cr:forsterite laser and likely converts into timing noise of its output pulses (see power noise spectra of the Yb:glass pump laser and the Cr:forsterite laser in Fig. 3b)). The integrated timing jitter among the two lasers is 20 fs in a bandwidth from 1 Hz to 100 kHz. This is sufficient to keep the pulses overlapped to a large enough fraction at all times such that a measurement of Δf_0 and F_0 is feasible. However, Fig. 3 strongly suggests that by eliminating the high frequency excess power noise and thereby the timing noise on the Cr:forsterite laser a timing jitter of 2 fs in a 100 kHz bandwidth is possible. This would be highly desirable because it immediately reduces the phase-noise of the mentioned heterodyne beat signals as will be discussed in the following paragraphs. Routes towards a reduced power noise on our Cr:forsterite laser are the use of a quieter pump laser or employing a broad-

[1] A Yb:glass double cladding fiber laser

band noise reduction feedback for our present pump laser, which is an ongoing project.

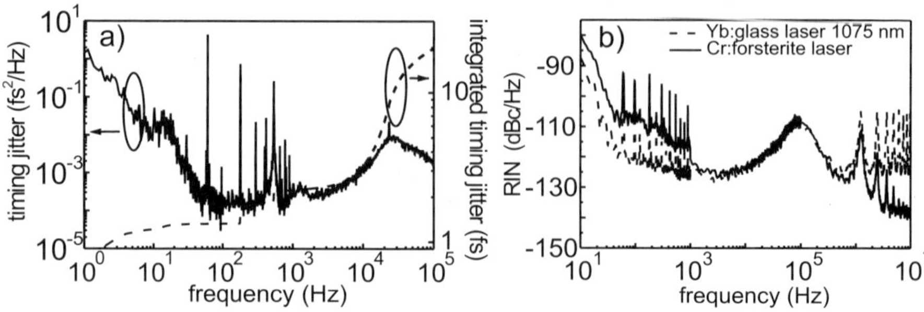

Fig. 3. a) Out-of-loop spectral density of the timing jitter between the Ti:sapphire laser and the Cr:forsterite laser with the feedback loop engaged (solid line, left scale). Integrated timing jitter versus integration bandwidth (dashed line, right scale). b) Relative intensity noise spectral density of the Cr:forsterite laser (solid line) and its pump laser (dashed line).

With the repetition rates of the two lasers locked, we are able to measure the difference of the lasers' carrier-envelope-offset frequencies Δf_0. We pick the overlapping portions of both lasers around 1.1 μm wavelength with dichroic beamsplitters. Subsequently the Cr:forsterite frequency comb around 1.1 μm is offset by 75 MHz by diffracting the beam with an AOM that is driven at f_{AOM}=75 MHz. This allows us to generate a heterodyne beat with frequency $f_b = \Delta f_0 + f_{AOM}$ on a low-noise InGaAs photodiode provided the pulses are temporally overlapped by rotating a thick quartz plate in the Ti:sapphire beam path. The estimated power that contributes to the beat signal is 10 μW from each laser. A radio frequency spectrum of the signal at f_b is shown in Fig. 4 (left panel). Its signal-to-noise ratio is 35 dB in 300 kHz resolution bandwidth. In order to phase-lock $f_{0,Cr}$ to $f_{0,Ti}$ (i.e. $\Delta f_0 = 0Hz$) we phase-lock f_b to f_{AOM} by mixing f_b and f_{AOM} and feeding back the filtered and amplified mixer output to the AOM that controls the pump power of the Cr:forsterite laser. This approach makes it possible to circumvent problems related to the low-frequency noise floor of the photodetector in our feedback loop. As a test of whether we really created a continuous frequency comb, we counted f_b at a 1 s gate time and derived from this a time record of the deviations of Δf_0 from 0 Hz. Fig. 5 shows that the deviations are on the few mHz scale, this result actually being limited by the resolution of our frequency counter. This proves that our combined frequency comb is readily suitable for precise optical frequency measurements. To characterize the tightness of our phase-lock, we measure the phase noise spectrum $S_\phi(f)$ of f_b which is equal to the phase-noise of Δf_0 and the relative phase-noise of the equally spaced optical frequency combs. The result is shown in Fig. 5.

The spectrum again shows a strong excess noise contribution at few tens of kHz that is likely due to the timing jitter among the lasers which broadens f_b by introducing both amplitude and phase-noise. Integration of the phase noise spectrum $\int_{10MHz}^{f_l} S_\phi(f)$ shows that 1 rad is reached at f_l =27.3 kHz, i.e. the coherence time of the lock between our lasers is only 36 μs. In other words, at this preliminary stage bunches of ≈15,000 successive pulses from the two lasers are truly phase-coherent.

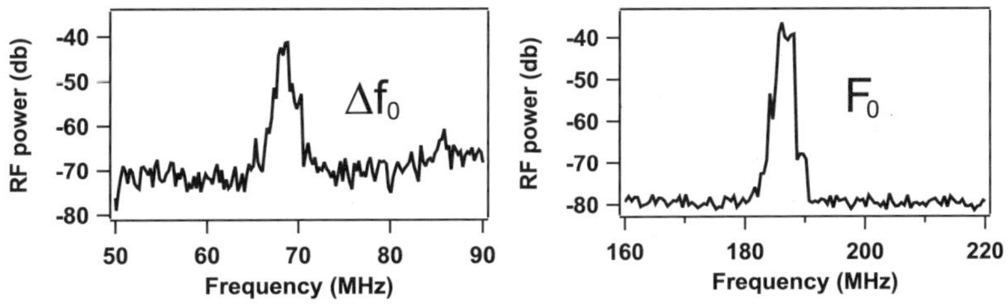

Fig. 4. Left panel: RF-spectrum of the Δf_0 signal. Right panel: RF-spectrum of the F_0 signal. The resolution bandwidth for both plots is 300 kHz.

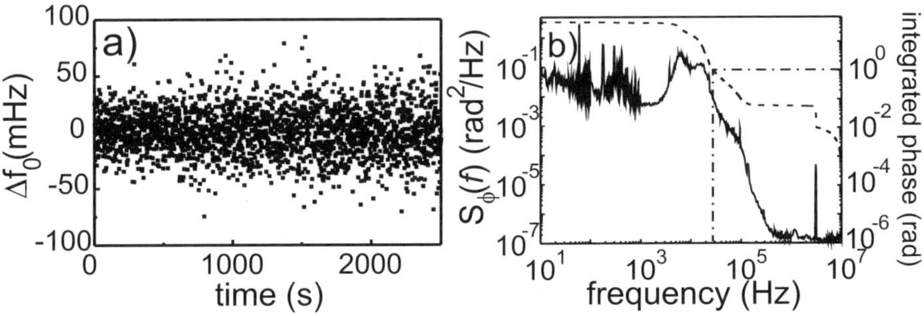

Fig. 5. Left panel: record of consecutive measurements of Δf_0 at 1 s gate time. The standard deviation is 20 mHz (counter resolution limited). Right panel: Phase-noise spectrum of the signal at f_b (also that of Δf_0), equal to the relative phase-noise of the equally spaced optical frequency combs (solid line). The dashed line is the integral over $S_\phi(f)$ mentioned in the text. The dashed-dotted line indicates where this integral reaches 1 rad.

To facilitate an absolute frequency measurement or ultimately control the carrier-envelope phase of a combined field, however, $F_0 = |2f_{0,Cr} - f_{0,Ti}|$ needs to be measured. To do so, we employ the conventional "f-2f" self-referencing method. A portion of the Cr:forsterite laser is frequency-doubled in a type-I phase matched BBO crystal and beat against the spectrally matching portion of the fundamental Ti:sapphire spectrum to generate a signal at

F_0 (see radio-frequency spectrum in Fig. 4, right panel). The power contributing to this beat is ≈1 mW for each laser, a signal-to-noise ratio exceeding 40 dB in 300 kHz resolution bandwidth is achieved. Preliminary attempts to also phase-lock F_0 have failed so far, most likely due to the fact that the timing jitter between the lasers writes too much phase-noise onto the F_0 signal.

3 Conclusion

In conclusion we have tightly phase-locked the repetition rates of a Ti:sapphire and a Cr:forsterite femtosecond laser at 433 MHz. Subsequently we have measured the offset Δf_0 among the now equally spaced frequency combs and are able to stabilize it to 0 Hz. For the first time to our knowledge we have thereby created a broadband, high-powered and phase-coherent frequency comb from two lasers with different gain media. A heterodyne beat between the second harmonic of the Cr:forsterite laser and the fundamental spectrum of the Ti:sapphire laser has given us access to F_0, the carrier envelope offset frequency of the combined frequency comb. The system readily allows us to perform absolute optical frequency measurements within the combined spectrum by simultaneous counting all relevant RF-frequencies. Combining possible second harmonic generation, sum and difference frequency generation [10] as well as broadening in nonlinear fibers [8] our approach has great potential for phase coherent optical frequency synthesis from the ultraviolet through the visible, the telecommunications band all the way to the mid-infrared.

We thank Lennart Robertson and Ma Long-Sheng from BIPM for the crucial loan of their pump laser.

References

1. Jones, D.J., Diddams, S.A., Ranka, J.K., Stentz, A., Windeler R.J., Hall, J.L., Cundiff, S.T., Science **288**, 635 (2000)
2. Morgner, U., Ell, R., Metzler, G., Schibli, T.R., Kärtner, F.X., Fujimoto, J.G., Haus, H.A., Ippen, E.P., Phys. Rev. Lett. **86**, 5462 (2001)
3. Ramond, T.M., Bartels, A., Diddams, S.A., Hollberg, L., Opt. Lett. **27**, 1842 (2002)
4. Baltuska, A., Udem, Th., Uiberacker, M., Hentschel, M., Goulielmakis, E., Gohle, Ch., Holzwarth, R., Yakovlev, V.S., Scrinzi, A., Hänsch, T.W., Krausz, F., Nature **421**, 611 (2003)
5. Udem, Th., Reichert, J., Holzwarth, R., Hänsch, T.W., Phys. Rev. Lett. **82**, 3568 (1999)
6. Holzwarth, R., Udem, Th., Hänsch, T.W., Knight, J.C., Wadsworth, W.J., Russell, P.St., Phys. Rev. Lett. **85**, 2264 (2000)
7. Bartels, A., Kurz, H., Opt. Lett. **27**, 1839 (2002).

8. Thomann, I., Bartels, A., Corwin, K.L., Newbury, N.R., Hollberg, L., Diddams, S.A., Nicholson, J.W., Yan, M.F., Opt. Lett. **28**, 1368 (2003)
9. Bartels, A., Diddams, S.A., Ramond, T.M., Hollberg, L., Opt. Lett. **28**, 663 (2003)
10. Foreman, S.M., Jones, D.J., Ye, J., Opt. Lett. **28**, 370 (2003)

Synthesis of Optical Frequencies and Ultrastable Femtosecond Pulse Trains from an Optical Reference Oscillator

A. Bartels, T. M. Ramond, S. A. Diddams, and L. Hollberg

National Institute of Standards and Technology, 325 Broadway, Boulder, CO 80305, USA albrecht@boulder.nist.gov

1 Introduction

Recently, atomic clocks based on optical frequency standards have been demonstrated [1,2]. A key element in these clocks is a femtosecond laser that downconverts the petahertz oscillation rate into countable ticks at 1 GHz. When compared to current microwave standards, these new optical clocks are expected to yield an improvement in stability and accuracy by roughly a factor of 1000. Furthermore, it is possible that the lowest noise microwave sources will soon be based on atomically-stabilized optical oscillators that have their frequency converted to the microwave domain via a femtosecond laser. Here, we present tests of the ability of femtosecond lasers to transfer stability from an optical oscillator to their repetition rates as well as to the associated broadband frequency comb. In a first experiment, we phase-lock two lasers to a stabilized laser diode and find that the relative timing jitter in their pulse trains can be on the order of 1 femtosecond in a 100 kHz bandwidth. It is important to distinguish this technique from previous work where a femtosecond laser has been stabilized to a microwave standard [3,4] or another femtosecond laser [5]. Furthermore, we extract highly stable microwave signals with a fractional frequency instability of 2×10^{-14} in 1 s by photodetection of the laser pulse trains. In a second experiment, we similarly phase-lock the femtosecond laser to an optical oscillator with linewidth less than 1 Hz [6]. The precision with which we can make the femtosecond frequency comb track this reference oscillator is then tested by a heterodyne measurement between a second stable optical oscillator and a mode of the frequency comb that is displaced 76 THz from the 1 Hz-wide reference. From this heterodyne signal we place an upper limit of 150 Hz on the linewidth of the elements of the frequency comb, limited by the noise in the measurement itself.

2 Synthesis of ultrastable pulse trains and microwave signals referenced to an optical oscillator

The different femtosecond lasers we employ (referred to with index i=1,2) have been described elsewhere [7, 8]. One of the lasers employs microstruc-

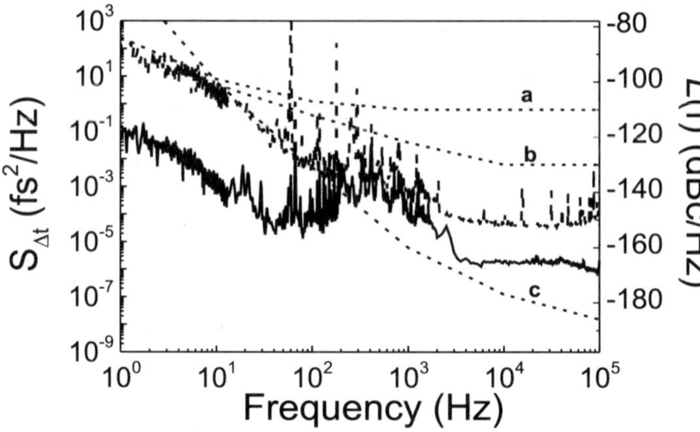

Fig. 1. Timing- and phase-noise spectrum at 1 GHz of the laser pulses $L_{opt.}$(f) relative to the optical reference oscillator (solid line) and phase-noise spectrum at 1 GHz of the microwave signals $L_{mic.}$(f) extracted from the lasers by photodetection (dashed line). For comparison the phase-noise at 1 GHz of a high-quality quartz oscillator, a synthesizer and a sapphire microwave oscillator are shown as dotted lines a, b, and c, respectively.

ture fiber to create an octave-spanning spectrum, while the second laser emits such a spectrum directly. Their repetition rates $f_{R,i}$ are both ≈1 GHz. We phase-lock the carrier-envelope offset frequencies $f_{0,i}$ (see references [9, 10] for experimental details) to a synthesizer that is referenced to a hydrogen maser. Subsequently, we heterodyne a stabilized single-frequency laser diode (f_{LD} ≈456 THz) with the neighboring components of both frequency combs (with mode numbers n_i, both ≈456,850) to generate beat signals at frequencies $f_{b,i} = f_{LD} - (f_{0,i} + n_i \times f_{R,i})$. These beat signals are also phase-locked to synthesizers. The repetition rates of both lasers are thus independently phase-locked, i.e. synchronized to f_{LD} as $f_{R,i} = n_i^{-1} \times (f_{LD} - f_{0,i} - f_{b,i})$. It can be shown that, within our measurement resolution, the pulse trains now derive their frequency stability entirely from the optical frequency reference. In fact, the phase-locked frequency spectra of $f_{0,i}$ and $f_{b,i}$ can both be at the millihertz linewidth level, indicating excellent coherence for the entire octave spanning comb. To characterize the phase-noise of the pulse trains relative to the common optical reference oscillator we use an optical nonlinear cross-correlation technique (described e.g. in reference [5]). The measured single-side-band phase-noise spectrum $L_{opt.}$(f) (also given as timing noise spectrum $S_{\Delta t}(f)$) of the pulse trains at 1 GHz is displayed in Fig. 1 (solid line). It is approximately 30-50 dB lower than that of high quality synthesizers and quartz oscillators. It is also comparable to the best available sapphire microwave oscillators at high frequencies with a much better stability at low frequencies ≤ 10 Hz. The excess-noise in the range from 0.1 - 1 kHz is attributed to uncompensated mechanical vibrations in the lasers and the measurement system. From $L_{opt.}$(f) we extract the timing jitter of the pulse trains relative to f_{LD} and find it to be 0.45 fs in a passband from 1 - 100 Hz, increasing to 1.5 fs in the full available bandwidth of 100 kHz. Although no specific measures towards vibration isolation of the lasers have been taken, our result is comparable to what has been achieved previously by employing microwave-domain phase-locking techniques but with use of elaborate vibrational damping and techniques for common mode rejection [5]. We attribute this to the fact that we are effectively phase-locking the repetition rate at its ≈456,000th harmonic instead of using a low order microwave harmonic. The Allan deviation of the repetition rates as calculated from $L_{opt.}$(f) is σ_y(1 s)= 8×10^{-16} in τ =1 s gate time, indicating excellent stability (see Ref. [11] for the conversion between L(f) and $\sigma_y(\tau)$). This limit is set by residual mechanical vibrations outside of our feedback loops.

As a test of the stability that can ultimately be reached for microwave signals that can be extracted from an optical atomic frequency standard, we photodetect the pulse trains referenced to f_{LD} and measure their phase-noise $L_{mic.}$(f). The result is shown in Fig. 1 (dashed line). $L_{mic.}$(f) is about two orders of magnitude greater than $L_{opt.}$(f). This indicates that the photodetection and/or the subsequent electronic processing still corrupt the initial quality of the laser repetition rates. Research to identify the cause of the signal degradation is underway. Amplitude-noise to phase-noise conversion and pointing fluctuations of the lasers have been identified as possible candidates [12]. The higher phase-noise also shows up in the Allan deviation of the 1 GHz microwave signals. With σ_y(1 s)=2×10^{-14} their fractional instability is an order of magnitude higher than that of the laser repetition rates.

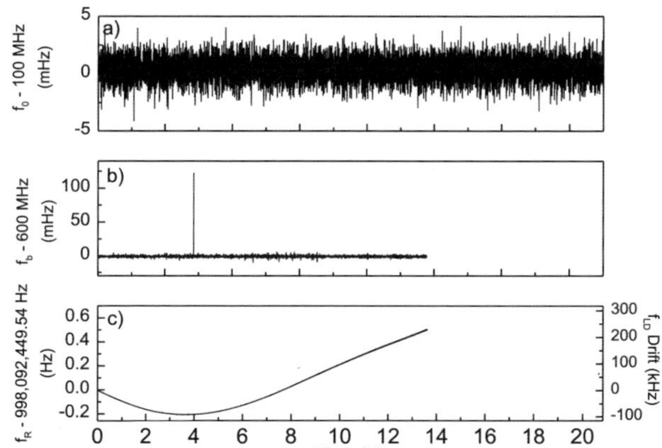

Fig. 2. Long term phase-locking of the broadband femtosecond laser. a) offset of f_0 from 100 MHz. b) offset of f_b from 600 MHz. c) time record of the offset of the laser repetition rate from 998,092,449.54 Hz (left scale) translated into the drift of f_{LD} (right scale).

3 Long-term stability of the femtosecond laser based optical clockwork

The laser that emits a broadband continuum directly circumvents many of the common problems with spectral broadening in microstructure fibers regarding long-term operation. This laser enables us to establish a phase-coherent optical-to-microwave link that operates continuously for a period approaching one day. As a demonstration, we count f_0, f_b and f_R simultaneously with frequency counters at 10 s gate time. The offset of the counter readings from their preset values is displayed in Fig. 2 for all three channels. The dataset shows uninterrupted and hands-off operation of the f_0-lock for more than 20 hours until we turned the system off (recent experiments have even proven operation for >50 h). The phase-lock on f_b operated for ≈14 h with only one detected cycle-slip at ≈4 h, i.e. ≈10 h of cycle-slip free data are contained. The reason for the failure at ≈14 h is not clear, although it is likely related to the stabilization of f_{LD} and not the femtosecond laser itself. As f_R actually presents a measurement of f_{LD} relative to our hydrogen maser, the time record of the offset of f_R from 998,092,449.54 Hz has been multiplied with the mode number n=456,857 to give the temporal drift of the Fabry-Perot cavity used to stabilize f_{LD} on the right axis of the graph in Fig. 2c). The cycle slip in the f_b feedback loop does not appear in the f_R record because the effect of the 120 mHz excursion in f_b results in an error of 260 nHz in f_R which is below our measurement limit. These data demonstrate that we are able to maintain a phase-coherent, cycle-slip-free link between an optical oscillator and the repetition rate of our laser for 10 h. In other words, we have the ability to count 1.6×10^{19} optical cycles at 456 THz without ever losing track of the oscillation. This ability is important for measurements where long averaging and/or recording times are necessary to attain useful data as well as to test the ultimate stability limit of optical frequency standards.

4 Generation of an array of narrow linewidth oscillators with the femtosecond frequency comb

While previous work has reported tight phase-locking of f_0 [13], to our knowledge there has not been direct verification of the linewidth of the individual modes of the stabilized femtosecond frequency comb. Here, we describe such a test that verifies the linewidth of the comb modes can be made at least as narrow as 150 Hz. Such an array of narrow linewidth oscillators can be used for spectroscopic references and frequency measurements, or even for time- or frequency-domain spectroscopy. Using techniques described above and in the references, the femtosecond comb created from nonlinear broadening in microstructure fiber is phase-locked with millihertz precision to a 532 THz (563 nm) dye laser having sub-Hz linewidth [6]. With the octave-spanning comb thus stabilized to the 532 THz laser, we can make a check of the linewidth of the other comb elements by heterodyning a second comb tooth against the 456 THz (657 nm) light from the stabilized diode laser mentioned above. The spectrum of this beat, as measured with an RF spectrum analyzer, is plotted as the lower trace in Fig. 3. The observed 150 Hz linewidth implies that the modes of the femtosecond laser comb can track the very narrow 532 THz laser at this same level. However, it is likely that the modes of the comb are even narrower, as the 10 m of optical fiber that transports the 456 THz light to the femtosecond laser and the linewidth of the diode laser itself also contribute to the measured linewidth. Independent measurements place an upper limit of 100 Hz on the 456 THz diode laser linewidth, and the extra noise introduced by the 10 m optical fiber also introduces broadening at the 100 Hz level, as we measured and show in the upper trace of Fig. 3. While we cannot yet state a lower limit for the linewidth of the comb teeth, these data suggests that it could be well below 100 Hz.

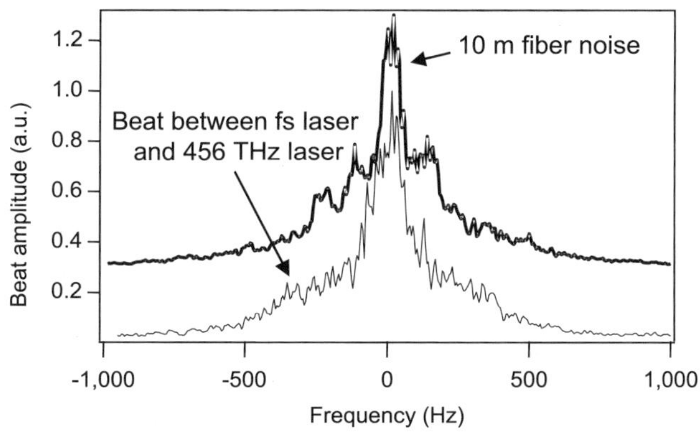

Fig. 3. Lower trace: Heterodyne beat between one mode of the optically referenced femtosecond laser and a 456 THz stabilized CW laser diode. The 150 Hz linewidth of this beat includes the additive noise from a 10 m long optical fiber (upper trace) that transport the 456 THz light to the femtosecond laser.

5 Conclusion

In conclusion, we have synchronized two femtosecond lasers to an optical reference oscillator with a timing jitter as low as 0.45 fs in a passband from 1 - 100 Hz. The method presented provides the means to exploit the stability of low-noise optical frequency standards for the synthesis of microwave signals with a stability superior to current microwave sources. Yet limited by excess noise due to the photodetection process we have shown that microwaves with a fractional frequency instability at the 10^{-14} level in 1 s gate time can readily be extracted. We are able to maintain a phase-coherent, cycle-slip-free link between an optical oscillator and the repetition rate of our laser for periods approaching one day. Finally, we have further shown that we can make the linewidth of the elements of the femtosecond mode comb reproduce that of an optical frequency standard at the level of $\leq$150 Hz (3×10^{-13}). Tight control of the octave-spanning optical comb means that we can now envision the phase-coherent transfer of not only the stability, but also the linewidth of a very narrow optical oscillator to several hundred thousand comb elements spanning the visible and near infra-red spectrum.

We are very grateful to C.W. Oates, E.A. Curtis, J.C. Bergquist, S. Bize, and R.S. Windeler for their contributions to these experiments.

References

1. Diddams, S.A., Udem, Th., Bergquist, J.C., Curtis, E.A., Drullinger, R.E., Hollberg, L., Itano, W.M., Lee, W.D., Oates, C.W., Vogel, K.R., Wineland, D.J., Science **293**, 825 (2001)
2. Ye, J., Ma, L.-S., Hall, J.L., Phys.Rev. Lett. **87**, 270801 (2001)
3. Spence, D.E., Dudley, J.M., Sleat, W.E., Sibbett, W., Opt. Lett. **19**, 481 (1994)
4. Clark, T.R., Carruthers, T.F., Matthews, T.J., Duling, I.N., Electronics Letters **35**, 720 (1999)
5. Shelton, R.K., Foreman, S.M., Ma, L.-S., Hall, J.L., Kapteyn, H.C., Murnane, M.M., Notcutt, M., Ye, J., Opt. Lett. **27**, 312 (2002)
6. Young, B.C., Cruz, F.C., Itano, W.M., Bergquist, J.M., Phys. Rev. Lett. **82**, 3799 (1999)
7. Bartels, A., Dekorsy, T., Kurz, H., Opt. Lett. **24**, 996 (1999)
8. Bartels, A., Kurz, H., Opt. Lett. **27**, 1839 (2002).
9. Diddams, S.A., Udem, Th., Vogel, K.R., Oates, C.W., Curtis, E.A., Windeler, R.S., Bartels, A., Bergquist, J.C., Hollberg, L., in: Laser Frequency Stabilization, Standards, Measurement, and Applications, J. L. Hall, J. Ye, eds., Proceedings of SPIE **4269**, 77 (2001)
10. Ramond, T.M., Bartels, A., Diddams, S.A., Hollberg, L., Opt. Lett. **27**, 1842 (2002)
11. Allan, D.W., IEEE Transactions on Ultrasonics Ferroelectrics and Frequency Control **34**, 647 (1987)
12. Ivanov, E.N., Hollberg, L., Diddams, S.A., Proceedings of 2002 IEEE International Frequency Control Symposium, 435 (2002)
13. Fortier, T.M., Jones, D.J., Ye, J., Cundiff, S.T., Windeler, R.S., Opt. Lett. **27**, 1436 (2002)

Generation of 14.8-fs Pulses in a Spatially Dispersive Amplifier

C. P. Hauri, M. Bruck, W. Kornelis, J. Biegert, and U. Keller

Swiss Federal Institute of Technology (ETH Zürich), Physics Department, Institute of Quantum Electronics, Zürich, Switzerland chauri@phys.ethz.ch

1 Introduction

The rapid advance of ultrashort amplified laser technology into a mature research tool has inspired its widespread use. Such laser pulses, with durations of less than 10 fs, are exclusively generated by using gas-filled capillaries with subsequent chirped-mirror compression [1]. Direct amplification of ultrabroadband laser pulses to high energies (millijoule region) still poses a challenging problem. Thermally induced lensing, gain-narrowing and phase aberrations during amplification reduce the ability to obain ultrashort pulses at a higher energy level with a good beam quality. These problems have been addressed in a number of variants, such as cryogenic cooling of the gain material, spectral filtering and phase control via adaptive shaping or specially designed mirrors and pulses shorter than 16 fs have been reported [2] [3]. Another promising approach, first proposed by Danailov *et al.* [4] employs a spatially dispersed seed beam inside an amplifier to decouple the amplification of competing wavelengths from each other. With this an artificial inhomogeneous broadening is induced in the Ti:sapphire, which is normally homogeneously broadenend, which makes it possible to tailor the gain profile. The effective gain factor of each wavelength component is only determined through the dedicated pump intensity profile. As a result, gain-competition between different wavelength components, and red-shifting of the spectrum can be eliminated provided that the individual wavelength channels are sufficiently decoupled from each other inside the gain medium. The implementation of such a dispersive amplifier has been previously reported [5], but compression to yield amplified ultrashort pulses was never demonstrated.

2 Dispersive amplification

2.1 Setup

Our setup consists of a Ti:sapphire oscillator, preamplifier and the dispersive amplifier. The Kerr-lens modelocked Ti:sapphire oscillator (Femtolaser) delivers 11-fs pulses at a repetition rate of 76 MHz. After single pulse selection down to 1 kHz and stretching to 200 ps, the pulses are sent through an acousto-optic programmable dispersive filter (Dazzler, Fastlite) to pre-compensate for higher-order dispersion in the amplifiers. The home-built pre-amplifier yields pulse energies of 200 μJ, with a reduced spectral width of 40 nm FWHM due to gain-narrowing.

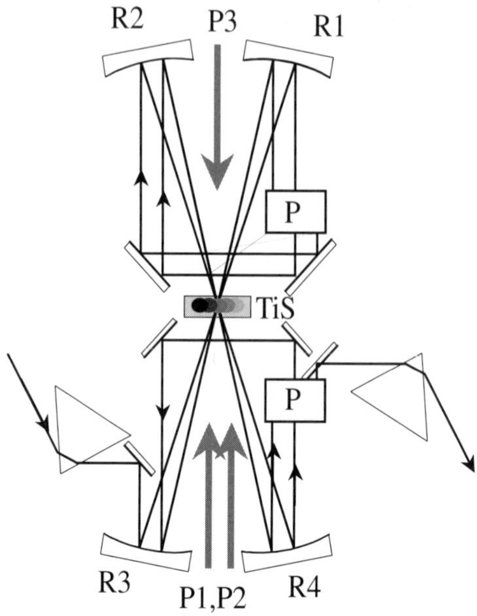

Fig. 1. Experimental setup. R1,R2,R3,R4: Focusing mirrors (ROC -750 mm); P:periscope assembly for mirroring spectral components;Ti:sapphire crystal with spatially separated spectral components; PR1,PR2: Brewster-cut prisms; P1,P2,P3:Pump beams.

The total spectral throughput of the pre-ampllifier is given by the spectral bandpass of the dielectric mirrors (110 nm, centered at 800 nm). The power density in the spectral wings is more than two orders of magnitude smaller than around the center wavelength. The dispersive amplifier is used subsequently to preferentially amplify these wings, whereas the central part of the spectrum does not experience any further gain. In the four-pass dispersive amplifier, shown in Fig. 1, the incoming collimated seed beam is sent through a Brewster-cut fused silica prism (FS) and the angularly dispersed spectral components are mapped into a line of foci inside the Ti:sapphire crystal, placed in the Fourier plane, which guarantees the best spectral resolution. The multi-pass geometry is designed to yield a real image in the focusing plane after each pass. After four passes, the angularly dispersed beam is sent through a second prism that reverses angular dispersion, before being compressed in a grating compressor. Depending on the spatial pump intensity profile, independent amplification of each wavelength component is possible and gain-narrowing can be overcome. A pump geometry employing three independent pump beams, each beam focused down to a circular spot of 230 μm inside the elliptical seed focus of dimensions of 200 μm and 1.3 mm (spatially dispersed dimension), is used. The pump beams are positioned in the gain medium such, that they exclusively overlap with the spectral wings of the seed pulse. The pump beams impose a spatially inhomogeneous thermal load in the gain medium, which could potentially make proper reversal of the angular dispersion difficult and result in a spatial chirp. To circumvent this problem, we cryogenically cool the crystal to 90 K and keep the number of passes low.

2.2 Pulse measurement

We measured the pulse duration by spectral-phase interferometry for direct electric field reconstruction (SPIDER) [6] and used the Dazzler to manually compensate for higher-order phase aberrations.

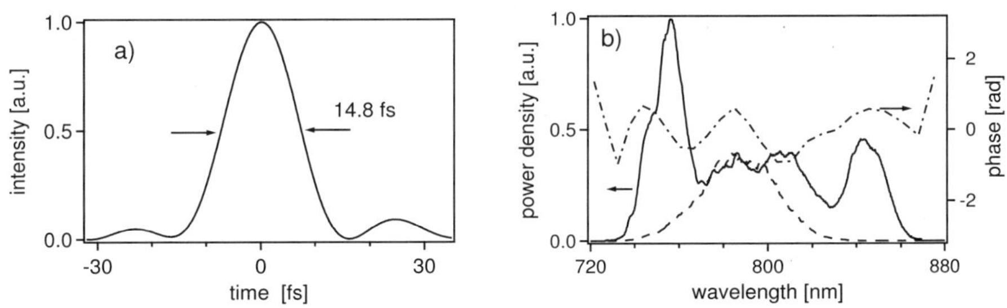

Fig. 2. a) Reconstructed temporal pulse profile showing a FWHM duration of 14.8 fs. **b)** Amplified pulse spectrum (solid line) and spectral phase (dash-dotted) after four passes through the dispersive amplifier. The dashed line shows the spectrum of the pre-amplified seed pulse.

Figure 2a shows the reconstructed temporal pulse shape of the compressed, amplified pulse, with a duration of 14.8 fs (FWHM). The pulse energy was 450 μJ. The associated spectral phase (dash-dotted line) and the spectrum (solid) are shown in Fig. 2b. The pulse spectrum entering the dispersive amplifier (dashed) shows a spectral width (FWHM) of only 40 nm due to serious gain-narrowing from the preamplifier. Optimizing the overlap of the three pump beams with the spectral wings of the dispersed seed, allows the previously narrowed spectrum to be broadened to more than 110 nm supporting a transform-limited pulse duration of 14.6 fs.

2.3 Numerical modelling

The amplification process in the dispersive amplifier was numerically modeled, with results shown in Fig. 3. The simulation accounts for spatially-distributed pumping, non-overlapping spectral components of the seed, and the gain cross-section of the Ti:sapphire modeled according to Eggleston *et al.* [7]. Starting from the measured pump (Fig. 3a) and seed (Fig. 3b, dashed line) parameters, the successively amplified spectra from passes 1 to 4 are shown in Fig. 3b. After the first pass through the gain medium, the input spectrum (dashed) experiences a slightly stronger amplification at shorter wavelength than at the center at 840 nm, where no amplification is observed. During the next three passes the spectrum changes significantly, leading to considerable amplification of the spectral wings. The simulation agrees well with the experiment as shown for the simulated spectrum after four passes and the measured spectrum (Fig. 2b, solid)

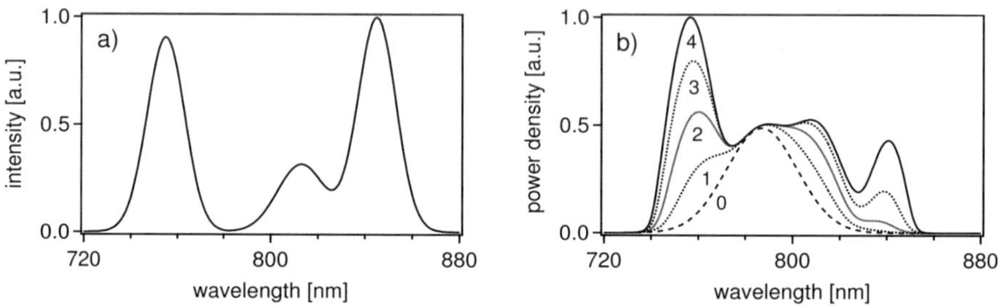

Fig. 3. a) Pump intensity profile in the Ti:sapphire plotted versus corresponding seed wavelengths.**b)** Evolution of the spectra in the dispersive amplifier from pass 1 to 4 predicted by the numerical simulation. The output spectrum of the preamplifier is labeled as pass number 0 (dashed).

3 Conclusions

We have for the first time demonstrated direct amplification and subsequent compression using a dispersive imaging multipass amplifier combined with a conventional preamplifier able to generate 14.8 fs pulses with an energy of 450 μJ at 1 kHz. A key adavantage lies in the decoupled spectral shaping capabilities, being an ideal source to generate ultrabroadband spectra with well-defined energy content. Our simulation shows excellent agreement with measured results, which allows for the confident predictions on further energy scaling and spectral shaping. With a combination of hollow fiber compression and subsequent dispersive amplification, we expect to overcome the energy limitations imposed by the hollow fiber approach.

This work was supported by the Swiss National Science Foundation and by the "Bundesamt für Bildung und Wissenschaft, Schweiz, Projekt BBW Nr. 02.0434".

References

1. M. Nisoli, S. Stagira, S. De Silvestri, O. Svelto, S. Sartania, Z. Cheng, G. Tempea, C. Spielmann, F. Krausz, Toward a terawatt-scale sub-10-fs laser technology, IEEE J. Sel. Top. Quantum Electron. 4, 414 (1998)
2. K. Yamakawa, M. Aoyama, S. Matsuoka, H. Takuma, C.P.J. Barty, D. Fittinghoff, Generation of 16-fs, 10-TW pulses at a 10-Hz repetition rate with efficient Ti:sapphire amplifiers, Opt. Lett. 23, 525 (1998)
3. E. Zeek, R. Bartels, M.M. Murnane, H. C. Kapteyn, S. Backus, G. Vdovin, Adaptive pulse compression for transform-limited 15-fs high-energy pulse generation, Opt. Lett. 25, 587 (1999)
4. M.B. Danailov, I.P. Christov, A novel method of ultrabroadband laser generation, Opt. Commun. 77, 397 (1990)
5. J. Faure, J. Itatani, S. Biswal, G. Cheriaux, L.R. Bruner, G.C. Templeton, G. Mourou, A spatially dispersive regenerative amplifier for ultrabroadband pulses, Opt. Commun. 159, 68 (1999)
6. C. Iaconis, I.A. Walmsley, Spectral phase interferometry for direct electric-field reconstruction of ultrashort optical pulses, Opt. Lett. 23, 792 (1998)
7. J.M. Eggleston, L.G. DeShazer, K.W. Kangas, Characteristics and kinetics of laser-pumped Ti:sapphire oscillators, IEEE J. Quantum Electron. 24, 1009 (1988)

A Sub-10-Femtosecond Terawatt-Scale Ti:sapphire Laser System

József Seres[1], Enikö Seres[1], Christian Spielmann[2], and Ferenc Krausz[1,3]

[1] Institut für Photonik, Technische Universität Wien, Gusshausstrasse 27/387, 1040 Wien, Austria jozsef.seres@tuwien.ac.at
[2] Physikalisches Institut EP1, Universität Würzburg, Am Hubland, 97074 Würzburg, Germany
[3] Max-Planck-Institut für Quantenoptik, Hans-Kopfermann-Str. 1, D-85748 Garching, Germany

Summary. A three-stage, 1-kHz amplifier system has been developed delivering sub-10-fs pulses with a peak power of 0.3 TW. Passive and active spectral amplitude and phase control allow preserving a bandwidth of 120 nm (FWHM) up to multi-millijoule energy levels and temporal compression of the broadband pulses close to their Fourier limit. Future improvements to increase the peak power beyond 1 TW will be discussed.

1 Introduction

Development of high-peak-power laser systems based on the chirp pulse amplification (CPA) concept [1] has opened the way to generate coherent x-ray radiation and attosecond pulses by the technique of high-order harmonic generation (HHG). The ultrashort duration of such pulses benefits both the conversion efficiency [2] and the confinement of the emission into an isolated sub-femtosecond pulse [3]. For Ti:sapphire based amplifier systems a repetition rate of 1 kHz is a good compromise between pulse energy and repetition rate [4-13] and the bandwidth of the amplifier material can support sub-10-fs pulses. We describe an amplifier system producing multi-millijoule sub-10-fs pulses for the first time.

2 Description of the amplifier system

The laser system (Fig. 1) contains three, multi-pass Ti:sapphire amplifier stages and is seeded by a Ti:sapphire oscillator (Femtosource Compact Pro, Femtolasers GmbH). The first stage implements 9 passes and is pumped by 10-mJ pulses after splitting the output of a Q-switched, frequency-doubled Nd:YLF Laser (Thomson CSF 621D) at a 1-kHz repetition rate. The other half of the output is used for pumping the second stage. The first stage delivers pulses with an energy of 1 mJ and gain narrowing is partially compensated [9,13] by thin film dielectric filter inserted into the first eight passes.

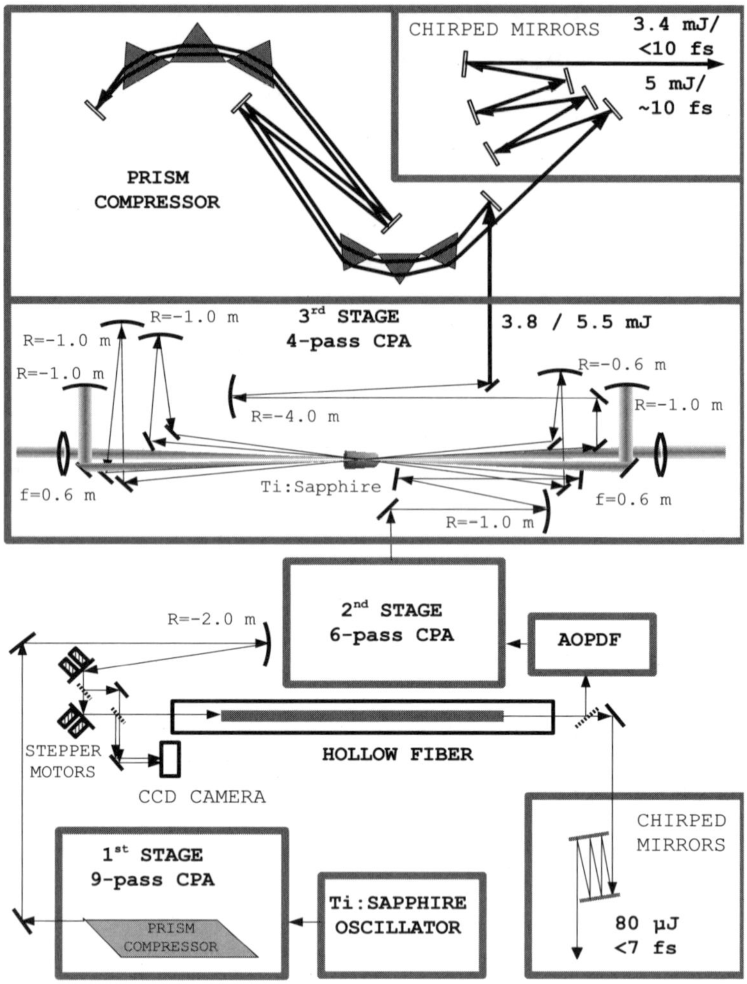

Fig. 1. Schematic draw of the laser system

After amplification the pulses have a bandwidth of 60 nm. With combination of an LAK16A double prism compressor [13] and specially designed chirped mirrors we were able to recompress the pulses close to their Fourier limit of about 20 fs.

In order to provide sufficient bandwidth for generating sub-10-fs pulses, the pulses are launched into a hollow-core fiber [17]. The direction of the laser beam is monitored at two different positions with a CCD camera (Fig. 1). Deviations of the ideal beam direction are corrected with a pair of motorized mirror mounts controlled by a computer-assisted feedback loop. The coupling into the fiber was designed to guarantee long term stability. The spectra of the pulses are broadened to nearly 140 nm (FWHM). Employing such a broad seed spectrum for the final amplifier stages ensures an output bandwidth broad enough to support sub-10-fs pulses. The task of the second 6-pass amplifier stage is to compensate for the losses introduced by the spectral broadening in the fiber and the subsequent shaping with a newly-developed acousto-optic programmable dispersive filter [15] (AOPDF) (DAZZLER, Fastlite), which is inserted before this stage. The new prototype AOPDF is able to control the amplitude and phase of spectral components over a wavelength range as broad as 350 nm with an efficiency [16] as high as 40 %. To avoid spectral narrowing, thin film dielectric filters are inserted into the first 4 passes and the input spectrum is pre-shaped with AOPDF.

Finally, the pulses are amplified in a 4-pass amplifier stage pumped by 22-mJ from a Q-switched frequency-doubled Nd:YLF Laser (Photonics GM-30-527). To minimize thermal aberrations the Ti:sapphire crystal is cooled to 150 K by a closed-loop cryostat (Cryotiger, APD Cryogenics Inc.). The pulses after the third stage have been compressed in a prism compressor consisting of two sets of fused silica triple prisms. Higher order dispersion terms are pre-compensated by specially designed chirped mirrors placed before the second stage and appropriate programming of the AOPDF. With the optimized compressor we have been able to recompress the amplified output pulses down to 9.4 fs (Fig. 2) at an energy of 3.4–3.5 mJ. This pulse width has been evaluated from a SPIDER analysis [18] and confirmed by an interferometric autocorrelation measurement.

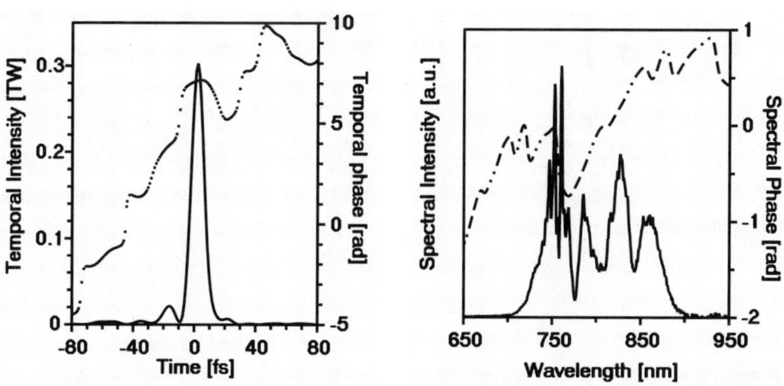

Fig. 2. Pulse shape and spectrum at the output of the 3-stage amplifier system (solid lines) with the corresponding phases (dashed lines)

3 Improvement of the booster amplifier stage

Up to now, the third amplifier stage is pumped by only one pump laser keeping beam diameter of the pump and seed beam in the crystal to be 0.6 mm and 0.4 mm, respectively. Using this configuration the pulse energy is boosted from 0.5 mJ to 3.7-3.8 mJ (Fig. 3). We are currently working on a further upgrade using two pump lasers instead of one. The pump power is 40 W now at 1 kHz repetition rate and it deliveres pulses with energy of 5.5 mJ (Fig. 3). Further important task to keep the output spectrum as smooth as possible to eliminate side pulses. The spectrum we reached recently at 5.5 mJ energy level is presented in Fig. 3. We estimated the maximum obtainable pulse energy without compromising the pulse duration. In our calculations we assumed beam diameters of 0.9 mm and 0.6 mm for the pump and seed, respectively and our calculation predicts sub-10-fs pulses with energy larger than 10 mJ and peak power of near 1 TW.

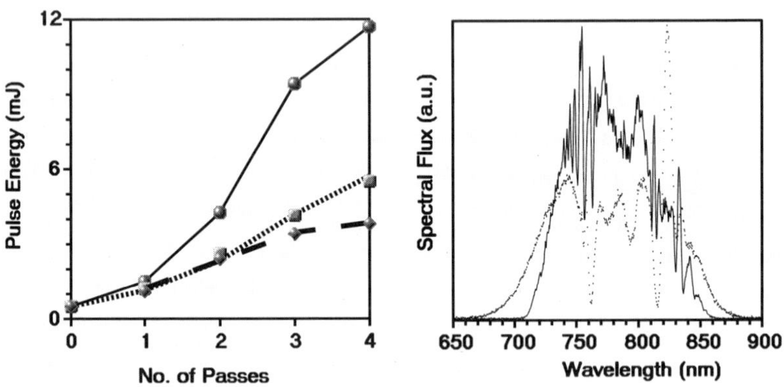

Fig. 3. Measured and calculated pulse energy evaluation in the 3rd amplifier stage using one (dashed line, 20 W) or two (dotted line, 38 w) pump lasers and the predicted available pulse energy (continuous line, 45 W). Right side graph presentes the spectrum after hollow-core fiber (dotted line) and the 3rd stage (continuous line) at 5.5 mJ output pulse energy.

4 Summary

We have developed a Ti:sapphire amplifier system delivering sub-10-fs pulses with a peak power of 0.3 TW in a near diffraction-limited beam. The new source is expected to allow the generation of powerful high-order harmonic radiation up to photon energies in the water window and extend attosecond time-resolved spectroscopy into this important spectral range. The envisaged upgrade of the system to peak powers exceeding 1 TW will open the way to few-cycle relativistic electron-light interactions.

Acknowledgement. This work has been sponsored by the Austrian Science Fund grant F016 (ADLIS), the BMBF project 13N7922, grant of P15382 (phase-control) and Z63 (Wittgestein). J. Seres acknowledges support from a Lise Meitner fellowship M739-N08. Ch. Spielmann acknowledges support from the DFG grant SP 687/1-1.

References

[1] P. Maine, D. Strickland, P. Bado, M. Pessot, G. Mourou, IEEE J. QE **24**, 398 (1988)
[2] T. Brabec and F. Krausz, Rev. Mod. Phys. **72**, 545 (2000).
[3] M. Hentschel *et al.*, Nature **414**, 512 (2001).
[4] S. Backus, Ch. G. Durfee III, G. Mourou, H. C. Kapteyn, M. M. Murnane, Opt. Lett. **22**, 1256 (1997)
[5] S. Santania, Z. Cheng, M. Lenzner, G. Tempea, Ch. Spielmann, F. Krausz, K. Ferencz, Opt. Lett. **22**, 1562 (1997)
[6] Y. Nabekawa, Y. Kuramoto, T. Togashi, T. Sekikawa, S. Watanabe, Opt. Lett. **23**, 1384 (1998)
[7] Y. Nabekawa, T. Togashi, T. Sekikawa, S. Watanabe, S. Konno, T. Kojima, S. Fujikawa, K. Yasui, Opt. Express **5**, 318 (1999)
[8] V. Bagnoud and F. Salin, Appl. Phys. B **70**, S165 (2000)
[9] M. Hentschel, Z. Cheng, F. Krausz, Ch. Spielmann, Appl. Phys. B **70**, S161 (2000)
[10] E. Zeek, R. Bartels, M. M. Murnane, H. C. Kapteyn, S. Backus, G. Vdovin, Opt. Lett. **25**, 587 (2000)
[11] S. Backus, R.Bartels, S. Thompson, R. Dollinger, H. C. Kapteyn, M. M. Murnane, Opt. Lett. **26**, 465 (2001)
[12] Y. Jiang, T. Lee, W. Li, G. Ketwaroo, Ch. G. Rose-Petruck, Opt. Lett. **27**, 963 (2002)
[13] Z. Cheng, F. Krausz, Ch. Spielmann, Opt. Commun. **201**, 145 (2002)
[14] J. Seres, A. Müller, E. Seres, R. Herzog, K. O'Keeffe, M. Lenner, D. Kaplan, Ch. Spielmann, F. Krausz, Opt. Lett. (2003) accepted
[15] F. Verluise, V. Laude, Z. Cheng, Ch. Spielmann, P. Tournois, Opt. Lett. **25**, 575 (2000)
[16] E. Seres, R. Herzog, J. Seres, D. Kaplan, Ch. Spielmann, Opt. Express **11**, 240 (2003), `http://opticsexpress.org/abstract.cmf?URI=OPEX-11-3-240`
[17] M. Nisoli, S. De Silvestri, O. Svelto, Appl. Phys. Lett. **68**, 2793 (1996)
[18] C. Iaconis, I. A. Walmsey, IEEE J. Quantum Electron. **35**, 501 (1999)
[19] C. Le Blanc, P. Curley, F. Salin, Opt. Commun. **131**, 391 (1996)

Direct Generation of 12-fs Pulses from a Ti:sapphire Chirped-Pulse Amplification System

Hideyuki Takada, Masayuki Kakehata, and Kenji Torizuka

The National Institute of Advanced Industrial Science and Technology (AIST), 1-1-1 Umezono, Tsukuba, Ibaraki 305-8568, Japan h.takada@aist.go.jp

Summary. We demonstrate a direct generation of ~12-fs pulses from a Ti:sapphire chirped-pulse amplification system utilizing a broadband pulse stretcher, broadband gain-narrowing compensators, broadband high energy mirrors, a dispersion compensator, and a broadband pulse compressor.

In these several years, the pulse duration of Ti:sapphire chirped-pulse amplification (CPA) system has been reduced. Yamakawa, et.al. demonstrated the generation of 20 fs, 100 TW pulses [1]. However, it is still a challenging problem to use the wide gain bandwidth of Ti:sapphire in a high peak power amplifier efficiently. To obtain shorter pulses, we developed a broadband pulse stretcher/compressor [2, 3], a broadband gain-narrowing compensator [2, 4], a broadband high energy mirror [2, 4], and an adaptive dispersion compensator [3]. Utilizing these devices, we demonstrate a direct generation of ~12-fs pulses from a Ti:sapphire CPA system.

The ~8-fs pulse train from a mode-locked Ti:sapphire laser with chirp mirrors and a pair of prisms is sent to the pulse stretcher to be stretched to ~300 ps. We adopt a pulse stretcher with a telescope, which consists of a concave mirror and a convex mirror. Since the aberration of this telescope is relatively small, the high-order dispersions of this stretcher are nearly reverse of those of the compressor. Therefore this stretcher is suitable for the CPA system including multipass amplifier chain having relatively small material dispersions. The size of the concave mirror is 500 mm(W) x 100 mm(H) x 80 mm(D). The size of the convex mirror is 300 mm(W) x 10 mm(H) x 50 mm(D). The number of grooves of the grating for the pulse stretcher is 1200g/mm. The ruled size of the grating is 220 mm(W) x 110 mm(H). The incidence angle of the grating is 55 degrees. The distance between the concave mirror and the grating is 800 mm. The curvatures of the concave mirror and the convex mirror are 998 mm and 519 mm, respectively. The distance between the concave mirror and the convex mirror is 499 mm. In case of an Offner-type telescope, centers of curvatures of a concave mirror and a convex mirror coincide at a single point [5], but those of the concave mirror and the convex mirror, which are components of our pulse stretcher, do not coincide at a single point. Based on our calculation utilizing ray tracing [2,3], if an Offner-type telescope is utilized in the pulse stretcher with wide spectral range (>200 nm), it is difficult to decrease the aberration of the Offner-type telescope to a negligibly small value. The results of our calculation suggest that optimization of a radius curvature and a position of a convex mirror is useful for decreasing an aberration of a telescope for a pulse stretcher.

The chirped pulse train from the pulse stretcher passes through a dispersion compensator with a spatial light modulator (SLM) for reduction of the residual dispersion of the CPA system. The dispersion compensator consists of two gratings, two concave mirrors, and the SLM. The number of grooves of the gratings is 200 g/mm. The radius curvature of the concave mirrors is 0.4 m. A width of a pixel of the SLM is 100 μm. The number of the pixels of the SLM is 128. The bandwidth of the compensator is ~320 nm.

The chirped pulse from the dispersion compensator is amplified by the 8-pass preamplifier with the gain narrowing compensator with 15 Al_2O_3/SiO_2 layers [2,4]. We proposed a method of a gain-narrowing compensation using multiple dielectric layers providing sufficient compensation for a multi-pass amplifier [2]. We decided the incident angle, the polarization of the incident beam, the materials for its coatings, the number of its layers, and the surface material of the its layers so that the layers alleviate the gain-narrowing well for the total gain of 10^9 and the transmitting number of the layers of 8. The repetition rate of the amplifiers is 10 Hz. The preamplifier is pumped with ~50 mJ, 532-nm pulses, and amplifies the chirped pulse to ~1 mJ.

The chirped pulse from the preamplifier is amplified by the 6-pass power amplifier. The power amplifier is pumped with ~1.2 J, 532-nm pulses. For first 2-pass amplification, protected gold mirrors are used, and for last 4-pass amplification broadband high energy mirrors [2,4] are used. To generate shorter (~ 10 fs) TW pulses, we need broadband mirrors ($\Delta\lambda$~ 200 nm at λ_0~ 800 nm) with practical tolerance to high fluence (> 1 J/cm^2) and relatively low dispersion. Therefore, we developed a broadband high energy mirror, consisting of broadband coatings (TiO_2/SiO_2) and optimized high damage threshold coatings (ZrO_2/SiO_2) for decreasing the intensity on the broadband coatings. The chirped pulse is amplified to ~300 mJ.

The pulse compressor compresses a portion (~10 mJ) of a ~300 mJ, ~300-ps chirped pulse to ~12 fs. The pulse compressor consists of a pair of two parallel gratings. The number of grooves of the gratings for the pulse compressor is 1200 g/mm. The optical path length between the gratings is ~36 cm. The ruled size of the first grating is 110 mm(W) x 110 mm(H). The ruled size of the second grating is 220 mm(W) x 110 mm(H). The incidence angle of the first grating is ~58.33 degrees.

We measure the phase of the recompressed pulse without control of the dispersion compensator utilizing spectral phase interferometry for direct electric-field reconstruction (SPIDER) [6]. Based on the results, we control the phase of the dispersion compensator to cancel the residual dispersion of the recompressed pulse. Fig. 1 shows the measured phase (dotted line) of the recompressed pulse without control of the dispersion compensator, the measured spectra (solid line) and the measured phase (dashed line) of the recompressed pulse with control of the dispersion compensator. This shows the residual dispersion of the recompressed pulse is reduced remarkably from ~710 nm to ~900 nm. Fig. 2 shows the measured autocorrelation trace (solid line) and the autocorrelation trace (dotted line) calculated from the measured spectra and the measured phase with control of the dispersion compensator. The measured autocorrelation trace and the calculated autocorrelation trace agree well each other. Fig. 3 shows the reconstructed pulse calculated from

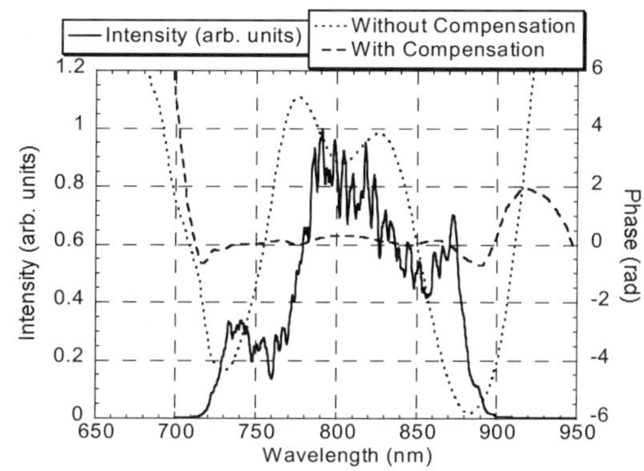

Fig. 1. Spectra and Phases of the recompressed pulse.

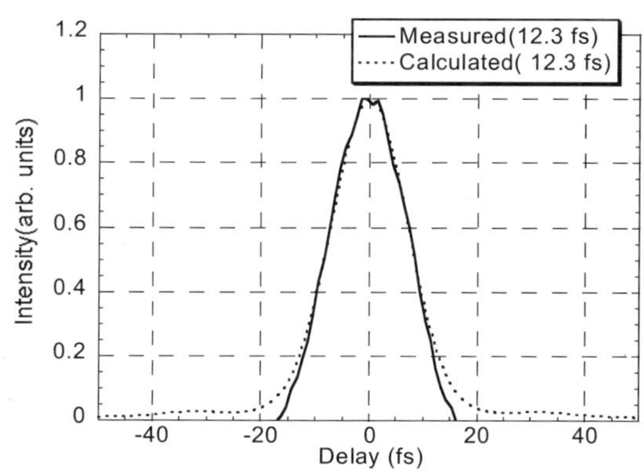

Fig. 2. Measured and calculated autocorrelation traces of the recompressed pulse with control of the dispersion compensator.

the measured spectra and the measured phase. The pulse width of the reconstructed pulse is ~12 fs. We believe that our Ti:sapphire CPA system generates >10 TW, ~12-fs pulses with a pulse compressor in a vacuum chamber.

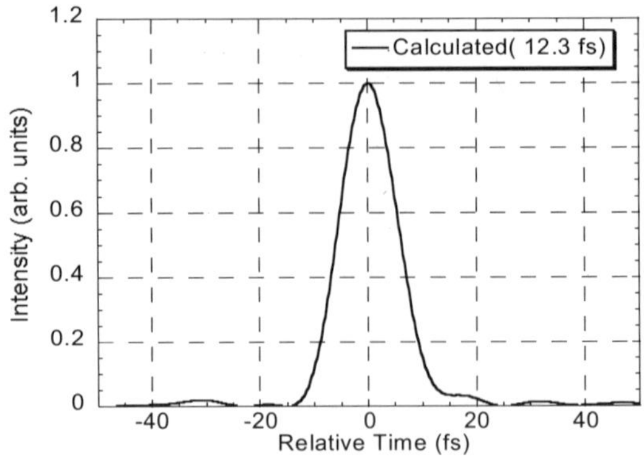

Fig. 3. Reconstructed pulse calculated from the measured spectra and the measured phase of the recompressed pulse with control of the dispersion compensator.

References

1. Yamakawa, K., Aoyama, M., Matsuoka, S., Kase, T., Akahane, Y., Takuma, H.: 100-TW sub-20-fs Ti:sapphire laser system operating at a 10-Hz repetition rate. Opt. Lett., **23**, 1468-1470 (1998).
2. Takada, H., Kakehata, M., Torizuka, K.: Study on design of 10 TW, 10 fs class Ti:sapphire laser. Rev. Laser Eng., **27**, 341-345 (1999) (in Japanese).
3. Takada, H., Kakehata, M., Torizuka, K.: Large-ratio stretch and recompression of sub-10-fs pulse utilizing dispersion managed devices and a spatial light modulator. Appl. Opt., **B74**, S253-S257 (2002).
4. Takada, H., Kakehata, M., Torizuka, K.: Broadband high-energy mirror for ultrashort pulse amplification system. Appl. Opt., **B70**, S189-S192 (2000).
5. Offner, A., U. S. patent 3,748,015(1971).
6. Iaconis, C., Walmsley, I. A.: Self-referencing spectral interferometry for measuring ultrashort optical pulses. IEEE J. of Quantum Electron., **35**, 501-509 (1999).

Generation of 3.8-fs Pulses Through Adaptive Cascaded Hollow Fiber Compression

J. Biegert[1], B. Schenkel[1], U. Keller[1], C. Vozzi[2], M. Nisoli[3], G. Sansone[3], S. Stagira[3], S. De Silvestri[3], and O. Svelto[3]

[1] Swiss Federal Institute of Technology (ETH Zürich), Physics Department, Institute of Quantum Electronics, Zürich, Switzerland biegert@phys.ethz.ch
[2] National Laboratory for Ultrafast and Ultraintense Optical Science - INFM,Dipartimento di Fisica, Politecnico, Milano, Italy
[3] INFM - Dipartimento di Fisica, Universitá degli Studi, Milano, Italy

1 Introduction

An important pre-requisite for a number of fundamental physics experiments, such as the generation of single attosecond pulses by high-order harmonic generation [1], and the investigation of absolute phase effects in nonlinear processes [2], is the availability of high-peak power laser pulses in the sub-4-fs regime. In the past decade, various sources have become available that are capable of producing supercontinua in the visible and near-infrared spectral region. However, it was only recently that a high-energy supercontinuum from two cascaded hollow fibers with bandwidths exceeding 500 THz was generated [3]. The remaining challenge is then to compress these enormous bandwidths to finally yield an *isolated* ultrashort optical pulse.

Successful compression to 4.5 fs of the supercontinuum output of a single hollow fiber was demonstrated using chirped mirrors and thin prisms for dispersion compensation [4]. With the same continuum generation technique, 5 fs pulses were obtained by using chirped mirrors [5], or a sole spatial light modulator (SLM) [6, 7]. With a non-collinear optical parametric amplifier (NOPA) pulses as short as 4 fs with a pulse energy of 0.5 μJ were generated with a more complex dispersion compensation technique employing chirped mirrors, gratings and a deformable mirror [8].

2 Adaptive pulse compression

2.1 Setup

In this article, we present the generation of 3.8-fs pulses with energies up to 15 μJ by compressing the high-energy supercontinuum output of two cascaded hollow fibers. Dispersion compensation is achieved by measuring the pulse with spectral-phase interferometry for direct electric-field reconstruction (SPIDER) [9], and then applying appropriate feedback to a liquid crystal spatial light modulator (SLM) to iteratively compress the pulse. Figure 1

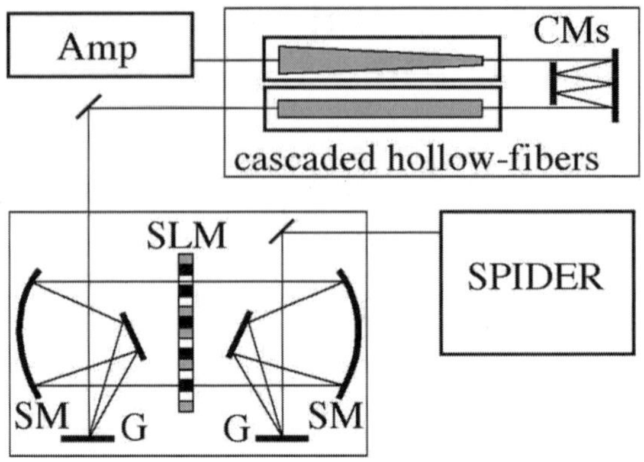

Fig. 1. Schematic of the experimental setup: Amp, Ti:Sapphire laser amplifier, 25 fs, 0.5 mJ, 1kHz; CM's, chirped mirrors; G, 300 l/mm diffraction gratings; SM, spherical mirror, ROC=600 mm.

shows our experimental setup: 25 fs pulses with a pulse energy of 0.5 mJ, produced by a 1-kHz Ti:Sapphire laser-amplifier system, enter the first Argon-filled capillary. The gas pressure (0.3 bar) in this fiber was chosen such, that after chirped mirror compression (5 bounces, 30 fs^2 each), 10 fs pulses emerge with negligible wings. Sending these pulses (300 μJ) into a second Ar-filled hollow fiber finally leads to a high-energy supercontinuum (100 μJ, 500 THz) with excellent spatial characteristics. Note, that the advantage of using two cascaded hollow fibers over a single one lies in the generation of a supercontinuum with more spectral energy in the short wavelength part of the spectrum. The emerging beam is then collimated and sent into the pulse shaper [10] consisting of a 640 pixel liquid crystal SLM (Jenoptik, pixels are 97 μm wide with a 3 μm gap), two 300 lines/mm gratings and two 300 mm focal length spherical mirrors (4-f-setup). We found, that for the large beam diameters of 6.4 cm that we used, alignment for normal incidence on the spherical mirrors (see Fig. 1) proved to be very critical in order to yield a good beam quality. The major disadvantage of this setup are the considerable losses caused by the gratings for such an enormous bandwidth: with the pulse energy at the entrance of the pulse shaper being 100 μJ, it was measured to be only 15 μJ at the exit.

2.2 Pulse measurement

For such a short pulse, diligent pulse characterization is prudent. We employed SPIDER, optimized for sub-10-fs pulses; see Ref. [11] for details. Furthermore, custom-designed 400 μm thin ultrabroadband dielectric beamsplitters, optimized for ultralow dispersion over bandwidths spanning from 450 nm to 1000 nm, were used. Over this large bandwidth, the reflectivity of these beamsplitters varies from 50% to 47% and the group delay dispersion varies from 0 fs^2 to 1 fs^2. For sum-frequency mixing, a 30 μm thin β-barium-borate (BBO) crystal, cut for type II interaction, was used. Hence, for type-II sum-frequency mixing of a broadband input pulse with the quasi-cw upconversion spectral slices, more than 50% conversion efficiency for a bandwidth spanning from 530 nm to 1030 nm (FWHM of the efficiency curve) is ensured. The spectral phase was then reconstructed from the measured SPIDER trace and used to iteratively compress the pulse: Compression is started by writing an initially flat phase on the SLM. Next, the measured spectral phase is inverted and added to the currently applied phase. Then the pulse is measured again and the cycle repeated should the pulse be too long. With this technique, we find that typically 5 iterations were necessary to yield the shortest pulse. The spectral phase of the supercontinuum has been observed to be very stable and remarkably constant for several hours. This is in excellent agreement with earlier measurements performed on phase fluctuations in hollow fibers [12] and is an important prerequisite for the successful application of this compression technique for everyday use.

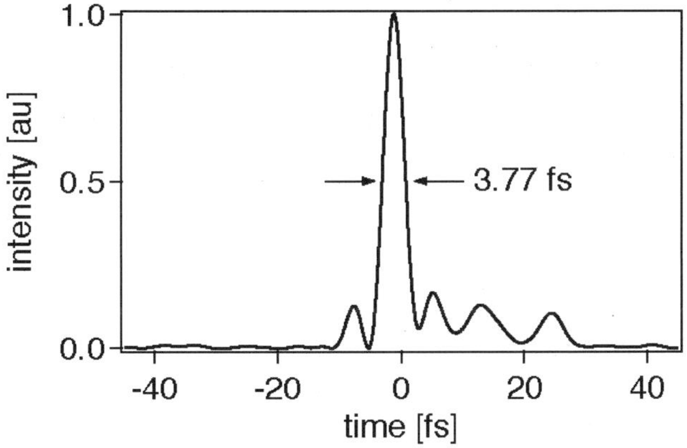

Fig. 2. Reconstructed temporal pulse profile showing a FWHM of 3.77 fs.

Figure 2 shows the temporal profile of the shortest pulse obtained so far with our technique. The full-width-at-half-maximum (FWHM) is measured to be (3.77 + 0.11) fs with a pulse energy of 15 μJ. The spectrum spans a bandwidth of about 270 THz (see Fig. 3) and has been cut at 566 THz by inserting a knife-edge into the spectrally dispersed beam — inside the pulse shaping apparatus. The reason for this is two-fold: Because of the lack of a strong enough up-conversion signal from the shorter wavelength region, the spectral phase remained undetermined in the cut region. The second reason is the strong scattered fundamental light which has to be kept from interfering with the SPIDER-signal in the spectrometer and therefore from decreasing the modulation depth. We expect to solve both of these problems in the future by using cross-correlation-SPIDER. Figure 3 shows the spectral phase to be

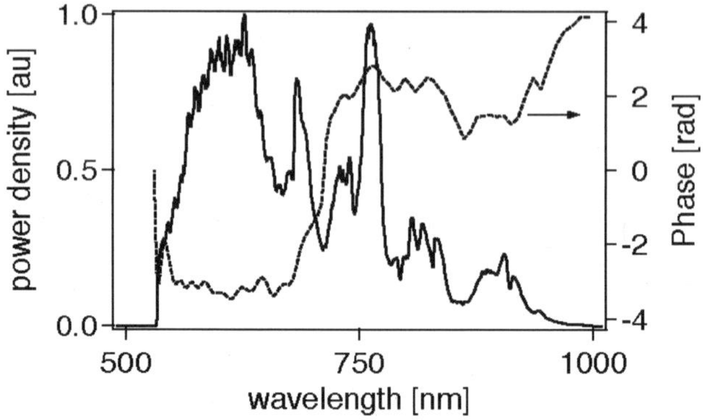

Fig. 3. The measured pulse spectrum is shown in solid. The spectrum has been cut at wavelengths below 530 nm (above 566 THz) and spans a bandwidth of 270 THz. The reconstructed spectral phase is shown dashed.

essentially flat over the entire bandwidth except for a near 2π-step at 711 nm (422 THz). The phase-step in conjunction with the strongly modulated spectrum accounts for the pre- and after-pulses in the temporal domain (Fig. 2). Due to high-frequency noise locally accumulated over a few iterations, compression failed in this spectral region and even though we tried to eliminate this phase shift by adding one more iteration, we were unsuccessful. Due to the steep phase transition however, the pulse duration was only slightly influenced and the main impact is on the pulse contrast. Additionally, we achieved clean compression to about 4.2-fs with slightly narrower spectra. We believe that careful application of a smoothing algorithm between the individual iterations could resolve this problem. Aside from technical challenges, this hints at the clear possibility of getting "cleaner" and shorter optical pulses.

2.3 Error analysis

Pulse parameters are given in most papers without any quantification of the deviation from the quoted values. We believe that it is prudent for such ultrashort pulse measurements to state the possible error sources and to quantify their contributions. E.g., within the SPIDER retrieval algorithm, the linear phase owing to the time delay between the two pulse replicas has to be subtracted from the SPIDER interferogram phase. The determination of this time delay is crucial because it has the most significant influence on measurement errors of the reconstructed pulse duration [13]. We therefore performed an error estimation on the measured pulse duration by evaluating the precision of the calibration of this time delay. The value of the delay retrieved from the interferogram of the two replicas was measured to be τ

= 179.81 fs. Comparing the linear phase of the above measurement with the linear phase measured several hours before, we found that the long term stability is good enough to yield a value within $\Delta\tau = 0.01$ fs of the original delay. Performing the reconstruction with $\tau + \Delta\tau$ yields a FWHM pulse duration of 3.81 fs.

The influence of random noise on the pulse reconstruction was estimated by fitting a straight line to the linear phase caused by the delay. The standard deviation of the delay was found to be $\sigma_\tau = 0.07$ fs. This corresponds to an error of 0.11 fs in pulse duration. From our experience with SPIDER and our particular implementation of it, we find that this delay parameter is typically reproducible to within better than 0.1 fs over several hours of operation.

The influence of finite bandwidth sum-frequency mixing on the measured pulse duration was investigated: Correcting the measured SPIDER-trace for a finite crystal bandwidth yields a FWHM pulse duration of 3.79 fs. Another possible error source is the measurement of the spectrum: After performing a careful calibration of the spectrometer, we have found that shifting the measured spectrum by $\pm$ 0.1 nm results in a change of the measured pulse duration of only 0.03 fs. Lastly, the influence of noise on the spectrum was simulated by adding 1% rms Gaussian noise onto the spectrum, which is significantly more than experimentally observed. We find that this only yields a 1% rms error in pulse duration.

3 Conclusions

We have demonstrated the generation of pulses as short as 3.77 fs with pulse energies of 15 μJ by adaptive compression of the supercontinuum output of two cascaded hollow fibers using only a SLM for dispersion compensation. To the best of our knowledge, these are the shortest pulses generated in the visible and near-infrared spectral region by any technique. This also corresponds to the broadest bandwidth that has ever been controlled using a SLM. Compared to the OPA [9], this technique provides more than an order of magnitude higher pulse energies therefore being a potential source for nonlinear optics experiments. We expect to generate even shorter pulses with higher pulse energies in the near future.

This work was supported by the Swiss National Science Foundation and by the "Bundesamt für Bildung und Wissenschaft, Schweiz, Projekt BBW Nr. 02.0434" (B. Schenkel, J. Biegert, and U. Keller), by the EU Human Potential Programmes, contracts N. HPRI-CT-2001-00148 and N. HPRN-CT-2000-00133 (ATTO); and by INFM PRA-CLUSTERS (M. Nisoli, G. Sansone, S. Stagira, C. Vozzi, S. De Silvestri, and O. Svelto).

References

1. M. Hentschel, R. Kienberger, Ch. Spielmann, G. .A. Reider, N. Milosevic, T. Brabec, P. Corkum, U. Heinzmann, M. Drescher, and F. Krausz. Attosecond metrology. *Nature*, 414:511, 2001.
2. G. G. Paulus, F. Grasborn, H. Walther, P. Villoresi, M. Nisoli, S. Stagira, E. Priori, and S. De Silvestri. Absolute-phase phenomena in photoionization with few-cycle laser pulses. *Nature*, 414:182, 2001.
3. M. Nisoli, G. Sansone, S. Stagira, S. De Silvestri, O. Svelto, and C. Vozzi. Ultrabroadband continuum generation by hollow fiber cascading. *Appl. Phys. B*, 75:601, 2002.
4. M. Nisoli, S. De Silvestri, O. Svelto, R. Szipocs, K. Ferencz, C. Spielmann, S. Sartania, and F. Krausz. Compression of high-energy laser pulses below 5 fs. *Opt. Lett.*, 22:522–524, 1997.
5. G. Cerullo, S. De Silvestri, M. Nisoli, S. Sartania, S. Stagira, and O. Svelto. Few-optical cycle laser pulses: From high peak power to frequency tunability. *IEEE J. Sel. Topics Quant. Electronics*, 6:948, 2000.
6. N. Karasawa, L. Li, H. Suguro, H. Shigekawa, R. Morita, and M. Yamashita. Optical pulse compression to 5.0 fs by use of only a spatial light modulator for phase compensation. *J. Opt. Soc. Am. B*, 18:1742, 2001.
7. A. Weiner, D. Leaird, G. Wiederrecht, M. Banet, and K. Nelson. Spectroscopy with shaped femtosecond pulses:styles for the 1990s. In *SPIE Proceedings 1209*, pages 185–195, Los Angeles, CA, 1990.
8. A. Baltuska, T. Fuji, and T. Kobajashi. isible pulse compression to 4fs by optical parametric amplification and programmable dispersion control. *Opt. Lett.*, 27:306, 2002.
9. C. Iaconis and I. A. Walmsley. Spectral phase interferometry for direct electric-field reconstruction of ultrashort optical pulses. *Opt. Lett.*, 23(10):792–794, 1998.
10. A. M. Weiner. Femtosecond pulse shaping using spatial light modulators. *Rev. Sci. Instrum.*, 71:1929, 2000.
11. L. Gallmann, D. H. Sutter, N. Matuschek, G. Steinmeyer, U. Keller, C. Iaconis, and I. A. Walmsley. Characterization of sub-6-fs optical pulses with spectral phase interferometry for direct electric field reconstruction. *Opt. Lett.*, 24:1314–1316, 1999.
12. W. Kornelis, J. Biegert, J. W. G. Tisch, M. Nisoli, G. Sansone, C. Vozzi, S. De Silvestri, and U. Keller. Single-shot kilohertz characterization of ultrashort pulses by spectral phase interferometry for direct electric-field reconstruction. *Opt. Lett.*, 28(4):1–3, 2002.
13. M. E. Anderson, L. E. E. de Araujo, E. M. Kosik, and I. A. Walmsley. The effects of noise on ultrashort-optical-pulse measurement using spider. *Appl. Phys. B*, S70:85, 2000.

Generation and Characterization of 3.4-fs Optical Pulses with Over-One-Octave Bandwidth

Keisaku Yamane[1], Zhigang Zhang[1], Akira Suguro[2], Kazuhiko Oka[1], Ryuji Morita[1], and Mikio Yamashita[1]

[1] Department of Applied Physics, Hokkaido University, Kita-13, Nishi-8, Kita-ku, Sappro, 060-8628 Japan
[2] Technical Research Laboratory, Citizen Watch Co, Ltd, 840 Shimotomi, Tokorozawa, Saitama, 359-8511 Japan

Summary. We feedback-compensated for chirp of optical pulses with over one-octave bandwidth (495–1090 nm) generated by self-phase modulation in a single argon-filled hollow fiber, and generated 3.4-fs, 1.56 optical-cycle pulses.

1 Introduction

The generation of sub-5.0 fs optical pulses was achieved in recent years by several groups [1–4]. For further pulse shortening, we must generate optical pulses with ultrabroad-band spectra — such as over one-octave — and compensate for the chirp. However, chirp compensation of such broadband pulses is mainly limited by poor capability of a measurement apparatus. It is difficult to characterize such ultrabroad-band pulses by frequency-resolved optical gating (FROG) because of its limited bandwidth or low sensitivity. Although the spectral phase interferometry for direct electric-field reconstruction (SPIDER) technique has the potential of characterizing the spectral phase which extends to over one octave [5], it has a low sensitivity when measuring such ultrabroad-band (low spectral density) signals. We developed a modified SPIDER (M-SPIDER) technique, which greatly improved its sensitivity [6, 7].

On the other hand, it is difficult to compress broadband optical pulses with an over-one-octave bandwidth by use of conventional passive chirp compensators such as prism and/or grating combinations, chirped mirrors, because of their bandwidth limitation and high-order dispersion. Liquid crystal spatial light modulator (SLM) is a superior candidate for a programmable phase compensator of over-one-octave pulses. It has a transmission spectrum from 300 to 1500 nm. We previously demonstrated that a SLM phase compensator enables us to compress pulses to sub 5-fs by inputing individual orders of dispersion into the SLM [1]. In this paper, we report, for the first time to the best of our knowledge, 1.56 optical-cycle, 3.4-fs pulse generation through a single hollow fiber and only a programmable SLM by phase feedback with a significantly improved M-SPIDER.

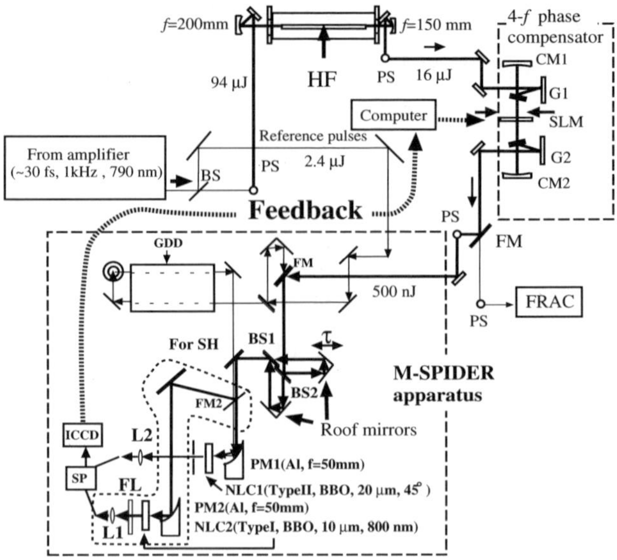

Fig. 1. Experimental setup. CM1,CM2, Al-coated concave mirrors;G1,G2, Al-coated reflective gratings; PS, periscopes; L1,L2, focusing lenses (fused-silica,f=50 mm); HF, Ar-filled hollow fiber; FM,flip mirrors; BS1,BS2, wedged beam splitters; FL,optical filter (cutoff: 500 nm); FRAC, fringe resolved autocorrelator.

2 Experimental

The experimental setup is shown in Fig. 1. The output beam from a Ti:sapphire laser amplifier system (pulse duration of < 30 fs, center wavelength at ~ 790 nm, repetition rate of 1 kHz) was divided into two beams with a beam splitter. One beam with pulse energy of 94 μJ was coupled into a hollow fiber with 34-cm length and 100-μm inner diameter, which was positioned in a chamber filled with argon. The chamber had two 1-mm-thick fused-silica windows. The other beam with pulse energy of 2.4 μJ was sent into the M-SPIDER as the external reference beam. After passing through the hollow fiber, the pulses (16 μJ energy) were injected into an active phase compensator containing a specifically designed liquid crystal SLM (Citizen Watch Co., pixel number of 648, pixel width of 97 μm, grayscale resolution of 192), two aluminum coated gratings (blazing wavelength of 500 nm, groove density of 150 lines/mm), and two concave mirrors (focal length of 350 mm). The output pulse (0.5 μJ energy) from the phase compensator was injected into the M-SPIDER.

In the M-SPIDER, the two replicas with a delay time τ were produced by a Michelson interferometer. The replicas were up-converted with the external reference chirped pulse, which passed through a 10-cm-long TF5 glass twice, in a 20-μm-thick β-barium borate crystal (BBO; Type II, 45°). The interferogram with a spectral shear Ω was measured by a spectrometer (Chromex 500is, 1200 lines/mm) with an intensified charge coupled device (ICCD, Andor Technology DH520; 1024×256 pixels). The synchronously gated ICCD was cooled to -25°C for increasing the signal-to-noise ratio.

The essential points for successful feedback are the accurate measurement and the accurate phase compensation. We made a lot of efforts on them. In order to accurately measure the phase $\phi(\omega)$, we analyzed the phase errors invoked by the inaccurately measured $\tau + \delta\tau$ and $\Omega + \Delta\Omega$. We found that the error $\delta\tau$ would introduce an additional second order dispersion $\delta\tau\omega^2/2\Omega$, while the error $\Delta\Omega$ would only result in an amplitude error of the phase $((1-\Delta\Omega/\Omega)\phi(\omega))$. Therefore, we make the following improvements to increase the accuracy of τ.

Improvement of the beamsplitters in the Michelson interferometer. We found that the SPIDER interferogram was spoiled by the interference between front and back surface reflections of the beam splitters. An anti-reflection coating on the backside should avoid this, however, it is difficult to make an anti-reflection coating for over-one-octave bandwidth. Therefore, we replaced them with wedged beam splitters with the thickness of 0.5-mm and a wedged angle of 0.5° (20 % reflection and 80 % transmission for s-polarization from 400 to 1300 nm: fused silica substrate). Consequently, the back reflection interference is completely eliminated.

Recalibration of the spectrometer. We calibrated the spectrometer at the sufficiently high accuracy (less than 0.05 nm in the whole spectral range of 300 to 460 nm) using a mercury lamp.

Subtraction of $\tau\omega$. In spite of the above calibration, we found that the calibrated wavelength contains a nonlinear error which may cause an error of the delay time τ in the order of a few femtoseconds As suggested by Walmsley [5], in order to determine the linear contribution $\omega\tau$ to the SPIDER signal, it is the best way to measure the second harmonic (SH) interferogram of the pules pair. We used the $\omega\tau$ term as a calibration trace and could remove the linear contribution $\omega\tau$ from the SPIDER inteferogram almost completely.

Increase of the SH signal intensity. The intensity of the SH interferogram was very week and it was difficult to measure over the same wavelength range as in the SPIDER measurement. The input pulse has a bandwidth from 490 to 1100 nm, corresponding to a SH signal from 245 to 550 nm. On the other hand, the SPIDER signal comes from 300 to 460 nm. The SH signal is too broad to cover the SPIDER wavelength range. Therefore, we initially reduced the pressure of Ar gas from 3.0 to 0.8 atm so that the spectral range of the significantly intensified SH interferogram is roughly the same wavelength range as in the SPIDER (Fig. 2(b)).

Additional optical path for SH interferogram. The crystal for the sum frequency generation in the SPIDER is cut for the type-II phase matching condition, which is insensitive to the bandwidth limitation, while the second harmonic generation (SHG) is achievable with type I. This asks for

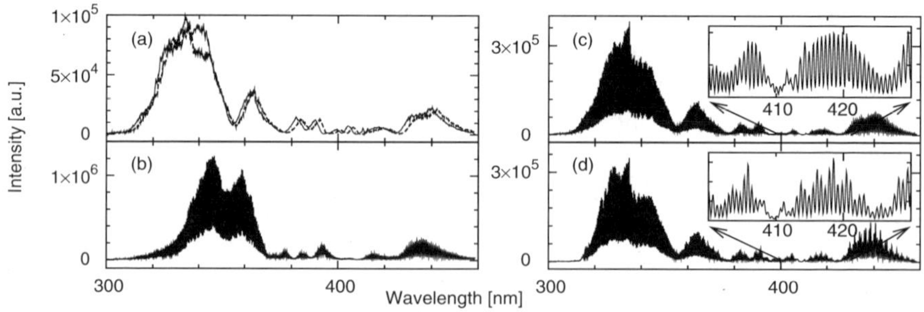

Fig. 2. Results of modified SPIDER measurements. (a) Replicas, (b) the second harmonic, (c) M-SPIDER signal interferogram before compensation. (d) M-SPIDER signal after 2nd feedback compensation.

a realignment of the crystal, the lens, the detector and the slit, for the best signal intensity. This alignment does not correspond to the same conditions as measuring the sum frequency waves of SPIDER. To avoid the perturbation of the SPIDER measurement and to get the reproducible signal, we built a separate optical path for the independent alignment and measurement of the SH interferogram (Fig. 1). This enabled highly reproducible measurements.

Selection of gratings in the active compensator. Since the pulse has an over-one-octave bandwidth, the first order diffraction of the gratings can overlap with the second order on the SLM (for example, the second order diffraction of 400-nm can superimpose on the first order of 800-nm). In the 4-f system, the second order diffraction can be modulated unexpectedly by the SLM and reduce the accuracy of phase compensation. It was found that the proper choice of a blazing wavelength could reduce the influence of the second order diffraction on the output from the compensator. For a spectrum of 490-1100 nm bandwidth, if the blazing wavelength is chosen to be 500 nm, the intensity of the second order diffraction can be reduced to only a few percents of the first order. In addition, the use of the longer focal length (f=200 mm in ref. 1 to f=350 mm) of two concave mirrors reduced the loaded bandwidth of each pixel (from $\Delta\lambda$=3.0 to 1.9 nm). and hence the load of phase compensation per each pixel.

Reduction of measurement error by feedback phase compensation Phase compensation with multiple feedbacks is very much demanded, especially in the case where the spectral shear Ω for each frequency component cannot be regarded as a constant. This could happen when the input pulse has a broader temporal width than the necessary ten percents of the reference pulse width [5]. The first feedback could make the input pulse narrow down such that the spectral shear Ω can be taken as a constant. Then the second feedback with such Ω can remove the residual errors. In general, n-multiple feedbacks rapidly decrease insufficiency of phase

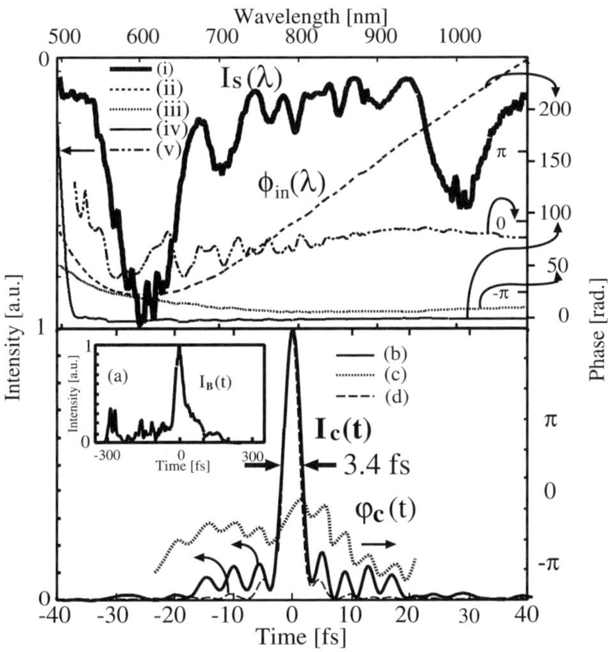

Fig. 3. Results of feedback compensation. (i) Intensity spectrum, spectral phases (ii) before, (iii) after 1st, and (iv),(v) 2nd feedback. (a) Temporal intensity before compensation. (b) Temporal intensity and (c) phase after 2nd feedback. (d) Transform-limited pulses.

compensation, which is due to any errors $\Delta\Omega$ of the spectral shear, according to $\phi^{(n)}(\omega) = |(\Delta\Omega/(\Omega + \Delta\Omega))^n|\phi^{(0)}(\omega)$ where $\phi^{(0)}(\omega)$ and $\phi^{(n)}(\omega)$ are the phases after and before n-times feedback compensation, respectively. In addition, the feedback can reduce the small phase calibration error of the compensator as well as the above Ω error.

3 Results

On the basis of the above improvement, we conducted the pulse compression experiment. First, we measured the SH spectral phase interference (Fig. 2(b)) and then the SPIDER one (Fig. 2(c)). Because the pulses have very broad bandwidths, it was necessary to take the signals by five rotations of the grating in the spectrometer , which were automatically controlled (exposure time was 0.1 s × 100). We found that the replicas are very similar over the whole range from ∼ 300 to ∼ 460 nm (Fig. 2(a)).

Figure 3 (i) is the measured spectrum (495-1090 nm) of the pulse output from the phase compensator, (ii) is the retrieved phase before feedback. This negative phase is fed back to the SLM, and the resultant reconstructed phase is curve (iii). After two times feedback compensation (Fig. 2(d); the corresponding SPIDER signal), the spectral phase was almost flat (Fig. 3(iv), and the enlarged is (v)), demonstrating the powerfulness for pulse compression of the feedback. The τ and $\Omega/2\pi$ confirmed in this experiment were 935 fs and 4.12 THz, respectively. The retrieved temporal intensity and phase of compensated pulses are shown in Fig.3(b),(c) (Fig.3(a); before compensation). We found that side-lobes were almost eliminated and the pulse duration was 3.4-fs, close to the 3.0-fs transform-limited pulse with small side-lobes(Fig.3(d)). This pulse duration corresponds to only 1.56 cycle, for the central wavelength of 655.4 nm. This was also confirmed by an independent measurement of fringe-resolved autocorrelation traces under the bandwidth limitation (500 - 980 nm) by a slit on the SLM.

4 Conclusion

We have demonstrated the successful pulse compression with negative phase feedback into a SLM phase compensator. The correct phase measurement with M-SPIDER for over-one-octave pulses was achieved by the improved hardware and retrieve process. The reconstructed pulse width is 3.4-fs in the mono-cycle region and the measured phase falls in the range much less than 2π. This research proved that the complicated spectral phase extending to over one octave could be compensated by direct negative feedback with only one SLM phase compensator.

References

1. N. Karasawa, L. Li, A. Suguro, H. Shigekawa, R. Morita and M. Yamashita, J. Opt. Soc. Am. B **18**, 1742-1746 (2001).
2. A. Cheng, G. Tempea, T. Brabec, K. Ferencz and F. Krausz, in *Ultrafast Phenomena XI*, (Springer-Verlag, Berlin, 1998), p. 8.
3. A. Baltuška, M. S. Pshenichnikov and D. A. Weiersma, IEEE J. Quantum Electron. **35**, 459-478 (1999).
4. A. Baltuška, T. Fuji, and T. Kobayashi, Opt. Lett. **27**, 306-308 (2002).
5. C. Iaconis and Ian A. Walmsley, IEEE J. Quantum Electron. **35**, 501-509 (1999).
6. M. Hirasawa, N. Nakagawa, K. Yamamoto, R. Morita, H. Shigekawa, M. Yamashita, Appl. Phys. B, **74**, S225-229 (2002).
7. R. Morita, M. Hirasawa, N Karasawa, S. Kusaka, N. Nakagawa, K. Yamane, L. Li, A. Suguro and M. Yamashita, Meas. Sci. Technol. **13**, 1710-1720 (2002).

Part II

Measurement and Shaping of Femtosecond Pulses

Acousto-Optic Spectral Filtering of Femtosecond Laser Pulses

D. Kaplan and P. Tournois

Fastlite, Ecole Polytechnique, 91128 Palaiseau

Summary. A programmable coherent filtering of a femtosecond laser pulse can be performed by a longitudinal acousto-optic interaction between the light pulse and a programmable acoustic waveform that describes the impulse response of the filter. Since the ratio between the speed of sound and the speed of light is of the order of 10^{-7}, coherent control of the optical pulses can be performed with electrical signals in the tens of megahertz range.

1 Introduction: Physical Concept

We present the operation and some applications of the Acousto-Optic Programmable Dispersive Filter (AOPDF) recently introduced [1–3] to control amplitude and phase of femtosecond laser pulses.

A schematic of the AOPDF is shown in Fig. 1. An acoustic wave is launched in an acousto-optic birefringent crystal, the tellurium dioxide crystal, by a transducer excited by a temporal RF signal. The acoustic wave propagates with a velocity V along the z-axis of the crystal and hence reproduces spatially the temporal shape of the RF signal. It is well known that two optical modes can be coupled efficiently by acousto-optic interaction only in the case of phase matching. If there is locally only one spatial frequency in the acoustic grating, then only one optical frequency can be diffracted at a position z. The incident optical short pulse is initially in fast mode 1 of the birefringent crystal. Every optical frequency ω travels a certain distance before it encounters a phase matched spatial frequency in the acoustic grating. At this position $z(\omega)$, part of the energy is diffracted in slow mode 2. The pulse leaving the device at mode 2 will be made up of all the spectral components that have been diffracted at various positions. Since the velocities of the two modes are different, each optical frequencies will see a different time delay $\tau(\omega)$ given by:

$$\tau(\omega) = \frac{z(\omega)}{v_{g1}} + \frac{L - z(\omega)}{v_{g2}}, \tag{1}$$

where L is the length of the acousto-optic interaction, v_{g1} and v_{g2} are the group velocities of mode 1 and 2.

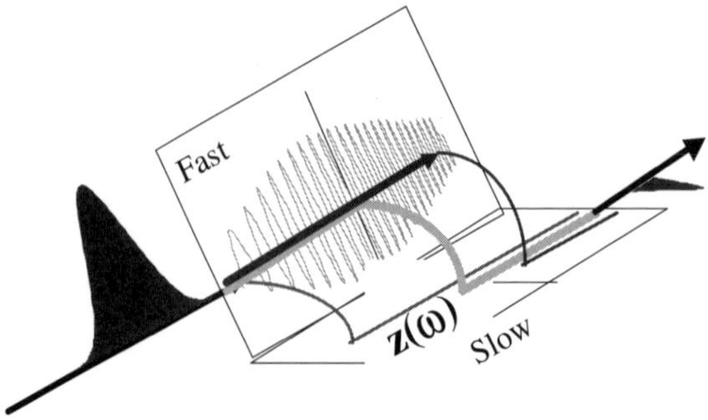

Fig. 1. Schematic of the AOPDF

The amplitude of the output pulse, or diffraction efficiency, is controlled by the acoustic power at position $z(\omega)$.The optical output $E_{out}(t)$ of the AOPDF is a function of the optical input $E_{in}(t)$ and of the electric signal $S(t)$. More precisely, it has been shown, [1] for low value of acoustic power density to be proportional to the convolution of the optical input and of the scaled electric signal $S(t/\alpha)$:

$$E_{out}(t) \propto E_{in}(t) \otimes S(t/\alpha), \tag{2}$$

where the scaling factor $\alpha = f/\nu$ is the ratio of the acoustic frequency f to the optical frequency ν. In the frequency domain, relation (2) can be written:

$$E_{out}(\omega) \propto E_{in}(\omega) \cdot S(\alpha\omega). \tag{3}$$

In this formulation, $S(t/\alpha)$ appears as the impulse response of a filter applied to the input optical pulse. By generating a proper function $S(t)$, one can achieve any arbitrary convolution with a temporal resolution given by the inverse of the available filter bandwidth.

This physical discussion qualitatively explains the principle of the AOPDF, but to obtain exact relations, we will next develop a first order theory of operation.

2 First order theory of the AOPDF

We consider an acousto-optic interaction in Paratellurite crystal TeO_2. The propagation directions of the optical and acoustical waves are in the P-plane which contains the [110] and [001] axis of the crystal. The acoustic wave vector K makes an angle θ_a with the [110] axis. The polarization of the acoustic wave is transverse, perpendicular to the P-plane, along the $[\bar{1}10]$ axis. Because of the strong elastic anisotropy of the crystal, the K vector direction and the direction of the Poynting vector are not collinear. The acoustic Poynting vector makes an angle β_a with the [110] axis. When an incident ordinary optical wave polarized along the $[\bar{1}10]$ direction with a wave vector k_o which makes an angle θ_o with the [110] axis, interacts with the acoustic wave, an extraordinary optical wave polarized in the P-plane with a wave vector k_d is diffracted with an angle θ_d relative to the [110] axis. To maximize interaction length for a given crystal length, in order to decrease the necessary acoustic power, we choose to align the incident ordinary beam with the Poynting vector of the acoustic beam, i.e. $\beta_a = \theta_o$. Figure 2 shows the k-vector geometry related to the acoustical and optical slowness curves, V_{110} and V_{001} are the phase velocities of the acoustic shear waves along the [110] axis and along the [001] axis respectively, n_o and n_e are the ordinary and extraordinary indices on the [110] axis and n_d is the extraordinary index associated with the angle θ_d.

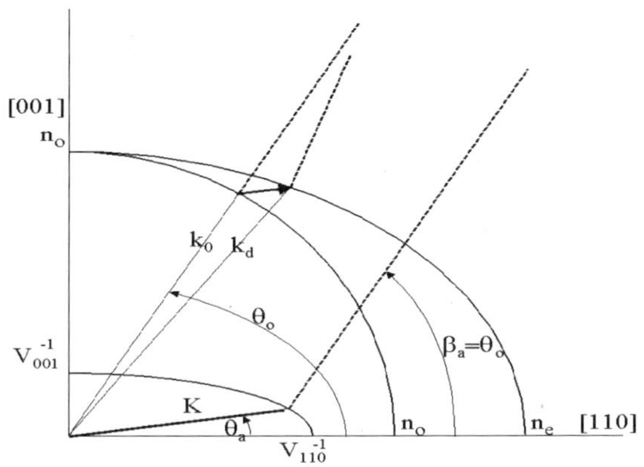

Fig. 2. Acoustic and Optic slowness curves and k-vector diagram.

The optical anisotropy $\Delta n = (n_e - n_o)$ being generally small as compared to n_o, the following relations can be obtained to first order in $\Delta n/n_o$:

$$\delta n = n_d - n_o = \Delta n \cos^2\theta_o, \qquad \theta_d - \theta_o = -\frac{\Delta n}{n_o}\cos^2\theta_o \tan(\theta_o - \theta_a), \tag{4}$$

$$K = k_o \frac{\Delta n}{n_o}\frac{\cos^2\theta_o}{\cos(\theta_o - \theta_a)} \implies \alpha = \frac{f}{v} = \Delta n \frac{V(\theta_a)}{c}\frac{\cos^2\theta_o}{\cos(\theta_o - \theta_a)}, \tag{5}$$

where c is the speed of light. Fig. 3 shows the evolution of α versus θ_o for θ_a=3,9° and θ_a=8° which are related to two different optimized crystal cuts at λ=0.8 μm for which Δn=0.147.

The single frequency solution of the coupled mode theory [4] allows to relate the diffracted light intensity to the incident light intensity and to the acoustic power density $P(f)$ present in the interaction area by the formula:

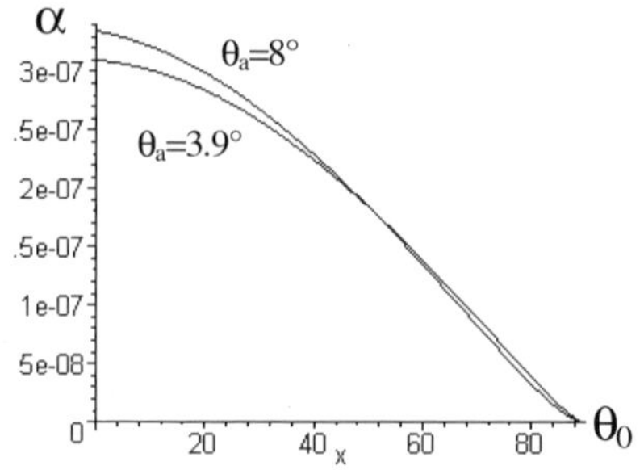

Fig. 3. Acoustic to optic frequencies ratio α versus optical incidence θ_o for $\theta_a = 3.9°$ and $\theta_a = 8°$.

$$I_d(v) = I_o(v)\frac{P(f)}{P_o}\frac{\pi^2}{4}\mathrm{sinc}^2\left[\frac{\pi}{2}\sqrt{\left(\frac{P(f)}{P_o} + \delta\phi^2\right)}\right] \tag{6}$$

with

$$P_o = \frac{1}{2M_2}\left[\frac{\lambda}{L\cos(\theta_o - \theta_a)}\right]^2 , \tag{7}$$

where $\delta\phi$ is an asynchronous factor proportional to the product of the departure δk from the phase matching condition (5) and of the interaction length along the acoustic wave vector K:

$$\pi\delta\phi = \delta k \cdot L\cos(\theta_o - \theta_a) = \frac{\Delta n}{n_o}k_o L\cos\theta_o\left[-\frac{\delta v}{v} + \delta\theta_o[2\tan\theta_o - \tan(\theta_o - \theta_a)]\right] , \tag{8}$$

L being the interaction length along the optical wave vector k_o, λ the wavelength of the light in a vacuum, ρ the density of the TeO_2 crystal, p an elasto-optic coefficient, and M_2 the merit factor given by:

$$M_2 = \frac{n_o^3[n_d(\theta_o)]^3[p(\theta_o, \theta_a)]^2}{\rho[V(\theta_a)]^3} \tag{9}$$

with $p = -0.17\sin\theta_a\cos\theta_o + 0.09\sin\theta_o\cos\theta_a$.

From (6), the ratio between the diffracted and incident light equals 0.5 for $\delta\phi = \pm 0.8$ when $P(f) = P_o$. So the spectral resolution and angular aperture become:

$$\left(\frac{\delta\lambda}{\lambda}\right)_{1/2} = \left(\frac{\delta\nu}{\nu}\right)_{1/2} = \frac{0.8}{\Delta n\cos^2\theta_o}\frac{\lambda}{L} \tag{10}$$

and

$$(\delta\theta_o)_{1/2} = \left(\frac{\delta\lambda}{\lambda}\right)_{1/2} \cdot \frac{1}{[2\tan\theta_o - \tan(\theta_o - \theta_a)]} . \tag{11}$$

If $\Delta\lambda$ is the incident optical bandwidth, the number of programming points N and the estimation P_N of the acoustic power density to maximally diffract the whole bandwidth will be:

$$N = \frac{\Delta\lambda}{(\delta\lambda)_{1/2}} = \frac{\Delta n}{0.8} L \cos^2\theta_o \frac{\Delta\lambda}{\lambda^2} \tag{12}$$

and

$$P_N = N P_o = \frac{1.25 \Delta n \cos^2\theta_o}{2 M_2 \cos^2(\theta_o - \theta_a)} \frac{\Delta\lambda}{L}. \tag{13}$$

Fig. 4 shows the variation of M_2 and N with θ_o when $\beta_a = \theta_o$ for the TeO_2 around $\lambda = 0.8$ μm for which $n_o = 2.226$ and $n_a = 2.374$ and ρ=5.99 g/cm^3. The M_2 curve goes through a maximum of M_2=280 mm^2/GW for $\theta_o = 58.5°$ ($\theta_a = 8°$).

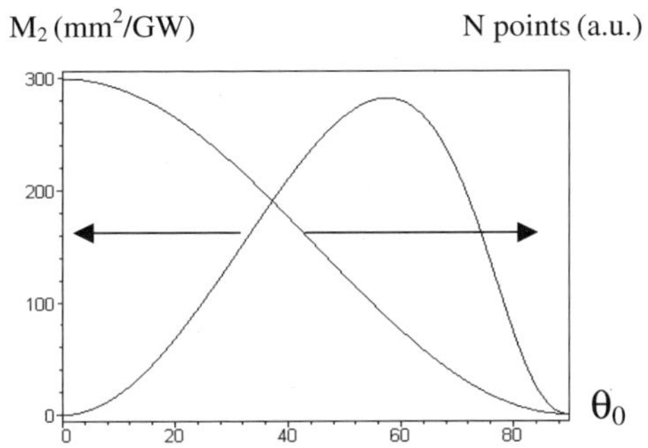

Fig. 4. Merit factor M_2 and number of points N versus θ_o in degrees.

3 Wide band and high resolution AOPDFs

The different applications of the AOPDFs call for two different cut optimizations of the TeO_2 crystal. When the goal is to control the spectral phase and amplitude in the largest possible bandwidth, to obtain the shortest possible pulse, the diffraction efficiency, hence the M_2 factor, has to be maximized. Fig.4 shows that this cut optimization is obtained for $\theta_o = 58.5°$ (i.e. $\theta_a = 8°$). When the goal is to shape the input pulse with a maximum number of independent points in a moderate bandwidth, the optimization is a trade-off between the spectral resolution and the diffraction efficiency. The crystal cut $\theta_o = 38.5°$ ($\theta_a = 3.9°$) allows to double the spectral resolution hence the number of programming points, while only loosing a third of the maximum diffraction efficiency.

The parameters for the Wide Band and High Resolution AOPDFs for a crystal length of 25 mm and λ=800 nm are given in Table 1.

	θ_a deg	θ_o deg	δn	$\theta_d - \theta_o$ deg	α 10^{-7}	M_2 $\frac{mm^2}{GW}$	$\Delta\lambda$ nm	$(\delta\lambda)_{1/2}$ nm	$n_o(\delta\theta_o)_{1/2}$ deg	N	P_o $\frac{mW}{mm^2}$
Wide Band	8	58.5	0.04	1.25	1.42	280	200	0.5	0.04	400	4.5
High Resolution	3.9	38.5	0.09	1.60	2.3	200	100	0.25	0.045	400	3.8

Table 1: Parameters for the Wide Band and High Resolution AOPDFs for a crystal length of 25 mm and λ=800 nm

Efficiencies larger than 50% in 100 nm and 30% in 300 nm bandwidth have been measured in Wide Band AOPDF configurations for an acoustic beam size of 5×5 mm^2 with an acoustic peak power smaller than 10 Watts.

4 Corrections of the dispersion induced by the paratellurite crystal

Since Paratellurite crystals are dispersive, the acoustic to optic frequency ratio α will depend on wavelength λ through the optical anisotropy Δn.

The relevant ordinary and extraordinary group optical indices, for a given θ_0, n_{go} and n_{gd}, are plotted on Fig. 5 versus optical frequency. As can be seen on this figure, the paratellurite dispersion becomes very large above 600 THz, i.e. below λ=0.5 μm. For small enough bandwidth $\Delta\lambda$, the dispersion of the crystal can be compensated by programming an acoustic wave inducing an inverse phase variation in the diffracted beam. This self-compensation is, however, limited by the maximum group delay variation given by:

$$\delta t_g = t_{gd} - t_{go} = (n_{gd} - n_{go})\frac{L}{c} = (n_{ge} - n_{go})\cos^2\theta_o\frac{L}{c}. \qquad (14)$$

More precisely, when the dispersion of the crystal is compensated by an adapted acoustic waveform, all the wavelengths in the optical bandwidth $\Delta\lambda$=λ_2 - λ_1 have to experience the same group delay time, i.e. the same group index, at length L of the crystal [3] and at the extrema of the bandwidth, we will have $n_{go}(\lambda_1) = n_{gd}(\lambda_2)$ as shown in Fig. 5. The maximum bandwidth $\Delta\lambda$ of self-compensation is plotted versus λ= $(\lambda_1 + \lambda_2)/2$ when θ_o= 38.5° and when θ_o= 58.5° on Fig. 6.

If the bandwidth of operation is larger than this maximum bandwidth $\Delta\lambda$ it is necessary to use an outside compressor. In the simplest implementation, this outside compressor will compensate mainly for second order phase deviations. One can then program the acoustic wave to compensate for higher orders. This leads to a new limit $\Delta\lambda_1 > \Delta\lambda$ associated to these higher orders compensation ($\Delta\lambda_1$ is also shown on Fig. 6).

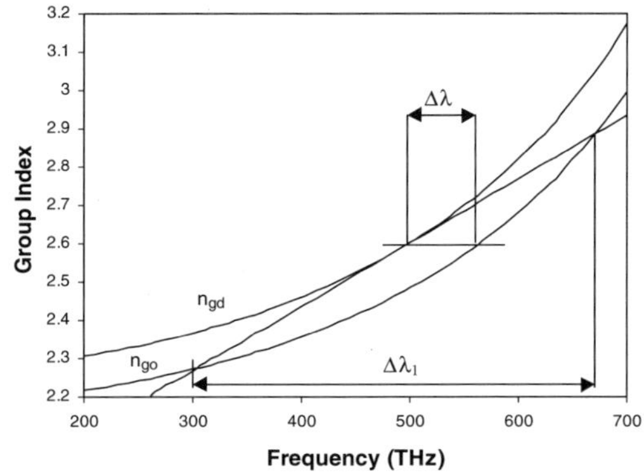

Fig. 5. Ordinary and extraordinary group indices versus optical frequency in THz for a given θ_o. The bandwidths of self compensation are: $\Delta\lambda$ without an external 2nd order compressor and $\Delta\lambda_1$ with an external 2nd order compressor.

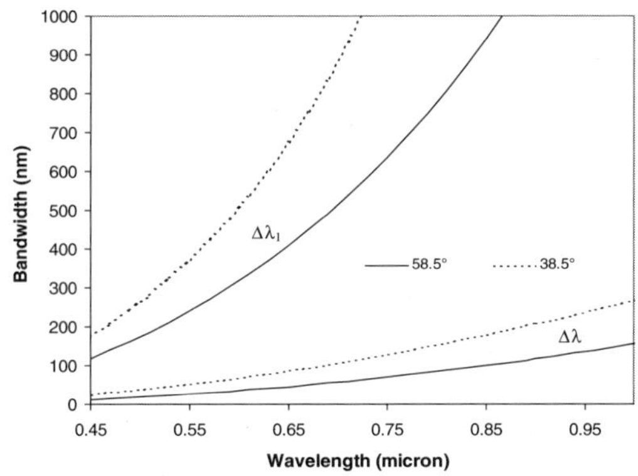

Fig. 6. Bandwidths of self compensation $\Delta\lambda$ and $\Delta\lambda_1$ versus λ for $\theta_o = 38.5°$ and $\theta_o = 58.5°$.

5 Advantages and limitations

The other existing devices used to shape the femtosecond laser pulses, e.g.: liquid crystal modulators, deformable mirrors or acousto-optic deflectors, are generally placed in the Fourier plane of a zero-dispersion line. In comparison the AOPDF pulse shaper is small, compact and geometrically stable. Being placed directly in-line between the oscillator and the amplifier, it requires only minor changes in an existing laser chain. The simultaneous spectral amplitude and phase controls are quantitative allowing single pass laser pulse optimization and generation of coherent multiple pulses. Moreover, the AOPDF crystal introduces no angular deformation and no spatial chirp. The refreshing time of acoustic waveform is fast enough to allow phase modulation experiments.

The main limitations of the AOPDF come from the paratellurite crystal. The crystal length limits the group time delay variation versus wavelength to 5 to 10 psec depending on crystal orientation. The maximum usable laser pulse repetition rate is also limited by the acoustic time length in the crystal to 20 to 40 kHz. Self focusing effects due to the nonlinear index of the crystal limits the maximum optical peak power to 100 MW/cm^2. The second order dispersion of the crystal limits the self compensation to a bandwidth of $\Delta\lambda$=70 nm at λ=800 nm, as shown on Fig. 6. If the AOPDF is placed between a stretcher and a compressor, however, the self compensation of higher orders extends to more than $\Delta\lambda$=500 nm at the same wavelength.

6 Optimization of amplified laser pulses

6.1 Pre-compensation

Given the power limitation of the crystal and the greater simplicity, most applications have opted for a pre-compensation scheme, i.e. inserting the AOPDF between the oscillator and amplifier. Usually the point of insertion is between the oscillator and stretcher where the beam diameter is minimal. The two principal goals are correction of gain narrowing and compensation of high order phase components not eliminated by the stretcher-compressor combination.

Correction of gain narrowing

The principle is to modify the intensity spectrum, in such a way that the intensity is minimal at the point of maximum gain. As an example, for an amplifier with 10 passes, compensation can be achieved over the full half width (3db) of the gain curve, if the dynamic range of intensity control reaches 30 db. In a pioneer experiment Verluise et al, [2] shown in Fig. 7 a doubling of the bandwidth was achieved. With proper phase correction a near Fourier transform limited pulse of 17 fs was obtained.

High order phase correction

For high power amplification, implying large compression ratios of 10000 or more, the problem of exact compensation becomes crucial. The AOPDF is well adapted for generation of fourth order or higher order corrections. This feature becomes particularly important because of the issue of pulse quality (peak to background contrast) in plasma generation experiments. A recent implementation of a 100 TW Ti:Sapphire amplifier chain at the Laboratoire d'Optique Appliquée [5] has used the AOPDF to reach 25 ps duration with a pulse contrast improved by an order of magnitude. The phase undulations, as measured by SPIDER, were kept below 0.15 radian in the whole spectral range.

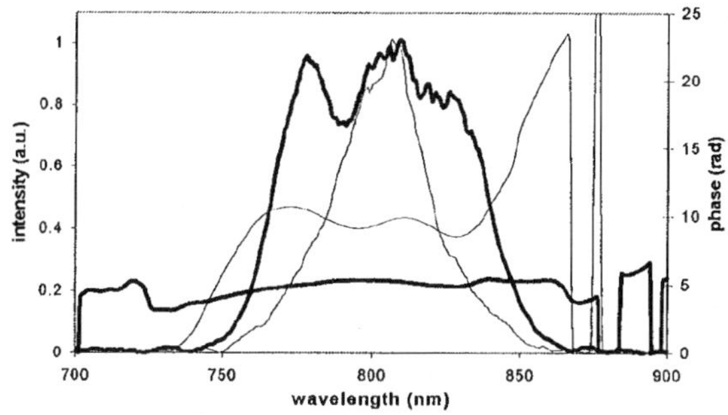

Fig. 7. Spectral Intensity before (thin curve) and after (thick curve) correction of the gain narrowing. Also shown are the corrected and uncorrected phase measured by FROG.

Correction procedures

The fast re-programmation time allows the use of sophisticated optimization algorithms. As an example, Ohno et al. [6]have matched the measured FROG pattern of their amplifier, with a theoretical 2D pattern computed for a Fourier transform limited pulse. A genetic algorithm has been made to converge to an optimal solution differing from theory by less than 1.6 % as measured by their quadratic weight function.

On the other hand, the phase introduced by the AOPDF is known with a high accuracy from the physical constants of the material, as opposed to other techniques where the phase depends on geometrical parameters of the set-up and has to be calibrated. If one disposes of a good phase measurement $\varphi(\lambda)$, it is then possible to program the opposite correction $-\varphi(\lambda)$ and obtain directly the desired flat phase. This has been done experimentally by Oksenhendler et al. [7]. The results are shown in Fig. 8.

6.2 Post compensation

The operation at low intensity is not always possible. An example application is the phase correction of pulses generated by hollow fiber spectral broadening. Recently Seres et al. [8] have shown that a properly designed AOPDF, with wide acoustic bandwidth can be used for that purpose. Because the optical bandwidth of order 200 nm was significantly larger than the limit for self compensation discussed above and because of the peak power limit, it was necessary to use an additional stretcher-compressor combination to operate the AOPDF. The AOPDF was used for higher order corrections. A spectral width of 160 nm was controlled and an optimal pulse duration of 7.6 fsec was achieved (Fig. 9). This set up was used as a pulse input to a terawatt-scale amplifier yielding 3.5 mJ, 9.4 fs FWHM, at 1 kHz repetition rate.

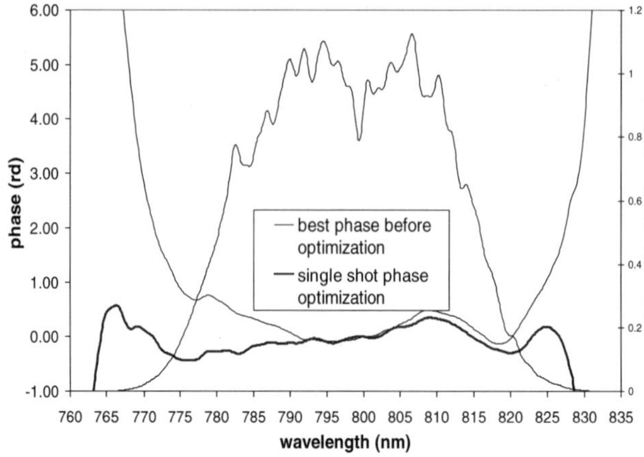

Fig. 8. Single shot phase correction: The thin curve represents the uncorrected phase. The thick curve represents the phase obtained by subtracting the phase error. Also shown is the power spectrum demonstating that adequate phase correction can be obtained even in the wings of the spectral curve.

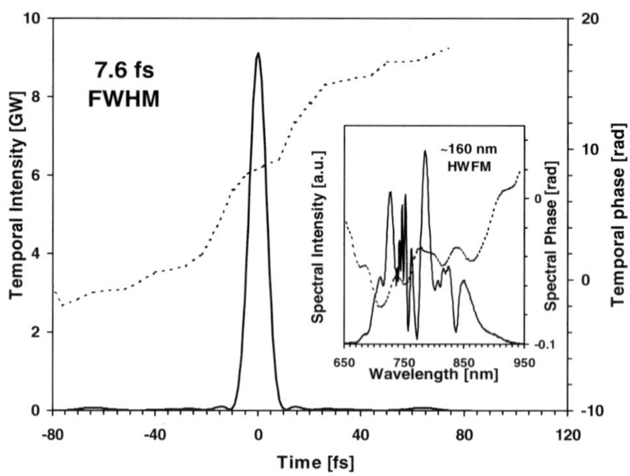

Fig. 9. Compression of hollow fiber output (solid line) with the corresponding phase (dashed line) measured by SPIDER. The spectral intensity and phase are shown in the inset.

7 Spectral phase measurements using the AOPDF

Because the phase introduced by the AOPDF is known with great accuracy, it is tempting to design a spectral phase measurement based on this device. The schematic below (Fig. 10), explains why and how this can be done.

Any spectral phase measurement method combines a linear optical filter, i.e. any linear optical combination (usually an interferometer geometry

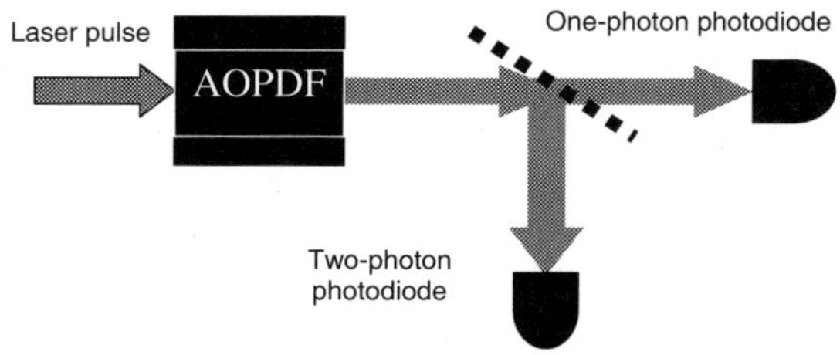

Fig. 10. Basic elements of a spectral phase measurement set-up.

), with a non linear element (crystal doubler, two photon diode,...). Other elements are necessary in some methods such as a spectrometer, a simple diode detector, etc. For instance, a typical SPIDER set-up [9], combines a narrow band filter, a splitter, a delay line, a crystal doubler and one or two spectrometers. The AOPDF alone can be used to emulate the effect of any linear optical filter. In particular, by summing multiple acoustic waveforms, on obtains multiple optical pulses with adjustable delays (within the limit set by the length of the crystal). Thus the combination of AOPDF and two-photon diode shown in the schematic contains the basic ingredients necessary for spectral phase measurements. If a particular method relies on knowing the pulse spectrum, this can be derived by Fourier transform spectroscopy using two pulses with variable delays.

As a first step interferometric autocorrelation has been investigated, using the AOPDF to generate two pulses with variable delays. One interesting feature is that the AOPDF can generate a true delay (i.e. first order derivative of the phase) rather than a path difference. One then obtains directly an interferometric pattern without oscillations ("baseband" measurement), without relying on averages as in traditional autocorrelators (Fig. 11).

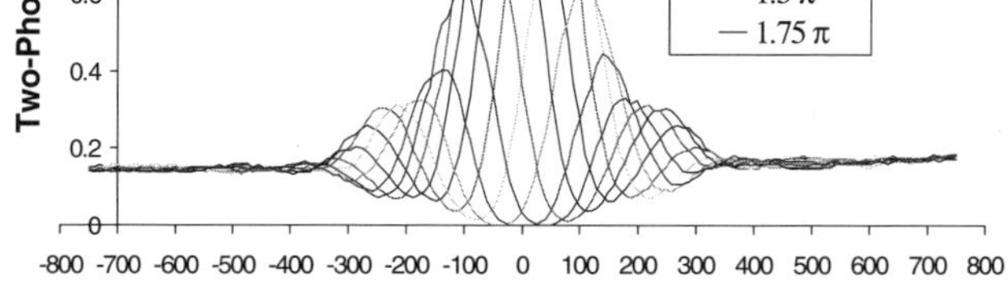

Fig. 11. Experimental "baseband" interferometric autocorrelation curves corresponding to different fixed phase between the two pulses.

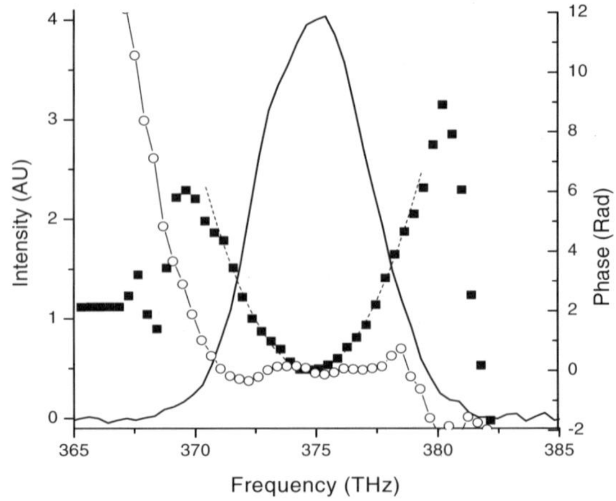

Fig. 12. Measured phase with the AOPDF implementation of the SPIDER technique. Open symbols show the initial flat phase. Closed symbols show the phase with a linear chirp added. The intensity spectrum is also shown.

By programming an additional quasi monochromatic pulse, Monmayrant et al. [10] have demonstrated an AOPDF version of the SPIDER method. The necessary spectral measurement where derived from Fourier transform spectroscopy, hence the denomination Time Domain SPIDER . Fig. 12 shows an example of measuring a second order phase component using this method.

Besides its simplicity, ease of alignment, and accuracy, the AOPDF measurement can dissociate the optical linear filter from the point at which the phase is measured. If an AOPDF is in-line in the experiment for pulse control or optimization, one can perform the measurement by inserting the two photon diode at the point of interest and adding the additional acoustic waveforms (usually smaller in intensity) that are required by the specific measurement algorithm. In particular, it can be envisaged that an AOPDF, installed at the **input** of an amplifier for pre-compensation, be used to measure the phase at the **output** of the amplifier or of the full experiment. This only requires that the phase changes due to the non linearities introduced in the amplifier by the additional acoustic waveforms can be considered negligible or can be determined independently.

8 Pulse shaping for coherent control and other applications

The AOPDF can be used with great simplicity in many experiments requiring spectral phase shaping. For instance, Reitze et al. [11] have controlled and optimized high harmonic generation using a pre-compensation configuration. Lee et al. [12] have applied the AOPDF control to luminescence studies.

Most interesting are the applications that require amplitude as well as phase control. For instance, the luminescence from a semiconductor quantum well structure excited by two pulses generated by the AOPDF has been studied [13]. The relative phase of the two pulses can be modified. Fig. 13 shows the luminescence response when the second delayed pulse is either in phase or out of phase with the first one. Efficient cancellation of the luminescence with a Pi shifted pulse is demonstrated.

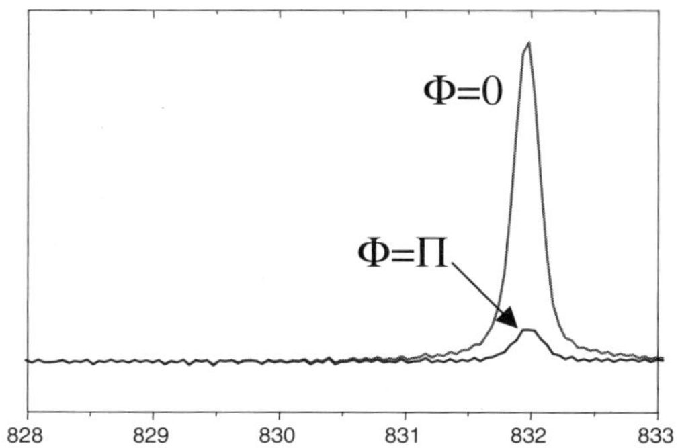

Fig. 13. Luminescence intensity as a function of wavelength in nm for two relative phases of the excitation pulses.

Another interesting application is the generation of fast rise time, flat top, picosecond long pulses for electron photo-injection (e.g. 10 psec flat pulses generated from a gaussian 100 fs pulse) [14]. If the spectrum of the pulse is gaussian, broadening with a simple second order dispersion will produce a gaussian time profile. Using the S-shaped spectral delay in the inset (i.e . adding fourth and higher even orders), one can generate the time profile of Fig. 14a, featuring a fast rise time but with a large overshoot. Trimming the spectral amplitude (i.e. reducing the amplitude on the edges of the spectrum), it is possible to generate the adequate time profile of Fig. 14b. Experiments are currently in progress to demonstrate this methodology experimentally.

9 Conclusion

We have reviewed the theory of operation of the AOPDF and investigated the main operational characteristics and limitations of this device. The AOPDF has now been used in a large variety of experiments and configuration settings. Its application range is now extending from the basic pulse optimization function to spectral phase measurements. We believe that there are many advantages in incorporating this device as a basic component of laser amplified chains. Its use may modify significantly the design of such systems by introducing simpler, not precisely matched, stretcher/compressor combinations,

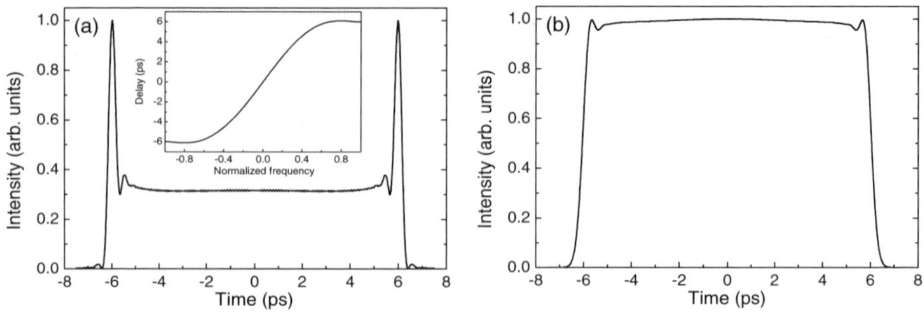

Fig. 14. Generation of flattop picosecond pulses. a) phase only control (spectral time delay shown in the inset); b) phase and amplitude control.

allowing real time corrections of pulse aberrations, eliminating the need for complex and difficult to align measurement instruments and permitting coherent control experiments to be performed without additional equipment.

References

1. P.Tournois, Opt. Commun. **140**, 245 (1997).
2. F.Verluise et al., Opt. Lett. **25**, 575 (2000).
3. D.Kaplan and P.Tournois, J.Phys.IV France **12**, Pr 5-69 (2002).
4. A.Yariv and P.Yeh, Optical Waves in Crystals (Wiley, New York, 1984).
5. M.Pittman et al., Appl..Phys.B **74**, 529 (2002).
6. Ohno et al., J.Opt.Soc.Am.B, **19**, 2781 (2002).
7. T.Oksenhendler et al., CLEO 2003, paper CWE1.
8. E. Seres et al., Opt. Express, **11**, 240 (2003).
9. I.Walmsley and V.Wong, J.Opt.Soc.Am B **13**, 2453 (1996)
10. A.Monmayrant et al., Opt. Lett. **28**, 278 (2003).
11. D.Reitze et al., *Ultrafast Phenomena XIII*, p. 57, Springer Series in Chemical Physics.
12. Lee et al., J.Chem.Phys, **117**, 9858 (2002).
13. Y.Lee et al., U. of Michigan, Private Communication.
14. M.Nisoli et al, CLEO 2003, paper CThM21.

Adaptive Femtosecond Quantum Control

T. Brixner, C. Dietl, G. Krampert, P. Niklaus, E. Papastathopoulos, T. Pfeifer, R. Selle, G. Vogt, D. Walter, C. Winterfeldt, and G. Gerber

Physikalisches Institut, Universität Würzburg, Am Hubland, 97074 Würzburg, Germany, Fax: +49-931-888-4906 gerber@physik.uni-wuerzburg.de

1 Introduction

Since the advent of quantum mechanics there has been a desire to understand in detail the behavior of quantum systems. This quest has been accompanied by the implicit dream to not only be able to observe in a passive way, but in fact also to actively control quantum-mechanical processes. The key question in quantum control is: Can one find external control parameters which guide the temporal evolution of quantum-mechanical systems in a desired way, even if this evolution is very complex?

Immediate applications of quantum control are found in many different branches of scientific and engineering research such as photochemistry, quantum optics, atomic and molecular physics, biophysics, solid-state physics, telecommunications or related quantum technologies such as quantum computing or quantum cryptography. In addition to these direct benefits, the successful implementation of quantum control concepts is also likely to provide new insights into the intricacies of the underlying quantum-mechanical dynamics.

Especially promising in this context is the implementation of adaptive control, with the basic idea of using a closed-loop setup in which the difficulties associated with complex quantum-mechanical Hamiltonians are circumvented [1]. With the help of learning algorithms and experimental feedback signals, it is possible to achieve automated control over complex systems without the necessity for knowing the underlying potential energy surfaces. Based on these experimental results, the control-parameter settings are optimized iteratively such that they adapt to the quantum-mechanical system.

The combination of adaptive quantum control with femtosecond laser spectroscopy – adaptive femtosecond quantum control – is a new research field which goes beyond "simple" observation, seeking to *control* chemical reactions by suitably shaped femtosecond light fields. This is done on a microscopic level by forcing the dynamical evolution of quantum wavefunctions into the desired direction. A number of recent review articles and books have treated the subject of quantum control from different perspectives [2–10]. Here we discuss the general scheme of adaptive control experiments and present a number of selected examples from our laboratory.

2 Experimental Scheme

An experimental setup of adaptive femtosecond quantum control is shown in Fig. 1. A femtosecond laser pulse shaper (right side of the figure) is used to impose specific spectral phase modulations. Details of our setup were published previously [11,12]. Briefly, the device consists of a zero-dispersion compressor in a 4f-geometry, which is used to spatially disperse and recollimate the femtosecond laser pulse spectrum. Insertion of a liquid-crystal display (LCD) in the Fourier plane of the compressor provides a mechanism for convenient manipulation of the individual wavelength components. By applying voltages, the refractive indices at 128 separate pixels across the laser spectrum can be changed, and upon transmission of the laser beam through the LCD, a frequency-dependent phase is acquired due to the individual pixel voltage values. In this way, an immensely large number of different spectrally phase-modulated femtosecond laser pulses can be produced.

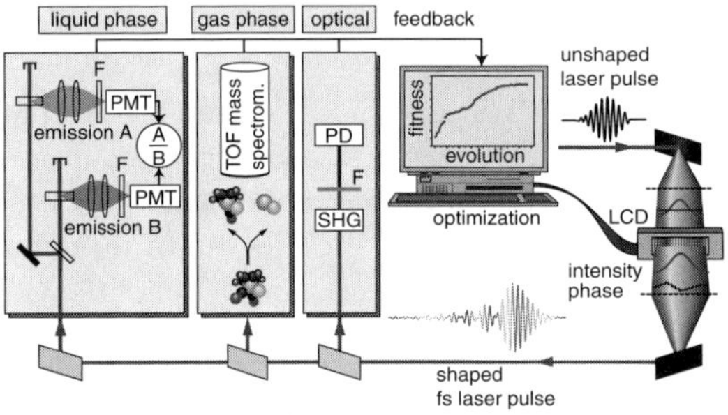

Fig. 1. Experimental setup. A femtosecond pulse shaper (right side) is used to generate phase-modulated laser pulses for closed-loop adaptive quantum control. Suitable experimental feedback signals are processed in a learning algorithm which iteratively improves the applied laser pulse shape until an optimum is reached.

The shaped laser pulses are then used in different types of adaptive quantum control experiments wherein experimental feedback signals guide an automated search for optimal electric fields within a learning algorithm. In general the optimal pulse is not known in advance and since the variational space of possible pulse shapes is so huge, scanning the complete parameter space is impossible. In order to overcome this problem, we use a learning loop to find optimized electric fields taking into account the experimental outcome. In an iterative process, the electric laser fields are improved with a computer algorithm until the particular optimization objective is reached and the feedback signals approach the user-specified goal. The global search method we use for this purpose is an evolutionary algorithm with crossover, cloning, and mutation procedures.

3 Selected Examples

3.1 Gas-Phase Control

In this realization of quantum control in the gas phase, the optimization objective is to maximize (or minimize) branching ratios within photodissociation reactions in complex molecules. The question is whether the evolutionary algorithm can find laser fields which preferably generate one photoproduct at the expense of another. Due to the very general implementation in a learning loop, many different classes of molecular photodissociation processes can be controlled. For example, we have shown bond-selective photochemistry in CH_2ClBr [13] and in lactic acid [14].

In the case of the model system CH_2ClBr, the question is whether in such a situation adaptive quantum control can be used to find shaped excitation fields (at a given central wavelength) which selectively break either one or the other carbon–halogen bond. Specifically, we attempted to break preferentially the stronger (C–Cl) bond versus the weaker (C–Br) bond. The fitness function in the evolutionary algorithm was therefore set to maximize the CH_2Br^+/CH_2Cl^+ ratio, and it could indeed be increased by 100% from 0.083 for bandwidth-limited laser pulses to 0.167 for the optimized pulse shape [13]. It can be shown that this effect is not simply a result of intensity changes in the laser pulse [13]. Rather it appears that the shaped pulse changes the fragmentation dynamics, and the control mechanism may involve manipulation of wavepacket motion on neutral dissociative PES which are not involved in photoproduct formation with bandwidth-limited 800 nm laser pulses. These results are an encouraging indication that even in systems such as CH_2BrCl where excited states are highly coupled, complex laser fields may be useful to achieve bond-selective photochemistry.

One important potential use of such a methodology is in chemical synthesis. With this technique, specific molecular groups could be separated from the parent molecule in a single step for further chemical synthesis, leaving the rest of the molecule intact. Thus the production of unwanted side-products in additional production steps can be avoided. In organic chemistry it is often desired to remove an acid group without destroying the remainder of the molecule. We used the organic molecule lactic acid ($CH_3CHOHCOOH$) as a prototype to illustrate this issue. Thus, the scope of the optimization was to efficiently convert lactic acid into ethanol by removing the carboxyl-group (COOH). The amount of the methyl-group fragment (CH_3) was used as a measure of an unwanted side-product arising after breakage of the "wrong" bond. We therefore optimized the branching ratio of the cleavage of the carboxyl-group (COOH) versus the cleavage of the methyl-group (CH_3). The $COOH^+/CH_3^+$ ratio could be increased from a value of 24 for unmodulated laser pulses by a factor of 5 up to a value of 130, obtained with the optimized laser pulse. Thus, an efficient chemical conversion from lactic acid to ethanol was realized by separating the acid group from the remaining intact molecule employing adaptive femtosecond pulse shaping.

3.2 Liquid-Phase Control

Probably the most intriguing initial motivation for quantum control was selective photochemistry in the liquid phase by which macroscopic amounts of chemical substances could be synthesized. While this dream has not yet been realized experimentally, a number of breakthroughs toward achieving this goal have been achieved. Our first approach to liquid-phase quantum control dealt with selective photoexcitation, monitoring a photophysical (rather than photochemical) observable. We have shown how light pulses can be optimized such that they selectively transfer electronic population within one specific complex dye molecule in solution [15]. This could be used for selective photoexcitation within mixtures of molecules.

For this, we investigated the dye molecule DCM and the organometallic complex $[Ru(dpb)_3](PF_6)_2$, where dpb $= 4,4'$-diphenyl-2,2'-bipyridin, both dissolved in methanol. After two photon excitation at 800 nm and the possibility for additional interaction of the excited species with the electric field of the shaped femtosecond laser pulse, the amount of excited-state population is recorded by monitoring the spontaneous emission signal from each of the two molecules. The objective here was to excite DCM while not exciting (or at least reducing the excitation of) $[Ru(dpb)_3]^{2+}$ [16]. This cannot be done by single parameter control schemes such as wavelength tuning, variation of the laser intensity or changes in the linear chirp of the laser pulses (see Fig. 2a-d).

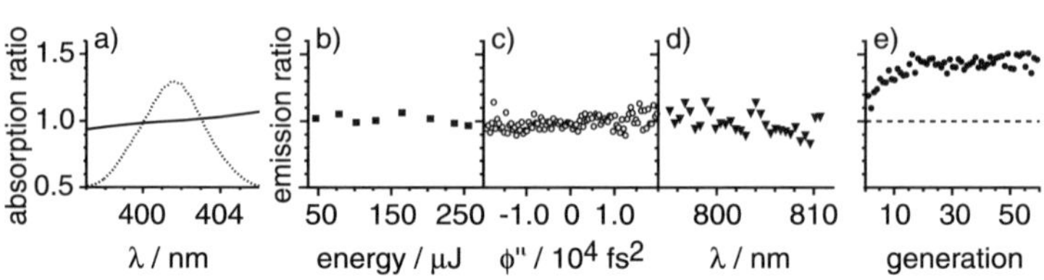

Fig. 2. Control of liquid-phase molecular excitation. a) The relative DCM/$[Ru(dpb)_3]^{2+}$ linear absorption ratio (solid line) is shown as a function of wavelength. The second-order power spectrum (dotted line) of a bandwidth-limited laser pulse is shown to illustrate possible two-photon transition frequencies. The DCM/$[Ru(dpb)_3]^{2+}$ emission ratio is plotted b) for varying pulse energies of unshaped laser pulses, c) for varying second-order spectral phase (i.e., linear chirp), d) for scanning a symmetric and rectangular window of 5 nm width over the laser spectrum, and e) for many-parameter phase-shaping as a function of generation number within the evolutionary algorithm.

Many-parameter quantum control, however, does provide a mean to achieve selective excitation of the two molecules. It is shown in Fig. 2e how the emission ratio evolves as a function of generation number within the evolutionary algorithm so that finally a 50% increase is observed. Thus it is possible to selectively excite one specific molecular species even within mixtures of molecules with identical absorption profiles. It should be emphasized that control is possible in the presence of complex solute–solvent interactions. The failure of the single-parameter schemes indicates that the control mechanism cannot be based on the initial excitation step (which is identical for the two molecules), but exploits the differences in the dynamical wavepacket evolutions on excited-state potential energy surfaces.

Further advance towards the control of a complex chemical reaction in solution is the manipulation of the geometrical structure of molecules. A model process that has attracted a lot of attention is the *cis-trans* isomerization reaction [17–20]. Irradiated by light of the proper wavelength, these molecules can undergo the *cis-trans* isomerization process with a certain probability. Controlling this kind of reaction would mean to suppress or enhance the isomerization process. Thereby the yield of the two isomers after excitation can be manipulated. An intensely investigated class of molecules that exhibit *cis-trans* isomerization are symmetrical cyanines (see [21] for a summary). In our experiment we employed adaptive femtosecond pulse shaping to steer the outcome of the photoisomerization reaction of 3,3-diethyl- 2,2-thiacyanine iodide (NK88) in the liquid phase. We excite the molecule dissolved in methanol with 400 nm laser pulses that are shaped in spectral phase and amplitude. The feedback is provided by transient absorption spectroscopy in the visible wavelength region. Using this scheme we can change the ratio of the *cis-* to *trans*-isomer compared to the yield of an unshaped femtosecond laser pulse of equal pulse energy by 20%.

3.3 Polarization Control

Physical systems such as molecules are three-dimensional objects. It would therefore be fascinating if not only scalar but also vectorial properties of light fields could be exploited to guide dynamical evolution. We recently developed the technique of femtosecond polarization pulse shaping by which the polarization state of light (i.e., its ellipticity and orientation) as well as the intensity and light oscillation frequency can be varied as functions of time within a single femtosecond laser pulse [22–25]. There are fundamentally new prospects which rely on the vectorial manipulation of light–matter interaction (for example, reaching enantiomer selectivity in quantum control [26–29]). This is done within a frequency-domain pulse shaper which contains two LCD layers instead of just one. Thereby it is possible to manipulate the phases of two mutually orthogonal polarization components. There are a number of experimental difficulties that need to be overcome, namely the polarization-dependent efficiency of optical gratings or the change of polarization states by any optical element between LCD and experiment. A major issue is also the accurate experimental characterization of such pulses which is more complicated than for linearly polarized light fields, but is possible using either dual-channel spectral interferometry or Jones-matrix analysis with experimental calibration [23]. General properties of these pulse shapes have been discussed [24], especially how "fast" polarization states can be changed within a single laser pulse.

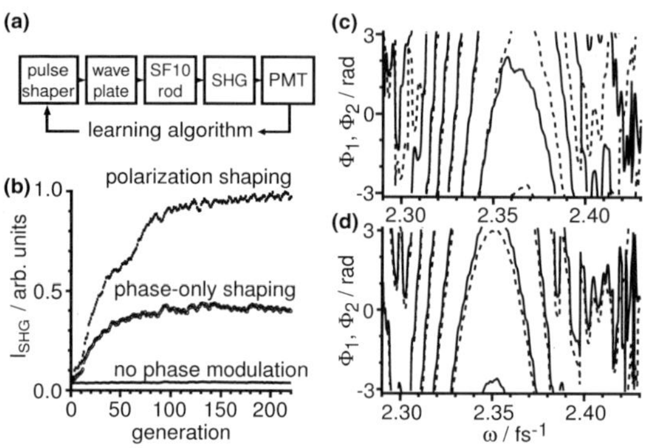

Fig. 3. (a) Schematic experimental setup for adaptive dispersion-and-polarization compensation experiment and (b) evolution of the SHG yield as a function of generation number for phase-and-polarization shaping, phase-only shaping, and unmodulated pulses. The resulting spectral phase modulations Φ_1 (dashed line) and Φ_2 (solid line) are shown (c) for polarization shaping and (d) for phase-only shaping.

To demonstrate the advantage of polarization shaping versus phase-only shaping, we carried out an experiment for adaptive polarization shaping [25]. The SHG yield generated in a BBO crystal was used as the feedback signal for the evolutionary algorithm (Fig. 3a). Installing a multiple-order wave plate, which is not designed for use with 800-nm femtosecond laser pulses, after the pulse shaper results in a frequency-dependent phase difference between the polarization components. This phase difference "messes up" the linear input polarization. Furthermore, a highly dispersive SF10 rod of 10 cm length is inserted into the beam path, introducing dominant second-order group-velocity dispersion, hence leading to a temporal broadening of the laser pulse. The optimal pulse shape therefore requires both specific chirping and polarization modulation. Two optimization strategies were pursued here: polarization shaping and phase-only shaping. In both cases, the SHG feedback signal clearly rises above the level of unmodulated laser pulses already in the first generation, and significant further improvements are accomplished in the following generations. However, the SHG intensity at the end of the polarization-shaping experiment is more than twice as high as in the linear phase-only-shaping experiment. Hence, with polarization shaping much better optimization results can be achieved. With phase-only shaping, it is possible to compensate for the material dispersion, but not for the polarization modulation of the wave plate. This additional compensation is achieved in the polarization shaping experiment. Hence, both the polarization state of a "distorted" laser pulse was "cleaned up" and material dispersion was compensated for.

3.4 Optimization of High-Harmonic Spectral Shape

High harmonic generation (HHG) has become a vivid field of research during the last decade and our understanding of the processes has reached a considerable extent for atomic systems [30]. The large bandwidth in the XUV and soft x-ray spectral region in conjunction with its coherence has been successfully exploited already, to yield pulses of approximately 500 attoseconds in duration. Such pulses provide a future tool to study the electronic dynamics of all kinds of atomic [31] and molecular systems.

While the large bandwidth is beneficial to attosecond pulse generation and observation of electronic dynamics, it is unfavourable for monitoring molecular dynamics. For this purpose, a small-bandwidth XUV source would clearly be advantageous. In particular if one is interested in ultrafast photoelectron spectroscopy, a high degree of monochromaticity of the XUV light is a very desirable property. In addition to spectroscopy, also zone-plate nanoscale microscopy using high-harmonic radiation can be improved by exploiting the same property [32].

Thus it seems reasonable to investigate whether it is possible to control the process of HHG in such a manner that just one particular harmonic order is generated while all the others are suppressed. Even beyond this, there is the question whether it is feasible to control the overall shape of the harmonic spectrum. We show here that this is indeed possible. In this sense, this work encompasses and surpasses an earlier study [33] which can be regarded as a special case of our results. This could open the door to a new field of quantum control, since one would be able to directly steer electrons on their natural time scale [34] with shaped attosecond XUV pulses.

Our setup, which was described earlier [35], consists of a 250 μm inner diameter argon gas filled hollow fiber and an electrostatically deformable membrane mirror built into a prism compressor. The laser pulse experiences self-phase modulation by traveling through the $\sim$60 cm long fiber and is then compressed to 20 fs using the prism compressor. The deformable mirror, replacing the commonly used symmetry mirror in the prism compressor, acts as a spectral phase modulator. By applying control voltages to each of its 19 electrodes, we are able to achieve different surface shapes that translate into different spectral phase functions of the laser pulse.

The shaped pulses are then focused into a second hollow core fiber, which has an inner diameter of only 140 μm and a total length of 10 cm, also filled with argon gas. To obtain a constant density in a part of this fiber, it is divided into three almost equally long pieces, which were rejoined with gaps of $\sim$0.5 mm on a v-shaped groove mount [36]. The end pieces were used to separate the vacuum from the gas-filled center part of the capillary. After exiting the fiber, the beam travels into a XUV monochromator equipped with an X-ray CCD detector. An aluminum filter was placed between the last capillary and the monochromator to remove the fundamental 800 nm radiation along with lower harmonics up to the ~17th order.

The evolutionary algorithm described above [11] is used to find the optimal pulse shape that produces the desired HHG spectrum. Fig. 4 shows an optimization experiment where our goal was to optimize a single harmonic while simultaneously suppressing neighbors. To define the fitness function, we

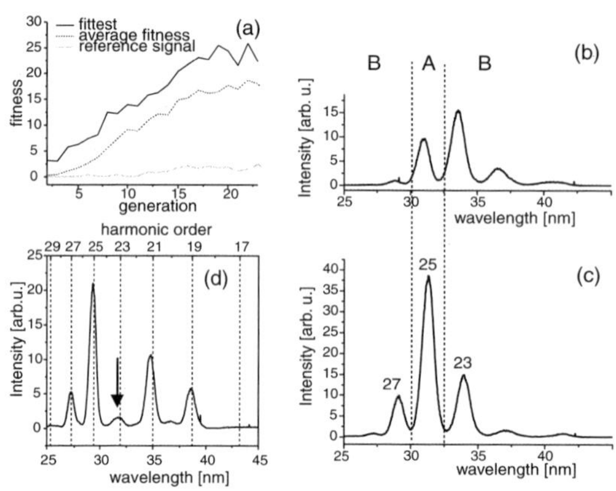

Fig. 4. Harmonic spectrum optimization. The fitness function (see text) was chosen such that a single harmonic (the 25th) was optimized whereas all other orders were suppressed or kept at low intensity. (a) Evolution of the fitness of the pulse shapes in each generation. (b) Reference spectrum before optimization. (c) Optimized spectrum. (d) High harmonic spectrum were a single harmonic order had been suppressed whereas all other harmonics are at about the same intensity level.

divided the harmonic spectrum into two regions as indicated in Fig. 4(b,c). Region A contains the harmonic order of choice, the other one (B) consists of the remaining part of the spectrum. The fitness function f was then written as $f = A^2/B$ to favor both enhancement of A and suppression of B. Fig. 4(a) displays the evolution of the fitness f during the course of the optimization. As a monitor of stability, we plotted the fitness of the pulse shape obtained with our reference mirror shape where all actuators are set to the same voltage level. At these voltage settings we manually adjusted the prism compressor for maximum harmonic signal (Fig. 4(b)). In Fig. 4(c) we show the result of the optimization experiment. It was indeed possible to enhance the 25th harmonic order while at the same time keeping the 23rd harmonic at approximately the same level and suppressing the 21st harmonic even further. In this run it was not possible to avoid an increase in the intensity of the 27th harmonic order. However, it is still much weaker than the 25th harmonic, which was enhanced by a factor of 4. In Fig. 4(d) we show that with a different pulse shape it is also possible to almost completely suppress one particular harmonic order.

4 Conclusion

Active control over quantum-mechanical phenomena is a fascinating perspective of modern physics. However, most of the early attempts at achieving chemical selectivity by the application of light fields were unsuccessful because of the complexity of the underlying photoprocesses. Following the theoretical suggestion of Judson and Rabitz [1], a closed-loop approach was developed by which quantum-mechanical processes can be controlled even in complex systems. The experimental outcome of light–matter interactions are used as a feedback within a learning algorithm which iteratively improves femtosecond pulse-shaper settings until an optimum is reached. This global optimization strategy does not require prior knowledge of the quantum system's Hamiltonian.

In molecular systems it is now possible to control gas-phase dissociation reactions. The example of CH_2ClBr demonstrated the ability for selective bond breakage even in complex molecules. Using lactic acid as a prototype system to illustrate an important step in organic chemical synthesis, we were able to break a desired bond without further fragmentation of the remaining molecule. A significant advance was also made in liquid-phase photoexcitation where solute–solvent interaction introduces additional complications. Nevertheless, it was possible to achieve selective excitation of specific molecular species by adaptive femtosecond pulse shaping. This technique was also employed to selectively steer a isomerization reaction of a complex molecule in solution. Further advances in this research area can be expected when the vectorial properties of the electric field are used. This is possible with the novel method of femtosecond polarization pulse shaping, by which quantum dynamics in three-dimensional systems can be influenced in a very general way on an ultrashort timescale.

The experiments in the gas phase and the liquid phase show control over the nuclear motion, but it is also possible to control the electronic motion. This is demonstrated experimentally by arbitrarily shaping high-harmonic spectra. XUV-photoelectron spectroscopy can immediately take advantage of the possibility to generate spectrally narrow coherent ultrashort XUV pulses. Future study in this direction will pave the way to coherent XUV spectral engineering, thus enabling us to control the electronic motion on an attosecond timescale.

References

1. R. S. Judson and H. Rabitz, Phys. Rev. Lett. **68**, 1500 (1992).
2. W. S. Warren, H. Rabitz, and M. Dahleh, Science **259**, 1581 (1993).
3. R. J. Gordon and S. A. Rice, Annu. Rev. Phys. Chem. **48**, 601 (1997).
4. S. A. Rice and M. Zhao, *Optical Control of Molecular Dynamics* (Wiley, New York, 2000).

5. H. Rabitz, R. de Vivie-Riedle, M. Motzkus, and K. Kompa, Science **288**, 824 (2000).
6. M. Shapiro and P. Brumer, in *Advances in Atomic, Molecular, and Optical Physics*, Vol. 42 of *Advances in Atomic, Molecular, and Optical Physics Series*, edited by B. Bederson and H. Walther (Academic Press, London, 1999), pp. 287–345.
7. T. Brixner, N. H. Damrauer, and G. Gerber, in *Advances in Atomic, Molecular, and Optical Physics*, Vol. 46 of *Advances in Atomic, Molecular, and Optical Physics Series*, edited by B. Bederson and H. Walther (Academic Press, London, 2001), pp. 1–54.
8. D. J. Tannor, *Introduction to Quantum Mechanics: A Time Dependent Perspective* (University Science Press, Sausalito, 2002).
9. M. Shapiro and P. Brumer, *Principles of Quantum Control of Molecular Processes* (Wiley, New York, 2003).
10. T. Brixner and G. Gerber, Chem. Phys. Chem. **4**, 418 (2003).
11. T. Baumert *et al.*, Appl. Phys. B **65**, 779 (1997).
12. T. Brixner, M. Strehle, and G. Gerber, Appl. Phys. B **68**, 281 (1999).
13. N. H. Damrauer *et al.*, Eur. Phys. J. D **20**, 71 (2002).
14. T. Brixner *et al.*, J. Mod. Opt. **50**, 539 (2003).
15. T. Brixner, N. H. Damrauer, B. Kiefer, and G. Gerber, J. Chem. Phys. **118**, 3692 (2003).
16. T. Brixner, N. H. Damrauer, P. Niklaus, and G. Gerber, Nature **414**, 57 (2001).
17. Y. Ohtsuki *et al.*, Chem. Phys. Lett. **369**, 525 (2003).
18. D. P. Aalberts, M. S. L. du Croo de Jongh, B. F. Gerke, and W. van Saarloos, Phys. Rev. A **61**, 040701 (2000).
19. F. Grossmann *et al.*, Europhys. Lett. **60**, 201 (2002).
20. D. G. Gruner and P. Brumer, J. Phys. Chem. **96**, 281 (1992).
21. Y. H. Meyer, M. Pittman, and P. Plaza, J. Photochem Photobio. A **114**, 1 (1998).
22. T. Brixner and G. Gerber, Opt. Lett. **26**, 557 (2001).
23. T. Brixner, G. Krampert, P. Niklaus, and G. Gerber, Appl. Phys. B **74**, S133 (2002).
24. T. Brixner, Appl. Phys. B **76**, 531 (2003).
25. T. Brixner *et al.*, J. Opt. Soc. Am. B **20**, 878 (2003).
26. M. Shapiro and P. Brumer, J. Chem. Phys. **95**, 8658 (1991).
27. Y. Fujimura *et al.*, Chem. Phys. Lett. **306**, 1 (1999).
28. M. Shapiro, E. Frishman, and P. Brumer, Phys. Rev. Lett. **84**, 1669 (2000).
29. L. González, D. Kröner, and I. R. Solá, J. Chem. Phys. **115**, 2519 (2001).
30. P. B. Corkum, Phys. Rev. Lett. **71**, 1994 (1993).
31. M. Drescher *et al.*, Nature **419**, 803 (2002).
32. M. Wieland *et al.*, Appl. Phys. Lett. **81**, 2520 (2002).
33. R. Bartels *et al.*, Nature **406**, 164 (2000).
34. R. Kienberger *et al.*, Science **297**, 1144 (2002).
35. T. Pfeifer, U. Weichmann, S. Zipfel, and G. Gerber, J. Mod. Opt. **50**, 705 (2003).
36. M. Schnürer *et al.*, Phys. Rev. Lett. **85**, 3392 (2000).

Measuring Spatial Chirp and Pulse-Front Tilt in Ultrashort Pulses Using Single-Shot FROG

Selçuk Aktürk, Mark Kimmel, Patrick O'Shea, and Rick Trebino

School of Physics, Georgia Institute of Technology, Atlanta, GA 30332-0430, USA
akturk@socrates.physics.gatech.edu

1 Introduction

Because their generation involves considerable spatio-temporal manipulations, ultrashort laser pulses commonly suffer from spatio-temporal distortions. The most common such distortions are spatial chirp and pulse-front tilt. Devices that use dispersive elements such as prisms and diffraction gratings introduce spatial chirp and pulse-front tilt (fig.1), only–in principle–to remove it afterwards. Unfortunately, even slight misalignments leave residual distortions in the output pulse. Since ultrashort pulses are very broadband, these distortions are usually noticeable and problematic.

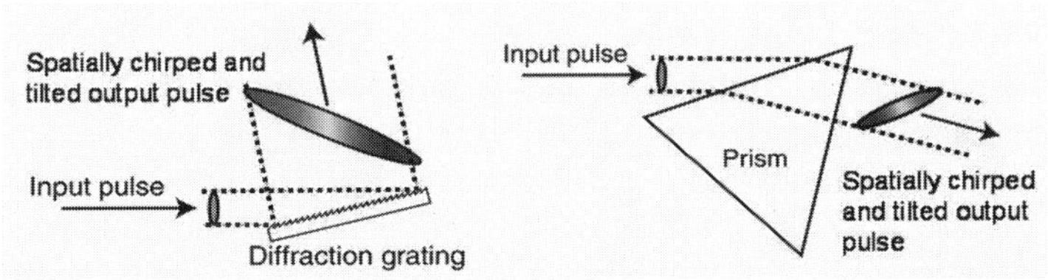

Fig. 1. Prism and gratings always introduce spatial chirp and pulse-front tilt.

2 Spatial Chirp in Ultrashort Pulses

Spatial chirp is the tendency for the wavelength to vary transversely across the beam. After two prisms in a pulse-compressor, the beam has considerable linear spatial chirp. While the next two prisms, in principle, remove this effect, in practice they typically do not completely do so. One cause of this distortion is that the first and last prism separations may not be equal. Using only two prisms and a mirror or mirrors to reflect the beam back on itself guarantees that the relevant prism separations are equal, but the beam may be diverging or converging while inside the device. As a result, the beam emerging from a pulse compressor is frequently contaminated with spatial chirp.

A window with a slight wedge, as is required for laser output couplers (to avoid feedback from the back surface), causes angular dispersion, which also imparts spatial chirp in the beam, and the further that the beam propagates from the optic the more spatial chirp it will have. In addition, even a simple plane-parallel window yields unavoidable spatial chirp when it is tilted. Thus, simply placing a pick-off mirror in the beam causes spatial chirp in the transmitted beam.

There is not a convenient diagnostic for spatial chirp. A spatially resolved spectral measurement, in principle, suffices, but aberrations in spectrometers can mimic this effect. Researchers have also used spatially resolved SPIDER [4, 7], but these interferometric methods are difficult to align and to keep aligned. SPIDER is also experimentally very complex and has within its apparatus a pulse stretcher, which requires very careful alignment or it will itself introduce spatio-temporal distortions.

Here, we report a simple device that measures spatial chirp. Any single-shot second-harmonic-generation frequency resolved optical gating (SHG FROG) device, including the extremely simple SHG FROG device we recently reported-GRENOUILLE [6]. Without a single modification, single-shot SHG FROG and GRENOUILLE yield the pulse spatial chirp–in addition to the intensity and phase vs. time. Specifically, the ordinarily symmetrical SHG FROG trace develops an asymmetrical shear (tilt) in the presence of spatial chirp, proportional to the spatial chirp (fig.2). Because no other effect is known to cause such asymmetry, this is a simple and clear indicator of spatial chirp [2]. Note that this result is independent of the pulse intensity and phase; the method is general.

Note that single-shot SHG FROG maps delay onto position and hence yields a plot of intensity vs. frequency and position, and a spatio-spectral diagnostic for spatial chirp involves a similar plot. The spatio-spectral plot develops a shear in the presence of spatial chirp. And so does the FROG trace. Indeed, we find that a spatially chirped pulse yields a single-shot FROG or GRENOUILLE trace with a shear that is approximately twice that of the spatial chirp when plotted vs. frequency and one half when plotted vs. wavelength. We show that the effects of spatial chirp may also be removed from the FROG trace, and the pulse intensity and phase can be determined in the usual manner.

2.1 Experiment

To introduce variable amounts of spatial chirp into a pulse, we modified the usual prism pulse compressor, placing mirrors between last two prisms, deflecting the pulse to two additional mirrors mounted on translation stage. By translating the latter two mirrors, we were able to align and misalign the compressor, obtaining positive, zero, or negative spatial chirp.

We performed pulse measurements for various amounts of spatial chirp using GRENOUILLE. We determined the spatial-chirp from the measured

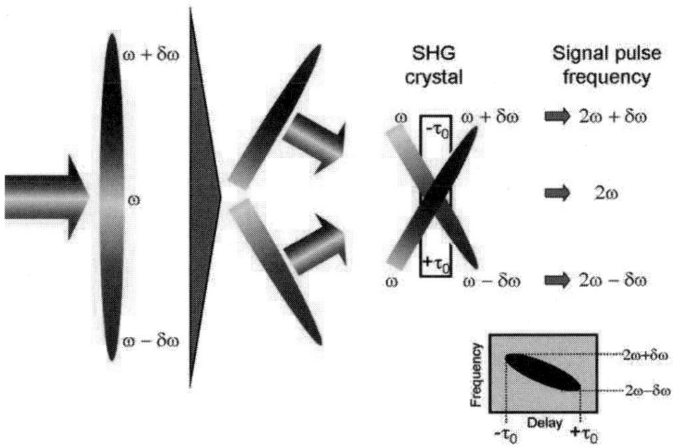

Fig. 2. GRENOUILLE measures spatial chirp.

GRENOUILLE trace from the linear slope of the trace (wavelength vs. delay). We also made independent measurements of the spatial chirp from a spatially resolved spectral measurement using a carefully aligned imaging spectrometer.

We calculated the slopes of both the GRENOUILLE traces and spatio-spectral plots (fig.3), and we find that the slope of this plot, that is, the ratio of the GRENOUILLE trace slope and the spatio-spectral plot slope, is 0.49 +/- .027. This measurement agrees very nicely with the theoretical value of $1/2\cos(\theta/2) = 0.4995$. We find that GRENOUILLE can measure spatial chirp with high sensitivity. Using this method, we were able to align our prism pulse compressor with a sensitivity (in prism separation) of 0.4 mm. With such accuracy, this device should provide a practical and reliable alignment of pulse compressors used in ultrafast laser laboratories.

3 Pulse-Front Tilt in Ultrashort Pulses

As with spatial chirp, the main source of pulse-front tilt is devices such as pulse compressors or stretchers. Propagation through a dispersive device causes angular dispersion, which also causes pulse-front tilt (Fig.1). Indeed, it has been shown that angular dispersion is fundamentally the same phenomenon as pulse-front tilt [3]. In principle, a second prism with its inner surface perfectly parallel to that of the first prism completely removes the angular dispersion and pulse-front tilt introduced by the first. However, even minor non-parallelism or beam divergence leaves residual angular dispersion and pulse-front tilt. As a result, beams used in ultrafast laser laboratories are frequently contaminated with pulse-front tilt (and angular dispersion).

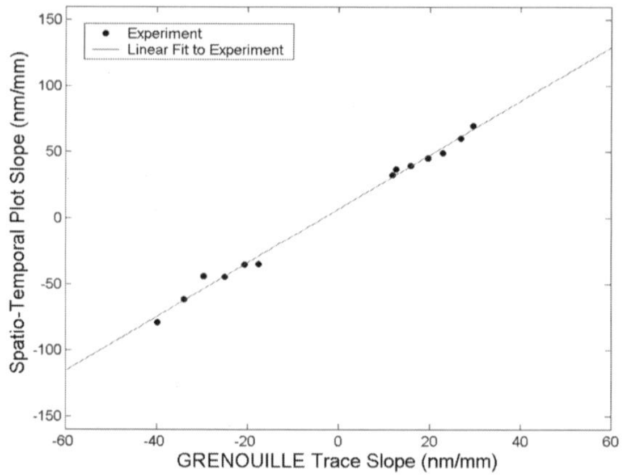

Fig. 3. Slopes of GRENOUILLE traces and corresponding spectrum vs. position slopes for various amounts of spatial chirp.

Pulses with pulse-front tilt are temporally broader than pulses without pulse-front tilt, and the assumption that temporal and spatial evolutions are independent ceases to be valid. Other problems arising due to pulse-front tilt include pulse frequency shifts and spatial profile variations [5].

Pulse-front tilt is not commonly measured, and only a few diagnostics have been proposed for it. The best existing method, "Spectrally resolved interference (SRI)" [9], is an interferometric method utilizing the spatial fringes formed when two beams cross at an angle. This method provides high precision measurement, but, like other interferometric measurements, it requires good temporal coherence and involves considerable labor for its alignment. Researchers have also used spatially resolved SPIDER [4, 7], but these interferometric methods are also difficult to align and to keep aligned. Adjustment of pulse-front tilt with a modified single-shot autocorrelator has also been suggested [8]but this method only indicates the presence of pulse-front tilt qualitatively; it does not provide quantitative measurement of it or yield its sign.

Here we show that GRENOUILLE also measures pulse-front tilt, and again it also does so without a single modification. This works because GRENOUILLE is a very symmetrical device, in which the pulses necessarily cross in space and time, typically at the center of the crystal, which can be made to coincide with the center of the CCD camera. Therefore, pulses without pulse-front tilt will generate their maximum-intensity second-harmonic signal at the center of the crystal (at zero relative delay). The effect of pulse-front tilt is then to move the traces off side of the crystal. And the shift from the center of the crystal is proportional to pulse-front tilt. Since no other effect is known to cause such a shift, GRENOUILLE traces unambiguously reveal and measure pulse-front tilt [1]. Note also that this result is independent of the pulse intensity and phase; the method is general.

3.1 Experiment

We have performed set of experiments by using a pulse compressor with one of the prisms placed on a rotation stage. With this setup, we can deliberately misalign and align the pulse compressor to get zero, positive, or negative pulse-front tilt. For all of the theoretical plots in this work, it was necessary to take into account the Gaussian behavior of the beam and its effects on the pulse-front tilt [10].

Figure 4 shows a comparison of the measured pulse-front tilt using GRENOUILLE the theoretical pulse-front tilt for a pulse emerging from our modified pulse compressor, plotted as a function of the prism angle of incidence. There is very good agreement between the two curves.

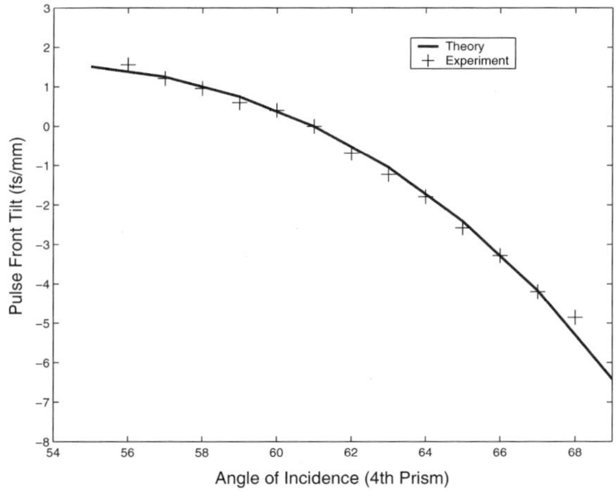

Fig. 4. Theoretically predicted pulse-front tilt and the experimentally measured pulse-front tilt using GRENOUILLE.

By using GRENOUILLE we obtained a sensitivity in the measurement of pulse-front tilt of 0.05 fs/mm. This is the most sensitive measurement of pulse-front tilt, to best of our knowledge. We were also able to align our prism pulse compressor with a sensitivity (in one prism angle of incidence) of 0.025 degrees.

4 Conclusion

In conclusion, we have demonstrated that the experimentally simple version of single-shot SHG FROG, GRENOUILLE, measures, not only the pulse temporal intensity and phase, but also the spatial chirp and pulse-front tilt. The GRENOUILLE trace determines the full spatio-temporal characteristics of a pulse with spatial chirp and pulse-front tilt.

Acknowledgement. This work was supported by National Science Foundation and a start-up grant from the Georgia Institute of Technology.

References

1. S. Akturk, M. Kimmel, P. O'Shea, and R. Trebino. Measuring pulse-front tilt in ultrashort pulses using grenouille. *Opt. Express*, 11:491–501, 2003.
2. S. Akturk, M. Kimmel, P. O'Shea, and R. Trebino. Measuring spatial chirp in ultrashort pulses using single-shot frequency-resolved optical gating. *Opt. Express*, 11:68–78, 2003.
3. Z. Bor, B. Zacz, G. Szabo, M. Hilbert, and H.A. Hazim. Femtosecond pulse front tilt caused by angular dipersion. *Opt. Eng.*, 32:2501–2504, 1993.
4. C. Dorrer, E.M. Kosik, and I.A. Walmsley. Self-referencing spectral interferometry for measuring ultrashort optical pulses. *JQE*, 35:501–509, 1999.
5. O.E. Martinez. Pulse distortions in tilted pulse schemes for ultrashort pulses. *Opt. Commun.*, 59:229–232, 1986.
6. P. O'Shea, M. Kimmel, X. Gu, and R. Trebino. Highly simplified ultrashort pulse measurement. *Opt. Lett.*, 26:932–934, 2001.
7. B. S. Prade, J. M. Schins, E. T. J. Nibbering, M. A. Franco, and A. Mysyrowicz. Spatio-temporal characterization of the electric field of ultrashort optical pulses using two-dimensional shearing interferometry. *Appl. Phys. B [Suppl.]*, 74:S209–S217, 2002.
8. Z. Sacks, G. Mourou, and R. Danielius. Adjusting pulse-front tilt and pulse duration by use of a single-shot autocorrelator. *Opt. Lett.*, 26:462–464, 2001.
9. K. Varju, A.P. Kovacs, G. Kurdi, and K. Osvay. High precision measurement of angular dispersion in a cpa laser. *Appl. Phys. B [Suppl.]*, 74:S259–S263, 2002.
10. K. Varju, A.P. Kovacs, K. Osvay, and G. Kurdi. Angular dispersion of femtosecond pulses in a gaussian beam. *Opt. Lett.*, 27:2034–2036, 2002.

Measurements of Ultracomplex and Ultraweak Pulses with FROG

Qiang Cao[1], Aparna P. Shreenath[1], Stephan Link[2], Xun Gu[1], Jing-yuan Zhang[1], Erik Zeek[1], Mark Kimmel[1], Rick Trebino[1], and Mostafa El-Sayed[2]

[1] School of Physics, Georgia Institute of Technology, Atlanta, GA 30332-0430, USA urlrick.trebino@physics.gatech.edu
[2] Laser Dynamics Laboratory, Georgia Institute of Technology, Atlanta, GA 30332-0400, USA

The full intensity-and-phase characterization of ultrashort laser pulses is now routine in scientific research. Techniques such as frequency-resolved optical gating (FROG) [1] have been widely applied to obtain such information, which is essential in many applications. However, what we have been able to measure so far are mostly simple laser pulses. There are many more *light* pulses in general, and they aren't always emitted by lasers, and they can be extremely complicated, very weak, and quite variable from shot to shot. Many more applications involve these "ill-conditioned" pulses, and the ability to measure these pulses will be an invaluable tool to many researchers. In this report we extend the FROG technique to the measurement of ultracomplex (supercontinuum) and ultraweak (bioluminescence) light pulses. Only minor modifications to the FROG apparatus and algorithm are required for these measurements.

In this report we extend the FROG technique to the measurement of ultracomplex (supercontinuum) and ultraweak (bioluminescence) light pulses. Only minor modifications in the FROG apparatus and algorithm are required for these measurements.

With the use of the microstructure fiber, ultrabroadband supercontinuum can now be easily generated using only a Ti:sapphire oscillator [2]. Many applications of the supercontinuum require that we know the phase of the light well. We have used cross-correlation FROG (XFROG) to measure the intensity and phase of the supercontinuum pulses from a 16-cm-long microstructure fiber [3]. To phase-match all the wavelengths in the supercontinuum, the nonlinear crystal (BBO) was rapidly dithered during the measurement, to allow access to a range of angles which corresponded to the entire supercontinuum bandwidth. We found that the continuum pulses had a time-bandwidth product of 4000, easily the most complicated pulses ever characterized, and also the retrieved spectrum was found to contain unstable ~1-nm-scale fine structure. Further single-shot spectrum measurements confirmed our findings, and the experiments agreed very well with theoretical predictions.

The newly revealed fine spectral structure and shot-instability of the supercontinuum pulses will inevitably present profound and often undesirable implications to the application of the light. For the many researchers who use this continuum, a very important question to ask is how we can create a cleaner supercontinuum that is free of or less plagued with these problems.

Our results have been instrumental in helping to understand the underlying spectral broadening mechanisms and in confirming recent advances in numerical simulations of supercontinuum generation in the microstructure fiber. Simulations using the extended nonlinear Schrödinger equation (NLSE) model, have matched experiments amazingly well. Although most microstructure-fiber supercontinuum experiments use 10–100 cm of fiber, simulations have revealed that most of the spectral broadening occurs in the first few mm of fiber [4]. Further propagation, which still slowly broadens the spectrum through less important nonlinear processes, such as Raman self-frequency shift, yields only increasingly unstable and fine spectral structure due to the interference of multiple solitons in the continuum spectrum.

This observation suggests that it would be favorable to use a short (< 1cm) length of microstructure fiber for supercontinuum generation, as the resulting continuum will still be broad, but short, more stable, and containing less fine spectral structure. Indeed, we successfully generated supercontinuum by pumping an 8-mm-long microstructure fiber with 40-fs Ti:sapphire oscillator pulses, and performed a similar crystal-angle-dithered XFROG measurement [5]. The results are shown in Fig. 1.

We note firstly that the temporal extent of the supercontinuum from the 8-mm-long microstructure is significantly shorter than the several-picosecond-long supercontinuum that is generated in longer fiber and, indeed, consists of series of sub-pulses that are shorter than the input 40-fs pulse. Meanwhile, the short fiber continuum has less complex temporal and spectral features than the continuum pulses previously measured from longer fibers. The spectral phase of the short fiber continuum is relatively flat, varying only in the range of 25 rad, while the long-fiber continuum is dominated by cubic phase spanning over 1000π rad [3]. In view of the complexity of the SC, Fig. 1 shows that there is good agreement between the retrieved spectrum (black) and that independently measured using a spectrometer and averaged over $\tilde{}10^7$ pulses (gray). The discrepancies are due to shot-to-shot instabilities in the SC, which smear out the spectrometer-measured spectrum, as we discussed previously [3].

Another development in our effort to extend FROG's capabilities is the measurement of ultraweak pulses, such as luminescence. Interferometric methods cannot measure most luminescence pulses, because, apart from being very weak, these pulses also suffer from variable absolute phase (requiring single-shot measurements), and poor spatial coherence (requiring small emission volume). However, the problem of variable absolute phase can be solved by using a FROG-type technique, whose signal is absolute-phase-invariant. And, to measure these ultraweak pulses, we can utilize optical parametric amplification (OPA), in which a strong bluer pump pulse not only gates the

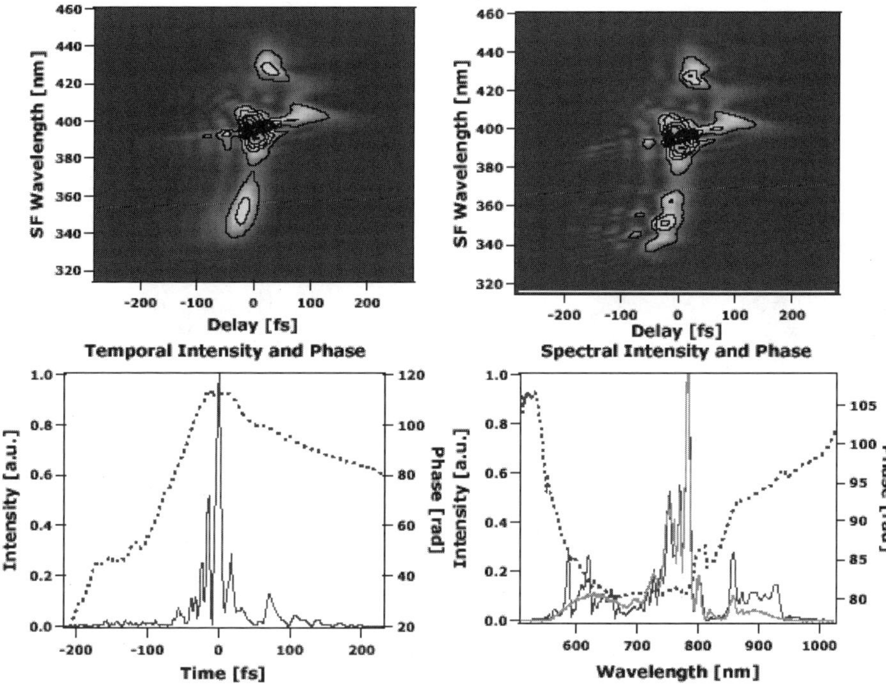

Fig. 1. XFROG measurement results of 8-mm-long microstructure fiber continuum: (a) measured trace, (b) retrieved trace, (c) temporal intensity and phase, and (d) spectral intensity and phase. The gray line in (d) is the spectrometer-measured spectrum averaged over many shots.

weak luminescence pulse in time, but also provides exponential gain to the weak luminescence pulse by a factor of up to 10^6. The phase of the luminescence pulse is retained in the process. The only requirements are that the gate (pump) pulse be measurable, bluer than, shorter than, synchronizable with, and have much higher intensity than the luminescence pulse. A natural choice would be the very same pulse that excites the luminescence, which has the same requirements!

We first tested the OPA XFROG technique [6] with an 80-fJ pulse, which was an attenuated, spectrally filtered slice of supercontinuum generated in a sapphire plate. In this measurement, variably delayed 5.8-μJ, 400-nm pulses gated the supercontinuum in a 1-mm-thick BBO crystal, yielding an OPA gain of about 150.

A modified XFROG retrieval algorithm, incorporating the OPA nonlinearity (the gate is now $G(t) = \cosh[g|E_{\text{gate}}(t)|]$, where g is the gain parameter which depends on crystal length), was used to retrieve the intensity and phase of the test pulse. Fig. 2 shows the measured and retrieved OPA XFROG traces, along with the retrieved intensity and phase. Another well established but less sensitive technique, sum-frequency generation (SFG) XFROG, was used to verify the results. The two measurements agree reasonably well, as does the independently measured spectrum.

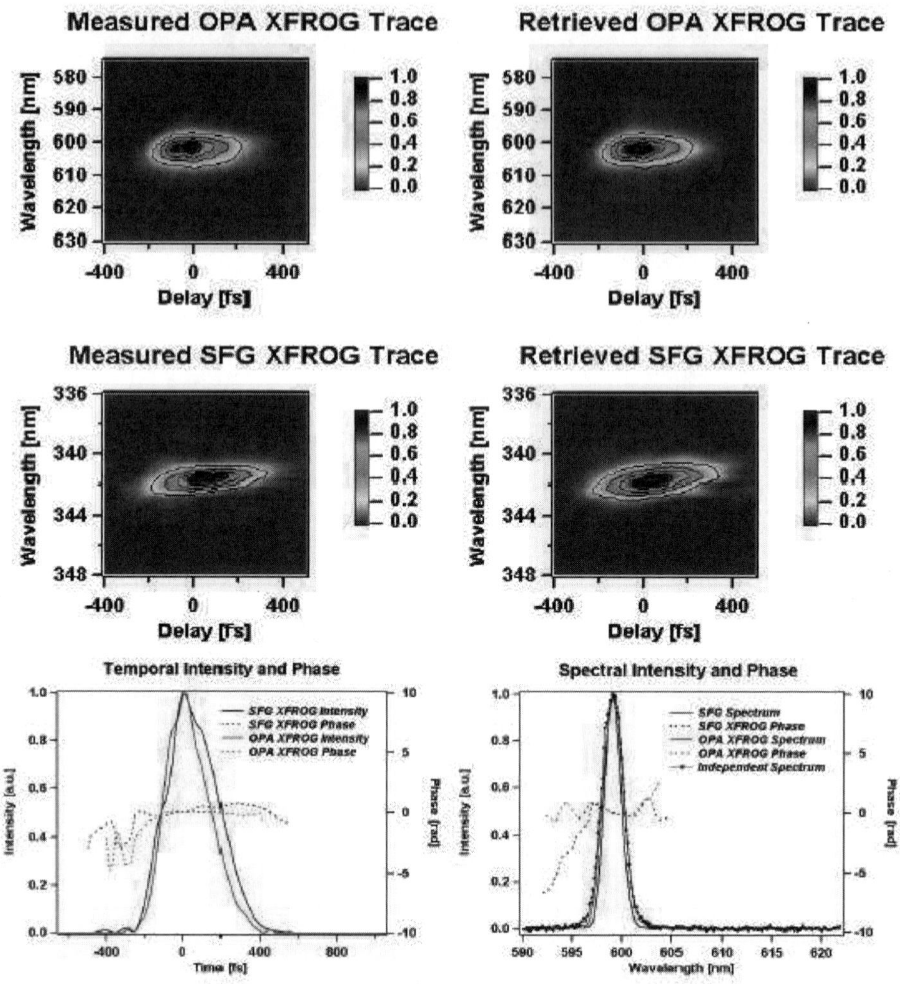

Fig. 2. The OPA XFROG measurement of 80-fJ pulses agrees well with that from SFG XFROG.

We applied the technique to measure a broadband section of the super-continuum, with its pulse energy attenuated to two different levels, 50 fJ and 500 pJ respectively. A non-collinear OPA (NOPA) geometry allowed us to phase-match the large signal bandwidth in these measurements. Both measurements yielded similar traces and retrieval results, as are shown in Fig. 3. There exists fine-scale structure in the retrieved temporal and spectral intensity profiles, similar to what was seen in the XFROG measurements of the microstructure-fiber supercontinuum [3]. We believe such fine structure in the OPA test pulse is also real. Its absence in a many-shot-averaged spectrometer measurement is due to shot-to-shot variations and spatial incoherence caused by multiple filaments in the continuum.

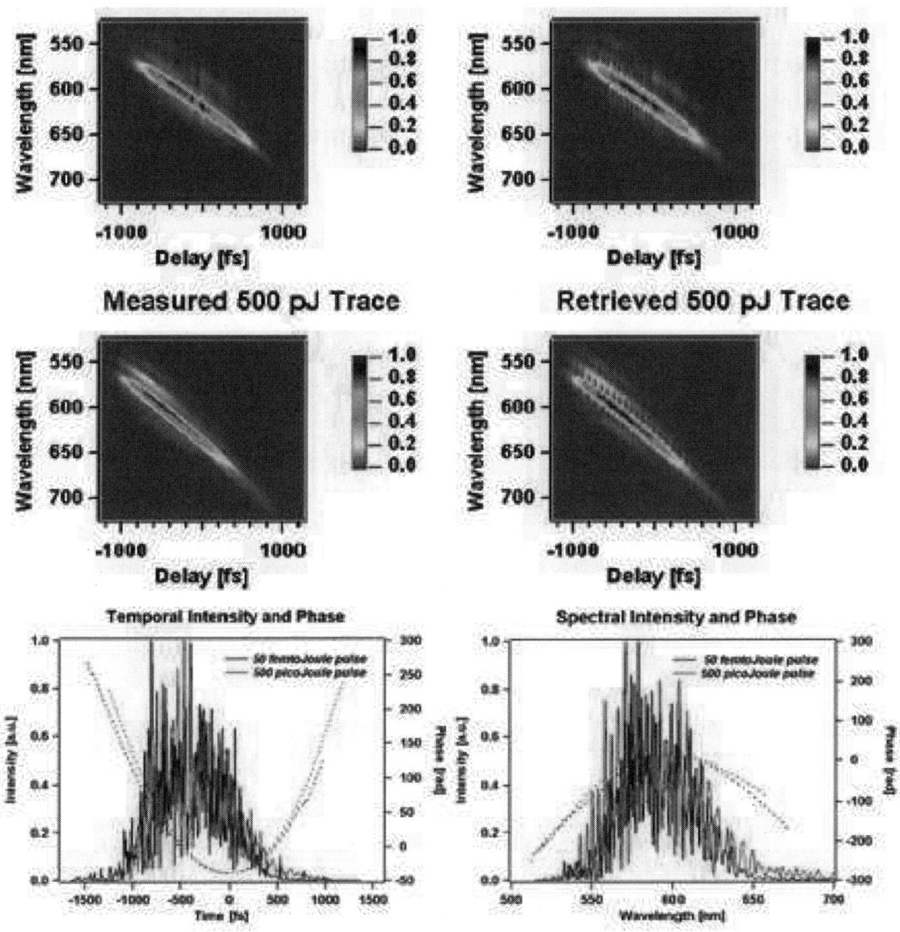

Fig. 3. OPA XFROG measurements of broadband 50 fJ (Gain ˜10^3) and, for comparison, 500 pJ (Gain ˜50) pulses. The fine-scale structure in the spectrum and intensity is real, although it varies from shot to shot.

To examine the sensitivity limit of the OPA XFROG technique, we attenuated a spectral slice of continuum to 50 aJ (˜150 photons), in simulation of a bioluminescence pulse, and we measured the pulse using the OPA XFROG technique in a 2-mm-thick BBO crystal, with a gain of about 5000. The results are shown in Fig. 4. In this high-gain measurement, amplified spontaneous noise, which is usually referred to as the optical parametric generation (OPG), constitutes an unwanted background, and limits the ultimate sensitivity of this technique. In the measurement of the 50-aJ pulse, the OPA signal was only about five times stronger than the OPG background. The measured trace also contained some intensity fluctuations from one delay step to the next, due to the variations in the pump pulse intensity and the inherent instability of a single-pass OPA process with high gain. Nevertheless, an OPA XFROG trace was successfully obtained, and retrieval of the intensity and phase was completed. The algorithm was able to smooth out the artifacts–as they do not correspond to any real pulse structure–and retrieved a reasonable pulse that had a FWHM of 170 fs.

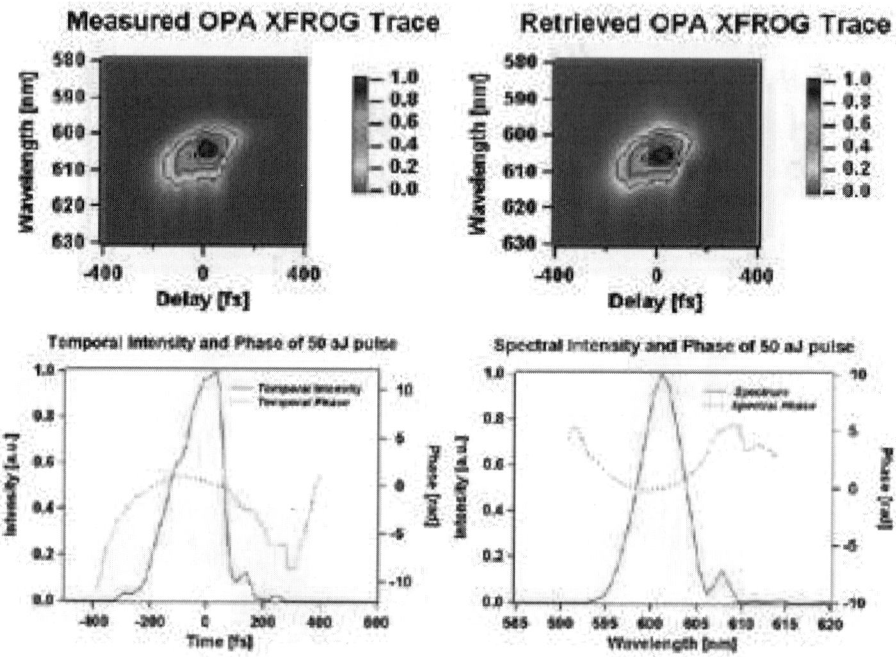

Fig. 4. OPA XFROG measurement of a 50-aJ pulse.

With the amplified gating provided by the OPA nonlinear optical process, we are now able to measure extremely weak, broadband, complex, spatially incoherent pulses with random absolute phase. This technique, which we call OPA XFROG, should be able to measure trains of fluorescence or spontaneous Raman pulses as weak as a few attojoules (i.e., a few photons).

In conclusion, the FROG technique has been extended to measure ultracomplex and ultraweak pulses. The apparatus remains quite simple, and modifications required to the FROG algorithm are minor.

This material is based upon work supported by the National Science Foundation under Grants No. ECS-0200223 and DBI-0116564.

References

1. Trebino, R.: Frequency-resolved Optical Gating: The Measurement of Ultrashort Laser Pulses. Kluwer Academic Publishers, Boston, Mass. (2002)
2. Ranka, J.K., Windeler, R.S., Stentz, A.J.: Visible continuum generation in air-silica microstructure optical fibers with anomalous dispersion at 800 nm. Optics Letters **25**, 25–27 (2000)
3. Gu, X., Xu, L., Kimmel, M., Zeek, E., O'Shea, P., Shreenath, A.P., Trebino, R., Windeler, R.S.: Frequency-resolved optical gating and single-shot spectral measurements reveal fine structure in microstructure-fiber continuum. Optics Letters **27**, 1174–1176 (2002)
4. Dudley, J.M., Gu, X., Xu, L., Kimmel, M., Zeek, E., O'Shea, P., Trebino, R., Coen, S., Windeler, R.S.: Cross-correlation frequency resolved optical gating analysis of broadband continuum generation in photonic crystal fiber: simulations and experiments. Optics Express **10**, 1215–1221 (2002)
5. Cao, Q., Gu, X., Zeek, E., Kimmel, M., Trebino, R., Dudley, J.M., Windeler, R.S.: Applied Physics B (in press) (2003)
6. Zhang, J., Shreenath, A.P., Kimmel, M., Zeek, E., Trebino, R., Link, S.: Measurement of the intensity and phase of attojoule femtosecond light pulses using Optical-Parametric-Amplification Cross-Correlation Frequency-Resolved Optical Gating. Optics Express **11**, 601–609 (2003)

Few-cycle Pulse Diagnostics Using Plateau Harmonics

Vladislav S. Yakovlev and Armin Scrinzi

Photonics Institute, Vienna University of Technology, Gusshausstrasse 27/387, A-1040 Vienna, Austria/EU, `scrinzi@tuwien.ac.at`

Summary. We show that the strength of the central electric field peaks in a few-cycle laser pulse can be recovered from a frequency-time image of the high harmonic spectrum generated in a gas volume. A simple observable is defined that provides a gauge of the carrier-envelope phase for pulse durations up to three optical cycles, corresponding to 8 fs FWHM at the Ti:sapphire wavelength of 800 nm.

The first experimental determination of the carrier-envelope phase φ_{CE} of a few-cycle laser pulse was achieved — up to an ambiguity of π — by analyzing the harmonic radiation generated in a noble gas [1]. The measurement was based on the observation that the harmonics near the cutoff sensitively depend on φ_{CE} [2], which for a chirp-free pulse with fundamental frequency ω_l is defined through

$$E(t) = f(t)\,\mathrm{Re}\,e^{i(\omega_l t+\varphi_{\mathrm{CE}})}, \tag{1}$$

where $E(t)$ denotes the laser electric field and the envelope $f(t)$ has its maximum at $t = 0$.

As cutoff harmonics represent only a small fraction of the full harmonic yield, their use for pulse diagnostics requires high quality of the spectra and, moreover, their diagnostic significance is limited to extremely short pulses with durations of around 2 laser optical cycles. A significant amount of additional information is contained in the plateau of lower harmonics. Indeed, plateau harmonics also strongly depend on φ_{CE} [3], but in view of the apparent irregularity of the dependence one must carefully investigate the underlying physical processes in order to associate a given plateau spectrum with a carrier envelope phase.

In the following we show, that the plateau keeps a rather detailed record of the electric field that generated it. Not only φ_{CE}, but also the heights of central field peaks can be recovered from the measured harmonic power spectrum. The recollision model of high harmonic generation [4, 5] finds another, most detailed confirmation in a time-frequency analyis of the harmonic field and the power spectrum. The field image reflected in the high harmonics is robust with respect to the spatial pulse profile and pulse propagation, provided the phase slip between fundamental and harmonics is kept small.

1 Time-frequency analysis of harmonic generation

Figure 1 shows the time-frequency analysis of the dipole response of a neon atom to a 5 fs laser pulse. The harmonic spectrum generated in a time window around t is obtained by multiplying the acceleration of the atomic dipole $\ddot{d}(t)$ with a Gaussian window function and taking the square of the Fourier transform

$$H(\omega, t) = \left|\mathcal{F}[\ddot{d}(t')e^{-(t-t')^2/T^2}]\right|^2. \tag{2}$$

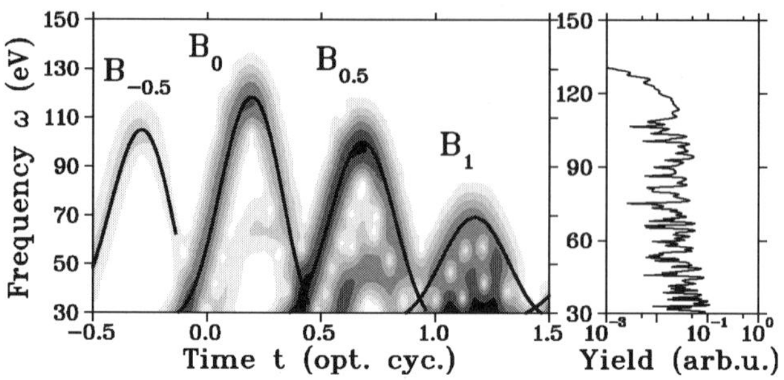

Fig. 1. Time-frequency plot $H(\omega, t)$ of the dipole response of a Ne atom to a strong short laser pulse. The gray shading indicates values from maximum (black) to 1/10 of maximum. Pulse duration is 5 fs FWHM with a sech pulse envelope at 800 nm wave length and peak intensity of $5 \times 10^{14}\, W/cm^2$. The box to the right shows the harmonic power spectrum. Solid lines in the time-frequency plot indicate harmonic energies obtained from the recollision model Eq. (3).

There is a trade-off between time- and frequency resolution, which depends on the width T of the time window. For the present purposes we have chosen $T = 0.05 \times 2\pi/\omega_l$, which restricts our analysis to harmonic energies $\omega \gtrsim 20\,\omega_l \approx 30\,eV$. Harmonics $\gtrsim 60\,eV$ are generated near the capture times according of the recollision model. These are given by [6]

$$\omega(t_s, t_c) \approx 1.3\, I_p + \frac{1}{2}\left[\int_{t_s}^{t_c} E(t)dt\right]^2, \tag{3}$$

where t_s and t_c denote start and capture times for a classical electron in the laser field. Each energy is generated twice during a laser half-cycle, corresponding to the "short" and "long trajectories" of the classical picture. Three-dimensional propagation effects favor contributions from the short trajectories, such that each laser half-cycle gives a single dominant contribution to the high harmonics. We label these radiation bursts by B_t with $t = -0.5, 0, 0.5, 1$ referring to the approximate times where the lower energies of the burst are generated. The intensity of each B_t is proportional to ionization at the starting times t_s of the classical trajectories, which precede recollision by typically $t_c - t_s \approx 1/2$ optical cycle. As ionization is strongest for the peak field at $t = 0$, the burst $B_{0.5}$ has the highest intensity.

2 Determining central field peaks from harmonics

The high frequencies in the dipole response contain a record of the laser field, which to a certain degree must reappear in the harmonic power spectrum, except that information on the harmonic phase is lost and therefore no absolute times but only time-separations of the electric field peaks may be recovered.

Propagation effects were simulated using the *ab initio* code of Ref. [7] for harmonic generation by an 800 nm, 5 fs FWHM pulse with peak intensity $5 \times 10^{14} W/cm^2$ during propagation through 2 mm of Ne at a pressure of 200 mbar. The focus of the Gaussian beam was placed in the center of the volume with a beam waste of 72 μm resulting in a confocal parameter of 2 cm. Under these conditions, the phase mismatch between laser and harmonics over the propagation length amounts to about 1/4 the optical period of the highest harmonics.

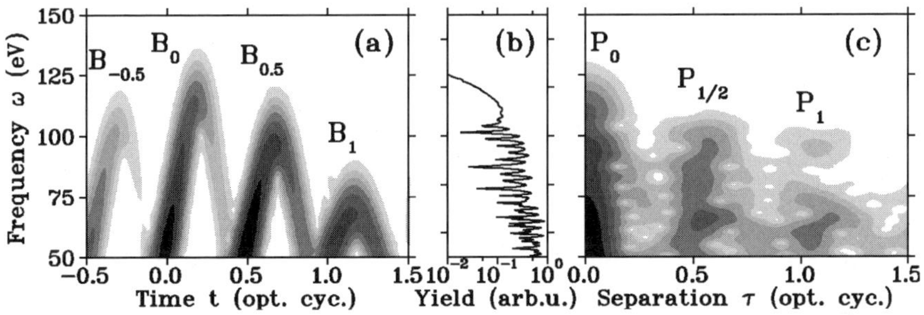

Fig. 2. Harmonic generation in a Ne gas volume: (a) time-frequency plot $H(\omega, t)$ of electric field on axis, (b) harmonic power spectrum, and (c) the frequency-time plot $T(\tau, \omega)$ of the power spectrum. (Logarithmic gray-scale from maximum (black) to 1/100 of maximum, pulse parameters as in Figure 1, see the text for the gas parameters.)

We analyzed the electric field on the propagation axis using Eq. (2) and the harmonic power spectrum $h(\omega)$ by an analogous frequency-time analysis

$$T(\tau, \omega) = \left| \mathcal{F}[h(\omega') e^{-(\omega-\omega')^2/\Omega^2}](\tau) \right|^2 . \tag{4}$$

The width of the Gaussian window function is set to $\Omega = 3\omega_l$ for optimal contrast. The results are shown in Figure 2. After three-dimensional propagation both, time-frequency plot of the dipole-field and harmonic power spectrum, are more regular than the corresponding single-atom quantities in

Figure 1, as the contributions of the classical short recollision trajectories are favored by phase matching (see [3,8,9] for a detailed discussion of this effect). The frequency-time analysis of the harmonic power spectrum shows distinct peaks at time-separations $\tau = 0$, $1/2$ and 1 optical cycles, which we label by P_τ. There is a close correspondence between the radiation bursts B_t in the analysis of the dipole and the peaks P_τ. The cutoff energy of P_0 coincides with the cutoff of B_0. The frequency dependence of $P_{1/2}$ and P_1 reflects the successive appearance of contributions with time separations 1/2 and 1, respectively. $P_{1/2}$ is predominantly due to the pair of bursts $(B_0, B_{0.5})$ and first appears at the maximal energy of $B_{0.5}$. A weak signal in P_1 is generated by the pair $(B_{-0.5}, B_{0.5})$, but the stronger contribution at lower frequencies is due to (B_0, B_1), due to the low intensity of $B_{-0.5}$.

3 Simple experimental observables

Based on the high harmonic images we can construct observables for phase-determination from the harmonic spectrum. For that we eliminate the τ-coordinate by defining the integrated peaks

$$P_\tau(\omega) = \int_{\tau-1/4}^{\tau+1/4} T(\omega, \tau')d\tau'. \tag{5}$$

We define the cutoff for each peak as the energy, where its height drops to 8% of the preceding maximum. Figure 3(a) shows the cutoff energies of $P_{1/2}$ and P_1 and the corresponding classical energies as a function of φ_{CE}, which agree within $\lesssim 10\%$. The deviations from the classical energies are brought about by interference of bursts of comparable height that contribute to $P_{1/2}$ and P_1, which blurs the cutoff. The ambiguity is more pronounced at phases, where the weaker burst B_t at the rising edge of the pulse contributes to $P_{1/2}$ and P_1, respectively.

Another experimental observable allows to extract φ_{CE} for longer pulses by defining the ratio between the P_1 and $P_{1/2}$ in the form

$$R = \int_{\omega_{0.5}}^{\omega_1} P_1(\omega)d\omega \;/\; \int_{\omega_{0.5}}^{\omega_1} P_{1/2}(\omega)d\omega. \tag{6}$$

The frequency range $[\omega_{0.5}, \omega_1]$ is given by the classical cutoff energies for $B_{0.5}$ and B_1, respectively, as obtained from Eq. (3) at $\varphi_{\mathrm{CE}} = 0$. The ratio R indicates, to which extent two bursts separated by $\tau = 1$ contribute to the harmonic spectrum. In figure 3(b) one sees a pronounced maximum near the cosine-pulse, which can be used to fix φ_{CE} for pulse durations up to 3 optical cycles ($\sim$ 8 fs).

A possible chirp of the pulse must be taken into account when deducing the electric field from the harmonic cutoff energies. A simple analysis for electric field $E(t) \approx E_0(1 - \alpha t^2)\sin(\omega_l t + \varphi_{\mathrm{CE}} + \gamma t^2)$ shows that two equally high

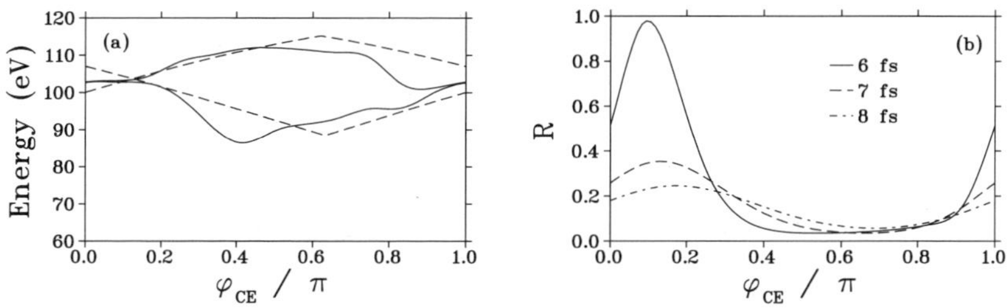

Fig. 3. Observables extracted from the frequency-time plots: (a) Cutoff energies for $P_{1/2}$ (upper curve) and P_1 (lower curve) and cutoff energies from the recollision model (dashed) as a function of $\varphi_{\rm CE}$. (b) Modulation of the intensity ratio R at period $\tau = 1/2$ and $\tau = 1$ for pulse durations between 6 and 8 fs FWHM; c.f. Eq. (6).

radiation bursts and the most strongly modulated cutoff region in harmonic spectra appear at $\varphi_{\rm CE} = \varphi_{\rm mco} \approx \pi/2 + \Delta\varphi + \gamma/\alpha$ ("mco" stands for "modulated cutoff"), while the smoothest cutoff appears at $\varphi_{\rm CE} = \varphi_{\rm sco} \approx \Delta\varphi + \gamma/\alpha$. The correction term $\Delta\varphi$ depends on the pulse intensity and on the ionisation potential, while dependence of $\Delta\varphi$ on pulse duration and shape is negligible. For the peak intensity of $5 \times 10^{14}\, W/cm^2$ numerical simulations give $\Delta\varphi = 0.35 \approx 20°$.

4 Discussion

The frequency-time analysis of the harmonic spectrum generalizes the phase-measurement method proposed by de Bohan *et al.* [2], where the structure of the harmonic spectrum near the cutoff was used to distinguish sine- from cosine-pulses. By exploiting a much larger part of the harmonic spectrum, the present method provides more detail on the pulse and in particular higher sensitivity to $\varphi_{\rm CE}$. While P_0 and $P_{1/2}$, which are responsible for the cutoff harmonics, are both related to bursts B_t at times t near the center of the laser pulse, whose height and intensity vary little with $\varphi_{\rm CE}$, the peak P_1 involves a third burst nearer to the slope of the laser pulse, which depends more strongly on $\varphi_{\rm CE}$. For that reason our method is applicable to longer pulses with durations up to $\sim 8fs$.

Several characteristic features of short pulse high harmonic spectra can be understood from time-frequency plots. Figure 2 shows that in any given range of harmonic frequencies, the separation between peaks is given by the time-separation of the two dominant bursts of radiation. As this separation increases with ω, the separation between peaks of the harmonic spectrum is not precisely $2\omega_l$. Another feature appearing in short pulse high harmonic spectra is an apparent phase-jump of modulations of the harmonic spectrum between groups of "odd harmonics" and "even harmonics". These jumps can be associated with the end of one of the two dominant bursts and takeover by a new pair of bursts. The most regular high harmonic spectra with peaks near the odd harmonic energies appear at $\varphi_{\rm CE} \approx \pi/2 + 0.35$, i.e. not exactly for the sine-pulse. Again, this is easily understood from the classical recollision energies: at $\varphi_{\rm CE} = \pi/2 + 0.35$ the two dominant radiation bursts reach the same maximal energy. The deviation from a sine-pulse $\varphi_{\rm CE} = \pi/2$ arises, since the collision energy does not depend on the instantaneous field, but on the field integral between ionization and collision.

The discussion above shows that the peak photon energies produced by the largest field peaks of a few-cycle pulse can be extracted from the high harmonic power spectrum. As these energies are strictly connected to the electric field strength by the recollision model, the basic pulse parameters can be recovered. For a full exploitation of the information contained in the harmonics, a fit to the frequency-time image can also make use of information on the spectral content of the laser pulse.

We are indebted to A. Baltuska and F. Krausz for numerous illuminating discussions. This work was supported by the Austrian Science Fund special research programs F011 and F016.

References

1. A. Baltuška, Th. Udem, M. Uiberacker, M. Hentschel, Ch. Gohle, R. Holzwarth, V. Yakovlev, A. Scrinzi, T. W. Hänsch, and F. Krausz. Attosecond control of electonic processes by intense light fields. *Nature*, 421:611, 2003.
2. A. de Bohan, L. P. Antoine, D. B. Milosevic, and B. Piraux. Phase-dependent harmonic emission with ultrashort laser pulses. *PRL*, 81:1837, 1998.
3. E. Priori, G. Cerullo, M. Nisoli, S. Stagira, S. De-Silvestri, P. Villoresi, L. Poletto, P. Ceccherini, C. Altucci, R. Bruzzese, and C. de Lisio. Nonadiabatic three-dimensional model of high-order harmonic generation in the few-optical-cycle regime. *PRA*, 61:063801, 2000.
4. P. B. Corkum. Plasma perspective on strong-field multiphoton ionization. *PRL*, 71:1994, 1993.
5. K. C. Kulander, K. J. Schafer, and J. L. Krause. *Proceedings of the Workshop, Super Intense Laser Atom Physics (SILAP) III, edited by B. Piraux (Plenum, New York)*, 1993.
6. M. Lewenstein, P. Balcou, M. Y. Ivanov, A. L'Huillier, and P. B. Corkum. Theory of high-harmonic generation by low-frequency laser fields. *PRA*, 49:2117, 1994.
7. N. Milosevic, A. Scrinzi, and T. Brabec. Numerical characterization of high harmonic attosecond pulses. *PRL*, 88:093905, 2002.
8. P. Salieres, A. L. Huillier, and M. Lewenstein. Coherence control of high-order harmonics. *PRL*, 74:3776, 1995.
9. M. B. Gaarde, F. Salin, E. Constant, Ph. Balcou, K. J. Schafer, K. C. Kulander, and A. L'Huillier. Spatiotemporal separation of high harmonic radiation into two quantum path components. *PRA*, 59:1367, 1999.

Part III

Measurement and Control of the Carrier-Envelope Phase

Carrier-Envelope Phase Stabilization of Modelocked Lasers

Tara M. Fortier, David J. Jones, Jun Ye, and Steven T. Cundiff

JILA, University of Colorado and National Institute of Standards and Technology, Boulder, CO 80309-0440 cundiffs@jila.colorado.edu

Summary. We present advances in carrier-envelope phase stabilization of mode-locked lasers within the scope of both optical frequency metrology and ultrafast science.

By drawing on the techniques of single-frequency laser stabilization and on improvements of ultrafast lasers, phase stabilization of Ti:sapphire (Ti:S) lasers has led to great advances in both the fields of optical frequency metrology and ultrafast phenomena [13]. In the former, it has resulted in the realization of an all optical atomic clock [5, 19] and in the latter it has allowed for waveform synthesis of ultrashort (~ 2 cycle) pulses [14]. In this paper, we report on both development of the technology of and experimental results for modelocked laser stabilization. Phase coherence measurements to characterize various noise sources that lead to contamination of the carrier-envelope phase are discussed. Using this highly phase stable laser, we discuss a lock-in based technique to measure phase fluctuations and extra-cavity changes in the carrier-envelope phase due to propagation through a dispersive material. We also present an octave spanning Ti:S laser, which allows for carrier-envelope phase stabilization without the use of external broadening in fiber.

1 Background

The spectrum of a modelocked laser is a frequency comb characterized by two radio frequencies (rf). One of these frequencies is the laser repetition rate, f_{rep}, which determines the comb spacing. The second is the offset frequency, f_0, which determines the absolute position of the comb. As a result the frequency of the nth comb line, ν_n, is represented by:

$$\nu_n = n f_{rep} + f_o \tag{1}$$

Here, n is a large integer multiplying f_{rep} up into the optical regime. Stabilization of both f_{rep} and f_0 results in an absolute optical comb with approximately 10^6 coherent lines spanning the visible to near IR. This provides a convenient secondary reference to which other oscillators, single frequency or pulsed, may be synchronized. Additionally, phase stabilization of the comb locks the evolution of the carrier-envelope (CE) phase of the pulse electric field, ϕ_{CE} [2,13]. ϕ_{CE} is defined as the phase difference between the peak of the carrier wave relative to the pulse envelope, and it's evolution is related to the offset frequency via

$$f_0 = \frac{1}{2\pi}\frac{d\phi_{CE}}{dt}. \tag{2}$$

The major advance in the Ti:S stabilization scheme is measurement of f_0 independent of a secondary optical standard. The measurement scheme uses the Ti:sapph laser spectrum for self-comparison by referencing harmonics at the spectral extremes, $|N(nf_{rep} + f_0) - M(mf_{rep} + f_0)|$ [17]. Here, the two harmonic numbers, N and M, are chosen such that spectral overlap is obtained, i.e., $N \times n = M \times m$. Interference between the two harmonics then directly yields the rf beat signal $(N - M)f_0$. A drawback of this measurement scheme, however, is that it requires a large bandwidth. For the simplest scheme, where $N = 1$ and $M = 2$, the required bandwidth is an optical octave, which has not typically been available from a Ti:S laser alone. Microstructure (MS) fiber technology provides a simple solution as it enables continuum generation using the pulse energy available directly from the laser output [16].

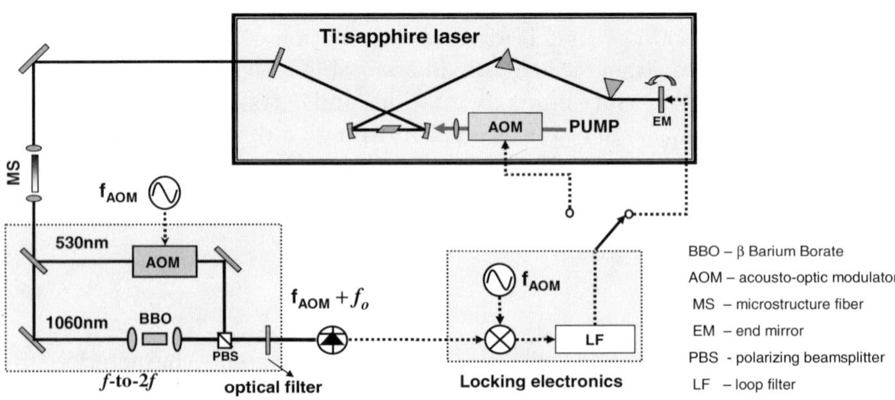

Fig. 1. Experimental apparatus of a carrier-envelope phase-stabilized fs Ti:S laser. The box in the lower left shows the ν–to–2ν interferometer used to measure f_0. An AOM is included in the ~530 nm arm of the interferometer to shift the frequency of one arm relative to the other such that f_0 may be set to zero.

Using the generated continuum the laser offset frequency is measured using a ν–to–2ν interferometer, which performs the necessary $N = 1$, $M = 2$ harmonic comparison (Fig. 1) [13]. The comb-offset frequency is measured as an rf signal that results from the optical heterodyne beat between the green portion (~530 nm) and the doubled near IR portion (~1060 nm) of the spectrum. Once the signal is measured, it is filtered, amplified and used in a feedback loop for comparison against a known reference. Negative feedback to stabilize f_0 is actuated by either tilting the laser end mirror after an intra-cavity prism sequence, or via amplitude modulation of the laser pump power (Fig. 1). For details about the feedback electronics and mechanisms and the repetition rate lock see Ref. [3].

The stability of the two characteristic frequencies is important in determining the quality of the optical comb. In metrology, the width of the comb lines presents a primary limit on precision. Given Eqn. 1, the dominant contribution to the optical linewidth results from fluctuations in f_{rep}. For ultrafast applications, however, it is small fluctuations in f_0 relative to f_{rep} that are important because they cause an accumulated phase error in ϕ_{CE}.

2 Carrier-envelope phase coherence

As mentioned above, the evolution of the CE phase is directly related to the laser offset frequency via Eqn. 2, yielding $\phi_{CE}(t) = 2\pi f_0 t + \phi_0$. If f_0 is stabilized to a frequency derived from the laser repetition rate, the value of f_0 fixes the pulse-to-pulse phase shift in the carrier-envelope phase, $\Delta\phi_{CE} = 2\pi\ f_0/f_{rep}$. The constant offset, ϕ_0, often termed the "absolute phase," determines the initial phase shift (at $t = 0$) between the carrier and the pulse envelope, making it an important parameter in field sensitive experiments. As a matter of course, fluctuations of f_0 are manifested as phase noise, $\Delta\phi(t)$, on ϕ_0. Therefore, for ultrafast experiments relying on the stability of the CE phase [4,14], knowledge of ϕ_{CE}'s stability is paramount as the dephasing of ϕ_0 determines the duration of a phase sensitive measurement. In this section, we present coherence time measurements of a stabilized Ti:S laser. These measurements, aside from determining the time scale over which the light pulses remain coherent, also give the quality of the servo system and aid in identifying the contributions due to the different noise sources within the stabilization loop.

Given the direct relationship between ϕ_{CE} and f_0, knowledge of the stability of the offset frequency directly yields that of the carrier-envelope phase. The stability of the offset frequency is determined from its lineshape. The noise analysis is straight forward since the spectrum of sidebands at frequencies relative to the carrier, ν, yield the power spectral density (PSD), $S_\phi(\nu)$, of the phase noise [18],

$$\Delta\phi_{rms}|_{\tau_{obs}} = \sqrt{2\int_{-\infty}^{-1/(2\pi\tau_{obs})} S_\phi(\nu)\ d\nu}. \tag{3}$$

Integration of the noise spectrum yields the total accumulated phase error on ϕ_{CE}, $\Delta\phi_{rms}$, due to frequency noise on f_0. Specifically, integration of $S_\phi(\nu)$

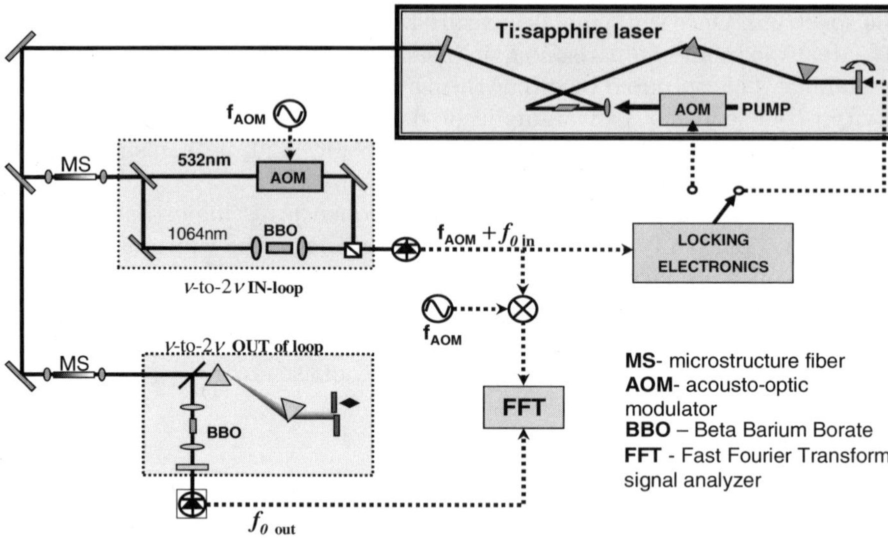

Fig. 2. Experimental setup showing how the coherence of ϕ_{CE} is measured. One interferometer is used to stabilize the laser while the second ν–to–2ν interferometer determines the phase coherence. The noise of the second interferometer is minimized by making the ν–to–2ν comparison as common mode as possible by using prisms for spectral dispersion.

up to an observation time, τ_{obs}, over which $\Delta\phi_{RMS}$ accumulates ~1 radian is generally taken to define the coherence time, τ_{coh}.

We determine the carrier-envelope phase coherence time of a stabilized Kerr-lens modelocked Ti:S laser capable of producing 10 fs pulses. The laser uses prisms for intra-cavity dispersion compensation [10]. The laser baseplate is temperature controlled and the laser is itself encased in a pressure sealed box. Negative feedback is obtained with a bandwidth of ~ 18 kHz via tilting the laser end mirror using a piezo-electric actuator (PZT). To perform an out-of-loop measurement of the offset frequency phase noise, we utilize two ν–to–2ν interferometers (Fig. 2) [8]. One interferometer stabilizes the laser, while the second determines the carrier-envelope phase noise from a phase sensitive measurment of f_0. The phase noise PSD is obtained by mixing f_0 down to base band where the noise sidebands are measured using a signal analyzer (see Fig. 2).

Figure 3 presents the results of the coherence measurement. A measurement of the unstabilized offset frequency and that of f_0 used for locking are included, comparison of the two indicates the noise suppression. The latter provides an in–loop phase noise measurement that is used only to determine the effectiveness of the stabilization circuitry. The difference between the out–of–loop and in–loop spectra yields the extra-cavity phase noise present within the stabilization loop (e.g., feedback electronics, the ν–to–2ν interferometer,

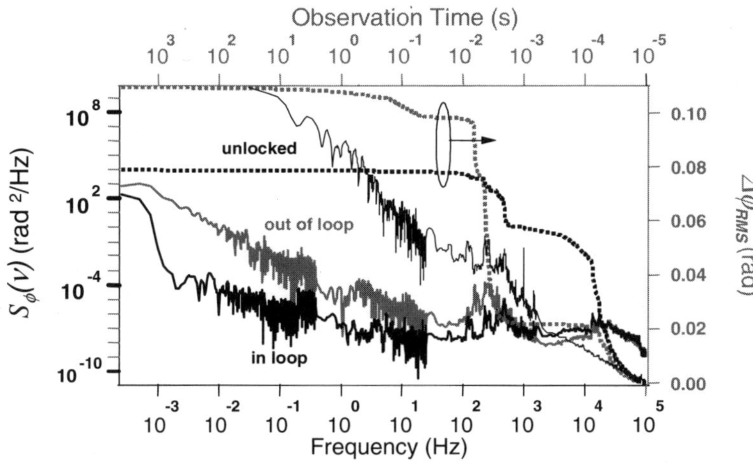

Fig. 3. (left axis)Phase power spectral density, $S_\phi(\nu)$, for the in-loop, solid line, and out-of-loop spectra, dashed line, of the comb offset frequency . (right axis)Integration of $S_\phi(\nu)$ yields the accumulated phase error as a function of observation time (top axis).

microstructure fiber, etc.) and the differential noise between the two loops. This noise is written onto the output of the laser, as the servo system uses the laser to compensate for extra-cavity noise. Integration of the phase noise PSD out–of–loop (in–loop), in Fig. 3 results in an accumulated phase error of 0.109 rad (0.08 rad) over the interval 102 kHz down to 488 mHz (resolution limited). Given that the out-of loop accumulated phase error is less than 1 rad, the lower frequency integration bound determines the coherence time, $\tau_{coh} = 1/(2\pi\ 244\mu\text{Hz}) = 652$ s.

Comparison between the unlocked and the in-loop phase noise PSD densities indicates a servo bandwidth $\sim$ 20 kHz. The in-loop PSD shows that additional phase noise results from the action of the servo loop at frequencies higher than 20 kHz, and that there is insufficient gain in the acoustic range ($\sim$100 Hz - 5 kHz). This frequency range is also responsible for the majority of the out–of–loop phase noise contribution. From Fig. 3, given that the out–of–loop accumulated phase error is less than 1 rad, the lower frequency integration bound determines the coherence time, $\tau_{coh} = 1/(2\pi\ 244\mu\text{Hz}) = 652$ s. This indicates that phase coherence is maintained for $>$ 65 billion pulses.

A major source of out–of–loop noise in the ν–to–2ν stabilization is the MS fiber, where amplitude noise on the laser output is converted to phase noise via the fiber nonlinear index of refraction [11]. This is detrimental when the direct output of the laser is to be used in an experiment, however it does not affect an experiment that uses the output of the microstructure fiber. A shortcoming of the dual interferometer method is that MS fiber is used in both interferometers, which because of common mode power fluctuations may

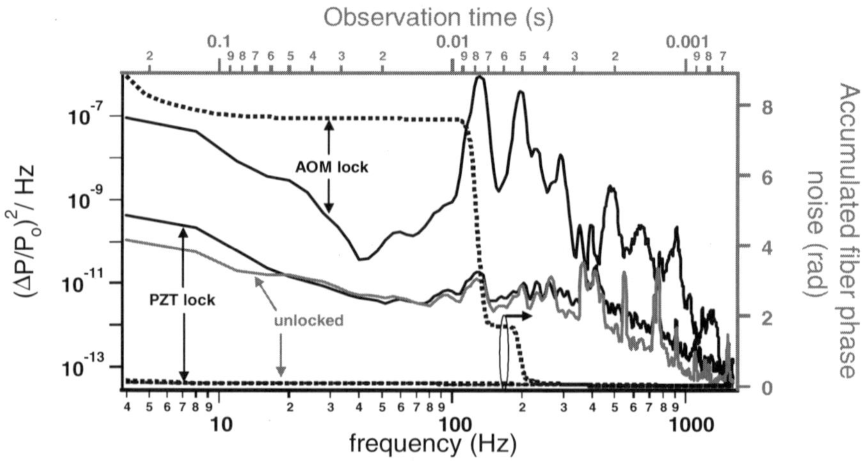

Fig. 4. (left axis) Phase noise spectrum for the unlocked laser, locked using an AOM in the pump beam and locked using the fast PZT. (right axis) Accumulated phase jitter obtained by integrating the spectra for the two locked cases.

result in a net cancellation of fiber generated noise in the out-of-loop measurement. To estimate the contribution of fiber noise, we measure the laser amplitude fluctuations and use them in conjunction with the amplitude-to-phase conversion factor for MS fiber [11]. This measurement is also used to compare the induced fiber noise by the PZT stabilization scheme described above to that obtained via modulation of the laser pump power [15]. Stabilization using the latter method is actuated by placing an AOM in the path of the pump beam for the Ti:S.

One drawback of modulating the pump power as a means for feedback to the Ti:S oscillator is the possibility of inducing fluctuations on the output power [12]. Figure 4 shows the amplitude noise PSD as well as the calculated accumulated fiber phase noise of the PZT versus AOM stabilized systems. A spectrum of the amplitude noise for the unstabilized laser is shown for comparison. As can be seen in Fig. 4, the PZT system contributes little additional amplitude noise during stabilization, whereas the opposite is true for the AOM stabilized laser. Integration of the AOM and PZT stabilized noise spectra from 8 Hz to 3.2 kHz yields the percent rms fractional laser power fluctuation, $(\Delta P/P_0)_{rms}$, to be 0.00473 and 8.34×10^{-5}, respectively. For a coupled laser power of 50 mW at 100MHz and an amplitude-to-phase conversion coefficient for MS fiber of 3784 rad/nJ [11], the amplitude noise would result in 8.78 rad of phase jitter for the AOM stabilized system. To connect back to the coherence measurement, the additional fiber noise contributed by the PZT stabilized system to the accumulated out-of-loop phase noise presented in Fig. 3 is determined to be 0.155 rad rms. This measured fiber noise then brings the total accumulated phase noise (fiber generated noise + out-of-loop phase noise (Fig. 3) to 0.264 rad.

3 Phase sensitive detection of ϕ_{CE}

In an ideal case, the phase of the f_0 signal includes the overall "absolute" phase ϕ_0. Since a phase-locked loop establishes a fixed phase, it seems that ϕ_{CE} is fully determined by the reference signal to which f_0 is locked, or that a phase sensitive measurement of the f_0 signal gives full knowledge of ϕ_{CE}. One limiting case is $f_0 = 0$ in which case ϕ_{CE} does not evolve pulse–to–pulse and thus is just given by ϕ_0. However, any ν–to–2ν interferometer is non-ideal and introduces an arbitrary phase shift that prevent this direct connection from being correct. This is illustrated in Fig. 5. The phase shifts in the ν–to–2ν interferometer arise from differences in the path length of the two arms and dispersion in any optical elements. (In principle an explicit interferometer is not needed if a chirp free pulse is used, however there is still unavoidable dispersion in the second harmonic crystal.)

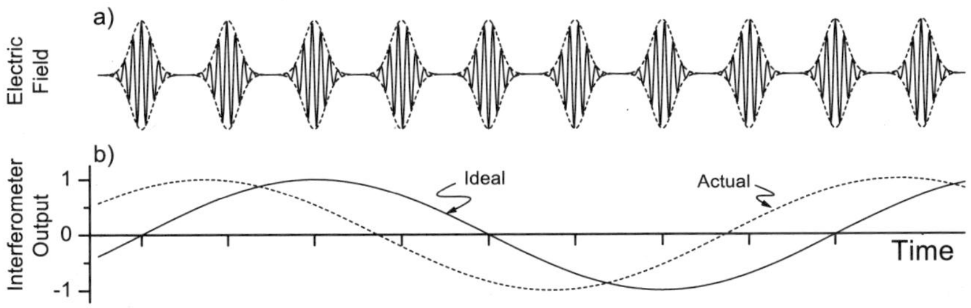

Fig. 5. Schematic showing relationship between pulse train and interferometer output. a) Pulse train showing pulse-to-pulse change in $\Delta\phi_{CE} = \pi/4$. b) Output of ν–to–2ν interferometer, solid line is the output of an ideal interferometer where zero signal is coincident with the pulse that has $\phi_{CE} = 0$. However, an actual signal from an interferometer has an arbitrary phase shift relative to the ideal signal.

Nevertheless it is still useful to make phase sensitive measurement of the f_0 signals in our dual interferometer setup [9]. With a dual-phase lock-in, it is possible to directly measure the phase difference between the reference channel and signal channel. As shown in Fig. 6, we phase lock f_0 to a reference frequency f_{ref} using one of the ν–to–2ν interferometers. As before, the acousto-optic modulator (AOM) in one arm facilitates locking f_0 at low frequencies as it offsets the beat note to higher frequencies. To be able to use a standard lock-in, we choose $f_{ref} \sim 100$ kHz. This signal is also provided to the reference channel of the lock-in. The output of the measurement ν–to–2ν interferometer is presented to the signal input of the lock-in. We then record the phase output of the lockin. We emphasize that this phase is a relative phase between the two channels, not ϕ_{CE}, and that ϕ_{CE} is evolving because $f_0 \neq 0$.

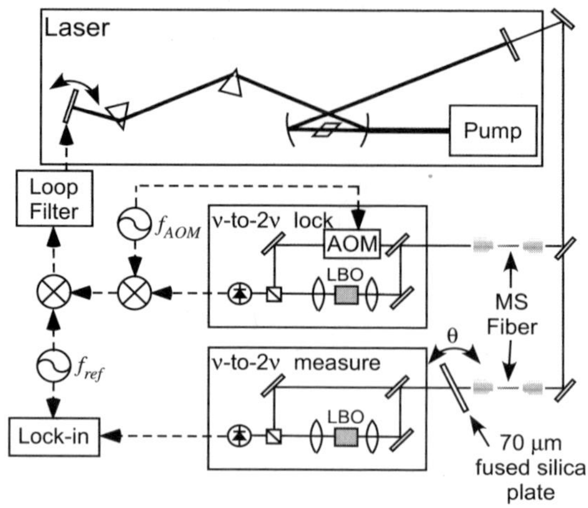

Fig. 6. Diagram of experiment used to monitor the relative ϕ_{CE} using a lock-in amplifier.

This measurement allows a direct visualization of the phase fluctuations. Figure 7 shows a 400 sec time record of the lock-in phase for an integration time of 100 ms. The standard deviation of the phase fluctuation is 3.8 deg rms (0.066 rad), with a maximum deviation of 20.8 deg (0.363 rad), which clearly shows that phase coherence is maintained. This result is related to the jitter measurement in section 2 by a Fourier transform where lower limit on the observation time is determined by the lock-in integration time. This measurement is more appropriate for measuring long time scale phase dynamics, where as the phase noise spectrum is better for characterizing short time scales.

This technique also promises to be very useful in searching for phase dependent processes. The extraordinary sensitivity of the phase sensitive detection performed by a lock-in allows very small signals to be retrieved. This is enabled by using a lock interferometer to establish a phase stable reference.

4 Extra-cavity adjustment of ϕ_{CE}

Using the same apparatus as in the previous section, we can demonstrate that the ν–to–2ν interferometer is able to track shifts in ϕ_{CE} and the capability of adjusting ϕ_{CE} outside the laser cavity by simple propagation through a dispersive optical element. Specifically, we insert a 70 μm thick fused silica

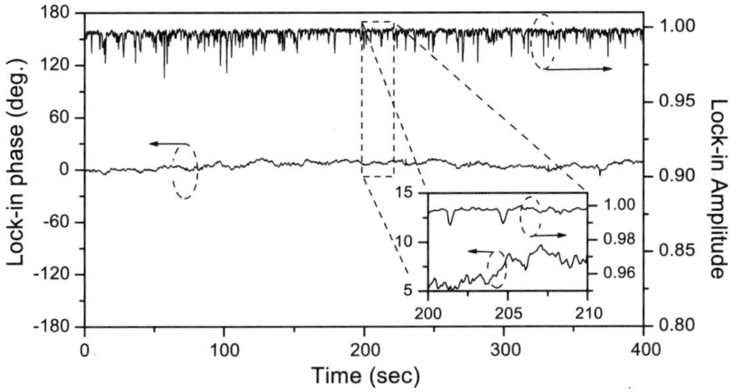

Fig. 7. Time record of the lock-in phase (lower trace, left axis) and amplitude (upper trace, right axis). The inset shows an enlargement to show that the amplitude and phase fluctuations are uncorrelated.

plate before the measurement interferometer. The plate is rotated to achieve a variable path length. The difference between group and phase velocities causes a shift in ϕ_{CE}. Because of the arbitrary phase shifts in the interferometers, we measure the change in the phase due to a change in the angle of the plate. We set the phase to be zero for normal incidence. The results are shown in Fig. 8. The curved line shows calculations based on the known dispersion of fused silica, which yields very good agreement with the experiment.

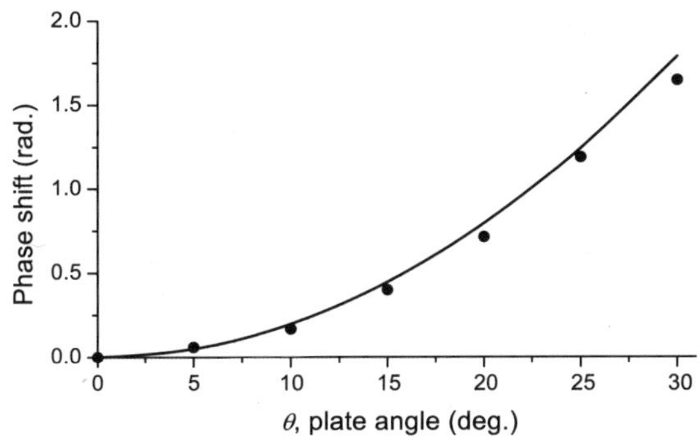

Fig. 8. Measured shift in ϕ_{CE} as a function of plate angle, θ, (circles). Line shows change ϕ_{CE} calculated from the Selemeier coefficients for fused silica

These results demonstrate that it is possible to impose a fixed shift in ϕ_{CE} and that it agrees well with the expected values. This capability will be useful as progress is made in experiments that are directly sensitive to ϕ_{CE} as it will allow systematic studies to be made. It also provides a means for correcting ϕ_{CE} using a feedback loop.

5 An octave spanning Ti:sapphire laser

Aside from the problems posed by fiber induced phase fluctuations, stabilization of Ti:S laser using MS fiber presents challenges to short pulse experiments because of fiber dispersion. Additionally, complexities in the fiber alignment often lead to loss of fiber coupling and degradation in the f_0 signal over time. This hinders optical frequency metrology since long term averaging is necessary to increase the measurement precision. Thus, a laser that directly generates an octave is preferable. In this section we present an octave spanning, conventional geometry Ti:S laser using intra-cavity prisms and negatively chirped mirrors. This has the advantage over a previously demonstrated octave spanning laser [6] in that the laser does not require precise intra-cavity dispersion compensation, nor does it require the use of an auxiliary space and time focus. We support the definition of octave spanning by demonstrating stabilization of the carrier-envelope phase using the bandwidth from the laser alone.

The octave spanning laser presented here is an x-folded cavity that utilizes CaF_2 prisms and commercially available negative chirped mirrors [7] for intra-cavity dispersion compensation (Fig. 9). The generation of intra-cavity continuum is obtained via optimization of self-phase modulation (SPM) in the laser crystal. The latter is obtained by strong misalignment of the curved mirrors away from the optimal cw position, which results in the production of a highly asymmetric and highly focused cw beam. When pumped with 5.5 W of 532 nm light the spectrum spans from 580 nm to 1200 nm (40 dB down from the 800 nm portion of the spectrum) with an average power of 400 mW (100 mW, cw) at a $\sim$ 88 MHz repetition rate [7].

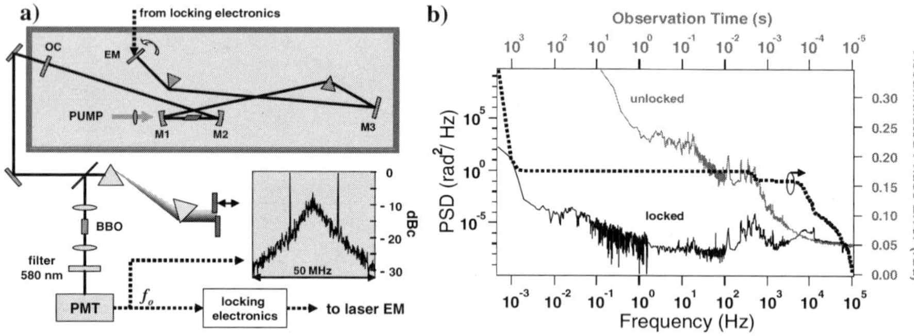

Fig. 9. (a) Experimental schematic of the Ti:S laser and the ν–to–2ν interferometer used for measurement of the laser offset frequency. The inset shows the rf spectrum of f_0 at a 100 kHz resolution bandwidth. (b) Phase noise power spectral density (PSD) of the stabilized f_0 versus frequency (left and bottom axes) and the integrated phase error as a function of observation time (right and top axes).

To demonstrate that the spectrum is indeed octave spanning, we measure f_0 using the laser output directly, i.e., without external broadening. This is done with a ν–to–2ν interferometer that uses prisms for spectral dispersion (not for compression) as shown in Fig. 9. The beat signal is detected using a fast photomultiplier tube (PMT) and yields a maximum signal–to–noise ratio of 30 dB at 100 kHz resolution bandwidth. This signal is then used to stabilize the laser using the PZT scheme described in section 2. Figure 9 presents the stabilization results. The phase noise PSD presented is an in-loop measurement of the offset frequency, which, as explained previously, may not reflect the total noise on the laser output. However, the use of the octave spanning laser eliminates MS fiber noise. Interferometer noise is also minimized by making the ν–to–2ν comparison as common mode as possible. As a result, the main contribution to the out-of-loop phase noise should stem from the feedback circuitry.

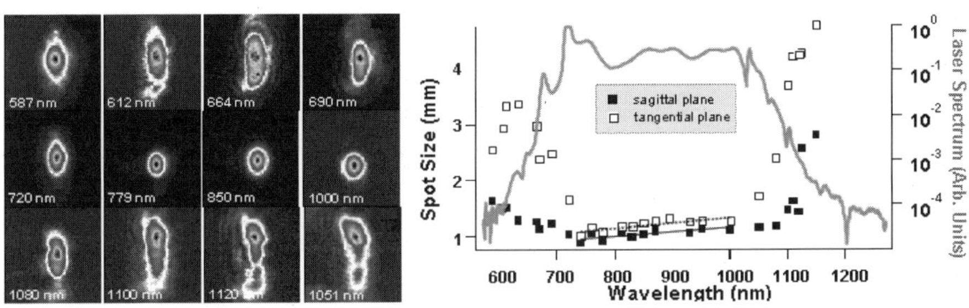

Fig. 10. The laser beam profile is displayed for selected wavelengths. A 10/90 knife edge fit was used to determine the spot sizes in the sagittal (filled squares) and tangential (empty squares) planes, displayed at the right of the figure. The solid and dashed lines are the respective fits to the expected $1/\lambda$ diffraction limit divergence. (Some of the diffraction rings observed for the larger beam modes may result from aperturing of the laser mirrors.)

An interesting aspect of the generated continuum is the laser beam spatial profile as a function of wavelength [1]. Because of the extreme breadth of the spectrum, light generated in the spectral wings is not resonant in the cavity and thus is not forced to obey the cavity transverse spatial mode conditions. This results in the production non-Gaussian modes (Fig. 10), which cause poor mode-matching between the ν and 2ν portions of the spectrum. The change in spot size of the beam as a function of wavelength, observed in Fig. 6, is the result of a sudden decrease in waist size for light in the wings of the spectrum, as shown in Fig. 11a. The beam waist is obtained from an M^2 measurement of the light at the output coupler. Because the beam parameters are derived using Gaussian beam propagation theory, they are simply meant to indicate a trend versus wavelength. The value of M^2 is obtained by comparing the ratio of measured divergence (Fig. 11b) with the

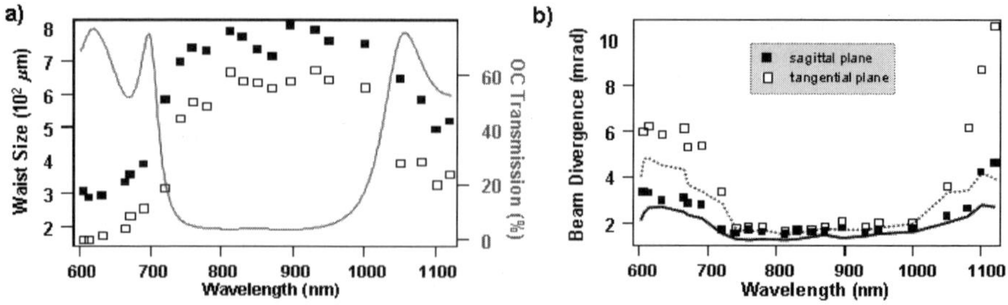

Fig. 11. (a) Measured beam waists and (b) measured beam divergence at the laser output coupler of the laser in the sagittal (filled squares) and tangential planes (empty squares) versus wavelength. Also show in (a), the output coupler transmission (right axis) is included to indicate the bandwidth of the resonant versus nonresonant modes. In (b) the solid and dotted lines are the calculated TEM_{00} beam divergences from the measured waist sizes [shown in (a)] in the sagittal and tangential planes, respectively.

divergence calculated for a TEM_{00} beam using the respective measured waists in Fig. 11a. This value indicates the strength of the non-ideal Gaussian mode propagation (higher order or non-Gaussian modes.)

6 Summary

This chapter has described a small contribution to remarkable advances that have been possible over the last few years by combining ultrafast lasers with frequency domain techniques for stablizing CW lasers. Further advances are opening new possibilities for controlling the electric field at optical frequencies.

Acknowledgments. The authors would like to acknowledge the contributions of J. Hall, and R. Windeler to this work. Funding is provided by NIST, NSF, DARPA and ONR.

References

1. S. T. Cundiff, W. H. Knox, E. P. Ippen, and H. A. Haus. Frequency-dependent mode size in broadband kerr-lens mode locking. *Opt. Lett.*, 21(9):662–664, 1996.
2. S. T. Cundiff. Phase stabilization of ultrashort optical pulses. *J. Phys. D-Appl. Phys.*, 35(8):R43–R59, 2002.
3. S. T. Cundiff, J. Ye, and J. L. Hall. Optical frequency synthesis based on mode-locked lasers. *Rev. Sci. Instr.*, 72(10):3749–3771, 2001.
4. P. Dietrich. Determining the absolute carrier phase of a few-cycle pulse. *Opt. Lett.*, 25(1):16–18, 2000.

5. S. A. Diddams, T. Udem, J. C. Bergquist, E. A. Curtis, R. E. Drullinger, L. Hollberg, W. M. Itano, W. D. Lee, C. W. Oates, K. R. Vogel, and D. J. Wineland. An optical clock based on a single trapped hg-199(+) ion. *Science*, 293(5531):825–828, 2001.
6. R. Ell, U. Morgner, F.X. Kärtner, J. G. Fujimoto, E. P. Ippen, V. Scheuer, G. Angelow, T. Tschudi, M.J. Lederer, A. Boiko, and B. Luther-Davies. Generation of 5 fs pulses and octave-spanning spectra directly from a ti:sapphire laser. *Opt. Lett.*, 26:373–375, 2001.
7. T. M. Fortier, Jones David. J., and Cundiff S. T. Octave spanning conventional geometry ti:sapphire laser. *Submitted to Opt. Lett.*
8. T. M. Fortier, D. J. Jones, J. Ye, S. T. Cundiff, and R. S. Windeler. Long-term carrier-envelope phase coherence. *Opt. Lett.*, 27(16):1436–1438, 2002.
9. T. M. Fortier, D. J. Jones, J. Ye, and S. T. Cundiff. Control over absolute carrier-envelope phase of ultrafast pulses: complete waveform synthesis. In *Nonlinear Optics*, Maui, 2002.
10. R. L. Fork, O. E. Martinez, and J. P. Gordon. Negative dispersion using pairs of prisms. *Opt. Lett.*, 9(5):150–152, 1984.
11. T. M. Fortier, J. Ye, S. T. Cundiff, and R. S. Windeler. Nonlinear phase noise generated in air-silica microstructure fiber and its effect on carrier-envelope phase. *Opt. Lett.*, 27(6):445–447, 2002.
12. F. W. Helbing, G. Steinmeyer, U. Keller, R. S. Windeler, J. Stenger, and H. R. Telle. Carrier-envelope offset dynamics of mode-locked lasers. *Opt. Lett.*, 27(3):194–196, 2002.
13. D. J. Jones, S. A. Diddams, J. K. Ranka, A. Stentz, R. S. Windeler, J. L. Hall, and S. T. Cundiff. Carrier-envelope phase control of femtosecond mode-locked lasers and direct optical frequency synthesis. *Science*, 288(5466):635–639, 2000.
14. G. G. Paulus, F. Grasbon, H. Walther, P. Villoresi, M. Nisoli, S. Stagira, E. Priori, and S. De Silvestri. Absolute-phase phenomena in photoionization with few-cycle laser pulses. *Nature*, 414(6860):182–184, 2001.
15. A. Poppe, R. Holzwarth, A. Apolonski, G. Tempea, C. Spielmann, T. W. Hänsch, and F. Krausz. Few-cycle optical waveform synthesis. *Appl. Phys. B*, 72(8):977–977, 2001.
16. J. K. Ranka, R. S. Windeler, and A. J. Stentz. Visible continuum generation in air-silica microstructure optical fibers with anomalous dispersion at 800 nm. *Opt. Lett.*, 25(1):25–27, 2000.
17. H. R. Telle, G. Steinmeyer, A.E. Dunlop, J. Stenger, D.H. Sutter, and U. Keller. Carrier-envelope offset phase control: A novel concept for absolute optical frequency control and ultrashort pulse generation. *Appl. Phys. B*, 69:327–332, 1999. STC.
18. J. Ye, S. T. Cundiff, S. Foreman, T. M. Fortier, J. L. Hall, K. W. Holman, D. J. Jones, J. D. Jost, H. C. Kapteyn, K. A. H. V. Leeuwen, L. S. Ma, M. M. Murnane, J. L. Peng, and R. K. Shelton. Phase-coherent synthesis of optical frequencies and waveforms. *Applied Physics B*, 74:S27–S34, 2002.
19. J. Ye, L.-S. Ma, and J. L. Hall. Molecular iodine clock. *Physical Review Letters*, 87(27):270801, 2001.

Frontiers of Carrier-Envelope Stabilization Schemes for Femtosecond Oscillators

Günter Steinmeyer[1,2], Florian W. Helbing[1], and Ursula Keller[1]

[1] Institut für Quantenelektronik, ETH Zürich Hönggerberg, CH-8093 Zürich, Switzerland

[2] *present address*: Max-Born-Institut für Nichtlineare Optik und Kurzzeitspektroskopie, Max-Born-Straße 2a, D-12489 Berlin, Germany
steinmey@mbi-berlin.de

1 Measurement of the carrier-envelope offset phase

With the advances of ultrafast optics towards ever shorter pulses, a regime has been entered where the relative phase between the carrier and the envelope of an ultrashort pulse starts to matter [1, 2]. The currently most widespread scheme to measure the carrier-envelope offset (CEO) frequency relies on heterodyning different harmonics from one and the same broadband spectrum, as originally proposed in [3]. The spectrum of a mode-locked laser consists of an equidistantly spaced frequency comb with an offset f_{CEO} at zero frequency,i.e. $f_i = f_{\mathrm{CEO}} + i f_{\mathrm{rep}}$. The comb spacing is given by the repetition rate f_{rep}. Taking any frequency component f_{2i} from the high frequency part of the comb and heterodyning it, e.g., with the second harmonic of a component f_i from the low-frequency part then isolates the CEO-frequency $f_{\mathrm{CEO}} = 2f_i - f_{2i} = f_{\mathrm{CEO}}$. While the f-to-$2f$ beat is the most straightforward scheme to extract the CEO-frequency it is also the most demanding scheme in terms of bandwidth as it requires an octave-spanning spectrum. Other laser harmonics can be used at slightly lower bandwidth requirements. For a more general treatment the reader is referred to [3, 4].

For measurement of the CEO-frequency, two different approaches have evolved. Either, one additionally broadens the laser spectrum by continuum generation in a microstructure fiber as in the first experimental realizations [5, 6], or a laser with extremely broad bandwidth is used directly to generate the beat note [7]. The latter approach may seem preferable at first sight. But given the spectral width of even the most broadband laser oscillators demonstrated to date, the power density at half an octave off pulse center is already several 10 dB lower than the peak power density. Therefore it is much more challenging to extract a beat note with sufficient signal-to-noise allowing for a cycle-slip free measurement or even for a stabilization of the CEO-frequency. So far, a successful phase lock without using external spectral broadening has only recently been reported [8]. The authors used a $2f$-to-$3f$ scheme, which relaxes the constraints on spectral bandwidth compared to the most commonly used f-to-$2f$ scheme.

On the other hand, the weak point of using microstructure fiber for extracavity broadening is amplitude-to-phase noise conversion processes. Amplitude-to-phase noise conversion has been reported for laser oscillators [9,10]. This coupling process is actually used to control the CEO frequency of an oscillator via the pump power. Various mechanisms have been suggested as the cause of this coupling. One mechanism is residual uncompensated cavity dispersion acting together with a shift of the laser spectrum [11]. Another process is coupling via the Kerr nonlinearity of the laser crystal [10], causing a peak power dependent shift of the roundtrip CEO phase of the cavity. Both processes cause conversion of pump power noise into fluctuations of the CEO frequency.

If the same amplitude-phase-coupling mechanisms also play a role in the white-light continuum process used for external broadening, one would also expect that a slight variation of the launched power in the fiber causes a CEO phase shift. For applications of the carrier-envelope stabilization it is important to characterize the order of magnitude of amplitude-to-phase coupling mechanisms in the microstructure fiber. Recently, a phase-lock of the CEO-frequency to an external resonator with only a few mrad residual jitter in a 0.01 Hz-1 kHz bandwidth has been

demonstrated [10]. This phase jitter corresponds to a timing jitter of only 10 attoseconds, which is clearly the lowest timing jitter ever reported for any kind of stabilization. This extremely small value has to be put into relation with spurious noise contributions from the microstructure fiber itself.

2 Differential CEO phase measurements

To address the issue of noise contributions from the microstructure fiber, we set up the differential measurement scheme depicted in Fig. 1. Our scheme is similar to the out-of-loop measurement on a stabilized oscillator [12], but only requires one microstructure fiber. Furthermore, our scheme does not require any servo control of the laser, which may adversely affect intracavity power stability. Our scheme generates two independent CEO beat notes, using only one microstructure fiber. For one beat note, the second harmonic of a supercontinuum component at about 1060 nm is heterodyned with another short-wavelength continuum component at 530 nm. For the second measurement branch, the second harmonic of a frequency slice from the long-wavelength laser spectrum at about 900 nm is heterodyned with a continuum component at 450 nm. The important point to realize is that in the former scheme both frequency components are generated in the microstructure fiber whereas the latter scheme measures the relative phase between a laser frequency component and a continuum component. Therefore the difference between both independent CEO phase measurement schemes is susceptible to a linear drift of the path length of the microstructure fiber as well as to amplitude-to-phase coupling effects inside the microstructure fiber.

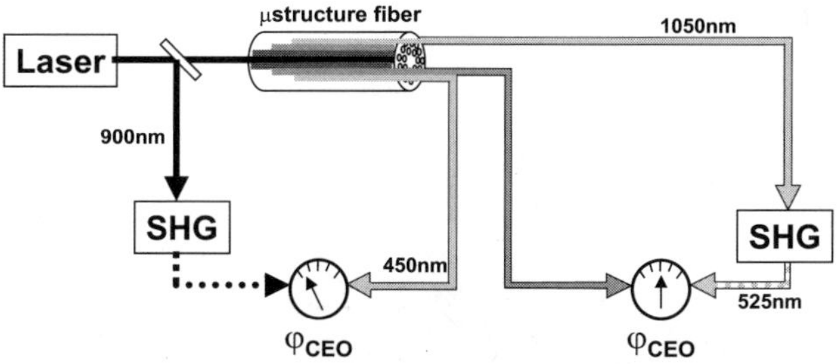

Fig. 1. Illustration of the differential CEO-phase measurement scheme. In one branch, a second harmonic and a fundamental frequency component of the fiber continuum are heterodyned. The second branch uses heterodyning of fundamental laser frequencies and the second harmonic from the fiber.

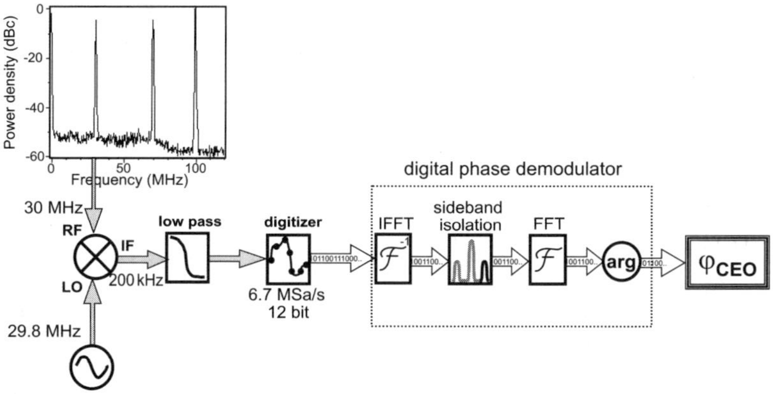

Fig. 2. Schematic illustration of the phase demodulation used to retrieve the CEO phase in the two branches of the set-up in Fig. 1. The measured beat signal is first transferred to an intermediate frequency of 200 kHz. The demodulation is digitally extracted from the IF signal employing the phase retrieval algorithm from [13].

As amplitude-to-phase conversion processes inside the microstructure are in the focus of our examinations, we must not use active control of the CEO frequency via modulation of the pump power. The power fluctuations induced by the feedback could potentially induce amplitude-to-phase conversion in the fiber by itself, which would defeat the purpose of our examination. Therefore, we operated the laser without any active CEO-frequency stabilization, but took measures to ensure a high degree of passive stability. The free-running laser exhibits a passive stability of the CEO frequency of a few 10 kHz over 1 minute. We extract the relative phase difference between our two measurement branches by first frequency shifting the two CEO beat notes from about 30 MHz to an intermediate frequency (IF) of 200 kHz (see Fig. 2). Both IF signals are digitized at 6.7 MSa/s with 12 bit resolution. Phase retrieval is then possible using the Fourier filtering technique described in [13]. The retrieved

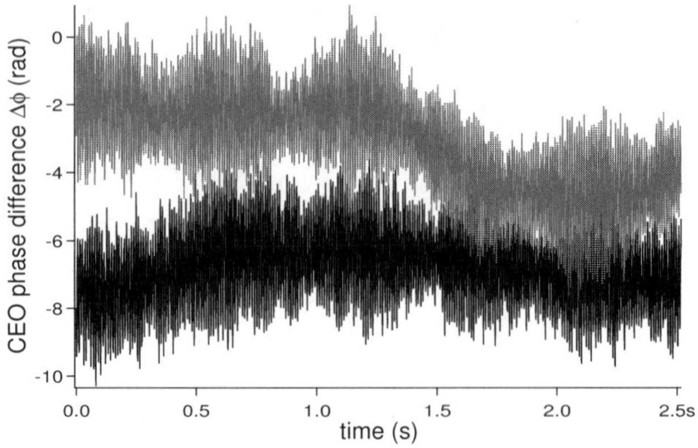

Fig. 3. Two traces of the phase difference $\Delta\varphi_{\mathrm{CEO}}$ without pump power modulation. The phase drifts exhibits a slow drift on the order of 1 rad/s. An additional oscillation with an amplitude of about one radian is caused by amplitude-to-phase coupling from a pronounced low frequency laser noise contribution.

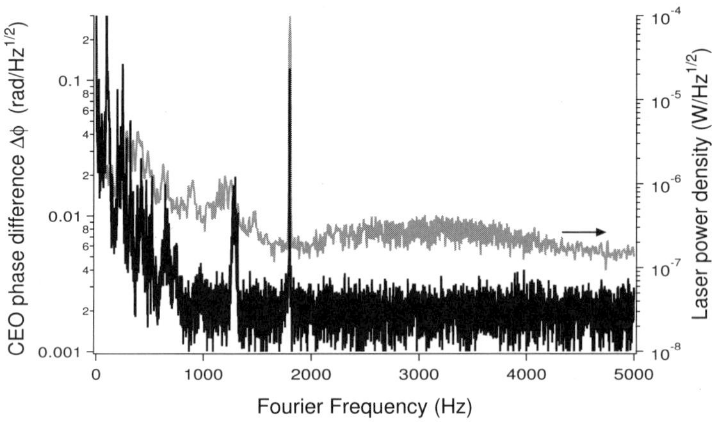

Fig. 4. Spectra of the differential CEO phase noise $\Delta\varphi_{\mathrm{CEO}}$ (black trace) in comparison to the noise spectrum of laser power (gray trace).

phases of both measurement branches are then subtracted from each other, which finally yields the differential CEO phase. Example measurements are displayed in Figs. 3 and 4. Figure 3 shows two independent measurements of the temporal evolution of the differential CEO phase $\Delta\varphi_{\mathrm{CEO}}$ for an observation time of 2.5 s. In these measurements, one can see a rapid modulation and a slow drift component, which is on the order of 1 rad/s. The rapid modulation is caused by amplitude-to-phase conversion processes. We attribute the drift to a slow variation of optical path lengths, both in the interferometers used for the detection of the CEO beats and also to a length drift of the fiber itself. Figure 4 shows a noise spectrum of a differential CEO measurement in comparison the amplitude noise spectrum of the laser. Taking into account that both measurements were not recorded exactly simultaneously, one can see a similarity of the noise distribution up to frequencies of about 1 kHz. The spectrum also shows a strong 1.8 kHz noise line, which goes back to active modulation of the pump power at this particular frequency. The strength of the differential CEO signal at 1.8 kHz allows for an estimation of the amplitude-to-phase coupling effect as 1 rad per mW change of the average laser output power. This translates to about 0.2 rad/mW in the microstructure fiber.

3 Conclusion

From the measurements, one can identify a linear and a nonlinear optical contribution to the differential CEO phase signal — interferometer drift and amplitude-to-phase coupling, respectively. From the differential CEO-phase signal induced by the 1.8 kHz pump modulation, it has to be concluded that this differential signal arises from the nonlinearity of the microstructure fiber. Nevertheless, this effect is orders of magnitude weaker than the influence of pump power modulation on the CEO frequency of the laser itself. To one part, the slower drift components stem from optical path length fluctuations in the two separate f-to-$2f$ interferometers. In our scheme an additional drift also comes from the fiber, e.g. induced by a slow temperature drift. We therefore expect the measured drift value of 1 rad/s to be an upper estimate for the severity of the drift problem. Optical frequency metrology employs frequencies on the order of 10^{15} rad/s. Therefore, the measured drift value would only play a role if the precision of frequency measurements advances well into the sub-10^{-15} regime.

In closing, it seems useful to point out that both contributions, interferometer drift and nonlinear amplitude-to-phase coupling, can be widely avoided. The nonlinear contribution vanishes if the same continuum is used for measurement of the CEO-frequency and for the experiment. One should therefore avoid to perform a CEO-sensitive experiment directly with a mode-locked laser and use additional spectral broadening only for measurement of the CEO-phase. However, this requires recompression of the continuum for CEO-sensitive experiments with short pulses.

The problem of amplitude-to-phase coupling effects in the microstructure fiber is also a further encouragement for directly obtaining the octave spectrum from a laser and totally avoiding external broadening. Nevertheless, even such an approach would not avoid corruption from a slow interferometer drift. Drift can be easily avoided by switching to a common-path interferometer, as first suggested by [14]. The idea here is to avoid splitting of the f and the $2f$ component into geometrically separated arms, e.g., as in a Mach-Zehnder interferometer. When the fundamental and the SHG are collinear, one only has to compensate for group delay dispersion between fundamental and SHG to create a temporal overlap of the two. In the experiments described by Kakehata, compensation for group delay effects could be suitably avoided by usage of an extremely thin optical crystal. For experiments with mode-locked oscillators, it may be necessary to compensate the group delay between fundamental and SHG by suitably designed chirped mirrors or similar devices.

The potential problems pointed out in our experiments can be avoided by suitable design. These measures may help to reach even higher precision in frequency metrology and advanced control of the CEO phase in extreme nonlinear optics.

References

1. G. G. Paulus, F. Grasbon, H. Walther, P. Villoresi, M. Nisoli, S. Stagira, E. Priori, and S. De Silvestri, *Nature* **414**, 182–184 (2001).
2. A. Baltuška, T. Udem, M. Uiberacker, M. Hentschel, E. Goulielmakis, C. Gohle, R. Holzwarth, V. S. Yakoviev, A. Scrinzi, T. W. Hänsch, and F. Krausz, *Nature* **421**, 611–615 (2003).
3. H. R. Telle, G. Steinmeyer, A. E. Dunlop, J. Stenger, D. H. Sutter, and U. Keller, *Appl. Phys. B* **69**, 327–332 (1999).
4. F. W. Helbing, G. Steinmeyer, and U. Keller, *IEEE J. Sel. Top. Quantum. Electron.* **9**, scheduled for publication May/June 2003
5. D. J. Jones, S. A. Diddams, J. K. Ranka, A. Stentz, R. S. Windeler, J. L. Hall, and S. T. Cundiff, *Science* **288** 635–639 (2000).
6. A. Apolonski, A. Poppe, G. Tempea, C. Spielmann, T. Udem, R. Holzwarth, T. W. Hänsch, and F. Krausz, *Phys. Rev. Lett. 85*, 740–743 (2000).
7. U. Morgner, R. Ell, G. Metzler, T. R. Schibli, F. X. Kärtner, J. G. Fujimoto, H. A. Haus, and E. P. Ippen, *Phys. Rev. Lett.* **86**, 5462–5465 (2001).
8. T. M. Ramond, S. A. Diddams, L. Hollberg, and A. Bartels, *Opt. Lett.* **27**, 1842–4 (2002).
9. A. Poppe, R. Holzwarth, A. Apolonski, G. Tempea, C. Spielmann, T. W. Hänsch, and F. Krausz, *Appl. Phys. B* **72**, 373–6 (2001).
10. F. W. Helbing, G. Steinmeyer, U. Keller, R. S. Windeler, J. Stenger, and H. R. Telle, *Opt. Lett.* **27**, 194–6 (2002).
11. L. Xu, C. Spielmann, A. Poppe, T. Brabec, F. Krausz, and T. W. Hänsch *Opt. Lett.* **21**, 2008–10 (1996).
12. T. M. Fortier, J. Ye, S. T. Cundiff, and R. S. Windeler, *Opt. Lett.* **27**, 445–7 (2002).
13. M. Takeda, H. Ina, and S. Kobayashi, *J. Opt. Soc. Am. B* **72**, 156 (1982).
14. M. Kakehata, H. Takada, Y. Kobayashi, K. Torizuka, Y. Fujihira, T. Homma, and H. Takahashi, *Opt. Lett.* **26**, 1436–8 (2001).

Direct Locking of the Carrier-Envelope Phase of Femtosecond Laser Pulses

Tae Jun Yu[1,2], Jong-Min Lee[2], Yong Soo Lee[1], Kyung-Han Hong[1], and Chang Hee Nam[1]

[1] Coherent X-ray Research Center and Dept. of Physics, KAIST, Daejeon 305-701, Korea
[2] Advanced Photonics Research Institute, K-JIST, Gwangju 500-712, Korea

1 Introduction

For the last several years, measurement and stabilization of the shot-to-shot carrier-envelope-phase (CEP) slip of mode-locked laser pulses have been intensively investigated. The CEP of femtosecond laser pulses can be measured using a nonlinear interferometry scheme, called f-to-$2f$ self-referencing interferometer [1,2], and amplified laser pulses with a low repetition rate have been explored using spectral interferometry techniques [3]. The CEP stabilization technique based on the f-to-$2f$ self-referencing interferometer has made a great impact on the optical frequency metrology because this technique allows the generation of a stabilized optical frequency comb in hundreds of THz range, which can be used as a simple but highly precise frequency standard [5]. The f-to-$2f$ self-referencing interferometer yields a heterodyne beat signal between the second harmonic (SH) of f-component and the $2f$ component in a one-octave spectrum, which is generated either directly from the mode-locked laser [4] or by the spectral broadening in optical fibers [6, 7]. The beat signal gives the carrier-envelope offset frequency f_{ceo}, given by $\Delta\phi/(2\pi\tau)$ with $\Delta\phi$ being the shot-to-shot CEP slip and τ the time interval between successive pulses, which can be used for the CEP stabilization.

The stabilization of the shot-to-shot CEP has been achieved by locking f_{ceo} to an external reference frequency f_{ref} using a phase-locked loop (PLL). Here f_{ceo} should be stabilized to zero frequency in order to make successive pulses have the same CEP, generating CEP-locked laser pulses. Since the PLL method intrinsically requires a non-zero f_{ref}, this problem has been solved by inserting an acousto-optic modulator (AOM) with a modulation frequency f_{AOM} at the blue (2f) arm of the f-to-$2f$ interferometer to shift the optical frequency of the blue arm relative to that of the red (f) arm [5]. As a result, the heterodyne beat with a frequency of $f_{ceo} - f_{AOM}$ may be measured at the modified f-to-$2f$ self-referencing interferometer. If f_{ref} is set exactly to f_{AOM}, f_{ceo} is logically locked to zero after stabilization. This is a complicated method that stabilizes the CEP in the frequency domain and, moreover, it is difficult to set the CEP at any arbitrary value. Here, we report a technique for direct CEP locking of femtosecond laser pulses, i.e., the generation of

CEP-frozen femtosecond laser pulses, using a simple time-domain feedback method combined with an f-to-$2f$ self-referencing interferometer.

2 Theory

The idea of our time-domain approach is based on the mathematical treatment of the heterodyne beat signal that appears in a conventional f-to-$2f$ self-referencing interferometer. Mode-locked femtosecond pulses are, first, spectrally broadened so as to have a one-octave spectrum after the propagation through an air-silica microstructure fiber (MSF). The electric field of the n-th pulse after the fiber can be described in the time and frequency domains as, respectively:

$$E_n(t) = A(t)\, e^{i[\omega_c t + \phi_{NL}(t) + \phi_n]}, \tag{1}$$

$$\tilde{E}_n(\omega) = \tilde{A}(\omega)\, e^{i\phi_n}, \tag{2}$$

where $A(t)$ and $\tilde{A}(\omega)$ are the temporal envelope and the complex spectral distribution of the of the n-th electric field, respectively. ω_c is the carrier frequency and $\phi_{NL}(t)$ is a nonlinear phase shift due to the self-phase modulation and other nonlinear effects in MSF. The CEP of the n-th pulse is given by $\phi_n = \phi_{n-1} + \Delta\phi_n$, where $\Delta\phi_n$ is the shot-to-shot CEP slip owing to the intracavity dispersion. $A(t)$, $\tilde{A}(\omega)$, and $\phi(t)$ are independent of n under the condition of the stable mode-locking and the fixed coupling efficiency in the MSF.

For the observation of the heterodyne beat, both the $2f$-component and the second harmonic of the f-component are spectrally filtered out by an infinitely narrow bandpass filter with a transfer function of $\delta(\omega - \omega_{2f})$. Resultant electric fields are given by

$$\tilde{E}_n^{2f}(\omega) = \tilde{E}_n(\omega) \cdot \delta(\omega - \omega_{2f}) \tag{3}$$

$$\tilde{E}_n^{f}(\omega) = -i\eta \exp(-\pi/2) \cdot \tilde{E}_n^{2}(\omega/2) \cdot \delta(\omega - \omega_{2f}) \tag{4}$$

where η is the second harmonic efficiency in terms of field amplitude and $\tilde{E}_n^{f}(\omega)$ contains a frequency-independent phase shift of $-\pi/2$ as predicted by Maxwell's equations. Then the electric fields in the time domain at two arms of the interferometer are given by inverse Fourier transformations of Eqs. (3) and (4):

$$E_n^{2f}(t) = \tilde{A}(\omega_{2f}) \cdot e^{i(\omega_{2f} t + \phi_n)}, \tag{5}$$

$$E_n^{f}(t) = -i\eta \cdot \tilde{A}^2(\omega_f) \cdot e^{i(\omega_{2f} t + 2\phi_n)}, \tag{6}$$

where ω_f is $\omega_{2f}/2$. The temporal overlap between $E_n^f(t)$ and $E_n^{2f}(t)$ generates an interference that depends on the CEP, ϕ_n. The heterodyne beat signal is given by

$$I_n(t) = |E_n^f(t) + E_n^{2f}(t)|^2 = \alpha + \beta \sin(\phi_n + \phi_0) \tag{7}$$

where α, β, and ϕ_0 are the dc offset,$|\tilde{A}(\omega_{2f})|^2 + \eta^2|\tilde{A}^2(\omega_f)|^2$, the amplitude of the pure interference signal, $2\eta \cdot |\tilde{A}^*(\omega_{2f}) \cdot \tilde{A}^2(\omega_f)|$, and an arbitrary phase shift, $Arg[\tilde{A}^*(\omega_{2f}) \cdot \tilde{A}^2(\omega_f)]$, respectively. Eq. (7) is dependent only on the sequence of pulse train, n, corresponding to the stream of time. As the CEP can be expressed as $\Delta\phi_n t/2\pi$, in which $\Delta\phi_n$ is not constant of n in an unstabilized laser, ϕ_n is given as a time-dependent phase, $\phi(t)$. The heterodyne beat signal is then given by $\alpha + \beta \sin(\phi(t))$ with an arbitrary phase shift ϕ_0. The zero CEP slip or constant $I(t)$ implies that the CEP is locked to a constant. After removing the dc offset signal α and obtaining the error signal, $I_{err}(t) = \beta \sin(\phi(t))$, we can directly use this as a feedback signal to keep $I(t)$ at a constant dc level.

3 CEP stabilization

The experimental setup for the CEP measurement and stabilization is illustrated in Fig. 1. Sub-10-fs pulses generated in a mirror-dispersion-controlled Ti:sapphire laser were launched into an MSF [6] to generate a one-octave spectrum, and the generated continuum was spectrally separated using a harmonic separator. The red (f) portion at the wavelength of 1064 nm was frequency-doubled using a 1-mm-thick KTP crystal and mixed with the blue ($2f$) portion at 532 nm. Since both beams have broad spectra around 532 nm, they are filtered by using two interference filters centered at 532 nm with a bandwidth of 10 nm. The delay arm in Fig. 1 ensured the temporal overlap of the two quasi-continuous pulses at 532 nm. The two beams were combined in a polarizing beam splitter to be spatially overlapped. The half-wave plates were used to correct a slight rotation of the polarization at the fiber. Finally, the beat signal of Eq. (7) was detected with an avalanche photodiode ($APD3$) in Fig. 1. The APD detected only the beat signal as the photodiode with a bandwidth of 100 kHz did not respond to the 100-MHz laser pulse train.

The dc component of the beat signal,α, should be constant in time, but it varies in time due to the fluctuations of fiber coupling efficiency and air flow. To remove the fluctuations in α, two APDs were additionally used. The time-dependent background signals detected in APD2 (proportional to $|\tilde{A}(\omega_{2f})|^2$) and APD3 (proportional to $\eta^2|\tilde{A}^2(\omega_f)|^2$) were subtracted from that of APD_{f-2f} to generate a pure ac beat signal, i.e., $\beta \sin(\phi(t))$. After confirming that the amplitude of this beat signal β did not change noticeably in time, we activated the feedback loop so as to keep the beat signal constant.

In the experiment, we first accurately adjusted an intracavity prism using a piezo-translator to set f_{ceo} close to zero and then turn on the feedback loop. Figure 2(a) shows the rf spectrum (resolution bandwidth of 10 kHz) of the beat signal when f_{ceo} is close to zero. Before the stabilization, a frequency jitter of about 30 kHz was normally observed, as indicated by Fig. 2(a).

For the active CEP-stabilization, the pump power, i.e., the intracavity pulse energy, was modulated using an AOM driven by the feedback loop. f_{ceo}

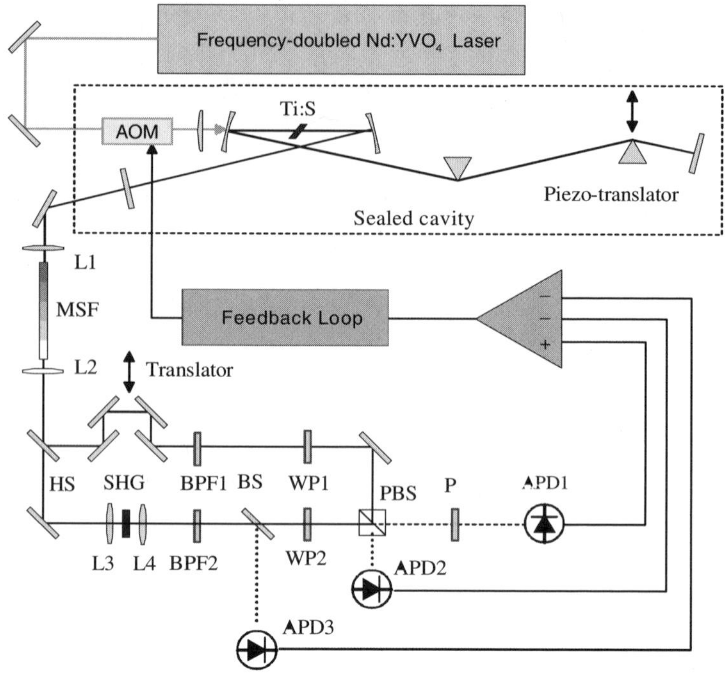

Fig. 1. Schematic diagram of CEP-locked Ti:sapphire laser. (AOM, Acoustic optic modulator; MSF, Microstructure fiber; HS, Harmonic separator mirror; BPF, Band pass filter with bandwidth 10 nm at 532 nm; BS, Beam splitter; WP, Half wave plate; PBS, Polarized beam splitter; APD, Avalanche photodiode)

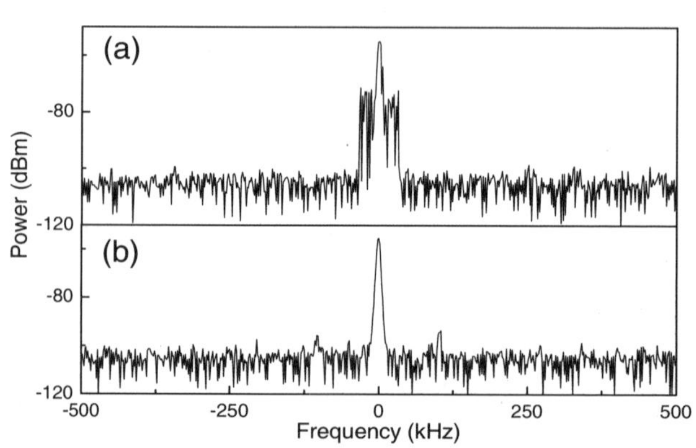

Fig. 2. Measured rf spectra of f-to-$2f$ heterodyne beat signals. (a) f_{ceo} is close to zero with a jitter of about 30 kHz (resolution bandwidth of 10 kHz), and (c) f_{ceo} is locked to zero using the feedback loop.

decreases linearly with the intracavity pulse energy with a slope of −5.834 MHz/nJ (or −0.316 rad/nJ in terms of $\Delta\phi$) as shown in Fig. 3. This slope shows that the intracavity-energy modulation of about 10 pJ (corresponding to the pump power modulation of about 0.2 mW) is enough for the full control of the beat signal, whose frequency is close to zero with a jitter of about 30 kHz. The AOM (with a modulation bandwidth of 11 MHz) and the feedback loop (with a bandwidth of about 100 kHz) were fast enough to immediately suppress the beat signal in the time domain. The feedback signal, loaded to the AOM while the feedback loop is turned on, is shown in Fig. 4(a) and the corresponding I_{f-2f} is shown in Fig. 4(b), where the amplitude fluctuation before the stabilization is an artifact coming from the data digitization in the measuring oscilloscope. Upon operation of the feedback loop, f_{ceo} quickly became zero, as shown as Fig. 2(b), where the small wings at around 100 kHz was the noise from the feedback loop, and the beat signal, shown in Fig. 4(a), became constant in time with a little jitter. These two observations clearly indicate that the zero CEP slip of the mode-locked pulse train (after the microstructure fiber in the present case) was achieved.

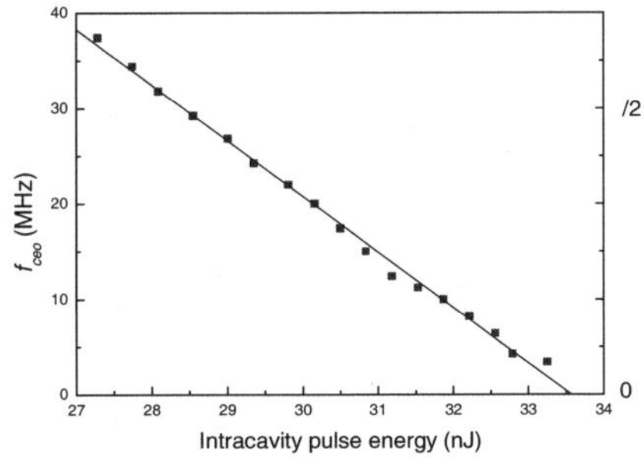

Fig. 3. Dependence of f_{ceo} on the intracavity pulse energy. Right axis represents the corresponding CEP slip.

To further confirm the CEP locking, we performed a quantitative analysis on the beat signals. Figure 4(d) represents the probability distribution of $I_{f-2f}(t)$, which shows that the beat signal after locking was confined close to the ground level with a near Gaussian distribution, whereas it oscillated from -1 to 1 (or $-\pi/2$to $\pi/2$ in terms of CEP) before locking. The probability distribution shows a minimal concentration at the ground level. The results reveal that f_{ceo} was locked to zero with a small random fluctuation, instead of locking to a small value as in the case of the PLL method that causes a regular CEP drift. The direct locking method does not allow such CEP drift and corresponding shot-to-shot CEP accumulation. From Fig. 3(d), the root-mean-square CEP jitter is calculated to be $\pm 0.02\pi$, which corresponds to a timing jitter of 30 attoseconds in the laser pulses with a center wavelength of 800 nm. Figure 5 shows that the locking of CEP was maintained for 10

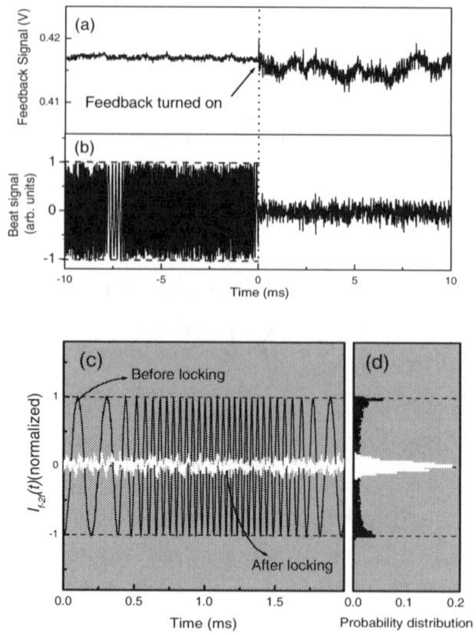

Fig. 4. CEP locking in the time domain. (a) Feedback signal loaded to the AOM in Fig. 1, (b) the evolution of the beat signal, with the feedback being turned on at t=0, (c) the beat signals before (black line) and after (white line) the locking process, and (d) the density distribution of the two beat signals before (black line) and after (white line) the CEP locking.

minutes without much change in phase jitter while the feedback signal was slightly reduced.

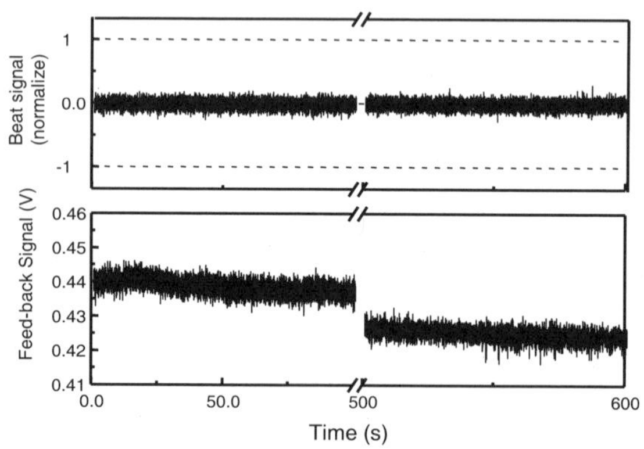

Fig. 5. Long-term stability of the CEP locking process.

This technique will be helpful for the construction of a CEP-controlled chirped-pulse amplification (CPA) laser because it allows the generation of laser pulses without any CEP slip from the laser oscillator. Using a laser oscillator based on our time-domain feedback method, a straightforward construction of a CPA system will be possible. In addition, the beat signal of the f-to-2f self-referencing interferometer can be locked to an arbitrary value (dc level), other than zero (the ground level), by intentionally adding a dc offset to the reference level of the feedback loop, which realizes the CEP locking to an arbitrary phase value. In the case of the PLL method, the CEP value cannot be easily adjusted since it has no mechanism to change the CEP value. The absolute value of the locked CEP, however, needs to be separately evaluated using a physical phenomenon such as high-order harmonic generation.

4 Conclusion

We proposed and demonstrated a new approach to directly lock the shot-to-shot CEP slip of mode-locked laser pulses to zero. The CEP locking in a sub-10-fs laser oscillator was clearly demonstrated with a CEP jitter of $\pm 0.02\pi$, corresponding to a timing jitter of 30 attoseconds.

Acknowledgement. This work was supported by the Ministry of Science and Technology of Korea through the Creative Research Initiative Program. We wish to thank R. S. Windeler for providing the air-silica micro-structure fiber.

References

1. H. R. Telle, G. Steinmeyer, A. E. Dunlop, J. Stenger, D. H. Shutter, and U. Keller, Appl. Phys. B **69**, 327 (1999).
2. K.-H.Hong, T. J. Yu, Y. S. Lee, C. H. Nam, and R. S. Windeler, J. Kor. Phys. Soc. **42**, 101 (2003).
3. M. Kakehata, H. Takada, Y. Kobayashi, K. Torizuka, Y. Fujihira, T. Homma, and H. Takahashi, Opt. Lett. **26**, 1672 (2001).
4. R. Ell, U. Morgner, F. X. Kartner, J. G. Fujimoto, E. P. Ippen, V. Scheuer, G. Angelow, T. Tschudi, M. J. Lederer, A. Boiko, and Luther-Davies, Opt. Lett. **26**, 373 (2001).
5. D. J. Jones, S. A. Diddams, J. K. Ranka, A. Stentz, R. S. Windeler, J. L. Hall, and S. T. Cundiff, Science **288**, 635 (2000).
6. J. K. Ranka, R. S. Windeler, and A. J. Stentz, Opt. Lett. **25**, 796 (2000).
7. T. A. Birks, W. J. Wadsworth, and P. St. J. Russel, Opt. Lett. **25**, 1415 (2000).

Ultrafast-Laser Stabilization with Application to Pulse Amplification by Use of Passive Optical Cavities

R. Jason Jones[1], Kevin Holman[1], Jun Ye[1], Eric Potma[2], and X. Sunney Xie[2]

[1] JILA, National Institute of Standards and Technology and University of Colorado, 440 UCB Boulder, CO 80309, email:rjjones@jilau1.colorado.edu
[2] Harvard University,12 Oxford Street, Cambridge, MA 02138

1 Introduction

Stabilization and control of the femtosecond laser is becoming increasingly important as novel applications utilizing the femtosecond comb are developed that require greater levels of precision. Improved stability is beneficial for both " "frequency domain" applications, where the relative quadratic phase between comb components, or "chirp", is unimportant (e.g. optical frequency metrology), as well as "time domain" applications where the pulse shape and/or duration is vital, such as in nonlinear optical interactions [1]. For both types of applications, minimizing jitter in the pulse train and noise in the carrier-envelope (CE) phase is often critical to achieve the desired level of precision. The stabilization of mode-locked femtosecond lasers has played a key role in recent advances in optical frequency measurement [2,3], carrier-envelope phase stabilization [4,5], all-optical atomic clocks [6,7] and coherent pulse synthesis [8]. Proper stabilization of ultrafast lasers can allow efficient coupling and temporary storage of ultrashort light pulses inside high finesse cavities [9]. This enables exciting possibilities for advancing external enhancement-cavity based techniques for short pulses, such as nonlinear frequency conversion [10], intracavity spectroscopy [11], and coherent pulse "amplification" [12,13] to name a few. A highly stable cavity may even itself serve as a frequency and phase reference for the pulse train [14]. This provides strong motivation to further improve tools for ultrafast laser control. Here we describe efforts in both active stabilization of ultrafast lasers and in storing these pulses in high finesse passive cavities. Results are given for pulse amplification when the single intracavity pulse is switched out with a Brag cell acting as a cavity dumper.

2 Stabilization of femtosecond lasers

To tightly lock a fs laser to a high finesse cavity and take advantage of the high signal to noise ratio of the recovered resonance information requires optimizing high bandwidth servo designs. An understanding of the dominant sources of noise in mode-locked lasers is important in determining the best choice of actuators such that corrections can be made without introducing more noise in another parameter of the laser. Like their single frequency counter-parts, mode-locked lasers are perturbed by mechanical and acoustic vibrations. However, because of the higher peak intensities and the nonlinear amplitude-phase coupling, the mode-locked lasers

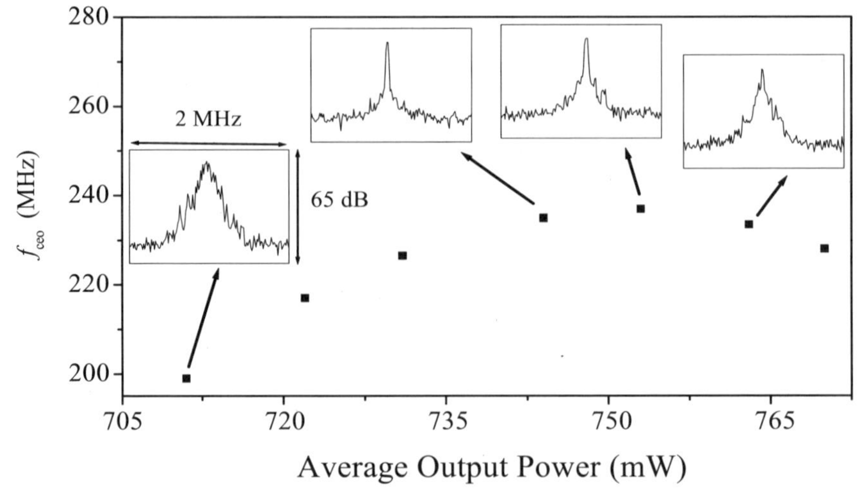

Fig. 1. Free-running linewidth of f_{ceo} near the turning point (i.e. $\Delta f_{ceo}/\Delta I \approx 0$) for the prism-less laser.

are much more susceptible to frequency noise induced by pump amplitude fluctuations. Previous work has attributed the dominant source of noise in the CE offset frequency (f_{ceo}) to power fluctuations, explained in terms of spectral shifts [15], self-steepening, and nonlinear refraction [16]. We have investigated intensity-related dynamics in both the repetition rate (f_{rep}) and f_{ceo} for laser systems incorporating prism-based and chirped mirror dispersion compensation designs [17]. An interesting sign-reversal in the dependence of f_{rep} and f_{ceo} on the laser power is observed and found to correspond to a shift of the laser spectrum. The point of reversal also coincides with minimum free running noise of both f_{rep} and f_{ceo} (see Fig.1). We find that in Ti:sapphire fs lasers with significant negative group delay dispersion (GDD), the dominant source of noise in f_{ceo} is due to power induced spectral shifts, in agreement with the findings of Xu et. al. [15]. This dependence is minimized with a decrease in the magnitude of the intensity dependent spectral shift and/or net cavity GDD, thereby minimizing the corresponding free-running linewidth of f_{ceo} and f_{rep}. An extremely broad bandwidth laser with near-zero net cavity GDD, such as that described by the dispersion-managed mode-locked model [20], is least susceptible to intensity fluctuations. In such cases other (smaller) noise mechanisms may dominate, and the use of power control to stabilize f_{ceo} may not be ideal.

3 External Cavity Amplification

An external passive optical cavity can be used to coherently superpose and temporarily store sequential pulses from a mode-locked laser. Equipped with a Bragg cell for cavity dumping, this approach can lead to an effective amplification process through decimation of the original pulse rate while preserving the overall coherence from the oscillator. The use of a passive cavity also enables amplification of short pulses where no suitable active gain medium may exist, such as in pulse trains generated in the infrared from difference-frequency mixing [18]. Unlike actively dumped laser systems, pulse energy is not limited by saturation of the gain or the effective saturable absorber needed for stable mode-locking. Instead, the linear response of the external cavity allows pulse energy to build up until limited by scattering loss and dispersion. The technique is applied to picosecond and femtosecond pulses.

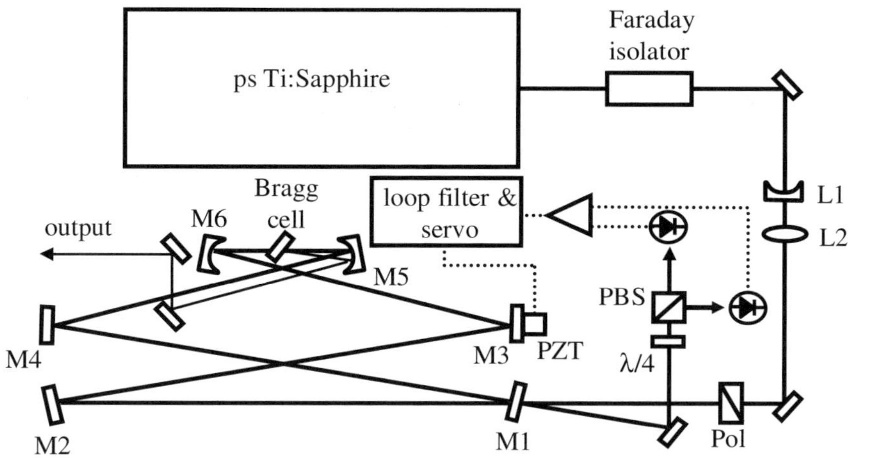

Fig. 2. Schematics of the passive optical amplifier. The Ti:sapphire laser is pumped by a 5W, 532 nm laser ; L1: Concave lens r = -75 mm; L2: Convex lens f = 125 mm; Pol: Polarizer; M1: output coupler 2% at 800 nm ; M2-M4: Low loss (¿ 30 ppm) high reflectors ; M5-M6: Low loss (>30 ppm) concave mirrors r = -100 mm ; Bragg cell: 3-mm fused silica cavity dumper; PBS: Polarizing beam splitter; $\lambda/4$: Quarter waveplate.

Initial experiments focused on amplification of ps pulses for immediate application in CARS microscopy. The use of ps pulses also allows us to separate out complications arising from intra-cavity dispersion and nonlinear effects from the higher peak intensities of femtosecond pulses. The layout of the amplifier is depicted in Fig. 2. The 835 nm output of a mode-locked 3.4-ps Ti:sapphire laser is fed into a six-mirror folded ring cavity. A mirror with 0.95% transmission is used as the coupling mirror. In order not to compromise the cavity finesse, all the other intra-cavity mirrors have reflectivity greater than 99.997%. The passive resonator incorporates a 3-mm fused silica Bragg cell placed under Brewster's angle, which acts as an acousto-optic cavity dumper. The FSR of the cavity is set to the 76 MHz repetition rate of the pump laser while the cavity resonances are actively locked to the average frequency of the laser. The error signal for the feedback loop is derived from the polarization sensitive detection scheme of Hänsch and Couillaud [19], and used to drive a PZT transducer attached to one of the cavity mirrors.

Results with 3.4 ps pulse durations emitted from a mode-locked Ti:sapphire laser are shown in Fig. 3. Based on measurements of the intra-cavity energy build up, a finesse of 349 is estimated, limited most likely due to loss in the input coupler.

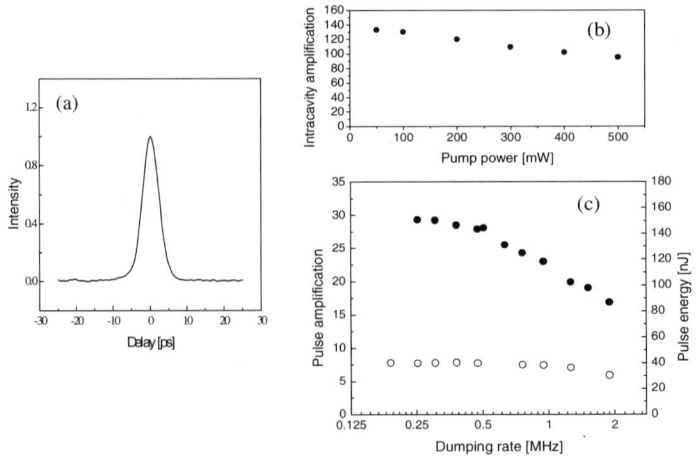

Fig. 3. Pulse amplification with the passive optical amplifier. a) Autocorrelation of 140 nJ amplified pulses at 475 kHz. Pulse width amounts to 3.4 ps, assuming a Gaussian envelope. b) Intra-cavity enhancement for various average powers of the incoming pulse train. c) Pulse amplification as a function of dumping frequency with single pass dumping efficiencies of 7.5% (open circles) and 30% (solid circles). Pulse energies (right axis) are obtained when seeding the amplifier with 400 mW from the Ti:sapphire laser. Note that dumping at frequencies beyond 2 MHz is limited by dumper electronics rather than by optical constraints.

Output pulse enhancements of 30 times are obtained at 253 kHz, yielding pulse energies greater than 150 nJ. The present pulse amplification is about three times better than what can be achieved by direct cavity dumping from a mode-locked laser while still maintaining relatively high repetition rates (hundreds of kilohertz). With optimization of the cavity finesse, we expect that amplifications greater than a hundred times are feasible, bringing pulse energies into the μJ range. This technique offers the additional practical benefits of being simple to implement with pre-existing commercial systems and providing all the pulse energy in a filtered spatial mode.

To efficiently couple sub-100 femtosecond pulses into an optical cavity with a finesse sufficiently high to build up pulse energy by several orders, the carrier and repetition frequency of the fs laser must be simultaneously stabilized to that of the cavity [14], and cavity dispersion must not distort the intracavity pulse. Stabilizing both degrees of freedom of the fs laser becomes increasingly important with shorter pulse durations and longer cavity lifetimes.

Preliminary results in enhancing low individual pulse energies for ≈75 fs pulses are shown in Fig. 4. The external cavity setup is similar to that shown in Fig. 2. The external enhancement cavity used incorporated specially designed negative GDD low loss mirrors to simultaneously compensate for the Bragg cell's 3 mm of fused silica and provide a high finesse. The input coupling mirror transmission was ≈0.8 %. From the buildup time of the pulses, a finesse of 440 is determined. An intracavity energy buildup of 163 is therefore expected, resulting in a single pulse amplification of approximately 65 for the current setup, given the 40% dumping

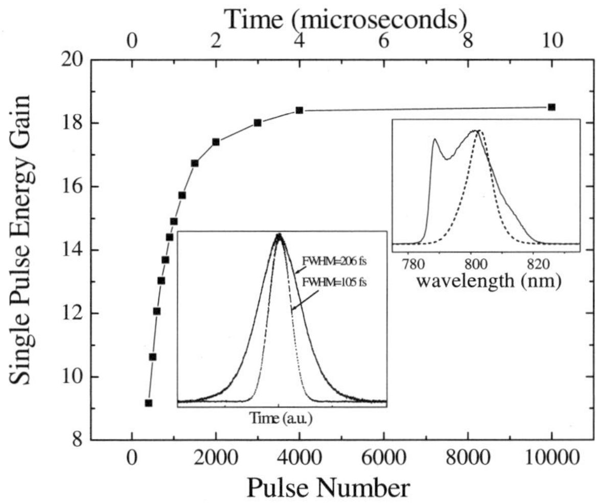

Fig. 4. External cavity amplification of single pulse energy versus number of pulses incident on cavity. Inset shows autocorrelation and spectrum of incident (dots) and dumped (solid) pulses.

efficiency of our Bragg cell. The negative dispersion mirrors were designed to only partially compensate for the total cavity dispersion. The remaining cavity GDD was estimated at +20-30 fs^2. The excess dispersion results in pulse broadening and a nonuniform filtering of the transmitted pulse spectrum, as shown in the inset of Fig. 4. The experimental results are in good agreement with independent numerical calculations predicting the observed transmission spectrum for a cavity with a net positive GDD of 26 fs^2. The observed amplification of only 18 times is therefore not surprising as the achievable pulse enhancement is limited by the lack of perfect resonance between the fs comb and the external cavity. Reducing the net cavity GDD by lowering the cavity air pressure will allow us to efficiently couple the entire pulse spectrum into the external cavity and enhance the pulse distortion free. The amount of external cavity amplification and minimum pulse duration achievable will crucially depend on the availability of low loss, broad bandwidth and dispersion tailored mirror technology.

References

1. T. Brabec and F. Krausz, Rev. Mod. Phys., **72**, 545 (2000).
2. T. Udem, J. Reichert, R. Holzwarth, and T. W. Hänsch, Phys. Rev. Lett. **82**, 3568 (1999).
3. S.A. Diddams, D.J. Jones, J. Ye, S.T. Cundiff, J.L. Hall, J.K. Ranka, R.S. Windeler, R. Holzwarth, T. Udem, and T.W. Hänsch, Phys. Rev. Lett. **84**, 5102 (2000).
4. D.J. Jones, S.A. Diddams, J.K. Ranka, A. Stentz, R.S. Windeler, J.L. Hall, and S.T. Cundiff, Science **288**, 635 (2000).

5. A. Apolonski, A. Poppe, G. Tempea, C. Spielmann, T. Udem, R. Holzwarth, T.W. Hänsch, and F. Krausz, Phys. Rev. Lett. 85, 740 (2000).
6. S.A. Diddams, T. Udem, J.C. Bergquist, E.A. Curtis, R.E. Drullinger, L. Hollberg, W.M. Itano, W.D. Lee, C.W. Oates, K.R. Vogel, and D.J. Wineland, Science **293**, 826 (2001).
7. J. Ye, L.-S. Ma, and J.L. Hall, Phys. Rev. Lett. **87**, 270801 (2001).
8. R.K. Shelton, L.S. Ma, H.C. Kapteyn, M.M. Murnane, J.L. Hall, and J. Ye, Science **293**, 1286 (2001).
9. R.J. Jones, et. al., Opt. Comm. **175**, 409 (2000).
10. G.T. Maker and A.I. Ferguson, App. Phys. Lett. **55**, 1158 (1989).
11. T. Gherman and D. Romanini, Opt. Ex. **10**, 1033 (2002).
12. R.J. Jones and J. Ye, Opt. Lett. **27**, 1848 (2002).
13. E. Potma, C. Evans, X.S. Xie, R.J. Jones and J. Ye, Opt. Lett. (In Press).
14. R.J. Jones and J.-C. Diels, Phys. Rev. Lett. **86**, 3288 (2001).
15. L. Xu, C. Spielmann, A. Poppe, T. Brabec, F. Krausz, T.W. Hänsch, Opt. Lett. **21**, 2008 (1996).
16. F.W. Helbing, G. Steinmeyer, U. Keller, R.S. Windeler, J. Stenger, and H.R. Telle, Opt. Lett. **27**, 194 (2002).
17. K.W. Holman, R.J. Jones, A. Marian, S.T. Cundiff and J. Ye, Opt. Lett. **28**, 851 (2003).
18. S. Foreman, D.J. Jones and J. Ye, Opt. Lett. **28**, 370 (2003).
19. T.W. Hänsch and B. Couillaud, Opt. Commun. **35**, 441 (1980).
20. Y. Chen, F.X. Kärtner, U. Morgner, S.H. Cho, H.A. Haus, E.P. Ippen, and J.G. Fujimoto, J. Opt. Soc. Am. B **16**, 1999 (1999).

Solid-State Carrier-Envelope Phase Detector

P. Dombi[1], A. Apolonski[1,2], G. G. Paulus[3], M. Kakehata[4], K. Torizuka[4], R. Holzwarth[5], Th. Udem[5], Ch. Lemell[6], J. Burgdörfer[6], T.W. Hänsch[4], and F. Krausz[1,5]

[1] Institut für Photonik, Technische Universität Wien, Gusshausstrasse 27/387, A-1040 Wien, Austria
[2] Institute of Automation and Electrometry, SB RAS, Novosibirsk 630090, Russia
[3] Texas A&M University Physics Department College Station, TX 77843-4242
[4] National Institute of Advanced Industrial Science and Technology (AIST), 1-1-1 Umezono, Tsukuba 305-8568, Japan
[5] Max-Planck-Institut für Quantenoptik, Hans-Kopfermann-Str. 1, D-85748 Garching, Germany
[6] Institut für Theoretische Physik, Technische Universität Wien, Wiedner Hauptstrasse 8, A-1040 Wien, Austria
dombi@tuwien.ac.at

Summary. A solid-state device is demonstrated, which is capable of measuring the carrier-envelope phase of ultrashort (4-femtosecond), low-energy (unamplified) laser pulses unambiguously. The measurement is based on photoelectron emission from a gold surface in the multi-photon regime.

1 Introduction

Atomic processes driven by few-cycle laser pulses depend on the timing of the oscillations of the electric field $E_{\mathrm{L}}(t)$ with respect to the pulse peak in addition to the carrier frequency ω_{L} and amplitude envelope $A_{\mathrm{L}}(t)$ [1]. With the electric field evolution given by $E_{\mathrm{L}}(t)=A_{\mathrm{L}}(t)cos(\omega_{\mathrm{L}}t+\varphi)$, this timing is quantified by the carrier-envelope phase φ. Even though this φ-dependence was observed experimentally a short time ago [2,3], the method relied on the interaction of high-energy ($\gg 1\ \mu$ J) pulses with atoms in gas phase, requiring a complex amplifier and vacuum system.

Following a recent proposal [4], here we demonstrate the first compact, solid-state detector of the carrier-envelope phase, based on a modified multiplier tube. This device is sensitive at nanojoule pulse energies, allowing complete characterization of light pulses from a laser oscillator for the first time.

In a femtosecond pulse train delivered by conventional mode-locked lasers φ is subject to a pulse-to-pulse shift, denoted by $\Delta\varphi$, that undergoes a substantial jitter. The novel "self-referencing" or "f-to-$2f$" technique made it possible to control the phase evolution in the typically multi-MHz train of pulses by stabilizing $\Delta\varphi$ [5,6]. However, an f-to-$2f$ interferometer usually used in this scheme is practically not suitable for measuring the absolute value of φ; only the value of the pulse-to-pulse phase shift $\Delta\varphi$ can be stabilized. Therefore a direct phase measurement technique is needed, being capable of accessing the actual value of φ at each pulse.

2 Simulation results

Simulations exploiting time-dependent density functional theory confirmed that the photoelectron yield of emission from a metal surface (modelled as a confined free-electron gas using the jellium model) exhibits a robust dependence on the absolute value of φ [4], provided that the pulse is short enough (the pulse length is below 6-7 fs, see Figure 1) and that the intensity of the light (ideally impinging on the metal with the E-field perpendicular to the surface) is not too high, i.e. the emission is dominated by multi-photon processes.

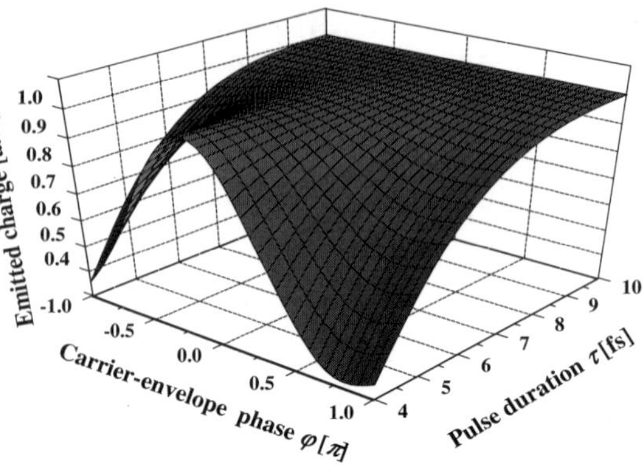

Fig. 1. Normalized photoelectron yield from a gold surface upon interaction with a Gaussian laser pulse with a central wavelength of 750 nm and a peak intensity of 5×10^{12} W/cm^2. The beam impinges at grazing incidence with 'P'-polarization.

Even though the effect is still observable at higher intensities, the change in the photoemission signal upon changing the carrier-envelope phase becomes significantly weaker in the tunnelling regime, and the phase calibration will also be different, i.e. the position of, say, the maximum of the signal on Figure 1 will be located at a different carrier-envelope phase value [4].

The observation that the carrier-envelope phase sensitivity is more pronounced in the multi-photon than in the tunneling regime is, at first glance, somewhat surprising. It may be explained by the fact that an ultrashort laser pulse is capable of supporting ionization pathways of different multi-photon orders, and the phase sensitivity arises from their interference.

3 Experimental results

For the experimental verification of these predictions we used our laser system delivering 4-fs, $\Delta\varphi$-stabilized, 3-nJ pulses at a central wavelength of 700-750 nm and a repetition rate of 24 MHz (the dashed box on Figure 2; for a detailed characterization of this laser system see [7]). We also utilized a modified electron multiplier tube (Hamamatsu R595) with the cathode substituted with a gold-coated nickel piece.

The 'P'-polarized laser beam impinged on the surface at an angle of incidence of 70 ° (Figure 2).

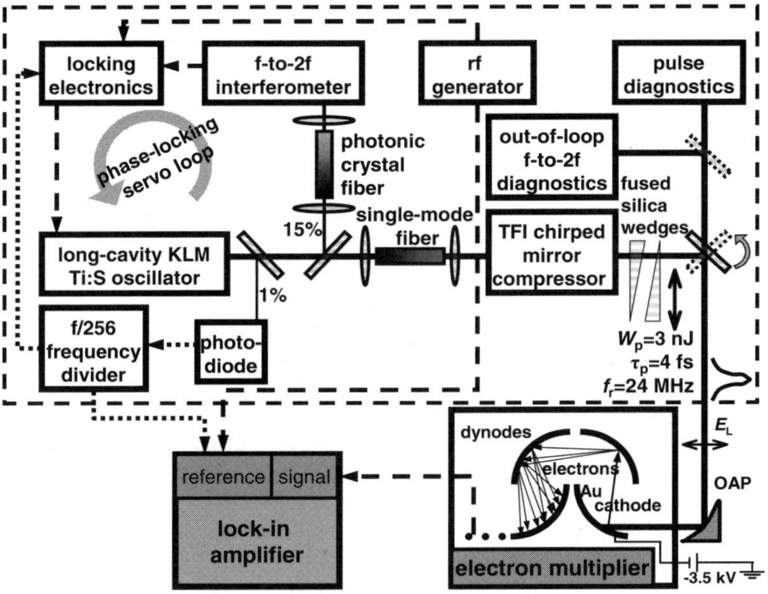

Fig. 2. Schematic layout. A fraction of the oscillator beam is split off to close the phase-stabilization servo loop, whilst the rest is spectrally broadened and compressed. The rate of change of φ is locked to $f_{ceo}=f_{ref}=(\Delta\varphi/2\pi)f_r$, with either f_{ref}=1 MHz (dashed-line electronic connections) or $f_{ref}=f_r/256$ (dotted-line electronic connections substituted, where applicable).

The photocurrent from the electron multiplier tube was found to exhibit power-law scaling with intensity: I^3, and, more importantly, a modulation at $f_{ceo}=f_{ref}$. This latter was detected by a lock-in amplifier referenced to f_{ref}. Since the lock-in amplifier might exhibit a coherent noise pick-up at f_{ref}=1 MHz, one needs a further, ultimate evidence for the φ-sensitivity of the photocurrent.

This evidence is provided by the observation that the in-phase component of the lock-in output evolved quasi-periodically upon increasing the propagation length of the pulses in a pair of extracavity wedges, as shown in Figure 3.

Shifting one of the wedges perpendicular to the beam means that the phase difference between the two periodic, electronic signals oscillating at f_{ref} and f_{ceo} (the latter now being the frequency of the weak 1 MHz signal superimposed on the photoemission signal measured at the output of the multiplier tube) also shifts continuously. Therefore a periodic behavior arises in the in-phase component of the lock-in output, just as depicted. This way one can exclude the possibility of coherent instrumental pick-up.

Moreover, the period length of 41 (+4/-3) μm retrieved from Figure 3 was found to be in excellent agreement with the calculated accurate carrier-envelope dephasing

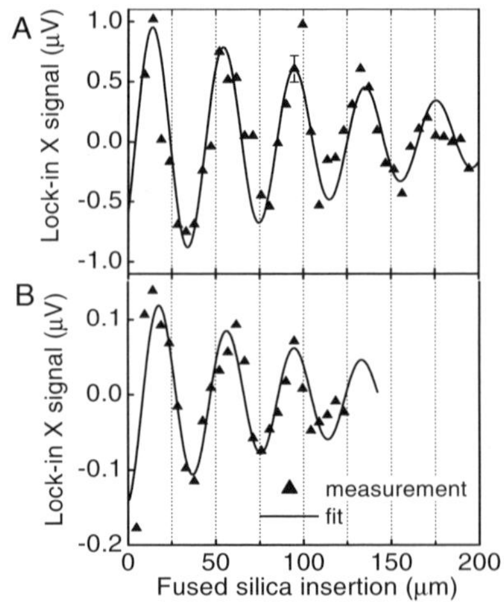

Fig. 3. The in-phase component of the photoemission signal as a function of material insertion in the beam, or, equivalently, as a function of the phase difference between the f_{ceo}-component of the photoemission and the reference signals. The lock-in amplifier was referenced to $f_{ref}(=f_{ceo})$ during the measurement. The amplitude drops rapidly due to the quick temporal broadening of few-cycle pulses upon propagation and the third-order phenomenon involved. The curve showing the more rapid decrease of amplitude corresponds to a shorter, possibly 3.7 fs-long pulse (B), whereas the other to a slightly longer, 4.5 fs-pulse (A). The substantially different amplitudes of the two curves can be attributed to experimental conditions not fully under our control, e.g. the exact position of the beam spot on the cathode surface lacking optical quality, or phase coherence fluctuations between the stabilization servo loop and the output beam etc.

length of 42 μm in fused silica. In this calculation we did not just rely on the approximate formula for the dephasing length as given in [1], but we evaluated its value accurately using the retrieved temporal pulse shape from our pulse diagnostic system [7], and propagating it through different amounts of fused silica. This way the effect of minor changes in the pulse envelope (caused by linear propagation) on carrier-envelope phase shift was also taken into account.

4 Conclusion

We presented experimental evidence for the φ-sensitivity of multi-photon-induced photoemission from a metal surface for the first time. This also permits the direct measurement of φ by a simple, compact, solid-state device. Along with standard pulse diagnostics the complete characterization and control of the electric field of the laser output can be carried out. Few-cycle, MHz-repetition-rate pulses with fully-characterized and fully-controlled electric and magnetic fields will permit the exploration of phase-sensitive strong-field processes at surfaces and in semiconductor nanostructures and the access to the sub-femtosecond electronic response of condensed systems.

This work was supported by the Austrian Science Fund grants F016, Z63 and P15382.

References

1. T. Brabec and F. Krausz, "Intense few-cycle laser fileds: Frontiers of nonlinear optics," *Rev. Mod. Phys.* **72**, 545-591 (2000).
2. G. G. Paulus, F. Grasbon, H. Walther, P. Villoresi, M. Nisoli, S. Stagira, E. Priori, S. De Silvestri, "Absolute-phase phenomena in photoionization with few-cycle laser pulses," *Nature* **414**, 182-184 (2001).
3. A. Baltuška, Th. Udem, M. Uiberacker, M. Hentschel, E. Goulielmakis, Ch. Gohle, R. Holzwarth, V. S. Yakovlev, A. Scrinzi, T. W. Hänsch, and F. Krausz, "Attosecond control of electronic processes by intense light fields," *Nature* **421**, 611-615 (2003).
4. Ch. Lemell, X.-M. Tong, F. Krausz, and J. Burgdörfer, "Electron Emission from Metal Surfaces by Ultrashort Pulses: Determination of the Carrier-Envelope Phase," *Phys. Rev. Lett.* **90**, 076403-1-4 (2003).
5. D. J. Jones, S. A. Diddams, J. K. Ranka, A. Stentz, R. S. Windeler, J. L. Hall, and S. T. Cundiff, "Carrier-Envelope Phase Control of Femtosecond Mode-Locked Lasers and Direct Optical Frequency Synthesis," *Science* **288**, 635-639 (2000).
6. A. Apolonski, A. Poppe, G. Tempea, Ch. Spielmann, Th. Udem, R. Holzwarth, T. W. Hänsch, and F. Krausz, "Controlling the Phase Evolution of Few-cycle Light Pulses", P*hys. Rev. Lett.* **85**, 740-743 (2000).
7. V.S. Yakovlev, P. Dombi , G. Tempea , Ch. Lemell , J. Burgdörfer , Th. Udem and A. Apolonski, "Phase-stabilized 4-fs pulses at the full oscillator repetition rate for a photoemission experiment" *Appl. Phys. B* **76**, 329-332 (2003).

Selection of Pulses Having a Constant Carrier-Envelope Phase from a Relatively Noisy Mode-Locked Oscillator

Masayuki Kakehata[1], Yoshihiko Fujihira[2], Yohei Kobayashi[1], Hideyuki Takada[1], Tetsuya Homma[2], Hideo Takahashi[2], and Kenji Torizuka[1]

[1] National Institute of Advanced Industrial Science and Technology (AIST), Central 2, 1-1-1 Umezono, Tsukuba, Ibaraki 305-8568, Japan
e-mail:kakehata-masayuki@aist.go.jp
[2] Shibaura Institute of Technology, 3-9-14 Shibaura, Minato-ku, Tokyo 108-8548, Japan

Abstract

This paper demonstrates a method that enables us to select pulses with the same carrier-envelope phase. This method, which selects pulses based on both the phase of the carrier-envelope offset beat signal and the pulse-repetition timing, is robust against large carrier-envelope offset fluctuations. This method relieves the required accuracy of the carrier-envelope offset beat stabilization for seeding the amplification system.

1 Introduction

Controlling the carrier-envelope phase (CEP) of high-intensity few-cycle laser pulses is a new approach to investigate and control nonlinear optical phenomena. As demonstrated, the beat signal between the fundamental and its second harmonic (f_{ceo}) was used as the feedback signal to control the slip of the CEP between pulses (CEO: carrier-envelope offset) of mode-locked oscillators [1, 2], and precise control of f_{ceo} were demonstrated [3, 4]. Single-shot measurement of the relative CEP was demonstrated by measuring the self-referencing spectral interferometry [5,6], and a passive-stabilization of the CEP of optical parametric amplifier was demonstrated by utilizing the phase relation between the pump pulse and seed pulse [7]. Recently, the CEP-controlled pulses were successfully amplified [8]. These researchers stabilized f_{ceo} and extracted pulses as seed pulses for amplification by dividing the pulse-repetition timing. Precise phase-locked-loop (PLL) control of f_{ceo} is required for pulse selection as long as the pulse selection signal is generated by dividing the repetition timing of the oscillator. Pulse selection by using the phase of the CEO beat signal was suggested in [2], but has not been demonstrated.

Here, we report selection of pulses having nearly a constant relative CEP from a Ti:sapphire oscillator that posseses large phase fluctuation of f_{ceo}. Our main point is that the requirement of the fine control of f_{ceo} is greatly relieved by utilizing the phase of the CEO beat signal to determine the pulse-selection timing. The scheme is based on the fact that the amplitude of the CEO beat signal reflects the temporal evolution of the relative CEP.

2 Experiment

Figure 1 presents schematic of the experimental setup. We controlled the CEO beat ($f_{\rm ceo}$) to $1/n$ of the pulse-repetition rate of the oscillator ($f_{\rm rep}$=80 MHz) using PLL electronics so that every n-th pulse possesses nearly the same CEP. We set $n = 8$ in this experiment ($\Delta\phi$=$2\pi(f_{\rm ceo}/f_{\rm rep})$=$2\pi/8$). The CEO beat was monitored by a f-to-$2f$ interferometer that is composed of a microstructured fiber and a SHG crystal. The error signal was used to control the angle of the end mirror of the prism-pair dispersion-compensated oscillator and the pumping power by using an EO modulator. The Ti:sapphire oscillator generates an average power of 250 mW with a spectrum width of ~30 nm FWHM centered around 790 nm. We used a spectrum analyzer, a frequency counter, and a vector signal analyzer

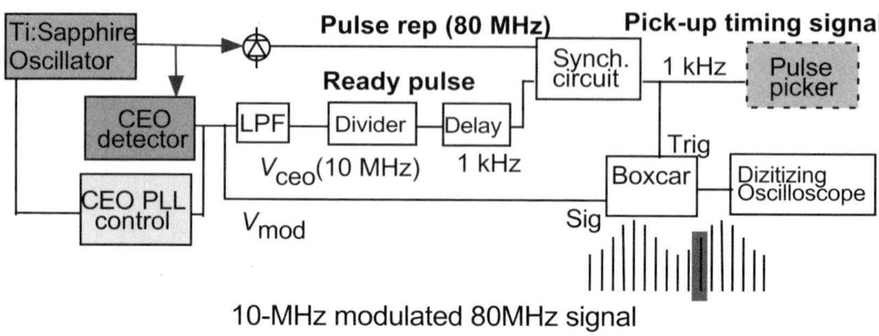

Fig. 1. Experimental set up of 1-kHz pulse selection and real-time monitoring of the relative CEP. LPF: low-pass filter

(Model 89441A, Agilent) to investigate the stability of $f_{\rm ceo}$. The linewidth of $f_{\rm ceo}$ was around 1 Hz, and the frequency counter shows the frequency fluctuation to be below 1Hz for 0.1-s time gate with the stabilization by EO and PZT. With stabilization only with PZT, the linewidth $f_{\rm ceo}$ becomes broader thant 1 Hz and the frequency counter shows the frequency fluctuation of over 10Hz for 0.1-s time gate. The power spectrum density (PSD) of the phase noise $S_\phi(\nu)$ and the accumulated phase jitter $\delta\phi_{rms} = (|\int_{\nu_{high}}^{\nu_{low}} S_\phi(\nu)d\nu|)^{\frac{1}{2}}$ are shown in Fig. 2 for stabilization with both EO and PZT. The stability of $f_{\rm ceo}$ is not high enough to pick up CEP-stabilized pulses by dividing the pulse-repetition timing. Because the amplitude of the CEO beat signal has information of the relative CEP, pulses that possess the same CEP can be selected even if the CEO fluctuation is large. We realized pulse selection as follows. First,we generated a 1-kHz ready pulse by dividing $f_{\rm ceo}$. The ready pulse follows the dynamic change of $f_{\rm ceo}$, therefore, it synchronizes with the timing at a certain relative CEP. The ready pulse, however, has timing jitter relative to the pulse train at 80 MHz because the phase fluctuation of $f_{\rm ceo}$ is large. The ready pulse therefore cannot be used as a pulse-selection signal. To generate a pulse-selection signal that synchronizes the pulse-repetition timing at 80 MHz, we introduced a synchronization circuit that generates electrical pulses (pick-up signal) synchronized to the first pulse that appears after the ready pulse. With this circuit, the pick-up

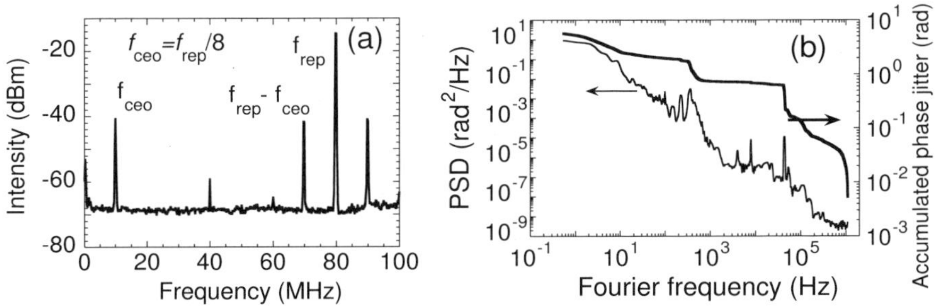

Fig. 2. The CEO beat signal; (a) Power spectrum of the beat signal ; (b) Power spectrum density of phase noise of CEO beat and the accumulated phase jitter.

signal perfectly synchronizes with the optical pulse timing but has timing jitter within $1/f_{\rm rep}$ relative to the ready pulse.

In order to monitor and evaluate the effectiveness of this method, we measured the amplitude of the beat signal ($V_{\rm mod}$) by a Boxcar integrator that is triggered by the pick-up signal. Figure 3 schematically shows the relation between the relative CEP and the amplitude. We set the delay of the Boxcar integrator to select pulses with half amplitude to measure the relative CEP change accurately . The synchronization between the laser pulse and the pick-up signal was confirmed by observing the oscilloscope trace. The range of relative CEP of π was determined by the peak-to-peak amplitude ($V_{\rm pp}$) measured with a CEO-uncontrolled oscillator and selecting the pulse by dividing the repetition rate. The relative CEP is calculated by

$$\phi_{\rm rel} = \arcsin(2V_{\rm meas}/V_{\rm pp}), \tag{1}$$

where $V_{\rm meas}$ is the Boxcar output voltage with the offset voltage $V_{\rm offset}$ subtracted.

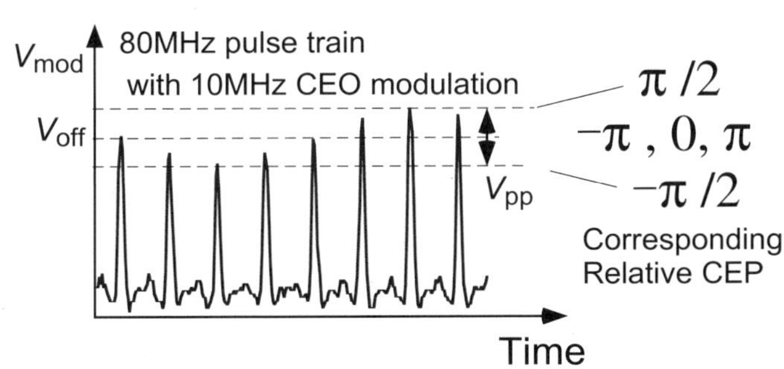

Fig. 3. Relation between the amplitude of the beat signal and the relative CEP in the temporal domain. The amplitude of the each pulse has information on the relative CEP.

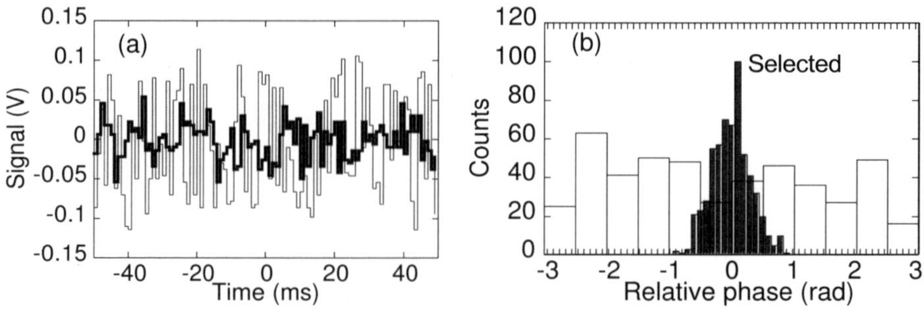

Fig. 4. (a) Temporal evolution of the amplitude of the beat signal for the free-running laser pulses (*thin line*) and the CEP-selected pulses (*thick line*). The offset voltage V_{offset} was subtracted. (b) Relative CEP distribution of the free-running laser pulses selected by dividing the repetition timing (*open bars*) and of the CEP-selected pulses (*filled bars*). The rms phase error was 0.4 rad by this measurement method.

Figure 4 (a) shows the temporal evolution of the V_{meas} sampled at 1 kHz. As shown in the figure, the distribution of the amplitude of the selected pulse localizes compared with the un-selected pulses. With the controlled oscillator and the pulse selection method, we observed synchronization of the triggering pulse with the phase of the V_{mod} (or V_{ceo}), therefore, the phase was calculated by eq. (1). In contrast, the relative phase was doubled for the non-selected result to make the phase distribution in the range of 2π because we did not observe synchronization of the selection signal with the V_{mod}. The distribution of the relative CEP is shown in Fig. 4 (b). Using the CEO-controlled oscillator with the CEO phase-pick-up method reduced the rms phase error to 0.4 rad. We also evaluated the CEP jitter by measuring the temporal jitter between the pick-up signal and V_{ceo} that is a sinusoidal signal with frequency f_{ceo}. The measured phase jitter was 0.3 rad , which was slightly smaller than the value measured by using the Boxcar integrator. This is because the amplitude fluctuation in the V_{mod} causes excessive errors in V_{meas} and increases the evaluated phase jitter.

We could select pulses with the same accuracy under very noisy conditions. By using the pulse-selection method for an oscillator controlled only with the PZT , we could select pulses with 0.3 rad rms error.

3 Discussions

To use this pulse-selection method for the amplification system, the time delay between the ready pulse and the pick up timing should be small. If the delay between the ready pulse and the pick up timing exceeds the coherence time of the CEO beat, the error of the relative CEP become large. We measured the dependence of the phase error as a function of the delay after the ready pulse as shown in Fig. 6. This figure presents results obtained with the PZT-controlled oscillator and PZT- and EO-controlled oscillator. The increase of the phase error is faster for the PZT-controlled oscillator, which is explained by the uncompensated high-frequency phase-noise component. Qualitatively, the phase error is determined by $\Delta\phi = 2\pi$ (f_{ceo}/f_{rep}) for small delay and by the accumulated phase error for large delay. This method can extract CEP-selected pulses from an uncontrolled oscillator if the delay is zero. In reality, however, the ratio f_{ceo}/f_{rep} should be stabilized to some extent because of the finite internal delay of the electrical circuit. Pulse selection accuracy can be improved by making $\Delta\phi$ small. We should note that the ratio f_{ceo}/f_{rep} should not necessarily be an integer in this pick up method.

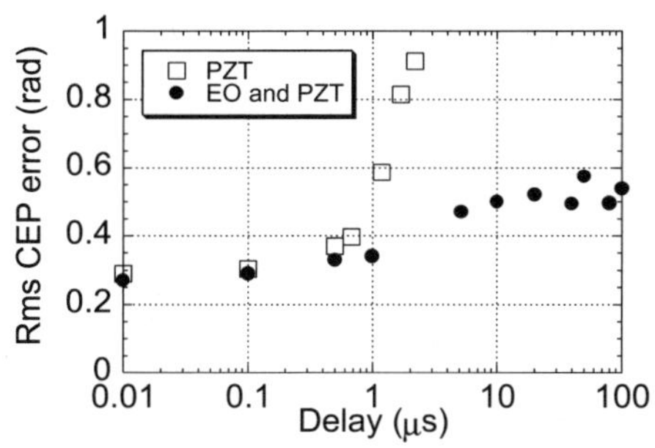

Fig. 5. Change of the standard deviation of the relative CEP as a function of the time delay.

The measured data are all in-loop results and the true value of the CEO can be affected by the phase fluctuation in the f-to-$2f$ interferometer. The phase fluctuation in the microstructured fiber caused by the intensity noise of the laser [9] should be minimized to correctly select the CEP-controlled pulses. Another issue is the optical path length drift in the interferometer. The drift of the optical path length in the f-to-$2f$ interferometer shifts the relative CEP, which is similar to the single shot measurement case [5,6]. Therefore, a drift of the f-to-$2f$ interferometer should also be compensated in a long term experiment as demonstrated in [8].

4 Conclusion

We have demonstrated selection of pulses having nearly a constant relative CEP from a Ti:sapphire laser oscillator by utilizing a pulse selection method that monitors both the phase of the CEO beat signal and the pulse timing. This method is useful for constant relative CEP pulse selection from oscillators with relatively large CEO phase noise.

Acknowledgement. The authors thank K. Okubo, H. Takamiya, and K. Nishijima for their technical assistance. A part of this study was financially supported by the Budget for Nuclear Research of the Ministry of Education, Culture, Sports, Science and Technology, based on the screening and counseling by the Atomic Energy Commission.

References

1. D. J. Jones S. A. Diddams, J. K. Ranka, A. Stentz, R. S. Windeler, J. L. Hall, and S. T. Cundiff , Science **288**, 635–639 (2000).
2. A. Apolonski, A. Poppe, G. Tempea, Ch. Spielmann, Th. Udem, R. Holzwarth, T. W. Hänsch, and F. Krausz, Phys. Rev. Lett. **85**, 740–743 (2000).
3. T. M. Fortier, D. J. Jones, J. Ye, S. T. Cundiff, and R. S. Windeler, Opt. Lett. **27**, 1436–1438 (2002).
4. F. W. Helbing, G. Steinmeyer, J. Stenger, H. R. Telle, and U. Keller, Appl. Phys. **B74**[Suppl.], S35-S42 (2002).
5. M. Kakehata, H. Takada, Y. Kobayashi, K. Torizuka, Y. Fujihira, T. Homma, and H. Takahashi, Opt. Lett. **26**, 1436–1438 (2001).
6. M. Kakehata, H. Takada, Y. Kobayashi, K. Torizuka, Y. Fujihira, T. Homma, and H. Takahashi, Appl. Phys. **B74**[Suppl.], S43– S50 (2002).
7. A. Baltuška, T. Fuji, and T. Kobayashi, Phys. Rev. Lett. **88**, 133901–133904 (2002).
8. A. Baltuška, Th. Udem, M. Uiberacker, M. Hentschel, E. Goulielmakis, Ch. Gohle, R. Holzwarth, V. S. Yakovlev, A. Scrinzi, T. W. Hänsch, and F. Krausz, Nature **421**, 611–615 (2003).
9. T. M. Fortier, J. Ye, S. T. Cundiff, and R. S. Windeler, Opt. Lett. **27**, 445–447 (2002).

Carrier Envelope Phase Measurement Using a Non Phase Stable Laser

K. O'Keeffe[1], J. Seres[1], E. Seres[1], H. Drexel[2], V. Grill[2], F. Krausz[1,3], and M. Lezius[1,2]

[1] Photonics Institute, Vienna Technical University, Gusshausstr. 27, A-1040 Vienna, Austria
[2] Ion Physics Institute, Innsbruck University, Technikerstr. 25, A-6020 Innsbruck, Austria matthias.lezius@uibk.ac.at
[3] Max-Plank-Institut für Quantenoptik, Hans-Kopfermann-Str. 1, D-85748 Garching, Germany

Ultrafast laser technologies permit the production of high power, well characterized yet variable optical waveforms in the few cycle regime. Laser systems can easily produce focused intensities in excess of $10^{15}\mathrm{W/cm^2}$, corresponding to electric fields stronger than the atomic Coulomb fields ($\approx 10^9\mathrm{V/cm}$), leading to strong field ionization. The liberated electrons move quasi- classically in the vicinity of the ion core and can be rescattered (ATI), or recombine after photon emission (HHG), or produce secondary electrons after electron impact (NSDI). Such electron - ion core interactions have been successfully studied in many experiments performed by the community. However, until most recently almost all experiments were conducted using pulses with durations longer than 10fs and therefore consisting of many optical cycles. With such pulses it is not possible to determine at which optical cycle ionization actually takes place or how many cycles later the electron recollides with the ion core. These limitations can be overcome by the use of few cycle pulses. Using techniques developed over the past few years it is now possible to obtain laser pulses with a duration of less than 5fs. In this so-called few cycle regime the carrier envelope phase (CE-phase) becomes an important measure in the description of the instantaneous electric field.

In order to investigate CE-phase effects of nonlinear processes it is necessary to have laser pulses consisting of approximately two optical cycles and to have the means of controlling, or monitoring the phase. Today, femtosecond oscillators exist which are capable of delivering pulses with a stable phase. However, the peak power of such oscillators pulse is insufficient to perform many strong field experiments. Then pulses must be amplified to become powerful enough. On the other hand, upon amplification various effects lead to an uncontrolled drift of the CE-phase. Fortunately the drift was observed to be relatively slow and may be monitored and compensated for by a secondary phase control loop. Such a system was realized in Vienna by Baltuska et *al.* [1]. Since the construction of this laser powerful, amplified, and phase stable pulses were available to observe phase effects in HHG and ATI [2]. Using this laser it was possible to adjust the CE-phase so that processes could be observed for different positions of the electric field relative to the pulse envelope, and to ultimately measure the absolute value of the phase [2]. While the laser technicians seem to have now control over most parameters of the laser pulse, we believe that it is by no means necessary to have a phase stable laser system in order to observe CE-phase effects like those predicted for example in [3, 4].

It is the purpose of the present work to measure phase effects in above threshold ionization using amplified pulses which are not phase stabilized. In particular we aimed to retrieve the phase for each laser shot from an unstabilized system. This appears at a first glance very similar to the measurement of Paulus *et al.* [5]. However there are distinct differences between that experiment and the one described here. In the experiment of Paulus *et al.* a stereo ATI measurement was performed to demonstrate that phase effects dominate in ATI electron ejection. Here we demonstrate that in combination with $f-2f$ interferometry the phase effects can be used to retrieve the value of the CE-phase for each laser shot. Note that $f-2f$ interferometry alone is insufficient to obtain the absolute value of the phase. Compared to Paulus et *al.* we have use here only one electron detector and record the electron yield for each laser shot into one particular direction. At the same time a value for the phase is retrieved by $f-2f$ interference. By correlation of both observables the

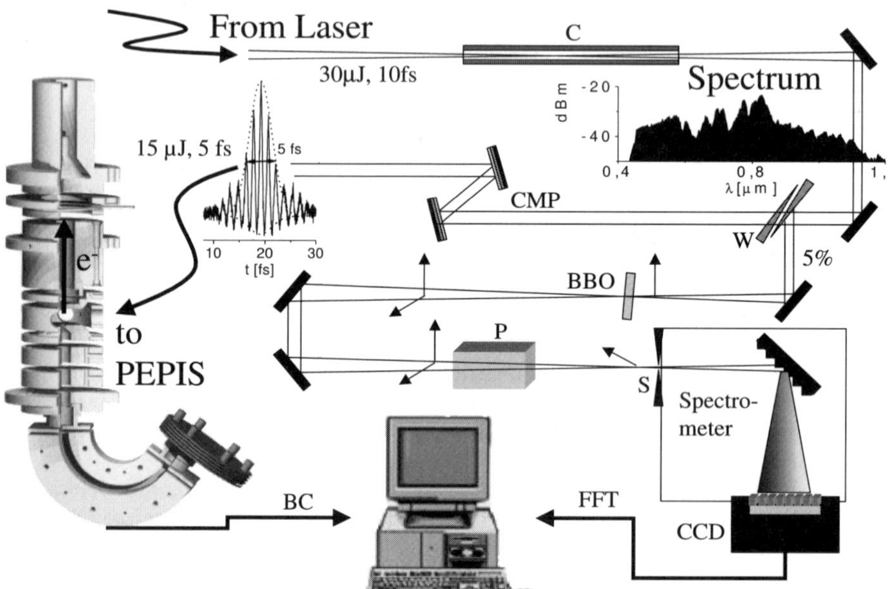

Fig. 1. Sketch of the experimental setup used for determining the carrier envelope phase. The laser pulses are spectrally broadened (see insert) up to one octave. Then the beam is split into two parts, a weak part that is used for $f-2f$ interferometry and an intense part that is compressed to 5fs and used to produce ATI electrons from a low pressure Xenon gas target in a photoelectron spectrometer (PEPIS, left side). Electron current and CE-phase from the $f-2f$ interferometer are stored and compared on a computer (see text).

is shown in Fig 1. We start with laser pulses that were produced from a 10 pass Ti:sapphire amplifier system with a repetition rate of 1kHz which is seeded with a 20fs Ti:sapphire oscillator. After the amplifier typical pulse energies are of the order of 1.3mJ. The beam passes through a prism compressor before being sent through a 1m long hollow fibre containing 1.2 bar Neon to achieve spectral broadening [6]. At the entrance of the hollow fibre the pulses have a power of approximately 1mW and a duration of 25fs. The throughput of the fibre itself is 40%. A beamsplitter with R=18% was placed after the fibre to attenuate the beam. The pulses are then compressed by a chirped multilayer mirror arrangement down to 7fs. Subsequently, pulses are further spectrally broadened by a second hollow fibre (C) to a one octave spanning spectrum for two reasons. On one side, the broad spectrum helps to achieve a 5fs pulse, on the other side it is necessary for the operation of the $f-2f$ interferometer. To generate the broadening a 7cm long hollow fibre with an inner diameter of 166μm was placed in an hermetic chamber filled with Krypton up to a pressure of 2.5bar. Pulses of 30μJ were sent into this hollow fibre, having a transmission efficiency higher than 40%. The generated spectrum extends from 400nm to over 1200nm (see insert in Fig. 1).

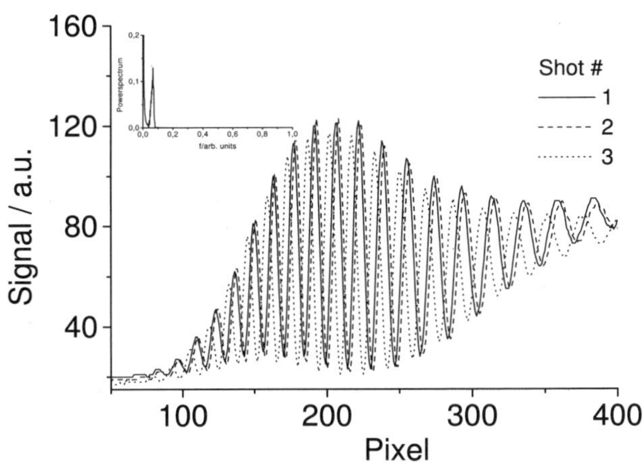

Fig. 2. Typical fringe pattern of the $f-2f$ interference obtained for three consecutive laser shots. Note the enormous contrast and number of fringes, which are necessary for a precision determination of the carrier envelope phase. The insert shows the corresponding power spectrum.

The laser beam leaves the second hollow fibre with 15μJ/pulse. 5% of this is used in the $f-2f$ interferometer [7]. For the interferometer the beam is focused into a 100μm BBO type II crystal. Since the fundamental beam is slightly chirped after the second hollow fibre there is a small delay between the second harmonic and the blue component of the fundamental. In the spectral overlap region between the fundamental and the generated second harmonic we can thus obtain a self-referencing interference. Numerical analysis of the interference fringes yields an (uncalibrated) value for the CE-phase. Since the second harmonic and the fundamental are at orthogonal polarizations the interference can only be seen if the ratio between both polarizations is adjusted correctly. By tuning the angle of a Glen-Thomson polarizer (P) the fringes can be maximized. After the polarizer the beam is analyzed by a high resolution spectrometer (ORIEL MS125). Attached to the spectrometer is a laser-clocked fast line-scan camera (CCD, BASLER L103) which is read for each shot by a computer. Fig. 2 shows the interference obtained for three consecutive laser shots. As can be seen the fringes of the interference pattern for each pulse are shifted relative to the fringes of subsequent pulses since the value of the CE-phase varies arbitrarily with each laser shot. Note the strong fringe contrast, which is much better than in [7]. The phase of each pulse is calculated by fast fourier transformation (FFT) of the spectral data. We are presently able to fourier-transform 512 data points within 0.5ms. It is thus possible to correlate photo-electron current to the FFT result at a 1kHz repetition rate. Further improvements in data processing speed can be expected.

The remaining 95% of the laser beam is compressed to 5fs using a pair of broadband multilayer chirped mirrors (CMP). A pair of thin wedges (W) made of fused silica is placed into the beam in order to fine tune the dispersion. In figure 1 a typical autocorrelation of the laser pulses was included. Autocorrelations have been measured with a special technique having minimum dispersion for one octave bandwidth. After compression 12μJ/pulse are available to perform the experiment. These short pulses are directed into a miniaturized photoelectron-photoion spectrometer (PEPIS, left side of Fig. 1). The spectrometer consists of two arms: A high resolution electron spectrometer based on a hemispherical analyzer (not used in the present experiment). The second arm of the PEPIS consists of a 500μm aperture followed by a time-of-flight spectrometer with a drift tube of 25mm in length. This arm can be used for ion and charge state detection, but also for electron current detection. The spectrometer is operated at a background pressure of $\approx 5 \times 10^{-10}$mbar. The laser beam was focussend into the ionization region of the PEPIS using a f=70mm spherical silver mirror. The target gas was 2×10^{-6}mbar of Xenon, which was introduced into the chamber by means of a precision capillary leak. Electron and ion signals were detected using fast, dual microchannel-plate detectors and then quantitatively evaluated with a 120ps minimum gate width sampling Boxcar (BC, Becker&Hickl PCS-150). Using this setup without any extraction voltages applied to the ionization region it is possible to detect the electron current per laser shot emitted to one side and to correlate it to the CE-phase. We have optimized the laser pulse duration by maximizing the electron current per laser shot with slight adjustments of the pair of wedges. The total electron yield is sensitive enough to the pulse duration so that in this way the shortest input pulses (further verified by autocorrelation) were obtained. For the CE-phase measurement 1 million laser shots were acquired within approximately 20 minutes. Over this time the conditions of the laser system do not change significantly. The result was added up in a histogram and a clear phase effect was immediately visible. Fig. 3 (upper panel) shows the resulting histogram of electron current versus CE-phase. We have checked the result with insertion of 0.7 mm glass into the beam. Positive dispersion caused a longer pulse duration so that phase effects vanished, as expected. As a further test we have calculated two additional histograms from the data. In the first histogram the measured electron signal was correlated to the CE-phase of the previous laser pulse and in the second histogram the same electron signal was correlated to the CE-phase of the next laser pulse. Since in our laser system the CE-phase of each pulse varies randomly with respect to the next laser pulse, no phase effect in these histograms is to be expected, which was indeed the case (see Fig. 3,

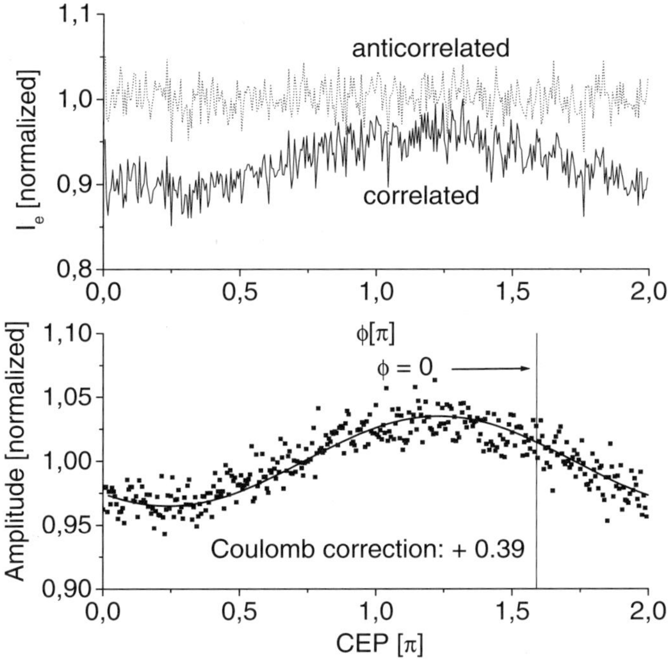

Fig. 3. Upper panel: Histogram of the electron signal adding up in dependence on the carrier envelope phase ϕ. Also shown is an anticorrelation obtained when the electron current of the actual laser shot is related to ϕ retrieved from the previous laser shot. Lower panel: The correlated signal divided by the anticorrelated signal yields the true dependence of the electron current on ϕ. The position where $\phi = 0$ can be retrieved if Coulomb correction ($\approx +0.39\pi$, see [2]) is taken into account.

upper panel). Therefore we conclude that the CE-phase can be observed with our setup and manifests as a sine-wave with a peek-to-peek variation of approximately 10% of the signal. This value agrees very well with a recent stereo-ATI experiment conducted in Vienna by G. Paulus et *al.* [2] on a stabilized laser system. The lower panel of Fig. 3 shows the ratio between the correlated and the anticorrelated histogram. A cosine curve can be immediately fitted to the data. It can be seen that the maximum electron yield occurs at phases between 1.15π and 1.25π, so that the position of $\phi = 0$ (e.g. *cosine-like pulse*, where the peak of the carrier coincides with the peak of the envelope and the electric field points towards the detector) of the CE-phase is situated at $\approx 1.59\pi$. For this evaluation we have taken a Coulomb correction factor of $0.39\pi (= 30^o)$ for low energy electrons (Paulus *et al.* [2]) into account.

In conclusion, we have demonstrated here an alternative method of determining phase effects using a non phase stable laser. The technique may be of interest to those who wish to observe phase effects in nonlinear processes but are unable or unwilling to use phase stable oscillators or multiple control loop techniques for amplified systems.

Acknowledgement. The authors would like to thank the Austrian Fund for the Advancement of Science (FWF, project P14447 and SFB F1616).

References

1. Baltuska A. et al. *Nature*, 421:611–615, 2003.
2. Paulus G. et al. *Physical Review Letters*, June 2003. submitted.
3. Cormier E. and Lampropoulos P. *European Physics Journal D*, 1998.
4. Dietrich P. et al. *Optics Letters*, 2000.
5. Paulus G. et al. *Nature*, 414:182–184, 2001.
6. Nisoli M. et al. *Optics Letters*, 22:522–524, 1997.
7. Kakehata M. et al. *Optics Letters*, 26(18):1436–8, 2001.

Stabilization and Characterization of the Carrier-Envelope Phase in Intense Few-Cycle Pulses

A. Baltuška[1], M. Uiberacker[1], E. Goulielmakis[1], Th. Udem[2], M. Hentschel[3], R. Kienberger[1], R. Holzwarth[2], T.W. Hänsch[2], and F. Krausz[1,2]

[1] Photonics Institute, Vienna University of Technology, Gusshausstrasse 27, A-1040 Vienna, Austria
[2] Max-Planck-Institut für Quantenoptik, Hans-Kopfermann-Strasse 1, D-85748 Garching, Germany
[3] Femtolasers Produktions GmbH, Fernkorngasse 10, A-1100 Vienna, Austria

Summary. We demonstrate a phase-stabilized 5-fs 0.1-TW amplified laser system that delivers pulses with exactly reproducible electric fields. Full temporal characterization of the resultant intense few-cycle light waveforms was achieved by combining conventional pulse characterization techniques with observing the signature of soft-X-ray emission.

Generation of high-intensity ultrashort laser pulses with a precisely controlled evolution of the electromagnetic field is a prerequisite for a number of applications in extreme nonlinear optics [1]. Control over the carrier-envelope phase (CEP), which describes how the intensity of the strongest field peak fits under the pulse envelope, should dramatically enhance the temporal resolution of attosecond metrology [2,3], improve stability of laser-driven sources of high-harmonic radiation [4], and enable production of identical isolated attosecond soft-X-ray bursts repeated with every laser shot [5]. The technique of CEP stabilization in mode-locked MHz—GHz laser oscillators was demonstrated a few years ago [6,7] and is now routine. However, its extension to TW-class amplifiers operating at 1—10-kHz repetition rates has not been achieved previously. Here, we present phase-controlled pulse amplification, based on active [5] phase stabilization and implemented in a standard multipass laser amplifier. Furthermore, we demonstrate a simple and intuitive technique for calibration of the actual CEP value. This method, in combination with a standard pulse-envelope measurement, allows us to characterize the precise shape of the electric field of the emitted light waveform.

The schematic of our CEP-stabilized amplified laser system is presented in Fig.1. To solve the problem with the phase control at a kHz repetition rate, we first stabilize the CEP of the Ti:sapphire modelocked seed oscillator (Femtosource Compact Pro, Femtolasers GmbH) operating at $f_{rep}\approx 80$ MHz. Following this primary phase control, the seed pulses are then injected into a 9-pass Ti:sapphire amplifier (Femtopower, Femtolasers GmbH), which is configured to minimize the total nonlinear phase shift accumulated throughout the process of pulse amplification. As the result, the remaining CEP pulse-to-pulse phase variations, acquired in the amplifier, are mainly caused by beam pointing instability. The latter leads to slight pathway changes for the pulse propagation through dispesive elements, such as the prisms of the pulse compressor. These fluctuations are attributed to air turbulence, mechanical vibrations, drift of the thermal lens, and thermal expansion of the laser base. The resultant CEP drift behind the amplifier is relatively slow, with a characteristic time scale of several seconds or higher, which enables us to track and compensate it by a secondary phase control loop.

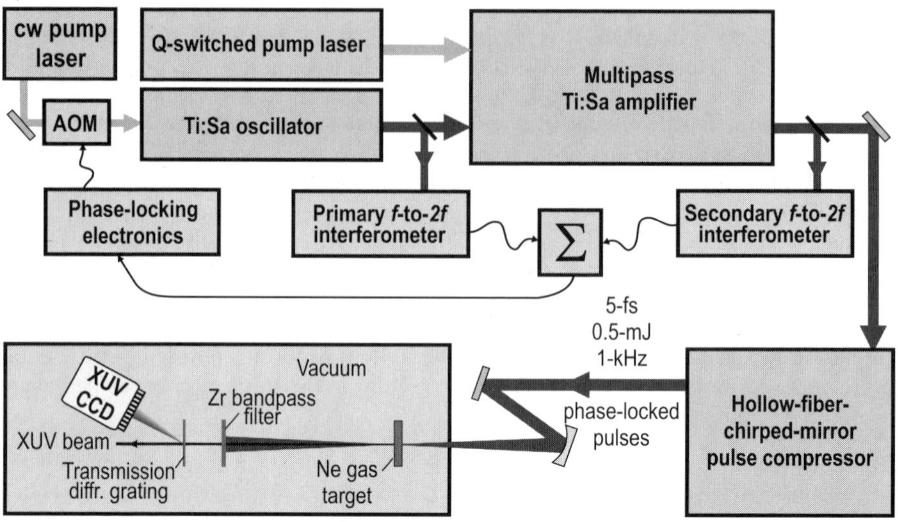

Fig. 1. Schematic of the experimental setup

The phase deviation at the output of the seed oscillator is monitored by a nonlinear interferometer that measures the interference beat between spectrally overlapping second-harmonic and fundamental components of the spectrum (the f-to-$2f$ arrangement [8, 9]). An octave-wide spectral broadening that is necessary to achieve the overlap between the harmonics is obtained in a 3-cm long piece of a photonic crystal fiber. The beat signal is then phase-locked to a reference oscillator (1/4 of the laser repetition rate) through a feedback that modulates the intensity of the pump laser beam with an acousto-optical modulator (AOM in Fig.1). The phase drift at the output of the amplifier is tracked by another f-to-$2f$ interferometer, which uses bulk sapphire as a medium for octave-wide spectral broadening, and is evaluated by Fourier transform analysis of the f-to-$2f$ spectral interference pattern [10]. The feedback proportional to the value measured in this second f-to-$2f$ device is used to adjust the phase of the reference frequency thereby allowing phase pre-compensation in the oscillator in order to keep the phase value steady at the output of the whole laser system.

Using such intense phase-stabilized driver pulses for the generation of soft X-rays, we demonstrate several remarkable effects that could not have been previously observed with conventional laser amplifiers. A portion of the output laser intensity was gently focused onto a target of atomic gas (neon), which results in photoionization, followed by acceleration of the photoelectrons, and emission of soft X-rays upon the electron re-collision with its parent atom. The resultant X-ray spectra for different driving laser pulses were examined with a simple spectrometer based on a transmission diffraction grating and an XUV CCD camera (Roper Scientific). Figure 2 shows the soft-X-ray spectra obtained with various CEP settings. The pulses were temporally broadened to 10 fs to sharpen spectral features through the entire spectrum up to the cutoff frequency (i.e. the highest frequency of the emitted X rays). The spectra shown by solid curves in Fig.2 (with phase stabilization) are distinctly different from those measured with a phase-unlocked laser (shaded contour) especially in the region near the cutoff.

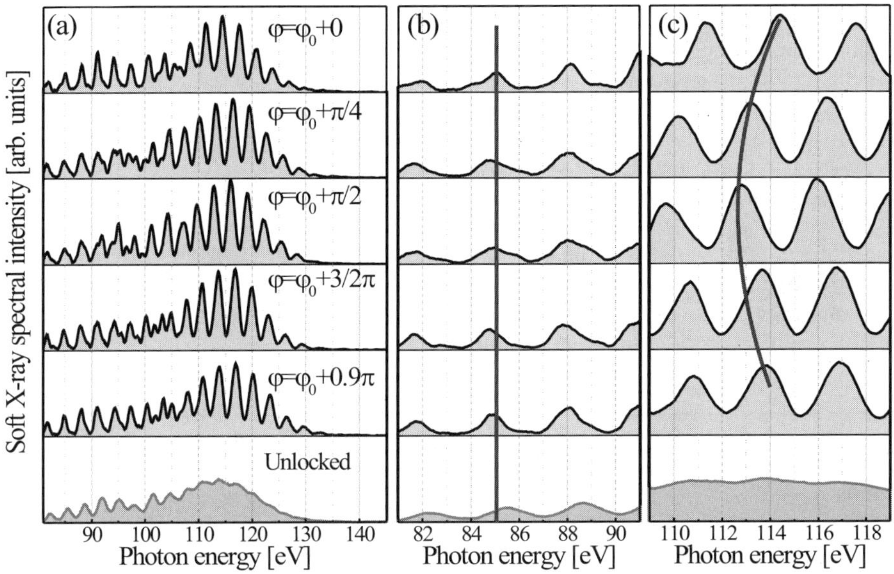

Fig. 2. Measured spectral intensity of soft-X-ray emission from ionizing atoms driven by 10-fs pulses. (a) Full-range spectra. (b,c) Close-ups on spectral regions with distinctly different spectral dependencies on the carrier-envelope phase.

A CEP-dependent shift of the harmonic peaks, clearly visible in our data, has been previously predicted by numerical simulations [11, 12]. The variation of the position in the spectrum indicates that the peaks that are visible in the few-cycle-driven harmonic spectra, strictly speaking, do not represent genuine harmonics, because the laser frequency is kept unchanged and its harmonics are expected to remain invariant as well. The shift of these "quasi-harmonics" is a direct consequence of the variation of the field amplitude within the light oscillation period, T_0, and the confinement of emission to within only a couple of oscillation cycles. The "genuine" harmonics at low frequencies (Fig.2b) are produced by many successive peaks of the laser field, whereas the "spurious" harmonics (Fig.2c), approaching the cutoff, are formed by very few field peaks. In this regard, it is interesting to note the anomaly in the energy interval of 95-105 eV. This "intermediate" region features quasi-harmonic peaks with a period of one laser photon energy. This is remarkable, since such periodicity corresponds to laser field peaks separated by T_0 rather than by the usual $T_0/2$ (i.e. by the time spacing between the nearest half cycles). In our understanding, the reason for this transient behavior is the involvement of three neighboring half cycles capable of generating X-rays at that frequency,—a condition that cannot be fulfilled at both lower and higher photon energies.

The use of even shorter, 5-fs driving laser pulses, that exhibit a correspondingly higher cycle-to-cycle variation of the electric field strength, allows us to identify directly the value of CEP, φ. This becomes possible because the neighboring X-ray spectral peaks broaden and overlap into a continuum toward the cutoff frequency. Since the X-ray photons at the cutoff energy can only be emitted after the interaction with the most intense electric field peak, pulses in which this field peak coincides with the envelope maximum (φ=0, the so-called "cosine" carrier wave) render smooth cutoff spectra. By contrast, the highest spectral modulation of the cutoff region corresponds to $\varphi=\pi/2$ (the "sine" carrier), since such a pulse carries two strong field peaks capable of overcoming the cutoff generation threshold. Consequently, the "sine"-carrier pulse yields two cutoff-energy X-ray bursts that in a spectral measurement manifest themselves by a pronounced modulation in the vicinity of the highest frequency.

Figure 3 illustrates the predicted difference between φ=0 and $\varphi=\pi/2$ in the case of a 5-fs driving laser pulse. Fig.3a and b show the markedly different cutoff spectra associated with these two phase settings, whereas Fig.3c and d depict the corresponding temporal profiles of the X-ray emission confined to a single (φ=0) and a double ($\varphi=\pi/2$) high-energy radiation burst.

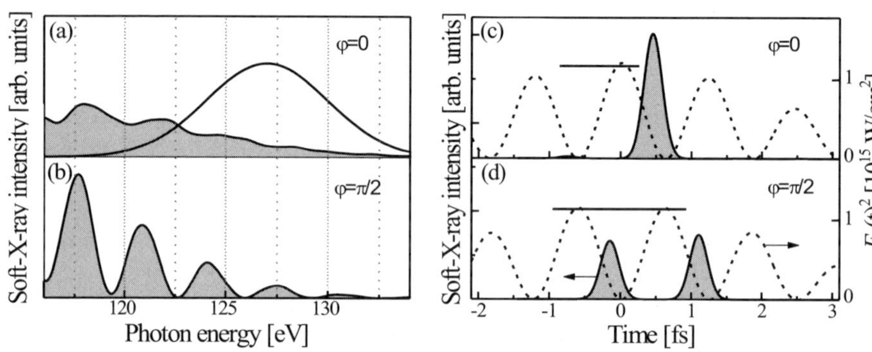

Fig. 3. Numerical simulations of few-cycle-driven coherent soft-X-ray emission from ionizing atoms. (a,b) Cutoff-range spectra for φ=0 and $\varphi=\pi/2$. (c,d) The solid curves depict the temporal intensity profile of the cutoff X-ray radiation filtered through a Gaussian bandpass filter with a full-width at half maximum of 7 eV (solid contour in (a)), whereas the dashed curves plot instantaneous laser field intensity $E_L(t)^2$. Horizontal bars in (c) and (d) indicate the intensity threshold required to produce the cutoff X-rays.

Figure 4 shows experimentally measured soft-X-ray spectra for φ=0 and $\varphi=\pi/2$ and pulse electric fields identified for these phase values. This method allows phase calibration with the ambiguity of $n\pi$, where the residual ambiguity in the determination of φ relates to the inversion symmetry of the interaction with the atomic gas medium. A π-shift in φ results in no change of the light waveform other than reversal of the direction of the electromagnetic field vectors. In an isotropic medium, such a phase flip cannot lead to a measurable consequence in the soft-X-ray emission studied here. CEP calibration without the field reversal ambiguity can be performed in photoemission experiments on both isotropic as well as anisotropic media by spatially and temporally resolving photoelectron trajectories. This capability has been demonstrated with our phase-stabilized amplifier in the measurements of X-ray driven electron photoemission [13] and above threshold ionization [14].

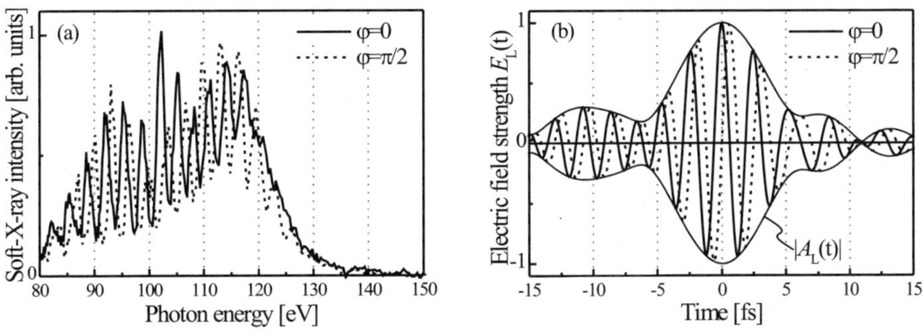

Fig. 4. Soft-X-ray generation with amplified pulses stabilized at two different CEP settings. (a) Measured spectral intensity of soft-X-ray emission from ionizing atoms driven by 5-fs pulses; (b) Electric field reconstruction for the corresponding CEP settings.

We believe that the developed phase-stabilized laser amplifier and the demonstrated technique of carrier-envelope phase measurement opens new technical possibilities to the field of extreme nonlinear optics and, in particular, should significantly benefit metrology and spectroscopy with attosecond pulses.

Acknowledgement. This work was sponsored by the Austrian Fund for the Advancement of Science (FWF, grants P15382, Z63, F016) and by the European ATTO network.

References

1. T. Brabec and F. Krausz: Rev. Mod. Phys. **72**, 545-591, (2000)
2. M. Hentschel, R. Kienberger, C. Spielmann, G. A. Reider, N. Milosevic, T. Brabec, P. Corkum, U. Heinzmann, M. Drescher, and F. Krausz: Nature **414**, 509-513, (2001)
3. R. Kienberger, M. Hentschel, M. Uiberacker, C. Spielmann, M. Kitzler, A. Scrinzi, M. Wieland, T. Westerwalbesloh, U. Kleineberg, U. Heinzmann, M. Drescher, and F. Krausz: Science **297**, 1144-1148, (2002)
4. M. Drescher, M. Hentschel, R. Kienberger, G. Tempea, C. Spielmann, G. A. Reider, P. B. Corkum, and F. Krausz: Science **291**, 1923-1927, (2001)

5. A. Baltuška, T. Udem, M. Uiberacker, M. Hentschel, E. Goulielmakis, C. Gohle, R. Holzwarth, V. S. Yakovlev, A. Scrinzi, T. W. Hänsch, and F. Krausz: Nature **421**, 611-615, (2003)
6. D. J. Jones, S. A. Diddams, J. K. Ranka, A. Stentz, R. S. Windeler, J. L. Hall, and S. T. Cundiff: Science **288**, 635-639, (2000)
7. A. Apolonski, A. Poppe, G. Tempea, C. Spielmann, T. Udem, R. Holzwarth, T. W. Hänsch, and F. Krausz: Phys. Rev. Lett. **85**, 740-743, (2000)
8. S. T. Cundiff, D. J. Jones, J. Ye, T. Cundiff, and J. L. Hall: Rev. Sci. Instrum. **72**, 3749-3771, (2001)
9. S. T. Cundiff: J. Phys. D: Appl. Phys. **35**, R43-R59, (2002)
10. M. Kakehata, H. Takada, Y. Kobayashi, K. Torizuka, Y. Fujihara, T. Homma, and H. Takahashi: Opt. Lett. **26**, 1436-1438, (2001)
11. G. Tempea, M. Geissler, and T. Brabec: J. Opt. Soc. Am. B **16**, 669-673, (1999)
12. E. Priori, G. Cerullo, M. Nisoli, S. Stagira, S. De Silvestri, P. Villoresi, L. Poletto, P. Ceccherini, C. Altucci, R. Bruzzese, and C. de Lisio: Phys. Rev. A **61**, 063801/1-8, (2000)
13. R. Kienberger, E. Goulielmakis, A. Baltuška, M. Uiberacker, Th. Westerwalbesloh, U. Kleineberg, U. Heinzmann, M. Drescher, and F. Krausz:'Reproducible generation and measurement of isolated sub-fs XUV pulses with phase-controlled few-cycle light', In: *Ultrafast Optics IV, Vienna, Austria, June 29 - July 3, 2003*, ed. by F. Krausz, G. Korn, P. B. Corkum, and I. A. Walmsley (Springer, 2003). See this volume, page 241.
14. F. Lindner, G.G. Paulus, H. Walther, A. Baltuška, E. Goulielmakis, M. Lezius, and F. Krausz: 'Measurement of the absolute phase of few-cycle laser pulses', In: *Ultrafast Optics IV, Vienna, Austria, June 29 - July 3, 2003*, ed. by F. Krausz, G. Korn, P. B. Corkum, and I. A. Walmsley (Springer, 2003). See this volume, page 209.

Measurement of the Absolute Phase of few-Cycle Laser Pulses

F. Lindner[1], G.G. Paulus[1,2,3], H. Walther[1,2], A. Baltuška[4], E. Goulielmakis[4], M. Lezius[4,5], and F. Krausz[1,4]

[1] Max-Planck-Institut für Quantenoptik, Hans-Kopfermann-Str. 1, 85748 Garching, Germany

[2] Ludwig-Maximilians-Universität München, Am Coulombwall 1, 85748 Garching, Germany

[3] Texas A&M University, Department of Physics, College Station, TX 77843-4242, USA

[4] Institut für Photonik, Technische Universität Wien, Gusshausstr. 27, A-1040 Wien, Austria

[5] Institut für Ionenphysik, Universität Innsbruck, Technikerstr. 25, A-6020 Innsbruck, Austria

. Recent developments of laser technology led to the generation of laser pulses with a duration (measured at full-width half maximum) of less than 5fs [1]. At a typical wavelength of 750nm, an optical cycle has a duration of 2.5fs; therefore, the pulses consist of only very few optical cycles (few-cycle pulses). Since the pulse envelope varies almost as fast as the electromagnetic field itself, the shape of the laser's field will significantly depend on the phase of the carrier frequency with respect to the envelope, the so-called *absolute* phase. The electric field *E(t)* of a laser pulse can be written as

$$E(t) = E_0(t)\cos(\omega t + \phi), \tag{1}$$

where $E_0(t)$ denotes the envelope of the pulse, ω the carrier angular frequency of the laser, and ϕ the absolute phase. The convention of choosing t = 0 at the maximum of the envelope is used. Thus, $\phi = 0$ corresponds to a *cosine-like* pulse with the absolute maximum of the electric field pointing to the positive direction. Accordingly, we speak of *sine-like* pulses for $\phi = \pm\pi/2$. Clear evidence of phase effects has been observed in photoionization with few-cycle pulses without phase-control [2] and in soft X-ray generation with phase-stabilized pulses [3]. However, no ultimate and unambiguous phase measurement has been performed so far.

The electric field of few-cycle pulses shows asymmetries directly connected with the absolute phase. Therefore, nonlinear photoionization is a possible approach to determine its value. In our experiments xenon atoms were ionized with peak intensities of nearly 10^{14}W/cm^2 of a phase-stabilized few-cycle pulses laser system [3]. Absorption of 7 photons is necessary in order to subdue the ionization threshold. As a matter of fact, processes of much higher order, i.e. absorption of photons in excess to those necessary for ionization can be observed and thus correspondingly high photoelectron energies. This phenomenon is known as above-threshold ionization (ATI) [4]. Our experimental approach to analyze this phenomenon relies on time-of-flight spectroscopy (see Fig. 1). A fast photodiode registers the arrival of the laser pulse and triggers two high-resolution (500ps) clocks (FAST 7886). Photoelectrons are generated inside a well-shielded ultra-high vacuum apparatus in the focus of the laser beam. Two micro-channel plates (diameter 18mm, Burle BiPolar TOF), each connected with one of the clocks, are placed at distances of 0.5m at opposite sides (left and right) of the focus. They detect photoelectrons emitted parallel to the polarization axis (*stereo-ATI spectrometer*). The respective time-of-flights are transferred to a computer and can be converted into kinetic electron energies.

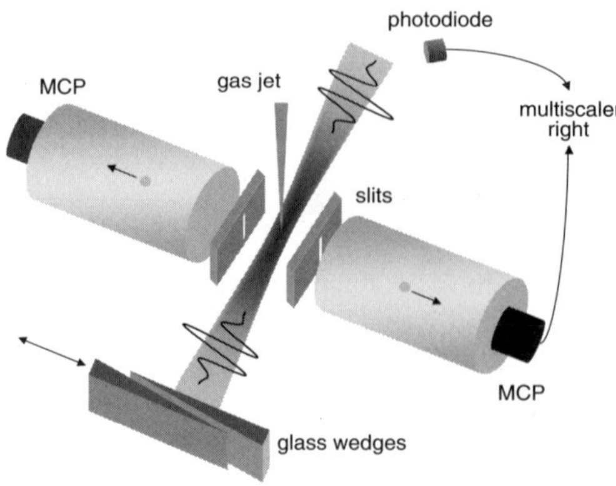

Fig. 1. Sketch of the experimental setup. Xenon atoms are ionized in the focus of the beam. The focal length is 250 mm, the pulse energy attenuated to 20 μJ. The laser polarization is parallel to the time-of-flight tubes. Slits of 250 μm width are used to discriminate a selected region of the focal volume. A fast photodiode and 2 micro-channel plates (MCP) allow determination of the time of flight of the electrons by means of two identical high-resolution clocks (only the right one is indicated in the figure).

Figure 2 shows a series of ATI spectra measured in the two directions corresponding to different absolute phases. Pulses with known phase differences $\Delta\phi$ can be realized by delaying the envelope with respect to the carrier. Since the carrier propagates with the phase velocity and the envelope with the group velocity of light, glass dispersion is one possibility to achieve this. At a central wavelength of 760nm adding 52μm of fused silica will change ϕ by 2π without affecting the pulse duration distinctly. Thus, two glass wedges, which can be shifted with respect to each other, allow adjusting any phase (see Fig. 1).

The spectra detected for different absolute phases significantly differ in many aspects. The total electron yield, though hidden in the semi logarithmic scale of the graphs, shows alternating maxima and minima in both directions. Even more striking is the high-energy part of the spectra, where the difference between emission to the left and to the right approaches one order of magnitude. One can verify that $\Delta\phi = \pi$ corresponds to a change from left to right while $\Delta\phi = 2\pi$ exactly reproduces the spectra. The agreement of the measured periodicity with that derived from the difference of group and phase velocity in the glass of the wedges is excellent, indicating that the phase can be changed and controlled with very high precision. On the other hand, for a phase assignment, or what is the same for determining the shape of the laser field, the theory of ATI needs to be taken into account. For a recent review see reference [5]. In short, both semiclassical and quantum mechanical

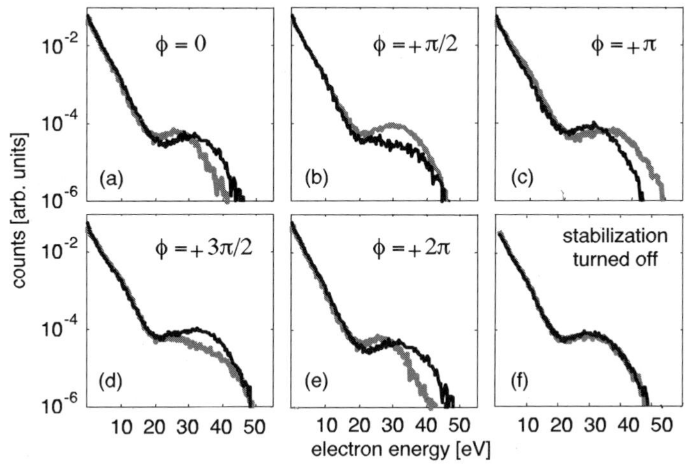

Fig. 2. Photoelectron spectra measured in the left (gray line) and right (black line) emission direction, for different absolute phases. The corresponding value of the attributed phase is indicated. When the phase-stabilization is switched off (f), identical spectra are measured in the two opposite directions as expected.

models [6] yield the basic result that the most energetic photoelectrons are mainly emitted in the direction of the peak of the electric field. Simply by inspecting the cutoff of the measured electron spectra in Fig. 2, one can conclude that the spectra in (a) and (e) correspond to a *cosine-like* pulse pointing to the right direction ($\phi = 0 \pm 2\pi$), whereas the spectra in (c) correspond to a *cosine-like* pulse pointing to the left direction ($\phi = \pm\pi$). Accordingly, the spectra in (b) and (d) correspond to an electric field having the same peak strength pointing to the left and to the right, i.e. to *sine-like* pulses. These results constitute the ultimate measurement of the absolute phase, since they remove the left-right ambiguity intrinsically connected with high-harmonic generation experiments [3]. Note that the measurement is fast, since many electrons per laser shot can be detected (more than 50 in each direction in these experiments). Furthermore, the target gas density is so low that the laser beam is not affected. Therefore, the measurement is non-invasive.

As a closing example for the power of the stereo-ATI scheme, we present results from another approach to change the absolute phase. The phase velocity in a laser focus is super luminal, while the group velocity is very close to the speed of light. This leads to the well-known phenomenon of the geometrical or *Gouy* phase. The absolute phase should change approximately in the same way as the Gouy phase. Hence, a phase shift of π is expected if the interaction region is moved from a few Rayleigh lengths before to a few after the focus (see also the electric field of Fig. 1). One possibility to observe this is to compare the electron yield in the two directions before and after the focus. For instance, if one detects dominant left emission before the focus, the opposite is expected after the focus. By varying the amount of glass by shifting the wedge, one can verify this for several absolute phases. The result of the respective measurement is shown in Fig. 3: the ratio of the total electron yield in the two directions is shown as a function of the glass thickness. The measurement

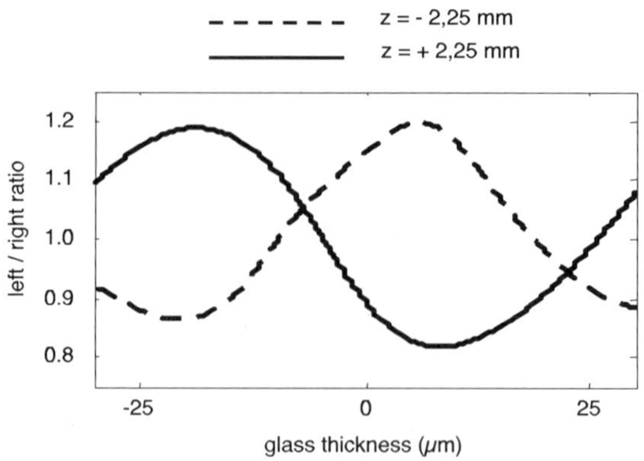

Fig. 3. Shifting the glass wedges changes the glass thickness and, as a consequence, the absolute phase of the pulse. The ratio (left/right) of the electron yield exhibits a clear oscillatory behavior. The measurement was performed before (dashed line) and after (solid line) the focus. The evident phase shift is due to the *Gouy* phase in the laser focus.

has been performed before (solid line) and after (dashed line) the focus. The two curves clearly indicate the expected phase shift.

Acknowledgements

This work was supported in part by the Austrian Fund for the advancement of Science (FWF, project SFB F1616).

References

1. Nisoli, M., De Silvestri, S., Svelto, O., Szipocs, R., Ferencz, K., Spielmann, Ch., Sartania, S., and Krausz, F.: Compression of high-energy laser pulses below 5 fs. Opt. Lett. **22**, 522-524 (1997).
2. Paulus, G.G., Grasbon, F., Walther, H., Villoresi, P., Nisoli, M., Stagira, S., Priori, E., and De Silvestri, S.: Absolute phase phenomena in photoionization with few-cycle laser pulses. Nature **414**, 182-184 (2001).
3. Baltuska, A., Udem, Th., Uiberacker, M., Hentschel, M., Goulielmakis, E., Gohle, Ch., Holzwarth, R., Yakovlev, V.S., Scrinzi, A., Hänsch, T.W., and Krausz, F.: Attosecond control of electronic processes by intense light fields. Nature **421**, 611-615 (2003).
4. Agostini, P., Fabre, F., Mainfray, G., Petite, G., and Rahman, N. K.: Free-Free Transitions following six-photon ionization of xenon atoms. Phys. Rev. Lett. **42**, 1127-1130 (1979).
5. Becker, W., Grasbon, F., Kopold, R., Milosevic, D., Paulus, G.G., and Walther, H.: Above-threshold ionization: from classical features to quantum effects. Adv. At. Mol. Opt. Phys. **48**, 35-98 (2002).

6. Milosevic, D., Paulus, G.G., and Becker, W.: High-order above-threshold ionization with few-cycle pulse: a meter of the absolute phase. Opt. Expr. **11**, 1418-1429 (2003).

Part IV

Extreme UV Radiation and Sub-Femtosecond Pulses

Quasi-Phase-Matching of High Harmonic EUV Generation at Very High Ionization Levels

E. A. Gibson[1], A. Paul[1], N. L. Wagner[1], R. Tobey[1], E. Gagnon[1], D. Gaudiosi[1], M. M. Murnane[1], H. C. Kapteyn[1], and I.P. Christov[2]

[1] JILA, University of Colorado, Boulder CO 80309-0440, USA
emily.gibson@colorado.edu
[2] Department of Physics, Sofia University, 1126 Sofia, Bulgaria

Summary. We experimentally demonstrate enhanced conversion efficiency for high harmonic generation in fully ionized gas using quasi-phase-matching. We report the first observation of high harmonic generation in Argon up to 180eV.

High-order harmonic generation (HHG) is a useful source of coherent light in the extreme ultraviolet to soft x-ray region of the spectrum. In HHG, an intense laser pulse is focused into a material, typically a gas. The extreme nonlinear interaction with the atoms results in generation of high-order harmonics of the fundamental laser that emerge as a coherent, low-divergence beam. The conversion efficiency, however, is limited by the difficulty of phase-matching the process, particularly at high ionization levels where the higher energy harmonics are generated.

One method to improve the conversion efficiency of HHG is to focus the fundamental light into a gas-filled hollow-core glass waveguide [1]. Phase-matching of the HHG conversion process can then be achieved by adjusting the gas pressure to balance the various contributions to the phase velocity of the light from the plasma, waveguide and neutral gas. Since these contributions are of opposite sign, phase-matching can be achieved and the harmonic signal increases quadratically over the interaction length, neglecting absorption. However, this method of phase-matching only works for low levels of ionization ($\sim 5\%$). At higher ionization levels, the plasma contribution becomes much greater than the neutral gas contribution - making phase-matching impossible. When the phase-velocities are not equal, the harmonic emission builds up periodically over a coherence length, and then interferes with out-of-phase light generated later along the length of the interaction. Since the highest harmonic orders are created at the peak of the pulse where the ionization level is greatest, this effectively limits the highest EUV photon energy that can be generated efficiently.

In recent work [2, 3], we demonstrated that by modulating the diameter of the waveguide, we could quasi-phase-match (QPM) the EUV conversion process up to ionization levels of $\sim 10\%$. The effect of the fiber modulations is to periodically change the driving laser intensity along the interaction length. Because the EUV phase is dependent on the driving laser intensity, the fiber modulations can both adjust the phase of the EUV light, and also suppress the EUV generation in certain regions of the fiber. This periodic dependence of the strength and phase of the nonlinearity enables QPM. In our past work, a large enhancement of the highest order harmonics was observed in Helium and Neon, corresponding to compensating for ionization levels below 10%. In this work, we show it is possible to quasi-phase-match HHG at dramatically high ionization levels of 100% in Argon and 35% in Neon. This work overcomes a critical limitation for extending HHG to higher energies where ionization destroys the build-up of signal. As a result of QPM, we enhance the harmonic emission between 160 - 180eV in Neon and Argon, which are the highest harmonics to be phase matched to date.

In this experiment, we focus light from a 1kHz Ti:sapphire laser system producing 23fs duration pulses with a typical pulse energy of 1.1mJ into either a straight or modulated 150μm diameter waveguide filled with Argon or Neon. The modulated fiber we use is shown in Figure 1. The modulations of the fiber inner diameter are approximately sinusoidal with a depth of $\sim$ 10% and a period of 0.25mm and the total fiber length is 2.5cm. The harmonic emission from the fiber passes through 0.4μm of Zirconium to block the fundamental light, and is spectrally dispersed by a Hettrick grazing-incidence EUV spectrometer and focused onto an Andor CCD camera. The HHG spectra are calibrated by recording the positions on the CCD of the Silicon and Boron absorption edges, at 99.9eV and 188.35eV respectively, to obtain the linear conversion from CCD position to wavelength.

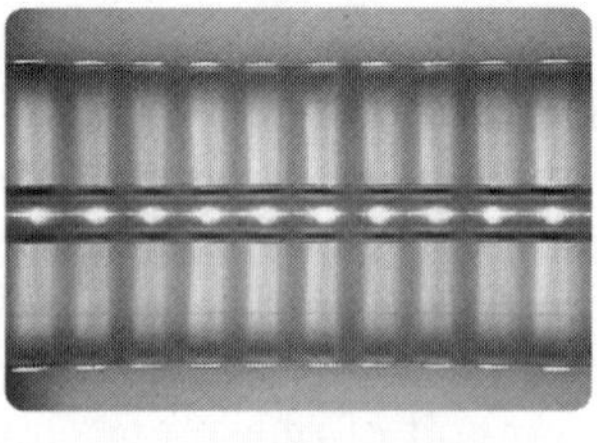

Fig. 1. *Optical microscope image of a 0.25mm period modulated waveguide.*

Figure 2 shows the harmonic emission from both a straight and a modulated waveguide filled with Neon. Both spectra are taken with the same driving pulse characteristics, gas pressure, fiber coupling efficiency, and CCD exposure time. The harmonics generated in the straight waveguide only extend to $\sim$ 135eV. For the higher harmonic orders, ionization creates a large a phase mis-match which greatly reduces the signal. In contrast, the harmonics from the modulated waveguide extend to $\sim$ 170eV. Note that for the higher harmonic orders, the modulated waveguide increases the signal by orders of magnitude over the straight fiber.

Figure 3 shows HHG emission in Argon from straight and modulated waveguides at low pressures. The modulated waveguide increases the HHG

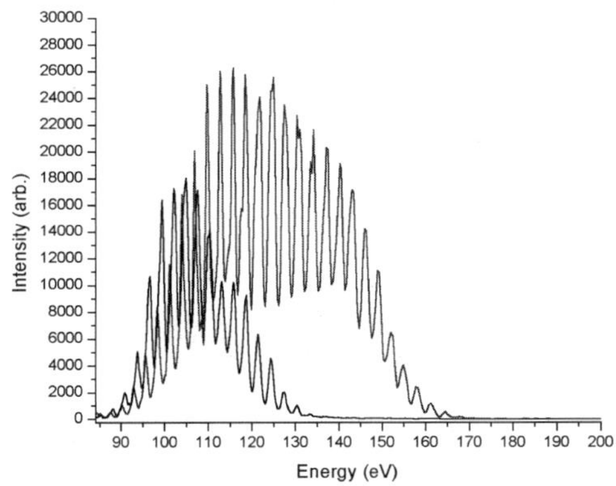

Fig. 2. *Harmonic emission from the straight waveguide (black) and the 0.25mm-period modulated QPM waveguide (gray) filled with 11 Torr of Neon showing the large enhancements of the highest harmonics due to QPM. The ionization levels near cutoff correspond to* $\sim$ *35%.*

flux by a factor of 2 to 3 over the entire spectrum. We observe HHG up to 180eV - the highest harmonic previously observed in Argon was 100eV using a Ti:Sapphire driving laser. [4] The use of the fiber geometry makes it possible for us to observe such high harmonics, because it reduces laser defocusing by the plasma. The Argon data was taken with a slightly higher laser intensity than the Neon case, explaining the higher cutoff energy.

Using the ADK theory after Coulomb correction, [5–7] we can calculate the level of ionization present at the time a given harmonic is generated by our laser pulse. The peak intensity of our driving pulse is obtained from our measurements of the pulse duration, input pulse energy, coupling efficiency, and fiber diameter. We also performed 1-D SAE quantum calculations which agree with the ADK model to within 5%. We calculate 35% ionization at the peak of the driving pulse of Neon and complete ionization of Argon, and 30% doubly ionized Argon. We can then calculate the amount of phase-mismatch from the following equation:

$$\Delta k = \frac{qu_{11}^2\lambda}{4\pi a^2} + P\eta N_{atm} r_e (q\lambda - \lambda/q) - \frac{2\pi(1-\eta)Pq}{\lambda}(\delta(\lambda) - \delta(\lambda/q)) \quad (1)$$

where λ, q, a, u_{11}, η, P, N_{atm}, r_e, and δ are the fundamental wavelength, harmonic order, waveguide radius, the first zero of the Bessel function J_0, ionization fraction, gas pressure in atmospheres, number density at 1 atm, classical electron radius, and the index of refraction of gas at 1 atm respectively. [8] From equation (1), we calculate Δk $\sim$ 26,000 m^{-1} for the cutoff

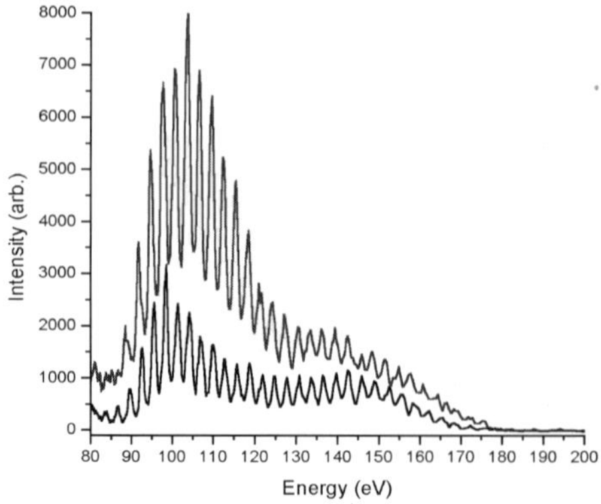

Fig. 3. *Harmonic emission from a straight (black) and 0.25mm-period modulated waveguide (gray) filled with low-pressure 7 Torr of Argon. QPM results in an enhancement in the HHG emission over the entire photon energy range. These are the highest harmonics ever observed in Argon gas. Near cutoff, the Argon gas is fully ionized.*

harmonic in Neon (109th order). For the 109th order in Argon, we calculate Δk $\sim$ 71,700 m^{-1}.

Quasi-phase-matching in a modulated waveguide can compensate for this amount of phase-mismatch. In a simple picture, the periodic change in the nonlinearity results in subtraction of the phase-mismatch between the fundamental and harmonic light by an effective k-vector, $K_m = 2\pi \mathrm{m}/\Lambda$, where m is an integer and Λ is the periodicity. [9] For a fiber with a 0.25mm periodicity, $K_1 = 25{,}400\ m^{-1}$. This value agrees well with the phase-mismatch calculated in Neon. The mismatch in Argon can be compensated for with higher order (m=3) QPM, explaining the enhancement in signal.

In conclusion, we have demonstrated quasi-phase-matched harmonic generation in Argon and Neon at very high ionization levels of 100% and 35% respectively. QPM increases the flux of all harmonic orders, and results in orders of magnitude increase in the highest harmonics. By using higher pulse energies, shorter modulation periods and longer modulated fibers, we expect to extend phase-matched harmonic emission to the water window region of the spectrum above 300eV.

References

1. Rundquist, A., *et al.*, Science **280**, 1412 (1998)
2. Christov, I.P., Optics Express **7**, 362 (2000)
3. Paul, A., *et al.*, Nature **421**, 51 (2003)

4. Wahlström,C.-G., *et al.*, Physical Review A **48**, 4709 (1993)
5. Krainov, V.P., JOSA B **14**, 425 (1997)
6. Spielmann, C., *et al.*, IEEE Journal of Selected Topics in Quantum Electronics **4**, 249 (1998)
7. Scrinzi, A., *et al.*, Physical Review Letters **83**, 706 (1999)
8. Durfee, C.G., *et al.*, Physical Review Letters **83**, 2187 (1999)
9. Fejer, M.M., *et al.*, IEEE Journal of Quantum Electronics **28**, 2631 (1992)

Enhancement and Tuning of High-Order Harmonic Generation Using Adaptive Pulse Shaping

D. H. Reitze[1], S. Kazamias[2], F. Weihe[2], G. Mullot[2], D. Douillet[2], F. Augé[2], O. Albert[2], V. Ramanathan[1], J. P. Chambaret[2], D. Hulin[2], and P. Balcou[2]

[1] Physics Department, University of Florida, Gainesville, FL 32601
reitze@phys.ufl.edu
[2] Laboratoire d'Optique Appliquée, ENSTA - Ecole Polytechnique, CNRS-UMR 7639, 91761 Palaiseau Cedex, France

1 Introduction

The optimization of high-order harmonic generation (HHG) processes continues to be an active area of investigation, not only for improving specific aspects of HHG (such as phase-matching, extending the cutoff frequency, generating tunable harmonics, and generating ultrashort harmonics), but also for developing a better understanding of fundamental HHG processes. While many intuitive schemes exist for improving HHG (such as chirping the laser pulse to tune the harmonic spectrum [1–3]), adaptive pulse shaping has also proven effective. The use of genetic algorithms coupled with pulse shaping has become a popular method in coherent control experiments [4]. Using sub-20 fs pulses, specific orders near the cutoff can be enhanced through shaping of the input pulse spectral phase [5]. Recent theoretical work using genetic algorithms have investigated the optimal focus, gas density, and interaction length to optimize the output flux, the pulse duration, and temporal coherence [6].

Here, we demonstrate tuning and enhancement of HHG in a free propagation geometry using shaped laser pulses synthesized using a learning algorithm. Compared with simple chirp tuning, both broader tunability and increased conversion efficiency is achieved in the 20-40 nm wavelength range. We theoretically investigate the mechanisms which produce the optimization and find that ionization coupled and specific quantum paths produce the optimized spectrum.

2 Experimental Description

A chirped pulse amplifier producing 6 mJ, 30 fs, 810 nm center wavelength pulses at 1 KHz was used in our experiments. An acousto-optic programmable dispersive filter (AOPDF) located before the first amplifier stage was used to modify the pulse spectral phase [7]. Pulses were focused using 1 m focal length into a 1 mm long gas cell containing 15 torr of argon within a high vacuum system. The focal point of the beam was located 2 mm before the cell entrance. An aperture was placed at the lens, which reduced the intensity at the focus to 3 x $10^{14} W/cm^2$, but improved the efficiency of harmonic generation process [8]. The resulting harmonics were filtered through two 150 nm thick Al filters and analyzed using a VUV/XUV spectrometer coupled to CCD detector. The absolute conversion efficiency was determined through a separate calibration of the spectrometer/detector. Approximately 100 shots were averaged for each spectrum. Prior to each optimization, the pulse temporal profile was measured using SPIDER.

A genetic algorithm was used to produce the desired harmonic output by selecting the best spectral phase, parameterized by 10 arbitrary values across the spectrum, bounded in the range 0 to 2π) summed with $\Phi^{(2)}$ and $\Phi^{(3)}$, the second and third order Taylor series phase coefficients. We chose this parameterization to provide a complex phase structure easily generated by the AOPDF. Each spectral phase was represented by an n-element bit (i.e., the spectral filter that determines the pulse phase). Each pulse was then tested by imparting spectral phase on the driving pulse and reading the harmonic spectrum from the CCD. Our goal was to tune a harmonic peak to a specific wavelength and simultaneously maximize the integrate HHG output. We therefore chose a fitness function that parameterized an individual harmonic peak as a Gaussian function specified by its center wavelength, height, and width. Optimizations were performed for a discrete set of wavelengths spanning two harmonic peaks. For each optimization, 41 individuals were tested. A mutation rate of 5% was used in all runs. Following application of the genetic operators, a new generation was tested, and this process continued until the algorithm ceased to improve.

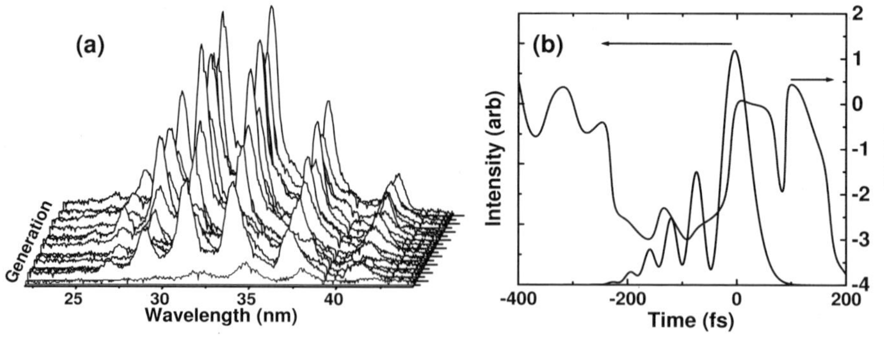

Fig. 1. (a) Optimization of the harmonic spectrum as a function of generation for tuning of the 27th harmonic (30 nm) to 31.16 nm, (b) temporal profile and phase of the laser pulse that produced the optimal output

3 Experimental Results

Figure 1 shows one example of an optimization, an attempt to tune the 27th harmonic (nominally at 30 nm) to 31.2 nm. The target wavelength is found early in the run, and the remaining generations are directed toward producing an enhancement of the peak height. Figure 1 also shows the laser pulse (intensity and phase) that produced the optimal output. The temporal phase is non-quadratic, indicating that the optimization occurs via a complicated means. Figure 2 displays the tuning curve for the entire set of optimizations. Almost complete tuning is obtained. This is substantially greater than the tuning range achieved by chirping the pulse (the left inset in Figure 2), where substantial EUV output observed only in a narrow (0.3 nm) band. The right inset in Figure 2 compares the spectrally integrated output harmonic flux using the GA against that of chirp tuning, plotted as a function of peak intensity of the laser field. A maximum conversion efficiency of 4 x 10_{-6} is found for the optimized laser fields, greater than that of the chirp tuning.

4 Theoretical Model

To gain an understanding of why the optimization works, we have developed a one-dimensional model of HHG using shaped laser pulses. Using the optimized laser fields as inputs, we evaluate the ionization rates in the medium using Ammosov, Delone and Krainov (ADK) rates. The spatial dephasing between the harmonic radiation and the laser-induced polarization that determines phase-matching is then found as:

$$\phi_q(z,t) = \alpha I(z,t) + \frac{q\omega z}{c}[\frac{n_e(z,t)}{2n_c} - \delta n_{at}] + q\arctan(\frac{z}{z_0}) \qquad (1)$$

where α is related to the specific quantum path [9], I is the laser intensity, q is the harmonic order, ω is the fundamental laser frequency, z is the propagation distance from the cell entrance, n_e is the electronic density, n_c is the critical density, n_{at} is the atomic dispersion term, z_0 is the Rayleigh range, and the final term is Gouy phase shift. From this, we find the complex time-dependent harmonic field at the exit of the medium:

$$\mathrm{E}_q(t) = \int_0^{l_{med}} |d_q(t)| e^{\frac{z-l_{med}}{2l_{abs}}} e^{i\phi(z)} dz \qquad (2)$$

Here d_q is the atomic dipole response, and $\phi(z,t) = \phi_q(z,t) + q\phi_L(z,t)$ where $\phi_L(z,t)$ is the phase of the laser field. The spectral shift is found as $\delta\omega = \partial\phi/\partial t$ and the harmonic spectrum is the magnitude of the Fourier transform of the E_q to the spectral domain.

Figure 3 displays the predicted harmonic spectrum corresponding to the optimization of Figure 1, as well as the theoretically predicted spectral shift

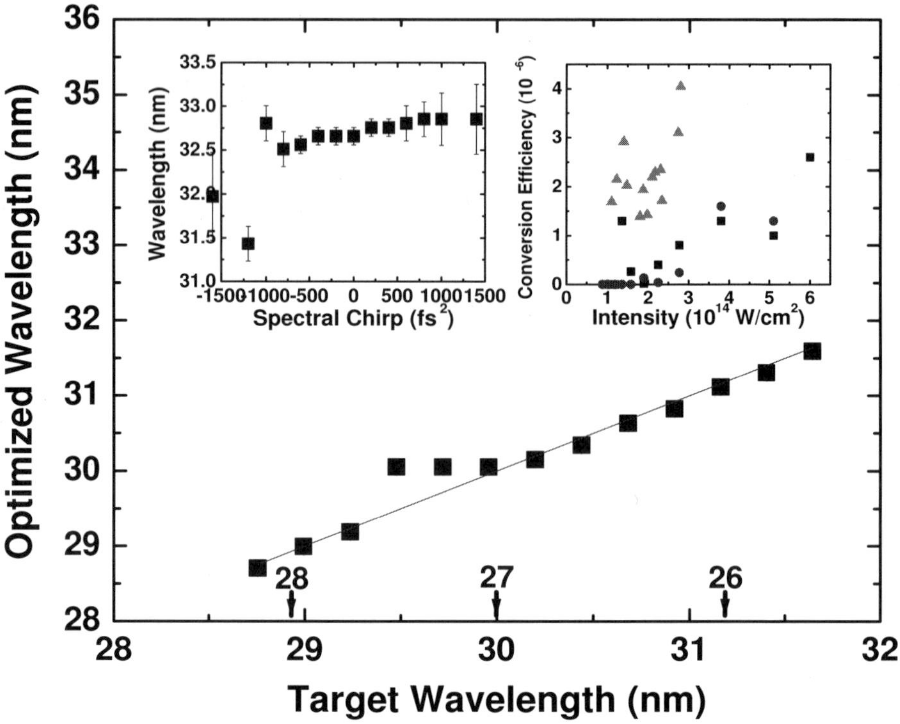

Fig. 2. Experimentally realized versus target wavelength. The straight line represents perfect tuning. The left inset shows the tuning curve achieved by simple chirp tuning. The right inset compares the conversion efficiency obtained from optimization (triangles), and positively (squares) and negatively chirped (circles) pulses

versus experimentally observed wavelengths. The major contribution comes from the second quantum path, indicating that it plays a larger role in the optimization for this specific wavelength. This was true for other optimizations as well. Furthermore, we find that ionization plays a key role. The spectral blueshift caused by the time-dependent phase shift of the driving laser field directly imprints on the harmonic phase [10]. When ionization is not included in the simulation, a negligible spectral shift is observed and first quantum path is the main contributor to the total harmonic intensity, inconsistent with the data. Also, the laser phase $\phi_L(z,t)$ is crucial in producing the magnitude of the shifts that we see in these experiments. In the absence of any temporal laser phase, the magnitude of the shift is much smaller than experimentally seen.

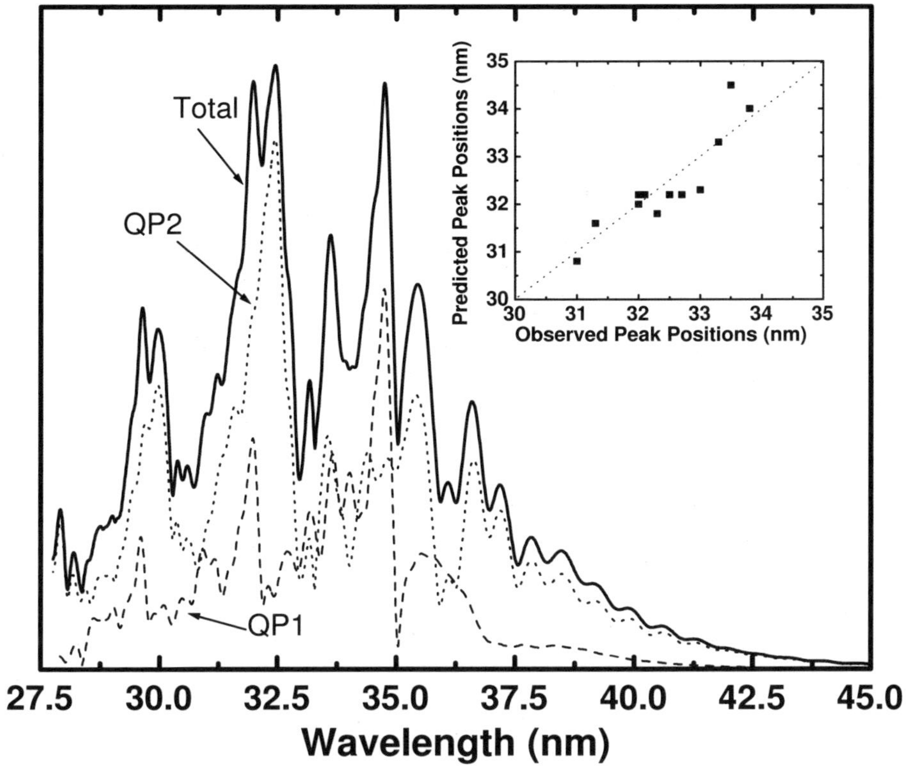

Fig. 3. Simulation of the optimized high-order harmonic spectrum (solid line). Harmonics corresponding to the first quantum path (dashed line) and the second quantum path (solid line) are also shown. Inset: Predicted versus experimentally observed spectral shifts obtained from the optimizing laser fields. The dotted line represents perfect correlation.

5 Conclusion

We have demonstrated tuning and enhancement of high-order harmonic generation in a free propagation geometry using shaped laser pulses synthesized using a genetic algorithm. Compared with simple chirp tuning, both broader tunability and increased conversion efficiency are achieved in the 20-40 nm wavelength range. Theoretical simulations reproduce the magnitude of the spectral shift and emphasize the role of ionization, specific quantum paths, and laser phase in the optimization.

References

1. Chang, Z., Rundquist, A., Wang, H., Christov, I. P., Kapteyn, H. C., Murnane, M. M.: Temporal phase control of soft-x-ray harmonic emission. Phys. Rev. A, **58** R30–R33 (1998)
2. Salières, P., Antoine, P., de Bohan, A., Lewenstein, M.: Temporal and spectral tailoring of high-order harmonics. Phys. Rev. Lett. **81**, 5544–5547 (1998)
3. Lee, D. G., Kim, J. H., Hong. K. H., Nam, C. H.: Coherent control of high-order harmonics with chirped femtosecond laser pulses. Phys. Rev. Lett. **87**, 243902-1–243902-4 (1992).
4. Judson R. S., Rabitz, H.: Teaching lasers to control molecules. Phys. Rev. Lett. **68**, 1500–1503 (1992)
5. Bartels, R. A., Backus, S., Zeek, E. Misoguti, L., Vdovin, G., Christov, I. P., Murnane, M. M., Kapteyn, H. C.: Shaped-pulse optimization of coherent emission of high-harmonic soft x-rays. Nature **406**, 164–166 (2000)
6. Roos, L., Gaarde, M. B., L'Huillier, A.: Tailoring harmonic radiation to different applications using a genetic algorithm. J. Phys. B: At. Mol. Opt. Phys. 34, 5041-5054 (2001).
7. Verluise, F., Laude, V., Cheng, Z., Spielmann, C., Tournois, P.: Amplitude and phase control of ultrashort pulses by use of an acousto-optic programmable dispersive filter: pulse compression and shaping. Opt. Lett. **25**, 575–577 (2000)
8. Kazamias, S., Weihe, F., Douillet, D., Valentin, C., Planchon, T., Sebban, S., Grillon, G., Augé, F., Hulin, D., Balcou, P.: High-order harmonic generation optimization with an apertured laser beam. Euro. Phys. J. D**21**, 353–359 (2002)
9. Lewenstein, M., Salières, P., L'Hullier, A.: Phase of the atomic polarization in high-order harmonic generation. Phys. Rev. A**52**, 4747–4754 (1995)
10. Altucci, C., Bruzzese, R., de Lisio, C., Nisoli, M., Stagira, S., De Silvestri, S., Svelto, O., Boscolo, A., Ceccherini, P., Poletto, L., Villoresi, P.: Tunable soft-x-ray radiation by high-order harmonic generation. Phys. Rev. A**61**, 021801(R)-1 - 021801(R)-4.

Multilayer EUV Optics for Applications of Ultrashort High Harmonic Pulses

Th. Westerwalbesloh, U. Kleineberg, Y.C. Lim, P. Siffalovic, M. Drescher, and U. Heinzmann

Faculty of Physics, University Bielefeld, D-33615 Bielefeld, Germany
uheinzm@physik.uni-bielefeld.de

Summary. Multilayers serve as tailored optical components for guiding, focussing, and monochromatization of high harmonics. A single harmonic order can be selected without pulse broadening thus enabling time-resolved photoelectron spectroscopy with high spectral and temporal resolution.

The proper conditioning of ultrashort extreme ultraviolet (EUV) pulses from high harmonic sources requires dedicated optical components which focus and spectrally filter the beam without compromising its temporal properties. Normal-incidence multilayer mirrors are ideally suited for this purpose. One objective of an optimized design is to maximize the spectral bandwidth for the characterization [1] and application [2] of attosecond EUV pulses. On the other hand, experiments with emphasis on a combination of spectral resolution and femtosecond time-structure require the selection of a single harmonic order from the harmonic comb.

Recently, EUV and soft x-ray multilayer mirrors have been developed for several wavelength regimes and for a number of applications in various fields of optical devices. Besides EUV lithography [3] applications of multilayer optics in other areas like astrophysics, Synchrotron Radiation optics, plasma diagnostics, x-ray laser and many more are reported. We have produced special EUV multilayer mirrors which are designed for broadband reflection or alternatively for spectral selection of individual high harmonics [4]. While broadband operation can be achieved by limiting the number of multilayer periods, high spectral selectivity requires a more advanced approach as discussed in the following. The spectral bandwidth $\Delta\lambda$ of the multilayer Bragg peak is limited by the number of contributing bilayers $N_{eff} = \lambda/\Delta\lambda$ and thus strongly dependent on the thickness ratio Γ between the absorber layer thickness and the bilayer thickness d [5]. A simulation of the peak reflectivity and bandwidth as a function of Γ based on a recursive Fresnel equation code and on optical constants published by Henke and Gullikson has been performed for multilayer mirrors of 30 bilayers and a bilayer thickness of 9.1...9.5 nm. For a photon energy of 70 eV we found by simulation that the Bragg peak bandwidth can be reduced to less than 3 eV (FWHM) almost without loss of peak reflectivity if the thickness ratio Γ is reduced to 0.2. Further reduction of Γ would result in a significant unwanted drop in peak reflectivity. Based on these simulations Mo/Si multilayers with $\Gamma = 0.2$ have been fabricated as a reasonable trade of of small spectral bandwidth and low peak reflectivity loss. A TEM image of the low-Γ multilayer with 12 bilayers deposited on a Si(100)

Fig. 1. TEM cross section of a low-Γ Mo/Si multilayer (12 bilayers, d = 10 nm, $\Gamma = 0.17$)

wafer substrate is shown in Fig. 1. The dark and light areas correspond to the Mo and Si layer, respectively. Note, that the thickness of the Mo layer is significantly overestimated in the TEM images due to diffraction contrast. The image shows that the Mo layers of the ion-polished sample are still partially polycrystalline. Measurements of the spectral bandwidth for a $\Gamma = 0.17$ multilayer in comparison with a $\Gamma = 0.33$ multilayer have been performed by using a laboratory EUV source. Two Mo/Si multilayers with 30 bilayers and d-spacing 10 nm, which differ in their gamma values (Γ = 0.17 and 0.33), were prepared on Si(100) wafers under identical deposition conditions. The reflectivity curves of both samples were measured in the wavelength range between 12 nm and 15 nm by using unpolarized radiation from a laser-induced plasma light source at an incidence angle of 45 deg with respect to the surface normal. The measured reflectivity curves (Fig. 2) show that the bandwidth of the reflection peak of the sample with low Γ value is appreciably smaller ($N_{eff} = 22$) compared to the other sample ($N_{eff} = 15$). Good agreement between the measurements and simulations have been achieved by assuming

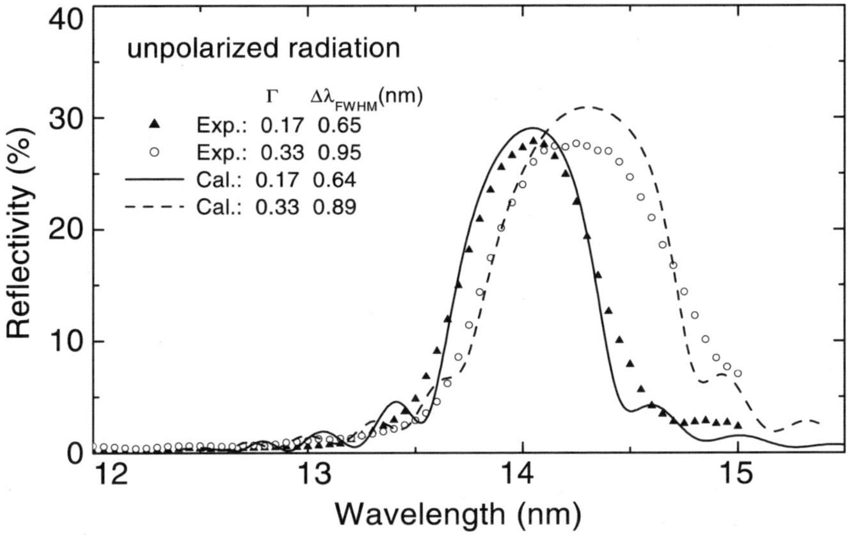

Fig. 2. Soft x-ray reflectometry measurements and simulations of a $\Gamma = 0.17$ vs. a $\Gamma = 0.33$ multilayer

a Debye-Waller roughness parameter of $\sigma = 0.55$ nm and $\sigma = 0.65$ nm for the $\Gamma = 0.17$ and $\Gamma = 0.33$ multilayer, respectively.

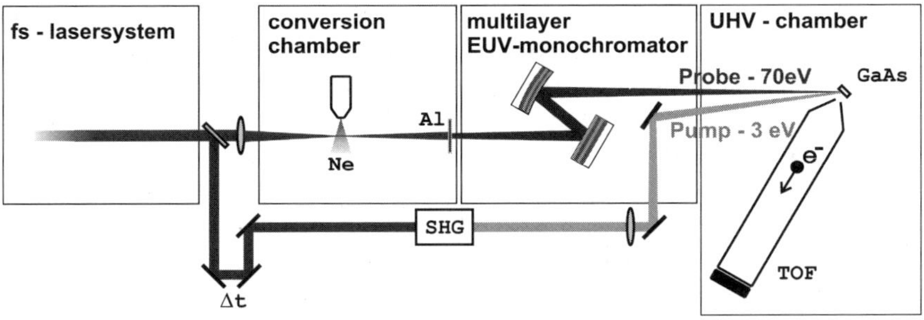

Fig. 3. Apparatus for vis-pump/EUV-probe photoemission

Utilizing a set of two low-Γ multilayer reflectors in series an apparatus for femtosecond time-resolved visible-pump / EUV-probe photoelectron spectroscopy on surfaces was built up (Fig. 3) [6]. Fig. 4 shows the photoelectron spectrum of Neon atoms (the solid GaAs target in Fig. 3 was replaced by an atomic Ne beam) measured by means of the time-of-flight spectrometer TOF. The measured bandwidth of 0.76 eV (FWHM) demonstrates the ability to select a single higher harmonic order with a suppression of adjacent harmonics to less than 5%. In order to characterize the time structure of the

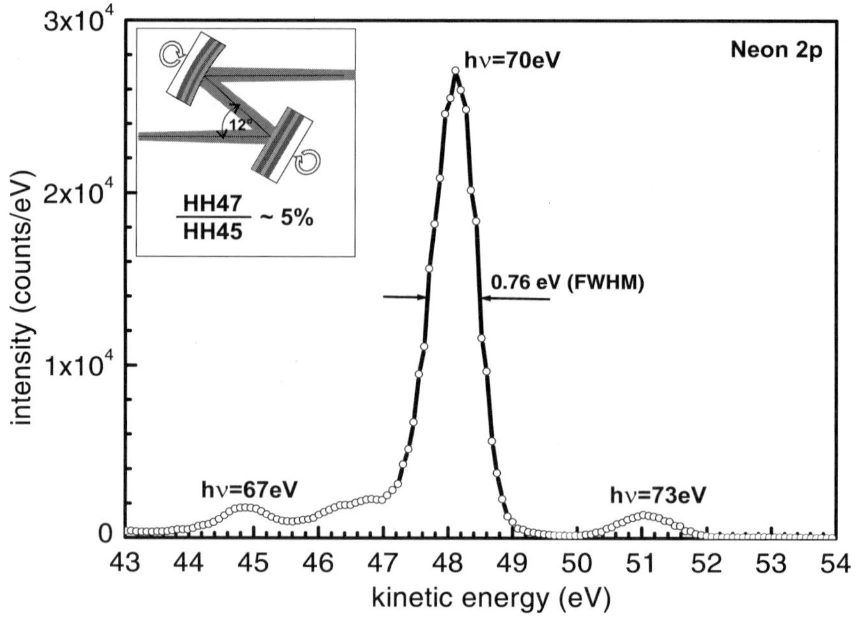

Fig. 4. Photoelectron spectrum of neon atoms measured behind the multilayer monochromator

EUV pulse at 70 eV photon energy, the hot electron dynamics on a Pt surface was studied in a visible-pump / EUV-probe photoemission experiment using the apparatus shown in Fig. 3. Fig. 5 displays the result of the visible-EUV cross-correlation measured at 1.5 eV above the Fermi energy of Pt. The width of 92.2 ± 6.5 fs yields an upper limit of the EUV pulse duration of 70 fs [6].

In a subsequent time-resolved experiment the charge carrier dynamics on a semiconductor surface was studied: the time evolution of kinetic energy shifts for Ga-3d core-level photoelectrons (Fig. 6) from p-doped GaAs(100) was probed following excitation with 3.1 eV femtosecond laser pulses [7]. The observed transient changes of the surface photovoltage in Fig. 7 reveal an electron transport from the bulk to the surface to occur within 500 fs after photoexcitation while a subsequent relaxation is determined to evolve on a time scale of 15 picoseconds. The high kinetic energies of the escaping electrons enable measurements at high pump-intensities of approx. 10 GW/cm^2 when a considerable background from multiphoton emission is already present but does not obscure the fast core-level photoelectrons. Owing to its extreme surface sensitivity the method supplements studies with a visible or UV probe which average over the entire space charge region.

Support by the DFG (SFB 613) is gratefully acknowledged.

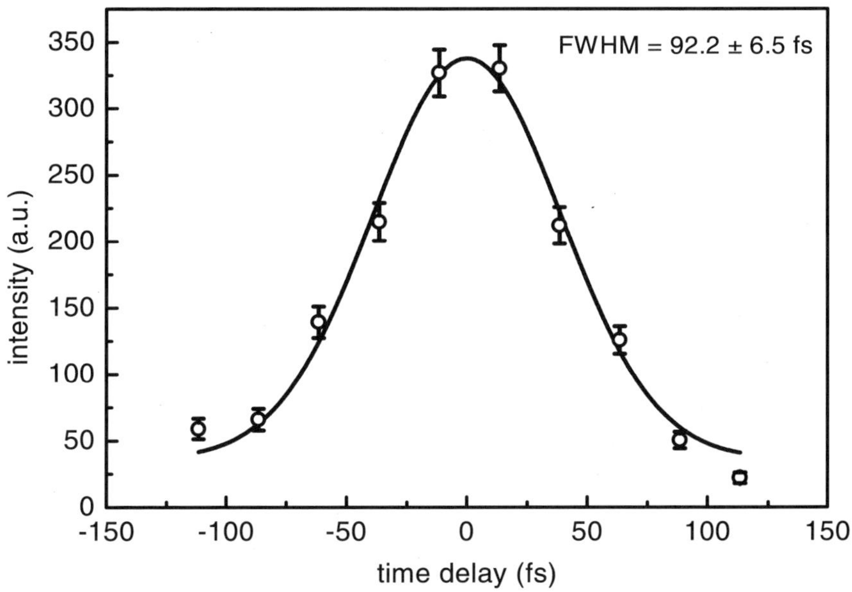

Fig. 5. Visible-EUV cross-correlation measured by means of hot-electron emission 1.5 eV above the Fermi edge of Pt.

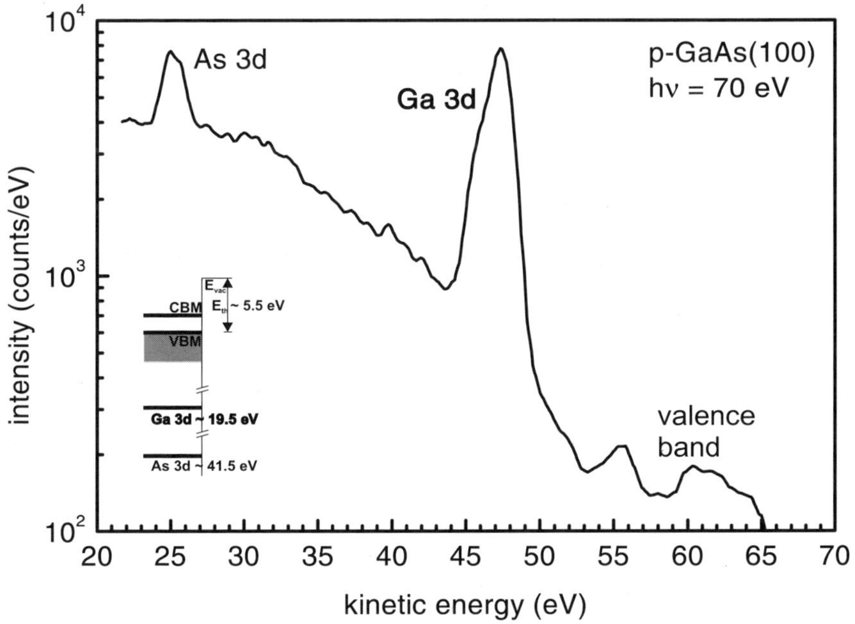

Fig. 6. GaAs photoelectron spectrum

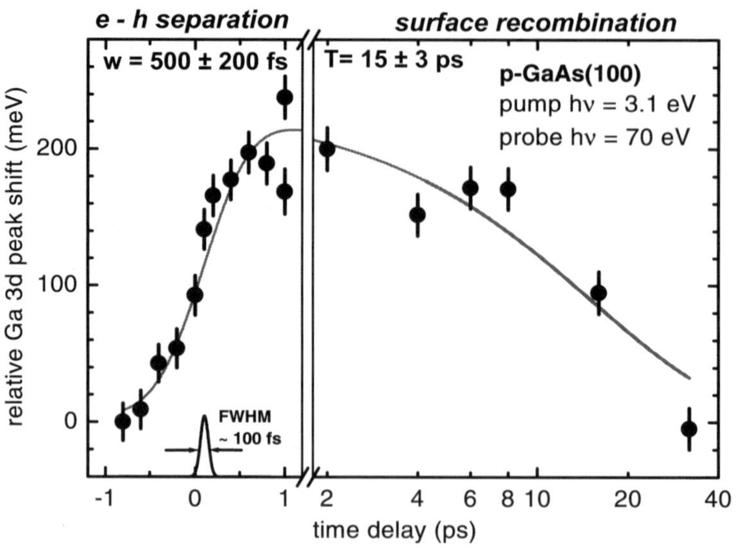

Fig. 7. Transient surface-photovoltage shift on p-doped GaAs mapped by observing the relative shift of the Ga-3d core level energy after photoexcitation with a laser-pump and a delayed EUV-probe pulse.

References

1. R. Kienberger, M. Hentschel, M. Uiberacker, Ch. Spielmann, M. Kitzler, A. Scrinzi, M. Wieland, Th. Westerwalbesloh, U. Kleineberg, U. Heinzmann, M. Drescher, F. Krausz: Steering attosecond electron wave packets with light. Science **297**, 1144 (2002)
2. M. Drescher, M. Hentschel, R. Kienberger, M. Uiberacker, Th. Westerwalbesloh, U. Kleineberg, U. Heinzmann, F. Krausz: Time-resolved atomic inner-shell spectroscopy. Nature **419**, 803 (2002)
3. U. Kleineberg, Th. Westerwalbesloh, W. Hachmann, U. Heinzmann, J.Tümmler, F. Scholze, G. Ulm: Effect of substrate roughness of Mo/Si multilayer optics for EUVL produced by UHV-e-beam evaporation and ion polishing. Thin Films (2003, in press)
4. Y.C. Lim, T. Westerwalbesloh, A. Aschentrup, O. Wehmeyer, G. Haindl, U. Kleineberg, U. Heinzmann: Fabrication and characterization of EUV multilayer mirrors optimized for small spectral reflection bandwidth. Applied Physics A **72**, 121(2001)
5. D.T. Attwood : "Soft X-Rays and Extreme Ultraviolet Radiation : Principles and Applications"; Cambridge University Press 1999
6. P. Siffalovic, M. Drescher, M. Spieweck, T. Wiesenthal, Y. Lim, R. Weidner, A. Elizarov, U. Heinzmann: Laser-based apparatus for extended ultraviolet femtosecond time-resolved photoemission spectroscopy. Rev. Sci. Instrum. **72**, 30 (2001)
7. P. Siffalovic, M. Drescher, U. Heinzmann: Femto-second time-resolved core-level photoelectron spectroscopy tracking surface photovoltage transients on p-GaAs. Europhys. Lett. **60**, 924 (2002)

Absolute-Phase Effects of Few-Optical-Cycle Light Pulses in Single-Shot High Order Harmonic Spectra

M. Nisoli[1], G. Sansone[1], S. Stagira[1], S. De Silvestri[1], C. Vozzi[2], M. Pascolini[3], L. Poletto[3], P. Villoresi[3], and G. Tondello[3]

[1] National Laboratory for Ultrafast and Ultraintense Optical Science INFM Dipartimento di Fisica, Politecnico, Milano (Italy)
mauro.nisoli@fisi.polimi.it
[2] INFM- Dipartimento di Fisica, Universita' degli Studi, Milano (Italy)
[3] Laboratory for Ultraviolet and X-Ray Optical Research - INFM D.E.I. - Universita' di Padova, Padova (Italy)

1 Introduction

Progress in femtosecond laser technology has led to the generation of laser pulses whose envelope varies on a time scale comparable to that of the electromagnetic field itself. Such pulses consist of less than two optical cycles [1]. Under these conditions, the phase of the carrier frequency with respect to the envelope (absolute phase) determines the variation of the laser electric field in time. Since all effects in strong-field laser interaction are driven by the electromagnetic field of the laser, the absolute phase is important for many different topics in laser physics. Experimental evidences of the role of the absolute phase have been obtained in strong-field photoionization [2], using pulses with random absolute phase, and in high-order harmonic generation, using phase-stabilized pulses [3]. In this work we report on single-shot measurement of harmonic spectra generated by few-optical-cycle pulses. The single-shot approach and the excellent stability of the driving pulses allow us to attribute the measured differences in the harmonic spectra to the absolute phase of the driving pulses, randomly changing from one pulse to the other. The experimental results have been interpreted in the framework of the strong-field approximation (SFA) model [4].

2 Experimental results

25-fs pulses were coupled into an argon-filled capillary consisting of a 60-cm-long fiber with an inner diameter of 0.5 mm at the entrance side and 0.3 mm at the exit side. The emerging pulses were then compressed by chirped mirrors. Typical pulse duration ranges from 5 to 7 fs. Single-shot, 1-kHz rate, characterization of the compressed pulses have been performed using spectral phase interferometry for direct electric field reconstruction (SPIDER) [5].

These measurements demonstrate the remarkable stability of the spectral phase and temporal profile of the compressed pulses. This is of paramount importance for the analysis of the single-shot harmonic spectra. In the experiments the harmonic beam is produced by focusing sub-10-fs laser pulses into a neon jet. To observe the harmonic emission we have used a high-resolution flat-field soft x-ray spectrometer and a high-resolution CCD detector. The spectrometer has been calibrated in order to provide an absolute measurement of the emitted photon flux. The gas jet, synchronized with the driving pulses, was operated at a 10-Hz repetition rate, with a 300-μs aperture time. The integration time of the CCD detector was set to 85 ms: in this way single-shot spectra can be measured.

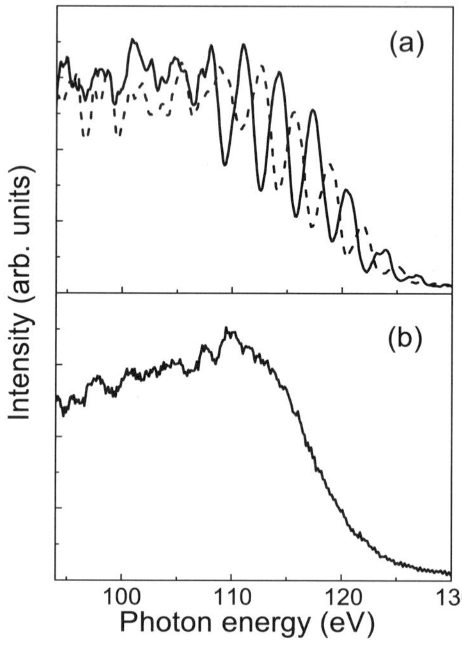

Fig. 1. Measured harmonic spectra in neon generated by 6-fs driving pulses: (a) single-shot harmonic spectra; (b) time-integrated harmonic spectrum corresponding to 100 XUV shots.

Figure 1(a) shows two single-shot harmonic spectra obtained in neon, for a laser peak intensity of $\sim 8.4\times10^{14}$ W/cm^2 at focus. The gas jet was located 2 mm after the laser focus. Figure 1(b) displays a harmonic spectrum measured in the same experimental condition, using a 10-s integration time, thus corresponding to 100 XUV shots. The single-shot spectra show discrete harmonic peaks, with different amplitude, up to the cutoff region. In the cutoff region, the harmonic peaks of the single-shot spectra corresponding to different driving pulses are frequency shifted, whilst in the plateau region (not shown in Fig.1) the harmonic peaks are not frequency shifted. This pulse-dependent frequency shift of the harmonic peaks explains why the time-integrated spectrum, which corresponds to several driving pulses with random absolute phases, appears as a structureless continuum. It is worth pointing out that the single-shot technique and the excellent stability of the compressed pulses guarantee that the only difference in the driving pulses is their absolute phase, which randomly changes from one pulse to the other.

3 Nonadiabatic single-atom response in the few-optical-cycle regime

The experimental results can be understood in the framework of the SFA. In particular the saddle-point equations allow one to gain a good deal of physical insight. Such equations are obtained requiring that the quasiclassical action, $S(\mathbf{p},t,\tau)$ is stationary, where: $S(\mathbf{p},t,\tau) = \int_{t-\tau}^{t} dt' \Big(I_p + [\mathbf{p} - \mathbf{A}(\mathbf{t}')]^2/2 \Big)$; $\mathbf{p}$ is the canonical momentum; $\mathbf{A}$ is the vector potential of the laser field; I_p is the ionization potential of the atoms of the gas medium; τ is the time spent by the electron in the continuum between the ionization instant and the recombination instant t. The complete set of saddle-point equations can be written as follows [4]:

$$\tau \mathbf{p} - \int_{t-\tau}^{t} dt' \, \mathbf{A}(t') = 0 \tag{1}$$

$$\frac{[\mathbf{p} - \mathbf{A}(t-\tau)]^2}{2} + I_p = 0 \tag{2}$$

$$\frac{[\mathbf{p} - \mathbf{A}(t)]^2}{2} - \frac{[\mathbf{p} - \mathbf{A}(t-\tau)]^2}{2} = \omega \tag{3}$$

where ω is the angular frequency of the emitted harmonic radiation. The solutions, $\mathbf{p}_s$, t_s and τ_s of the saddle-point equations are complex. Using the stationary phase approximation the Fourier transform of the dipole moment, $x(\omega)$, can be written as a coherent superposition of the contributions from the electron quantum paths corresponding to the saddle-point solutions.

$$x(\omega) = \sum_s C_s(\mathbf{p}_s, t_s, \tau_s) \exp[-iS(\mathbf{p}_s, t_s, \tau_s) + i\omega t_s] \tag{4}$$

where C_s is an amplitude determined by the saddle-point solutions. The sum (4) extends over the relevant saddle points with $0 < \mathrm{Re}(\tau) < T_0$, where T_0 is the optical period of the laser field. In the upper-plateau spectral region

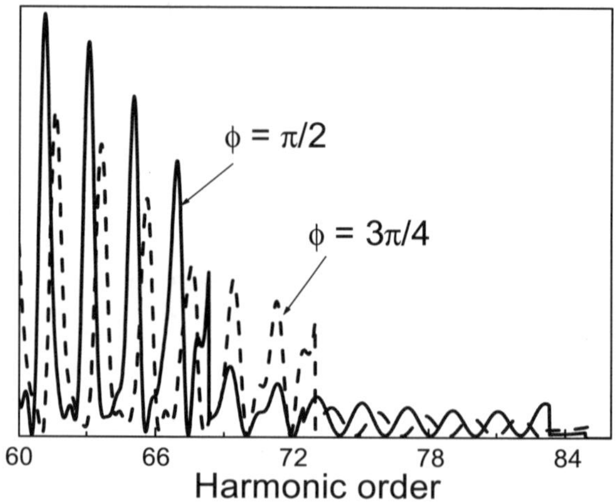

Fig. 2. Harmonic spectra in neon calculated using the saddle-point approximation as coherent superposition of the short quantum paths, for two absolute phases of a 6-fs driving pulse.

two quantum paths give the most relevant contribution to the dipole moment [6]. One trajectory has a short return time, close to half an optical period, whereas the second one has a return time close to one period. It has been demonstrated that phase-matching can provide a powerful method to identify the contributions of the individual quantum paths [6–9]. When the gas jet is located after the laser focus the short quantum paths are selected, whereas, when the gas jet is placed around and before the focus the contribution of the long quantum paths increases. In order to simplify the physical interpretation, we have placed the gas jet after the focus, so that the measured harmonic spectra (shown in Fig. 1) are completely dominated by the short path contributions. Figure 2 shows the harmonic spectra calculated using the saddle-point method as coherent superposition of the short electron quantum paths, for two different absolute phases. The spectra are characterized by the following features: the low-order harmonics (not shown in Fig. 2) are not influenced by the absolute phase of the driving pulses, and corresponds to odd harmonics of the fundamental wavelength; the highest photon energy region, which is generated around the peak of the driving pulses, shows discrete and well resolved peaks, whose spectral position is determined by the absolute phase, as experimentally observed. Therefore the absolute-phase dependent spectral shift is a single-atom effect.

The physical origin of the observed spectral shift of the harmonic peaks can be clearly understood as follows. The formation of the *harmonic* peaks

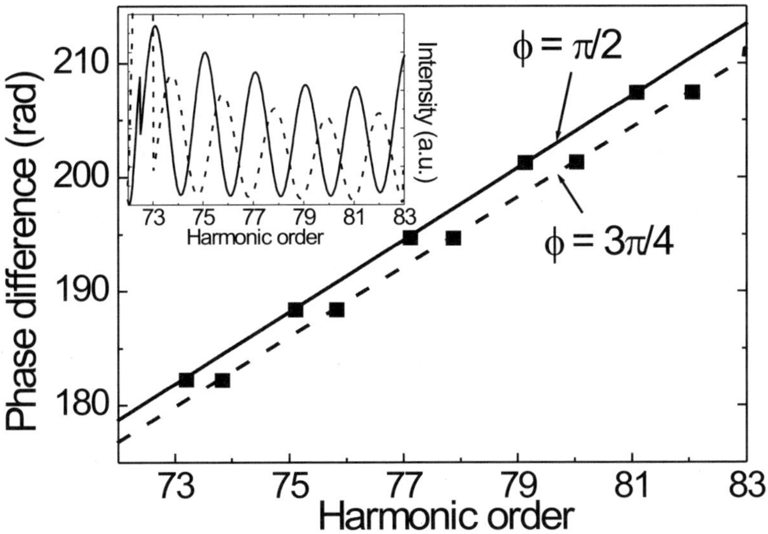

Fig. 3. Calculated difference, $\Delta\Theta$, between the phases associated to the short quantum paths for the two emission processes relevant to the upper cutoff region, for two absolute phases ($\phi = \pi/2$ and $\phi = 3\pi/4$). The squares give the spectral position of constructive interference. Inset: enlarged view of the cutoff region of the single-atom emission spectra shown in Fig. 2.

is due to the interference of the XUV spectra emitted periodically every half an optical cycle by the electron returning to the parent ions. In the upper cutoff spectral region the emission is produced by just a pair of recombination events. In this situation it is straightforward to calculate the spectral position of the *harmonic* peaks. We have calculated, using the saddle-point method, the phases associated to the individual (short) quantum paths. The interference between the radiation emitted during the two emission processes is constructive when the phase difference is $\Delta\Theta = 2m\pi$ ($m = 0, 1, 2, ...$). Figure 3 shows $\Delta\Theta$ as a function of the harmonic order for two values of the absolute phase of the driving pulses, namely $\phi = \pi/2$ and $\phi = 3\pi/4$. The squares reported on the curves mark the spectral position for constructive interference: the absolute-phase-dependent shift of the *harmonic* peaks is evident. The inset of Fig. 3 shows an enlarged view of the upper cutoff region of the calculated spectra reported in Fig. 2: the position of the *harmonic* peaks is obviously in agreement with the points reported on the $\Delta\Phi$ curves. The same kind of reasoning can be applied to the other spectral regions of the harmonic spectrum: in this case we have to take into account the condition for constructive interference among several emission processes.

4 Conclusions

In this work we have reported on *single-shot* measurements of harmonic spectra generated by few-optical-cycle pulses. Clear absolute-phase effects have been observed in the measurements. In particular, an absolute-phase-dependent frequency shift of the *harmonic* peaks has been measured. Such behavior can be ascribed to the nonadiabatic single atom response of the gas medium.

This work was partially supported by the European Community's Human Potential Programme under Contract No. HPRN-CT-2000-00133 (ATTO), and by INFM under the project CLUSTERS.

References

1. Schenkel, B., Biegert, J., Keller, U., Vozzi, C., Nisoli, M., Sansone, G., Stagira, S., De Silvestri, S., Svelto, O.: Generation of 3.8 fs pulses from adaptive compression of a cascaded hollow fiber supercontinuum. Opt. Lett., in press (2003).
2. Paulus, G.G., Grasbon, F., Walther, H., Villoresi, P., Nisoli, M., Stagira, S., Priori, E., De Silvestri, S.: Absolute-phase phenomena in photoionization with few-cycle laser pulses. Nature, **414**, 182–184 (2001)
3. Baltuska, A., Udem, Th., Uiberacker, M., Hentschel, M., Goulielmakis, E., Gohle, Ch., Holzwarth, R., Yakovlev, V.S., Scrinzi, A., Hansch, T.W., Krausz, F.: Attosecond control of electronic processes by intense light fields. Nature **421**, 611–615 (2003).
4. Lewenstein, M., Balcou, Ph., Ivanov, M.Y., L'Huillier, A., Corkum, P.B.: Theory of high-harmonic generation by low-frequency laser fields. Phys. Rev. A, **49**, 2117–2132 (1994).
5. Kornelis, W., Biegert, J., Tisch, J.W.G., Nisoli, M., Sansone, G., Vozzi, C., De Silvestri, S., Keller, U.: Single-shot kilohertz characterization of ultrashort pulses by spectral phase interferometry for direct electric-field reconstruction. Opt. Lett., **28**, 281–283 (2003).
6. Salieres, P., Carre, B., Le Deroff, L., Grasbon, F., Paulus, G. G., Walther, H., Kopold, R., Becker, W., Milosevic, D. B., Sanpera, A., Lewenstein, M.: Feynman's path-integral approach for intense-laser-atom interactions. Science, **292**, 902–905 (2001).
7. Salieres, P., L'Huillier, A., Lewenstein, M.: Coherence control of high-order harmonics. Phys. Rev. Lett., **74**, 3776–3779 (1995).
8. Bellini, M., Lynga, C., Tozzi, A., Gaarde, M. B., Hansch, T. W., L'Huillier, A., Wahlstrom, C. -G.: Temporal coherence of ultrashort high-order harmonic pulses. Phys. Rev. Lett., **81**, 297–300 (1998).
9. Lee, D. G., Shin, H. J., Cha, Y. H., Hong, K. H., Kim, J.-H., Nam, C. H.: Selection of high-order harmonics from a single quantum path for the generation of an attosecond pulse train. Phys. Rev A, **63**, 021801–1–4 (2001).

Reproducible Generation and Measurement of Isolated Sub-fs XUV Pulses with Phase-Controlled Few-Cycle Light

R. Kienberger[1], E. Goulielmakis[1], A. Baltuska[1], M. Uiberacker[1], Th. Westerwalbesloh[2], U. Kleineberg[2], U. Heinzmann[2], M. Drescher[1,2], and F. Krausz[1,3]

[1] Institut für Photonik, Technische Universität Wien, Gusshausstr. 27/387, A-1040 Wien, Austria.
[2] Fakultät für Physik, Universität Bielefeld, D-33615 Bielefeld, Germany.
[3] Max-Planck Institut für Quantenoptik, Hans-Kopfermann-Straße 1, D-85748 Garching, Germany.

Summary. We report on the use of intense phase-controlled few-cycle(5-fs @ 750 nm)pulses for the reproducible generation of high-order-harmonic-like XUV radiation near 100 eV and for measuring the temporal structure of the generated coherent XUV light in a two-colour ionization experiment. Our experiments reveal that control of the carrier-envelope phase of the few-cycle driver allows precise control of the temporal structure of the XUV emission filtered in the cut-off range. Depending on the phase setting, we can generate a single, isolated sub-femtosecond burst (with a cosine driver waveform) or a couple of bursts of comparable energy (with a sine waveform). Our two-colour photoionization measurement permits determination of the XUV pulse duration, the fractional energy carried by satellite pulses and the value of the carrier-envelope phase of the few-cycle driver laser pulse.

The generation of isolated sub-femtosecond XUV bursts by few-cycle-driven high-order harmonic generation has been reported before [1] but some of our peers expressed doubts as to what extent the XUV pulse energy is confined in a single burst rather than being distributed among several spikes [2]. Here we use, for the first time, phase-controlled laser pulses [3] for the control and simultaneous measurement of the temporal structure of XUV radiation at a sub-laser-cycle time scale and present results that conclusively corroborate the feasibility of isolated sub-femtosecond XUV pulse generation with few-cycle laser pulses.

The experimental setup employed for the generation of the XUV light is the same as described in Ref [3] except for the peak intensity of the phase-stabilized 5-fs 750-nm laser pulses, which is adjusted in the current experiments such that the highest-energy (cut-off) XUV photons are emitted near 100 eV within the high-reflectance band (93 ± 5eV) of our Mo/Si multilayer mirror used for filtering the cutoff radiation and focusing it (together with the driver laser pulses) into a neon gas target for the two-colour ionization experiments (Fig. 1). The ionization experiment is performed in the same excitation and detection geometry as described in Ref [4]. The photoelectrons

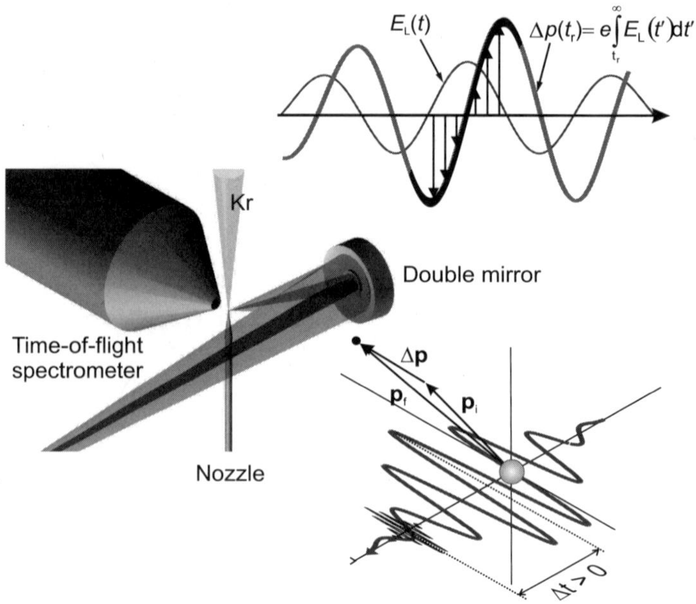

Fig. 1. Setup of the experiment and principle of the energy shift of the attosecond electron wave packet.

detached upon the absorption of an XUV photon pick up a momentum from the strong few-cycle laser field present during the XUV ionization [5,6]. If we detect the photoelectrons ejected parallel to the direction of the laser electric field (as we do in our study), this momentum transfer increases or decreases the initial momentum of the electron resulting in a shift of its final kinetic energy. This energy shift is maximum for the electrons "born" at the zero transition of the field. Our recent experiment [4] revealed that the cut-off XUV burst generated previously by the same few-cycle pulse approximately coincide with a zero transition of the driver field, allowing to time the instant of ionization to this point without having to introduce any extra delay between the XUV and laser pulse. Hence, we can focus both beams by the same mirror into the ionization target. Our previous experiments were carried out with laser pulses of random carrier-envelope (CE) phase, resulting in the simultaneous appearance of up- and downshifted features in the photoelectron energy spectra accumulated over many laser shots (because of equal number of laser pulses with a CE phase of $\varphi = \varphi_0$ and those having $\varphi = \varphi_0 + \pi$ contributing to the resultant data).

As depicted in Fig. 2, the situation changes dramatically with phase-stabilized pulses. With the CE phase fixed, i.e. the laser waveform precisely reproduced from shot to shot, we may expect to be able to isolate a clean up- or downshift, provided that the XUV radiaton is confined to a single burst of

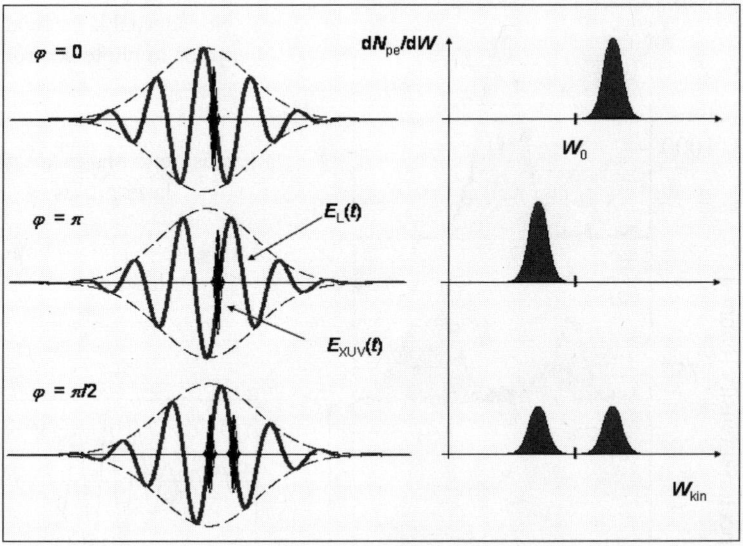

Fig. 2. Schematic of the energy shift depending on the absolute phase of the laser light wave.

sub-femtosecond duration. Depending on the direction of the detection, the XUV photoelectron spectrum is expected to be upshifted for $\varphi = 0$ owing to the production of a single XUV burst by a cosine laser waveform and downshifted for $\varphi = \pi$, or vice versa. As φ is gradually tuned away from zero a satellite burst spaced by the half laser oscillation period from the main burst is predicted to emerge, which would produce photoelectrons that suffer a shift of equal magnitude but opposite sign as compared to that induced by the main burst. For $\varphi = \pi/2$ (sine waveform) two XUV bursts of approximately equal energy are predicted, which should result in the simultaneous appearance of an up- and downshifted feature in the photoelectron energy distribution.

Fig. 3 summarizes our experimental results, which fully support the major conclusions of the above analysis. Panel (a) depicts the XUV photoelectron spectrum recorded in the absence of the laser field. It peaks at $\mathrm{W}_0 = \mathrm{h}\omega_{xuv} - \mathrm{W}_b \approx 70$ eV. Fig. 3(b) shows the spectrum obtained in the presence of the laser field without CE phase stabilization, reproducing our results reported in Ref. [4].

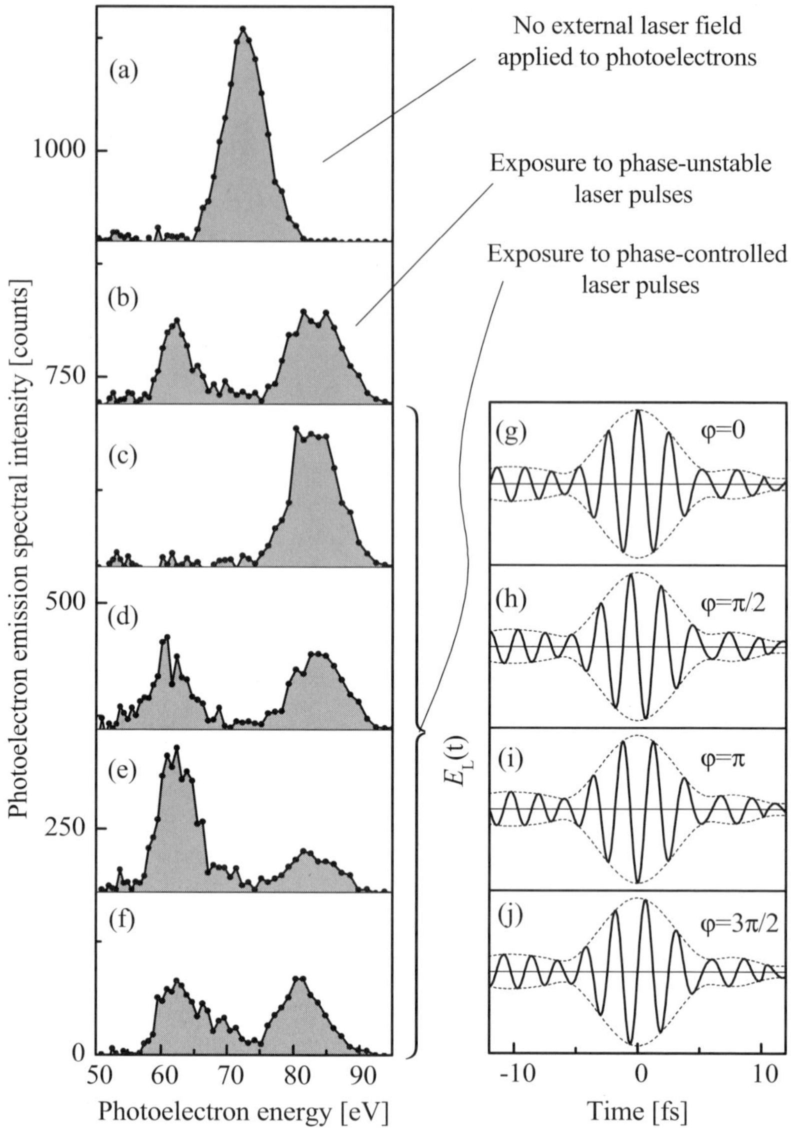

Fig. 3. Measured energy shift depending on the absolute phase of the laser light wave.

Panels (c)-(f) display photoelectron distributions in the presence of phase-stabilized laser pulses. For a cosine wave ($\varphi = 0$) we observe a single upshifted spectral feature (c), which turns into a single downshifted feature (e) for a cosine wave of reversed sign ($\varphi = \pi$). For $\varphi = +\ \pi/2$ or $-\ \pi/2$ (sine waveforms, panels (d) and (f)), we observe up-and downshifted features of equal magnitude.

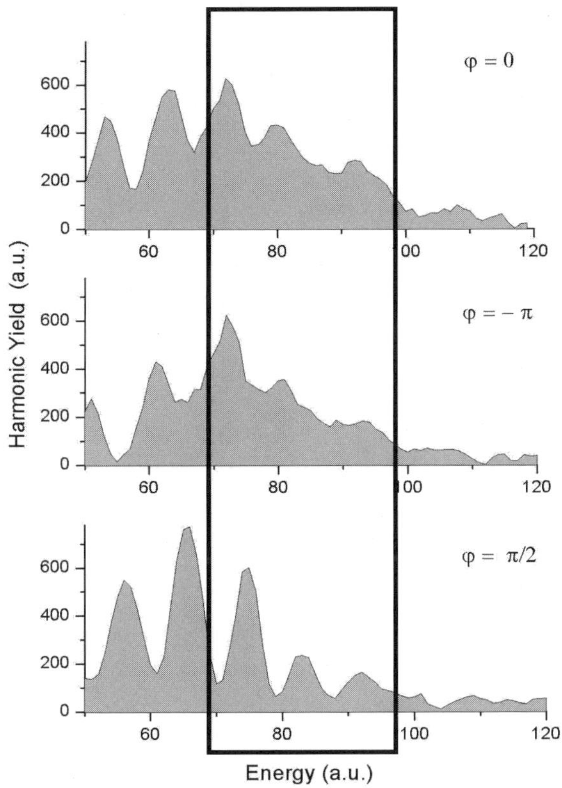

Fig. 4. Spectra of harmonic radiation measured with a transmission grating and an X-ray CCD camera.

In addition to the photoelectron spectra we measured the harmonic spectra by a grating spectrograph for each phase feature. Fig. 4 shows the results for CE phases $\varphi = 0$, $\varphi = \pi$ and $\varphi = +\ \pi/2$. The flat continuum-like structure in the cut-off region in the case of $\varphi = 0$ and $\varphi = \pi$ corresponds nicely to the occurrence of a single attosecond burst at the cosine feature of the driving (few-cycle) laser pulse at the harmonic generation. By the way, in this case the expression "harmonic" loses its meaning. The deep modulation in the case of $\varphi = \pi$ indicates the train/like structure of the harmonic radiation.

The negligible broadening of the shifted spectra in Fig. 3 (as combared to the unperturbed one in the top panel) implies an XUV pulse duration as well as a timing jitter of the XUV pulse with respect to the laser phase well below 1 fs. These result demonstrate for the first time the controlled, reproducible generation of single sub-femtosecond XUV pulses.

References

1. M. Hentschel, R. Kienberger, Ch. Spielmann, G. A. Reider, N. Milosevic, T. Brabec, P. B. Corkum, U. Heinzmann, M. Drescher, F. Krausz, *Attosecond metrology*, Nature **414**, 509-513 (2001).
2. H. C. Kapteyn, Laser Focus World **38**, 5, p. 8 (2002). Response: F. Krausz and P. B. Corkum, Laser Focus World **38**, 8, p.7 (2002).
3. A. Baltuska, Th. Udem, M. Uiberacker, M. Hentschel, E. Goulielmakis, Ch. Gohle, R. Holzwarth, V. S. Yakovlev, A. Scrinzi, T. W. Hänsch, F. Krausz, Nature **421**, 611-615 (2003).
4. R. Kienberger, M. Hentschel, M. Uiberacker, Ch. Spielmann, M. Kitzler, A. Scrinzi, M. Wieland, Th. Westerwalbesloh, U. Kleineberg, U. Heinzmann, M. Drescher, F. Krausz, *Steering Attosecond Electron Wave Packets with Light*, Science **297**, 1144–1148 (2002).
5. J. Itatani, F. Quéré, G. L. Yudin, M. Yu. Ivanov, F. Krausz, P. B. Corkum, *Attosecond streak camera*, Phys. Rev. Lett. **88**, 173903 (2002).
6. M.Kitzler, N. Milosevic, A. Scrinzi, F. Krausz, Th. Brabec , *Quantum Theory of Attosecond XUV Pulse Measurement by Laser Dressed Photoionization*, Phys. Rev. Lett. **88**, 173904 (2002).

Controlling High-Harmonic Generation via Molecular Alignment

D. Zeidler[1], J. Levesque[2], J. Itatani[1], K. Lee[1], P. Dooley[1], I. Litvinyuk[1], D. M. Villeneuve[1], and P. B. Corkum[1]

[1] National Research Council of Canada, 100 Sussex Drive, Ottawa, Ontario K1A0R6, Canada paul.corkum@nrc-cnrc.gc.ca
[2] INRS, Énergie et Materiaux, Varennes, Quebec, J3X 1S2, Canada

Summary. The yield of high harmonic generation with N_2 and O_2 molecules is enhanced when the molecules are aligned parallel to the laser field that produces the harmonics and suppressed for the perpendicular case, compared to randomly oriented molecules. The alignment is produced via creation of a rotational molecular wave packet with a non-ionizing 800-nm, 60-fs Ti:Sa pulse and probed at its revival via high harmonic generation with a delayed intense 800-nm, 30-fs pulse.

1 High harmonic generation in molecules

High harmonic generation (HHG) from atoms is well understood in terms of the three-step quasi-static model [1, 2]: (i) emission of a photoelectron wave packet by tunneling ionization, (ii) acceleration of the electron wave packet into the continuum by the laser field, and (iii) recombination with the ion that induces atomic dipoles to emit XUV photons. Each of these steps offers an opportunity to control HHG. In the case of HHG from molecules, alignment can affect all these steps: (i) The orientation of the molecule with respect to the laser field determines the ionization probability [3] and the initial condition of the electron wave packet launched by tunneling ionization [4], such as the initial spatial spread of electronic wave packet [6]. (ii) The electron motion in the continuum can be affected by Coulomb or dipole forces exerted by the ionized molecules [5]. (iii) The dipole induced by recombination can be influenced by the spatial distribution of the ground state wave function. So far, a number of articles has been published on high harmonic generation with adiabatically aligned molecules [7–10].

We show that the harmonics of all orders are sensitive to molecular alignment of N_2 and O_2 molecules. The ratio of the high harmonic signal between N_2 aligned parallel and perpendicular to the laser polarization can be as high as 4.2:1. We use dynamic field-free alignment [11] which occurs repeatedly as the wave packet evolves after the aligning pulse has passed. This enables us to perform experiments on field free molecules.

2 Experimental results

Fig. 1 shows a sketch of the experiment. Our Ti:Sa CPA laser system is capable of producing 30-mJ, 30-fs pulses at 800nm and a repetition rate of 50Hz. A Mach-Zehnder interferometer is inserted before the compressor to generate two pulses with mutual delay t_d.

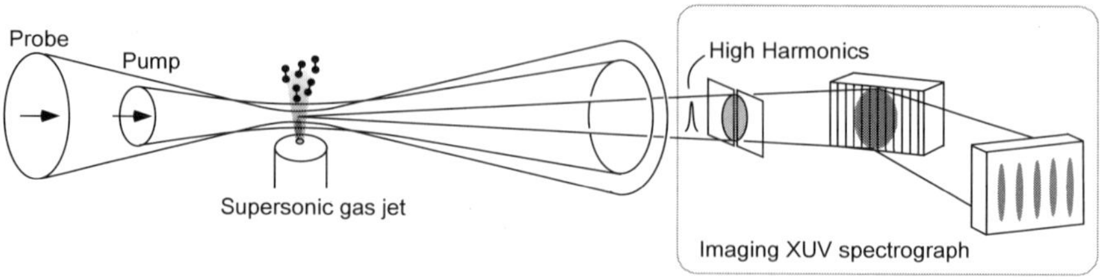

Fig. 1. Two time-delayed pulses, linearly polarized in the same direction, are focused (f = 500 mm) into a supersonic gas jet. The pump pulse ($\tau_{pump} = 60$ fs) creates a rotational wave packet. The second, delayed probe pulse ($\tau_{probe} = 30$ fs) generates high harmonics. The intensity of the pump pulse was kept just below the threshold of photoionization and HHG ($I \sim 10^{13}\text{W/cm}^2$). The peak intensity of the probe was set below the saturation ($I = 3 \times 10^{14}\text{W/cm}^2$) to avoid full ionization. The pump pulse is spatially clipped to ensure that its focal spot size is larger than that of the probe. The focal position is set 0.5mm below the orifice to realize low rotational temperature ($\sim$ 10 K for N_2) for better alignment. The high harmonic spectra are measured by an imaging spectrograph using a variable-spacing concave grating (1200 lines/mm) and an MCP with a phosphor screen. The images are captured by a CCD camera (exposure time 2 seconds, or 100 laser shots).

Fig. 2a depicts the evolution of the alignment parameter $\langle\cos^2\theta\rangle$ of N_2 directly measured by Coulomb explosion imaging (CEI) in an independent experiment [12]. As shown in Fig. 2b, high harmonics are strongly enhanced when N_2 is aligned parallel to the laser field ($t_d = 4.2$ ps), while they are suppressed when N_2 is aligned perpendicular($t_d = 4.6$ ps).

The revival structures for all observed orders of high harmonics were similar. The broader revival structures compared with CEI measurements suggest that the rotational temperature in the high harmonic case is lower than in the CEI case, leading to fewer J states involved in the wave packets.

Fig. 3 shows the enhancement and suppression ratios of the high harmonic intensities at the half revivals (Fig. 2b, delay B and C). These ratios range from 1.6-2.0 and 0.4-0.8, respectively. Maximum enhancement is observed at harmonic 21, where the intensity ratio of high harmonics from aligned N_2 with respect to perpendicularly aligned N_2 reaches ≈ 4.2. The dip in the relative HHG yields may be related to the interference effects discussed in [8]. The shape of the high harmonic plateau is predicted to show a local minimum when the wavelength λ_e of the recolliding electron satisfies the condition

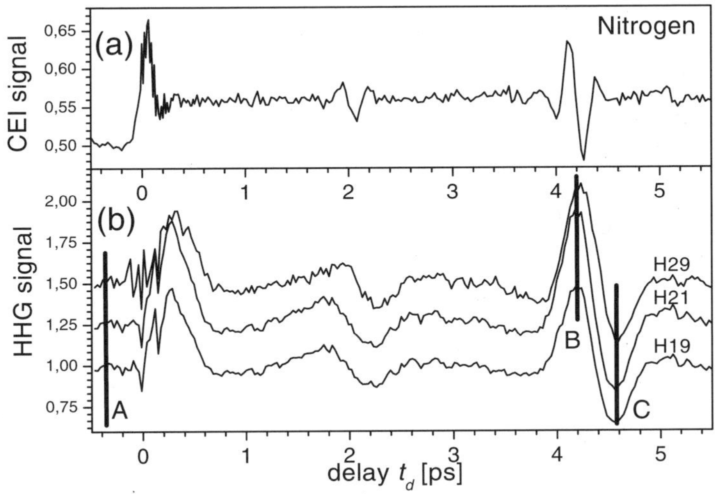

Fig. 2. Revival of rotational wave packets in N_2: (a) The degree of alignment $\langle\cos^2\theta\rangle$ directly probed by CEI. $\langle\cos^2\theta\rangle = 1, 1/2, 0$ correspond to perfect alignment, an isotropic distribution, and complete anti-alignment, respectively [12]. (b) Intensities of three different harmonics. These are normalized with respect to the unaligned molecules at $t_d < 0$ ps. Note that the curves for harmonic 21 and 29 are vertically offset.

for destructive interference $\lambda_e = 2R/(2n+1)$ (n is an integer and R the projection of the internuclear separation along the direction of the electron motion). Assuming N_2 fully aligned, this dip is expected around harmonics 29-31.

A comparison between N_2 and O_2 is most interesting since O_2 has an asymmetric electronic wave function and a different nuclear spin configuration. Fig. 4 shows the revival structures of O_2 under similar conditions as in the N_2 experiment. Pronounced revival structures are visible at 1/2 and 1/4 revivals ($\tau_d = 2.9, 5.8$ ps) both in the HHG and CEI data. In addition, small revivals can be seen at 1/8 and 3/8 of a rotational period ($t_d = 1.4, 4.4$ ps) in the high harmonic intensities since the rotational spectrum of O_2 contains only odd J states. The appearance of 1/8 and 3/8 revivals in the HHG data in contrast to the CEI data suggests that HHG does not simply depend on the degree of alignment, but is also sensitive to the phase of the J states. Compared to N_2, the contrast of the revival structure of O_2 is much smaller which is due to the lower polarizability of O_2 that makes O_2 harder to align.

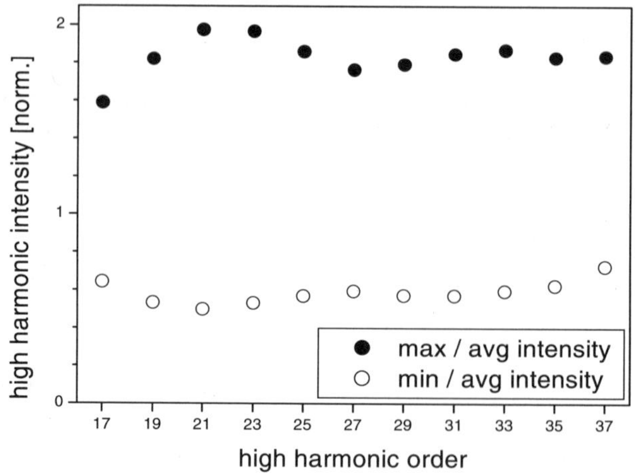

Fig. 3. Enhancement factor I_B/I_A(filled circle) and suppression factor I_C/I_A(open circle) normalized with respect to the nearly isotropic states for $t_d < 0$ ps for the case of N_2. Here I_A,I_B and I_C denotes the intensity at the delay positions A, B, and C as shown in Fig. 2b. The data shown here has been taken at a different run than shown in Fig. 2b, using laser intensities that were ≈ 1.5 times higher.

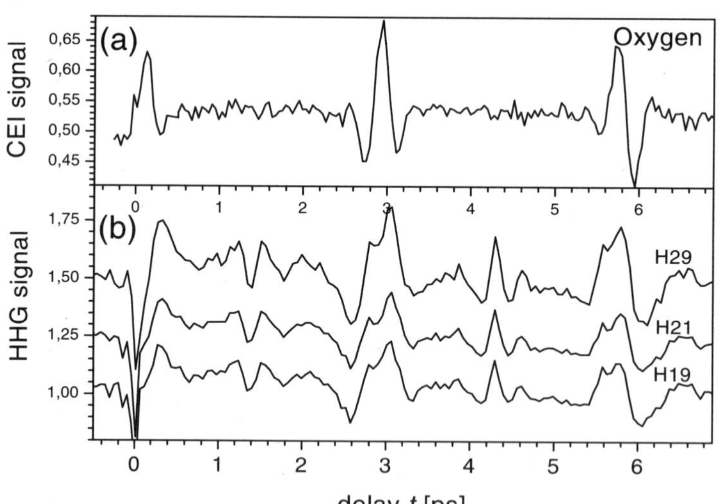

Fig. 4. Revival structures for O_2: probed by (a) CEI [13] and by (b) HHG. The curves for harmonic 21 and 29 are vertically offset.

3 Perspectives

Concluding, the evolution of the high harmonic intensity agrees well with the dynamics of rotational wave packets as directly probed by CEI. HHG promises to be an efficient method to detect the dynamics of molecular structure since the intensity of high harmonics scales as $\sim N^2$ (N being the number of atoms or molecules) under phase-matched condition. In addition, high harmonics can be used to probe the orientation of molecules since it is sensitive to the inversion symmetry of ionizing molecules.

The many external degrees of freedom of molecules will give us wide opportunities for controlling high harmonics. Quasi-phase matching may be used to enhance the conversion efficiency and to increase the cut off frequency of high harmonics. Rapidly spinning molecules may offer the opportunity to gate high harmonics in the time domain, thereby generating attosecond pulses.

Acknowledgement This work is in part supported by Photonic Research Ontario.

References

1. Corkum, P.B.: Phys. Rev. Lett., **71**, 1994 (1993).
2. Lewenstein, M., Balcou, Ph., Ivanov, M.Y., L'Huiller, A., Corkum, P.B.: Phys. Rev. A, **49**, 2117 (1994).
3. Tong, X. M., Zhao, Z. X., Lin, C. D.: Phys. Rev. A, **66**, 033402 (2002).
4. Bhardwaj, V. R., Rayner, D. M., Villeneuve, D. M., Corkum, P. B.: Phys. Rev. Lett. **87**, 253003 (2001).
5. Brabec, T., Ivavnov, M. Yu., Corkum, P. B.: Phys. Rev. A, **54**, R2551 (1996).
6. Niikura, H., Légaré, F., Hasbani, R., Bandrauk, A.D., Ivanov, M.Yu., Villeneuve, D.M., Corkum, P.B.: Nature, **417**, 917 (2002).
7. Hay, N., Velotta, R., Lein, M., de Nalda, R., Heesel, E., Castillejo, M., Marangos, J.P.: Phys. Rev. A., **65**, 053805 (2002).
8. Lein, M., Hay, N., Velotta, R., Marangos, J. P., Knight, P. L.: Phys. Rev. A, **66**, 023805 (2002).
9. Lein, M., Hay, N., Velotta, R., Marangos, J. P., Knight, P. L.: Phys. Rev. Lett., **88**, 183903 (2002).
10. Lein, M., Corso, P.P., Marangos, J. P., Knight, P. L.: Phys. Rev. A, **67**, 023819 (2002).
11. Rosca-Pruna, F., Vrakking, M. J. J.: Phys. Rev. Lett. **87**, 153902 (2001).
12. Litvinyuk, I.V., Lee, K.F., Dooley, P.W., Rayner, D.M., Villeneuve, D.M., Corkum, P.B.: Phys. Rev. Lett. **90**, 233003 (2003).
13. Dooley, P.W., Litvinyuk, I.V., Lee, K.F., Rayner, D.M., Spanner, M., Villeneuve, D.M., Corkum, P.B.: Phys. Rev. A, in print (2003).

Towards Attosecond Half-Cycle Pulses

Emil Persson[1], Severin Puschkarski[1], Xiao-Min Tong[2], and Joachim Burgdörfer[1]

[1] Institute for Theoretical Physics, Vienna University of Technology, A-1040 Vienna, Austria emil@concord.itp.tuwien.ac.at
[2] Kansas State University, Manhattan, KS 66506, USA

Summary. We propose to tailor a burst of the harmonics generated by a bichromatic field interacting with helium to produce either a single half-cycle pulse or a train of attosecond unidirectional half-cycle pulses. Such a field will be an important tool to pump, control and probe electronic states near the ground state.

1 Introduction

Ultra-short single half-cycle pulses (HCP) in a pump-probe setting can be used to impulsively tailor and detect wave packets [1, 2], and to retrieve quantum information [3]. Studies of periodically driven quantum systems, such as the Rydberg atom exposed to a train of unidirectional half-cycle pulses (u-HCP), have revealed the effect of dynamical localization [4, 5]: the driven system has a very small ionization rate and its wave function remains close to the initial state for a long time even though the driving field is not negligible compared to the binding energy of the atom.

A train of HCP's on a nanosecond scale can be produced by conventional pulse generators [6]. Sequences of 500fs HCP have been produced using GaAs wafers driven by short laser pulses [7]. For a HCP to be ultra-short on the electronic time scale, the pulse width τ_P must be short compared to the orbital period T_n ($T_n = 2\pi n^3$a.u. for Rydberg states). Shaping and manipulating electronic wave packets has up to now only been possible for Rydberg states $n \gg 1$ with long orbital periods.

The goal of the present paper is to discuss the feasibility of producing a train of u-HCP with a pulse width τ_p in the attosecond regime. This would allow to impulsely drive, shape and control electronic wave functions near the ground state. The shortest electromagnetic pulses presently available are obtained from high harmonics generated through the non-linear interaction (HHG) of a short high-intensity laser pulse with an atomic target [8, 9]. The electric field of such a pulse ("burst") still contains many cycles during the pulse length (Fig. 1.b). For a HCP, on the other hand, the electrical field contains less than one complete cycle under the envelope (Fig. 1.e). Here, we propose a strategy to combine the harmonics produced by HHG to form a train of attosecond u-HCP.

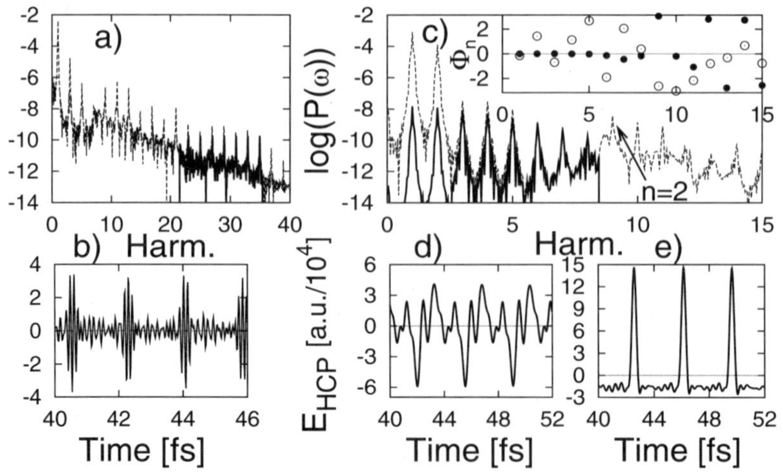

Fig. 1. Non-linear response of hydrogen interacting with a laser with $\lambda = 1064$nm. **(a)** Power spectrum (dashed line), filtered power spectrum (full line) and **(b)** resulting attosecond bursts for a single color field with $I = 7.0 \times 10^{13} W/cm^2$. **(c)** Power spectrum and phases ϕ_n (inset) of the harmonics for a two-color field with $I_A = 1.8 \times 10^{13}$, $A_R = 0.37$ and two different driving field phases **(i)** $\delta_2 = 0.56\,\pi$ (circles) and **(ii)** $\delta_2 = 0$ (filled circles). The power spectrum is only shown for (i) since the result for (ii) is similar. **(d, e)** resulting field for the phases (i) and (ii).

2 A strategy for producing u-HCP

In our analysis we study the non-linear single atom response for both hydrogen atoms and helium. We consider a driving field consisting of a fundamental frequency ω together with multiples of ω,

$$F(t) = f(t) \sum_{n=1}^{N} A_n \sin(n\omega t + \delta_n - \delta_1), \tag{1}$$

where $f(t)$ is an envelope function. To produce a train $E_{HCP}(t)$ of u-HCP, both odd and even harmonics from the HHG must be present. Therefore the driving field $F(t)$ must consist of at least two frequencies.

We start by studying a two color driving field consisting of ω and 2ω and choose the relative phase $\delta_2 = 0$. The high harmonic generation is calculated by means of the pseudo-spectral method [10], which has the advantage of practically exact bound states and stable long-time evolution due to quadrature methods used in a split-operator method. For the HHG-output to form u-HCP, two conditions must be met: (a) to avoid fluctuations around the peaks, the amplitudes of the harmonics should follow a Gaussian distribution (Fig. 1.c). (b) all phases ϕ_n of the harmonics should be aligned to zero (Figs. 1.d, e). Since the amplitude of the harmonic peaks decrease rapidly with the harmonic number, the fulfillment of (a) implies a trade-off between intensity (by giving the lower harmonics a larger weight) and narrow peaks. We assume that the adjustment of the amplitudes can be achieved by a combination of filters and propagation effects, and focus on optimizing the condition (b).

Using two colors and $\delta_2 = 0$ leads to two parameters subject to optimization, the intensity $I_A = A_1^2 + A_2^2$ and the relative amplitude $A_R = A_2/A_1$. Figure 2, left panel, displays the number of phases with $| \phi_n |< 0.1$ produced by the non-linear response of hydrogen at $\lambda = 1064$nm for different I_A and A_R. Due to the non-linear character of the interaction, a complicated pattern results. The phases ϕ_n and the output field E_{HCP} of the best case, which is at $I_A = 1.8 \times 10^{13} W/cm^2$ and $A_R = 0.37$ with eight phases aligned, are shown in Fig. 1.c and e. The width of the HCP's $\tau_P = 360as$, and the ratio $\tau_P/T = 0.10$ where $T = 3.5$fs is the period of the train of pulses.

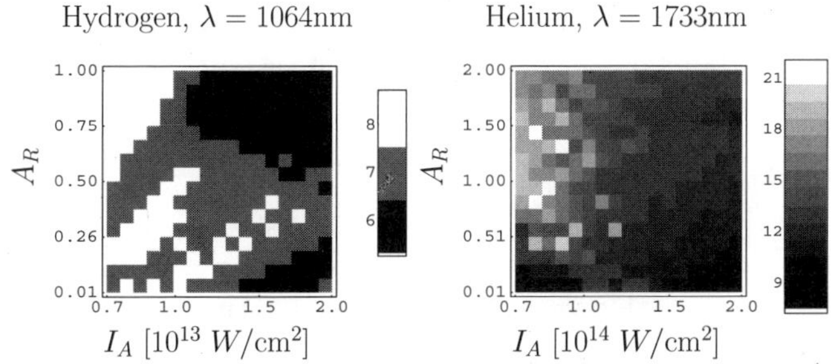

Fig. 2. Number of aligned phases of the non-linear response of H (left panel) and He (right panel) interacting with a two-color field λ, $\lambda/2$ with $\lambda = 1064$nm (left panel) and $\lambda = 1733$nm (right panel) at different I_A and A_R (for details see text).

We investigated the dependence of the phases ϕ_n of the harmonics on the phase δ_2 of the driving field (Fig. 1). The number of aligned phases start to decrease for $\delta_2 > 0.1\,\pi$. We note that the dependence of ϕ_n on δ_n might imply some possibility to compensate for phase mismatches produced by filters or propagation effects.

We further tested more than two colors in the driving field but we did not find any significant improvement.

3 Shorter u-HCP: Increasing the number of aligned phases

The fact that only eight phases could be successfully aligned is related to the maximum in the power spectrum around the ninth harmonic (Fig. 1.c), which corresponds to the energy difference between the hydrogen ground state $n = 1$ and first excited state $n = 2$. We can attribute the structure seen in the power spectrum and the misalignment of the phases above the eight harmonic to resonant effects caused by the excited levels.

One way of increasing the number of harmonic peaks below the $n = 2$ level is to increase the wavelength of the driving field. We studied the number of aligned phases for hydrogen interacting with a two-color driving field with $\delta_n = 0$ for $\lambda = 1733$nm as a function of I_A and A_R (not shown). Now, the first excited state lays close the harmonics 14 and up to 11 phases can be aligned. The best u-HCP field (not shown), found at $I_A = 1.26 \times 10^{13} W/cm^2$ and $A_R = 1.0$, has width $\tau_P = 430as$ and $\tau_P/T = 0.07$.

The number of harmonic peaks below the first excited level can be further increased by considering another atom with a larger energy gap between the ground state and the first level, such as helium. We treat helium in the single active electron approximation (SAE) using the pseudo-spectral method and a screening potential $V_{\text{SAE}}(r) = -(1 + \exp(-r/r_H))/r$. With $r_H = 0.4686$, the numerical energies (experimental values) are -0.9034(-0.9033), -0.1458(-0.1574), -0.1236(-0.1279), for the 1s, 2s and 2p states, respectively. Figure 2, right panel, shows the result from of a

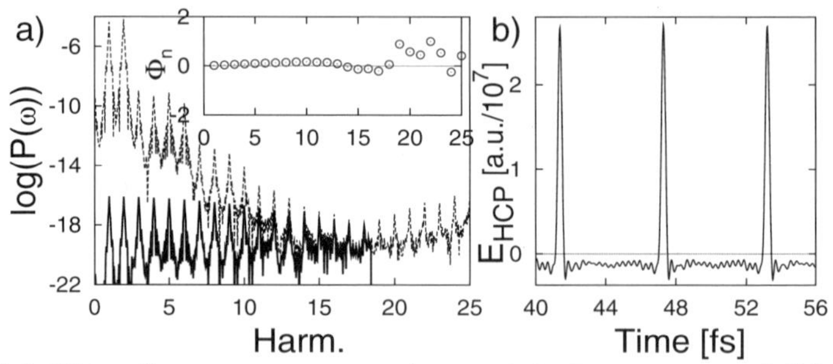

Fig. 3. **(a)** (filtered) power spectrum, phases of the harmonics, and **(b)** resulting train of u-HCP for He interacting with a two-color field with $\lambda = 1773$nm.

scan over I_A and A_R for a two-color field with $\lambda = 1773$nm interacting with helium. The energy of the first excited level corresponds to about 29 harmonics. Results for a field found at $I_A = 1.03 \times 10^{14} W/cm^2$ and $A_R = 1.25$, with about 18 harmonics aligned, are shown in Fig. 3. Its width $\tau_P = 280as$ and $\tau_P/T = 0.047$.

4 Controlling the length of the u-HCP trains

The concept above can be extended to long trains needed, for example, for the investigation of dynamical localization. For hydrogen interacting with a two-color field at $I_A = 2.5 \times 10^{13} W/cm^2$, we have found trains with 100 u-HCP pulses and a ground state depletion of 6%.

For pump and probe experiments, on the other hand, the pulses should ideally consist of only a single half-cycle pulse. We study a two color field with an envelope covering only two cycles of the fundamental frequency and different carrier-envelope phases ϕ_{CE} (Fig. 4). For $\phi_{CE} = 0$, we find a single half-cycle pulse with two low-amplitude side-peaks. Note that the peaks are about 2.5 times smaller than the main peak in the driving field and about 4.5 times smaller in the resulting field. For $\phi_{CE} \neq 0$ we find two peaks with virtually no additional peaks. The ratio between the two HCP can be adjusted by varying $\phi_{CE} \neq 0$. Since the distance between the two peaks is given by the frequency of the driving field, these peaks can be used as an attosecond u-HCP pump-probe field with variable strengths and time-delays.

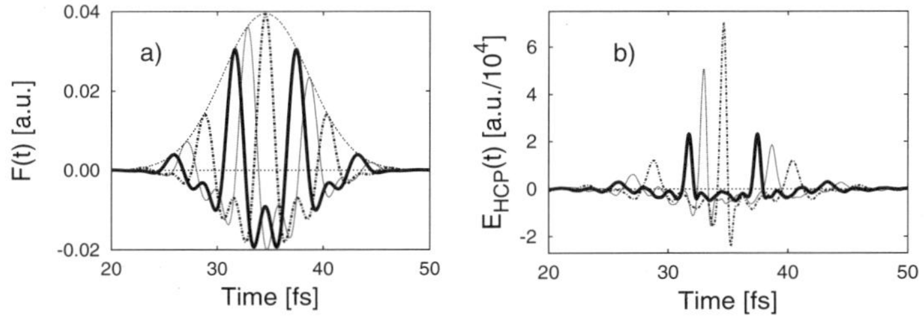

Fig. 4. **(a)** driving field, and **(b)** resulting field for a short two-color laser pulse with different carrier-envelope phases $\phi_{CE} = -\pi, -\pi/2$, and 0 interacting with hydrogen.

5 Conclusions

We theoretically analyze a protocol for forming a train of unidirectional half-cycle pulses in the attosecond regime. We obtain the train by combining harmonics obtained through high harmonic generation of a strong two-color laser field interacting with hydrogen atoms or helium in non-linear single-atom response. We assume the amplitudes of the harmonic peaks to be tunable and optimize the phases of the harmonics.

The two-color field with zero relative phase of the colors give promising results which should be experimentally accessible. Our findings indicate that it is possible to align phases of harmonics below the region where resonant effects caused by excited levels in the atoms strongly influence the heights and phases of the harmonics.

In our studies, we find **(a)** trains consisting of hundreds of unidirectional half-cycle pulses, which could be used to periodically drive an atom. We further find **(b)** both single half-cycle pulses and **(c)** trains consisting of only two pulses with variable strengths and delay between the pulses. Both the latter findings (b, c) imply possibilities for attosecond pump-probe experiments. The narrow pulse widths of the half-cycle pulses open up a way towards impulsively shaping, probing, and manipulating wave packets of low-lying electronic states.

This work is supported by FWF (Austria) under contract number SFB-016

References

1. D. Arbó et al., Phys. Rev. A **67**, 063401 (2003)
2. C. Wesdorp et al., Phys. Rev. Lett **87**, 083001 (2001)
3. J. Ahn et al., Phys. Rev. Lett. **86**, 1179 (2001)
4. S. Yoshida et al., Phys. Rev. **A**, 023408, 2000
5. E. Persson et al., (submitted Phys. Rev. A, 2003) and references therein
6. C.O. Reinhold et al., Phys. Rev. Lett. **79**, 5226 (1997)
7. R.R. Jones et al., Phys. Rev. Lett. **70**, 1236 (1993)
8. M. Hentschel et al., Nature **414**, 509 (2001)
9. P.M. Paul et al., Science **292**, 1689 (2001)
10. X.M. Tong, and S.I. Chu, Chem. Phys. **217**, 119 (1997)

Temporal Characterization of Attosecond Pulses

F. Quéré[1,2], J. Itatani[1], G. L. Yudin[1], and P. B. Corkum[1]

[1] Steacie Institute for Molecular Sciences, National Research Council of Canada, Ottawa, Ontario, Canada K1A 0R6 jiro.itatani@nrc.ca
[2] DSM-DRECAM-Service des Photons, Atomes et Molécules, CEA Saclay, 91191 Gif-sur-Yvette Cedex, France quere@drecam.cea.fr

1 Introduction

Two paths have so far been demonstrated for time-resolved experiments with attosecond resolution, using intense femtosecond lasers. The first one consists in using high-harmonic generation (HHG) in gases to produce attosecond x-ray pulses, which can then be used to trigger or to probe an ultrafast process [1]. In the second path, attosecond electron pulses, generated by strong-field ionization of molecules and correlated with the nuclear wave packet, are used as a probe of the nuclear dynamic [2,3].

Finding methods for accurate temporal characterization of attosecond x-ray pulses is crucial for the development of this new field of ultrafast science. A direct extension of the methods used for femtosecond pulses -such as nonlinear autocorrelation- turns out to be difficult for soft x-rays, although it might eventually become possible [4]. The difficulty arises from the lack of an efficient nonlinear process in this wavelength range.

The same problem was encountered more than a decade ago with the characterization of picosecond to femtosecond x-ray pulses. This was solved by cross-correlating the x-ray pulses with an intense near-infrared laser pulse -a so-called "dressing field"- of comparable or shorter duration, via atomic photoionization [5,6]. In this paper, we show how this pioneering approach can be generalized to the measurement of attosecond pulses, which are shorter than the optical period of the dressing field.

The common basis to most proposed [7–12] or demonstrated [13,14] attosecond measurement methods is as follows. Attosecond x-ray pulses can ionize atoms by single photon absorption. This ionization generates an attosecond electron wave packet in the continuum, which, far from any resonance in the continuum, is a replica of the attosecond field. Characterizing this wave packet thus provides direct information on the attosecond x-ray field. As we will demonstrate, strong near-infrared laser fields, such as femtosecond laser pulses generated by Ti-Sa lasers, provide an ideal tool to manipulate and characterize these attosecond electron wave packets. Atomic units are used throughout the paper.

2 Theory of atomic x-ray photoionization in a strong-laser field

We first analyze the connection between an attosecond x-ray pulse and the electron wave packet it generates in the continuum when ionizing an atom. We then study the effect of a strong low-frequency laser field on the photoelectron spectrum. We first use a quantum model which allows simple numerical calculations. We then detail a semiclassical model [15, 16], that provides an intuitive understanding of the quantum model, and is thus very helpful to design and understand attosecond measurement techniques. We finally present a graphic tool which is useful in the context attosecond measurements.

2.1 Electron wave packet replica of an attosecond field

We first consider the ionization of an atom by an attosecond field alone, and use the independent-electron approximation. According to first order perturbation theory, the transition amplitude at time T to the final continuum state $|\mathbf{v}\rangle$ with momentum $\mathbf{v}$, is given by

$$a_{\mathbf{v}}(T) = -ie^{-iWT} \int_{-\infty}^{T} dt\, \mathbf{d_v E}_X(t) e^{i(W+I_p)t} \,, \tag{1}$$

where $W = \mathbf{v}^2/2$ is the energy of the final continuum state. $\mathbf{E}_X(t)$ is the x-ray electric field, $\mathbf{d_v}$ is the dipole transition matrix element from the ground state to the continuum state $|\mathbf{v}\rangle$, and I_p is the ionization potential of the atom. Eq. (1) gives the connection between an attosecond pulse and the electron wave packets it generates in the continuum when ionizing an atom.

It shows that the continuum states get populated only during the x-ray pulse: once $|\mathbf{E}_X(t)| = 0$, $a_{\mathbf{v}}(T)$ no longer varies with T. In this respect, we can say that the temporal shape of the electron wave packet is directly related to that of the attosecond pulse. This is exploited in the attosecond streak camera (see part 3.2).

Taking $T = +\infty$ in Eq.(1) shows that the photoelectron spectrum is directly related to the attosecond pulse spectrum, both in phase and amplitude. The two spectra might differ in amplitude if $|\mathbf{d_v}|$ depends on $\mathbf{v}$. They differ in phase because of the $-WT$ phase term, which simply accounts for free-space propagation of the wave packet, and because of a possible phase dependence of $\mathbf{d_v}$ on $\mathbf{v}$. If this phase dependence is negligible, or is known either from theory or experiment, the spectral phase of the attosecond pulse can be directly deduced from the spectral phase of the electron wave packet. This direct connection is exploited in attosecond SPIDER (see part 3.1).

We now turn to the effect of a low-frequency laser field on this photoelectron spectrum.

2.2 SFA quantum model

In the case of x-ray photoionization in the presence of a laser field, we use the strong field approximation (SFA) [17] in addition to the independent electron approximation. This approximation consists in neglecting the effect of the ionic potential on the electron's motion after ionization. Despite its name, the SFA is expected to be valid even for moderate laser intensities, provided $\Omega_X \gg I_p$ (where Ω_X is the x-ray pulse central frequency). The sudden approximation [18, 19] can also be used when $\Omega_X \gg \omega_L$ is satisfied (where ω_L is the frequency of the laser field). In this situation, both approximations lead to the same results.

Within this approximation, at a time large enough for both the x-ray and the laser fields to vanish, the transition amplitude to the final continuum state $|\mathbf{v}\rangle$ with momentum $\mathbf{v}$, is given by

$$a_{\mathbf{v}} = -i \int_{-\infty}^{+\infty} dt\, \mathbf{d}_{\mathbf{p}(t)} \mathbf{E}_X(t) e^{-i\left[-I_p t + \int_t^{+\infty} dt' \mathbf{p}^2(t')/2\right]} . \tag{2}$$

$\mathbf{p}(t) = \mathbf{v} + \mathbf{A}(t)$ is the instantaneous momentum of the free electron in the laser field, $\mathbf{A}(t)$ being the vector potential of this field, such that $\mathbf{E}_L(t) = -\partial \mathbf{A}/\partial t$. $\mathbf{d}_{\mathbf{p}(t)}$ is the dipole transition matrix element corresponding to the momentum $\mathbf{p}(t)$ that the electron had at the time of ionization t. The integral in the exponential of Eq.(2) is the phase acquired by an electron with an initial momentum $\mathbf{p}(t)$ at time t, during its motion in the continuum from t to $+\infty$, in the presence of the laser field. Note that if there is no laser field, this equation reduces to first-order perturbation theory (Eq. (1)) with respect to the x-ray field.

A very useful form of this equation can be obtained if we use the slowly-varying envelope approximation for the intense laser pulse. For a linearly polarized laser field $\mathbf{E}_L(t) = \mathbf{E}_0(t)\cos(\omega_L t)$, one obtains

$$a_{\mathbf{v}} = -i \int_{-\infty}^{+\infty} dt\, e^{i\phi(t)} \mathbf{d}_{\mathbf{p}(t)} \mathbf{E}_X(t) e^{i(W+Ip)t} \tag{3}$$

with

$$\begin{aligned}
\phi(t) &= \phi_1(t) + \phi_2(t) + \phi_3(t) \\
\phi_1(t) &= -\int_t^{+\infty} dt U_p(t) \\
\phi_2(t) &= (\sqrt{8WU_p}/\omega_L)\cos\theta \cos\omega_L t \\
\phi_3(t) &= -(U_p/2\omega_L)\sin(2\omega_L t)
\end{aligned} \tag{4}$$

where $U_p(t) = E_0(t)^2/4\omega_L^2$ is the ponderomotive energy of the electron at the time of ionization, and θ is the observation angle, defined as the angle between $\mathbf{v}$ and the polarization direction of the laser field.

If $\mathbf{d}_{\mathbf{v}}$ is assumed to be independent of $\mathbf{v}$, comparing Eq. (3-4) with Eq. (1) shows that, in a given direction of observation θ, the only effect of the laser field is to induce a temporal phase modulation $\phi(t)$ on the electron wave packet generated in the continuum by the x-ray field.

This ultrafast electron phase modulation comprises three terms. $\phi_1(t)$ varies slowly in time (on the time scale of the laser pulse envelop), while $\phi_2(t)$ and $\phi_3(t)$ oscillates respectively at the laser field frequency and its second harmonic. $\phi_2(t)$ will be predominant over $\phi_3(t)$ for angles $\theta \ll \arccos(\sqrt{U_p/32W})$. In many cases $U_p \ll W$, so that $\phi_2(t)$ predominates for almost all angles except $\theta \approx \pi/2$. As opposed to $\phi_1(t)$ and $\phi_3(t)$, $\phi_2(t)$ depends on the final electron energy W. This dependence has a number of drawbacks for attosecond measurements, as we will see in part 3.

In the case of an x-ray pulse longer than the optical cycle of the dressing field, $\phi_2(t)$ and $\phi_3(t)$ induce a periodic phase modulation on the photoelectron wave packet : just as a periodic phase modulation applied on a light pulse induces satellites in its spectrum, this leads to the so-called side-bands in the photoelectron energy spectrum [5]. These side-bands can be used to cross-correlate ultrashort (ps to fs) x-ray pulses with the envelop of intense femtosecond laser pulses [5]. The slowly-varying phase term $\phi_1(t)$ is responsible for the ponderomotive shift [20] of these peaks in the photoelectron spectrum.

In some cases, the dependence of $|\mathbf{d_v}|$ on $\mathbf{v}$ might not be negligible. The laser field then acts both as a phase and amplitude modulator, through the temporal dependence of $\mathbf{p}(t) = \mathbf{v} + \mathbf{A}(t)$. This amplitude modulation could also be used for attosecond measurement (for instance to produce an amplitude gate for blind-FROG measurements, see part 3.3), but this possibility has hardly been explored so far.

2.3 Semiclassical model

The following semiclassical treatment gives an intuitive picture of x-ray photoionization in the presence of a laser field. Physically, the large difference in frequencies between the x-ray pulse and the laser pulse naturally divides the two-color ionization process in two steps: absorption of an x-ray photon followed by acceleration in the laser field [18, 19]. Classical mechanics can be used to describe the second step [15, 16].

We consider an electron ejected in the continuum by the x-ray field at time t_i, with an initial kinetic energy $W_0 = v_0^2/2 = \Omega_X - I_p$, where $\mathbf{v}_0$ is the initial velocity. Assuming $\Omega_X \gg I_p$, we neglect the influence of the ionic potential on the electron's motion in the continuum. Using classical mechanics, the time-dependent momentum $\mathbf{p}(t)$ in the continuum is given by

$$\mathbf{p}(t) = \mathbf{A}(t) + [\mathbf{v}_0 - \mathbf{A}(t_i)] \tag{5}$$

The first term on the right hand side describes the electron's quiver motion in the laser field and goes to zero as the laser pulse ends. The second term in brackets is determined by the initial condition $\mathbf{p}(t_i) = \mathbf{v}_0$ at the time of ionization. This term is the final drift velocity $\mathbf{v} = \mathbf{v}_0 + \mathbf{A}(t_i)$ measured after the laser pulse. The effect of the laser field on the drift velocity can be visualized by plotting $|\mathbf{v}|$ as a function of the observation angle θ [9, 13, 21]. For a given t_i, this distribution is a circle, which center is displaced from $|\mathbf{v}| = 0$ under the action of the dressing field (see Fig. 1, left panel).

The drift kinetic energy $W = \mathbf{v}^2/2$ is obtained by solving the second-order polynomial equation $\mathbf{v}_0^2 = (\mathbf{v} - \mathbf{A}(t_i))^2$, which leads for a linearly polarized laser-field to

$$\begin{aligned} W &= W_0 + 2U_p \cos 2\theta \sin^2 \omega_L t_i \pm \alpha \sqrt{8W_0 U_p} \cos\theta \sin\omega_L t_i \\ \alpha &= \{1 - (2U_p/W_0) \sin^2\theta \sin^2 \omega_L t_i\}^{1/2} \end{aligned} \tag{6}$$

As before, θ is the observation angle. For $U_p < W_0/2$, only the positive sign in Eq. (6) has a physical meaning.

Equation (6) shows how the energy change $\delta W = W - W_0$ induced by the laser field varies with the observation angle θ and the ionization time t_i. Almost all attosecond measurement methods presented in this paper can be understood from this simple equation. In Fig. 1, W is plotted as a function of $\omega_L t_i$ for three observation angles θ.

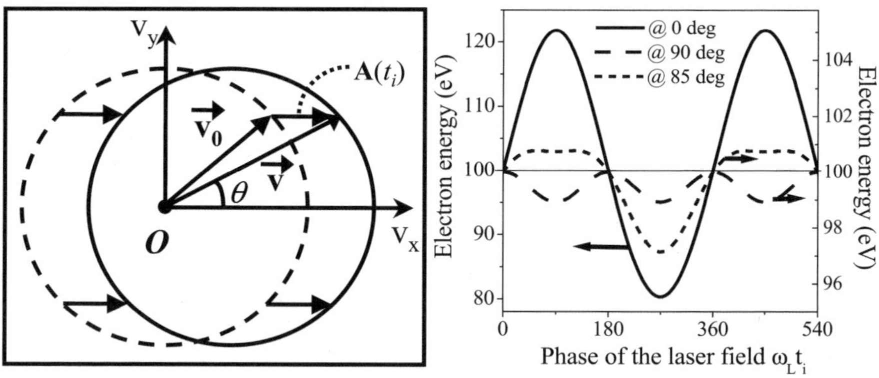

Fig. 1. Right panel: drift kinetic energy W of a photoelectron as a function of its ionization phase $\omega_L t_i$, for three observation angles θ (linearly-polarized dressing field, $\lambda = 800$ nm, $I = 9$ TW/cm^2, $U_p \approx 0.54$ eV, with $W_0 = \Omega_X - I_p = 100$ eV). These curves can be viewed as resulting from the oscillation with $\omega_L t_i$ of the velocity circle (left panel) along the laser polarization direction of the laser field.

This semiclassical model and the SFA quantum model are closely related [10]. If the x-ray pulse is short compared to the laser field period, a linear expansion of the phase modulation $\phi(t)$ with respect to time can be used in Eq. (3-4). Physically, a linear phase modulation on the electron wave packet leads to a shift of its spectrum, as in the case of optical pulses. This shift is given by $\delta W = W - W_0 = -\partial\phi/\partial t$, which leads to the same expression for W as in Eq. (6). Thus, the SFA quantum model and the semiclassical model are strictly equivalent in the limit of x-ray pulses much shorter than the laser period. The semiclassical model provides the intuitive basis to understand the underlying physics.

2.4 Wigner distribution of an attosecond electron wave packet

We define the Wigner distribution $\Upsilon_\theta(t, W)$ of the photoelectron wave packet in a given observation direction θ as

$$\Upsilon_\theta(t, W) = \frac{1}{2\pi} \int dE\ a_\theta^*(W - E/2)\ e^{-iEt}\ a_\theta(W + E/2) \tag{7}$$

where $a_\theta(W) = a_{\mathbf{v}}$ are the transition amplitudes given by Eq.(2). This is a particular case of time-frequency distribution [22] describing the electron wave packet. Comparing the Wigner distributions obtained with and without the dressing laser field enables to visualize the effect of this field on the photoelectron wave packet, and provides insight into the connection between the quantum and semiclassical model.

This is illustrated in Fig. 2. The attosecond pulse used in this calculation is Fourier-transform limited. The Wigner distribution in the absence of the dressing field therefore looks like an horizontal ellipse (Fig. 2(a)). In the central part of the pulse, the dressing field induces a positive chirp of the electron wave packet (Fig. 2(b)). On the edges, the sign of the chirp changes. This behavior is qualitatively reproduced by the semiclassical model (Fig. 2(b)). What misses in the semiclassical model, compared to the quantum model, are the interferences, in the spectral domain, between parts of the wave packet emitted at different times but having the same final energy. These interferences lead to strong modulations in the photoelectron spectrum, and appear as usual [22] as "islands" in the Wigner distribution. As the x-ray pulse gets longer, they eventually lead to the appearance of well-separated side-bands in the photoelectron spectrum.

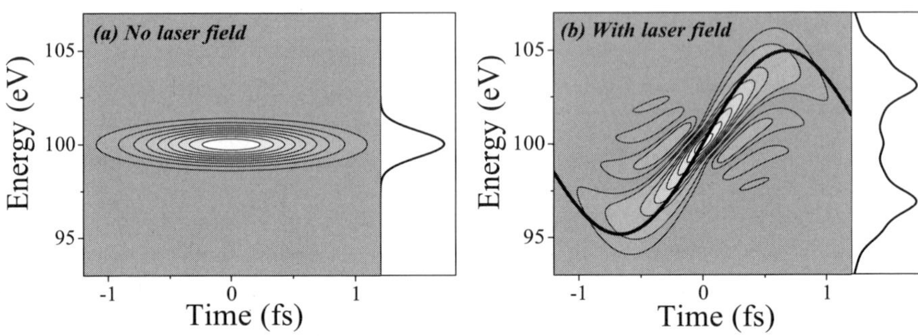

Fig. 2. Wigner distributions of the electron wave packet in the $\theta = 0$ direction, with and without the laser field. The panels on the right show the photoelectron energy spectrum (marginal of the Wigner distribution). $\mathbf{d_v}$ was assumed to be constant for this calculation. The Wigner distribution in (a) is therefore identical to that of the attosecond x-ray pulse (1.2 fs FWHM, with $\Omega_X - I_p = 100$ eV). The laser field has an intensity of 0.5 TW/cm^2 ($\lambda = 800$ nm, $U_p = 0.03$ eV), and its phase is 0 at the maximum of the x-ray pulse. The full line in (b) shows the photoelectron final energy W as a function of t_i, as deduced from the semiclassical model.

3 Attosecond measurement methods

We have shown that for electrons observed in a given direction relative to the laser polarization, a dressing low-frequency laser field essentially acts as an ultrafast phase - and in certain conditions, amplitude - modulator on the electron wave packet. Such a well-defined and controllable modulator enables to transpose techniques used for the characterization of femtosecond pulses to attosecond fields, using the electron wave packet as an intermediate tool (see in particular [23–25] for close optical analogues of the methods presented here).

Depending on the delay between the modulator and the field to be characterized, the phase modulation can either be used to shift the photoelectron spectrum - allowing SPIDER measurements to be carried out on the electron wave packet [12] - or be used to streak the energy of the attosecond electron wave packet -thus providing an attosecond streak camera [9]. By scanning the delay, it can also be used as a gate for blind-FROG measurements.

3.1 Attosecond SPIDER [12]

One method for the complete characterization of ultrashort near-visible light pulses is spectral phase interferometry for direct electric-field reconstruction (SPIDER) [26]. SPIDER measures the relative phases of the different frequency components of a light pulse, by using the spectral domain interferences between two replicas of this pulse, temporally delayed and slightly shifted in frequency.

SPIDER can not be directly applied to attosecond x-ray pulses, because no nonlinear optical process is available in this spectral range to induce the required frequency shift. However, it can be applied to the photoelectron wave packets generated by photoionization, since their spectra can be steered by a dressing laser field [12].

Two identical and time-delayed electron wave packets can be produced by ionizing an atom with two time-delayed replicas of the attosecond pulse to be characterized. Producing a time-delayed duplicate of an x-ray pulse with a large spectral bandwidth is experimentally difficult, but it could be achieved by using a dispersionless Michelson interferometer [27] or the zeroth order of a laminar grating [12]. By choosing appropriate observation angle and ionization phases, a dressing laser field then induces the required energy shift ΔW between these two wave packets. Such a controlled spectral shearing has already been demonstrated experimentally for a single electron wave packet [28].

For SPIDER measurements, it is essential to avoid inducing spectral distortions when shearing the twin pulses in frequency. For attosecond SPIDER, this implies that the attosecond pulse has to be short compared to the optical period of the dressing laser field. The phase modulation $\phi(t)$ can then be approximated by a linear expansion with respect to time. This leads to a shift $\delta W = -\partial\phi/\partial t$ of the photoelectron spectrum given by Eq.(6). The delay t_0 with the dressing field and the time separation τ between the twin pulses have to be chosen so that the two wave packets experience different energy shifts δW. Besides, the linear expansion of $\phi(t)$ is most accurate, and the distortion of the spectrum is the weakest, when $\partial^2\phi/\partial t^2 = \partial\delta W/\partial t = 0$ at $t = t_0$ and $t = t_0+\tau$. These two conditions determine the most adequate arrival times for the two attosecond pulses with respect to the laser field oscillation. For example, at $\theta=\pi/2$, this leads to $\omega_L t_0 = 0$ and $\omega_L \tau = 2\pi(n/2+1/4)$, with a corresponding energy shear of $2U_p$, while at $\theta=0$, $\omega_L t_0 = \pm\pi/2$ and $\omega_L\tau = 2\pi(n+1/2)$, with an energy shear of $\sqrt{32W_0U_p}$.

At $\theta = 0$, a residual distortion of the photoelectron spectrum always occurs, because of the dependence of δW (or $\phi_2(t)$) on W_0. This can lead to errors in the retrieved phase. However, such errors can be corrected by modifying the conventional SPIDER reconstruction algorithm, to take into account this systematic and predictable distortion of the spectrum.

Since there is no fundamental limitation to the temporal resolution of attosecond SPIDER, this method opens the route to the complete characterization of extremely short x-ray pulses [12]. The numerical results presented in Fig. 3 shows that this method should also allow measurement of currently accessible pulses: using parallel observation (θ =0) and an 800-nm 10-GW/cm^2 shearing field, the spectral phase of a 400-asec pulse centered at $\approx$100 eV is satisfactorily reconstructed. Note that the twin pulses are separated by 1.5 periods, to obtain a sufficient number of fringes in the photoelectron spectrum. Numerical calculations show that 400-asec pulses are near the upper limit for accurate reconstruction using 800-nm shearing fields. Beyond this duration, the photoelectron spectrum is distorted by the dressing field, and attosecond SPIDER loses its accuracy. We will now see how these distortions can be exploited for attosecond measurements.

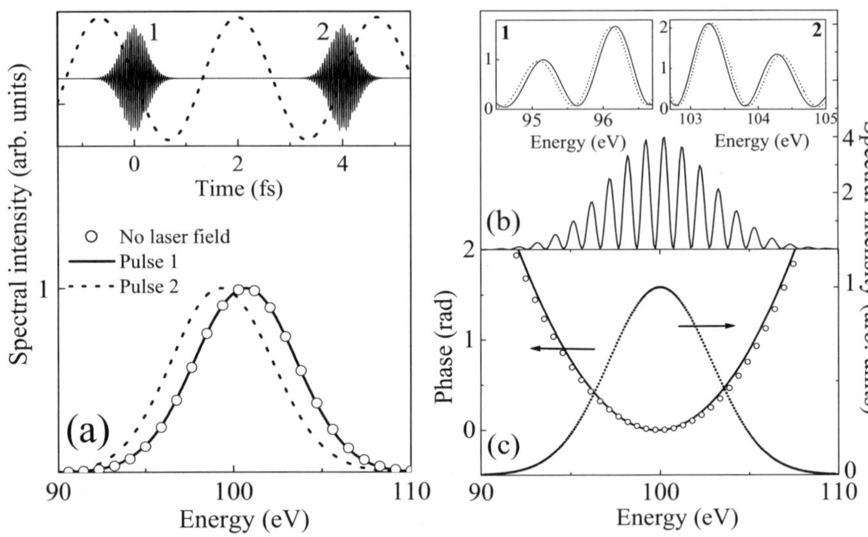

Fig. 3. (a) Photoelectron spectra at $\theta = 0$, produced by each of the two x-ray pulses shown in the top inset (400 asec chirped pulses, 6.5 eV spectral bandwidth, $\Omega_X - I_p = 100$ eV) in the presence of a laser field ($\lambda = 800$ nm, $I = 10^{-2}$ TW/cm^2, energy shear $\Delta W \approx 1.4$ eV). The open circles show the spectrum obtained in the absence of the laser field (shifted in energy for comparison with the sheared spectrum). The dashed line in the inset shows the laser field. (b) Photoelectron spectrum obtained when both pulses successively excite the atom in the laser field. Insets 1 and 2 compare the interference patterns obtained with unchirped (270 asec, full line) and chirped (400 asec, dashed line) pulses. (c) Comparison between the spectral phase retrieved from this SPIDER spectrum (circles) and the exact spectral phase $\varphi(W)$ (full line). The dotted line shows the spectral intensity.

3.2 Attosecond streak camera [9]

As shown in part 2.3, an electron generated by x-ray photoionization can be deflected by a strong laser field. Its energy and angular distribution depends on the phase of the laser field at the time of ionization. In analogy with conventional streak cameras, this phase dependence offers an opportunity to use the sub-cycle oscillation of the laser electric field as a time reference, to determine the duration of x-ray pulses shorter than the optical cycle [9].

There are many possible configurations of this attosecond streak camera, depending on the laser field polarization and on the measurements performed on the photoelectrons [9]. We will concentrate on the case where the laser field is linearly polarized and where the photoelectron spectrum is measured in a single direction. The method then consists in streaking the photoelectron in energy with the laser field. As opposed to the case of attosecond SPIDER, the best configuration, giving the maximum streaking speed, now corresponds to the attosecond pulse being synchronized with the dressing field so that $\left|\partial^2\phi/\partial t^2\right| = |\partial\delta W/\partial t|$ is maximum.

The change in the width of the photoelectron spectrum induced by the laser field is then directly related to the duration of the electron wave packet, i.e. to the duration of the x-ray pulse (see Fig. 2). This effect enables to characterize pulse longer than ≈ 100 asec [9]. If the pulse is chirped, the sign of the chirp can also be determined. For instance, let us suppose that the energy streaking speed $\partial\delta W/\partial t$ is negative. As it can be seen in Fig. 4(a-b), if the x-ray pulse is positively chirped, a laser field of appropriate intensity can then induce a narrowing of the photoelectron energy spectrum. On the opposite, for a positively chirped pulse, the spectrum will be widened whatever the intensity.

Beyond its width, the detailed shape of the streaked spectrum contains information on the exact temporal structure of the x-ray pulse, including high-order spectral phases (see Fig. 4(c-d)). Recent progresses in ultrashort visible pulses metrology have shown how to fully characterize a pulse using only the un-streaked pulse spectrum and one streaked spectrum [24]. This implies that the attosecond streak camera can in principle be used to fully characterize sub-laser-cycle attosecond pulses, which is a significant improvement compared to what has been proposed in [9].

The two methods we have described so far are restricted to attosecond fields shorter than the optical cycle of the dressing laser field. We will now describe a technique that does not suffer from this restriction.

3.3 Beyond the attosecond streak camera: attosecond FROG

Another widely used technique for the full temporal characterization of visible femtosecond pulses is frequency-resolved optical gating (FROG) [29]. It consists in decomposing the pulse to be characterized in temporal slices, and then measuring the spectrum of each slice. This provides a spectrogram, or FROG trace

$$S(\omega,\tau) = \left| \int_{-\infty}^{+\infty} dt G(t-\tau) E(t) \mathrm{e}^{i\omega t} \right|^2 \tag{8}$$

where $E(t)$ is the field of the pulse, $G(t)$ is the temporal gate used for the measurement, and τ is the variable delay between the gate and the pulse. The gate may either be a known function of the pulse, as in most implementations of FROG, or it may be an unrelated -and possibly unknown- function. This later case is known as blind-FROG [30]. An efficient algorithm, called principal component generalized projections algorithm (PCGPA) [30], then enables to extract both $E(t)$ and $G(t)$ from the spectrogram $S(\omega, \tau)$.

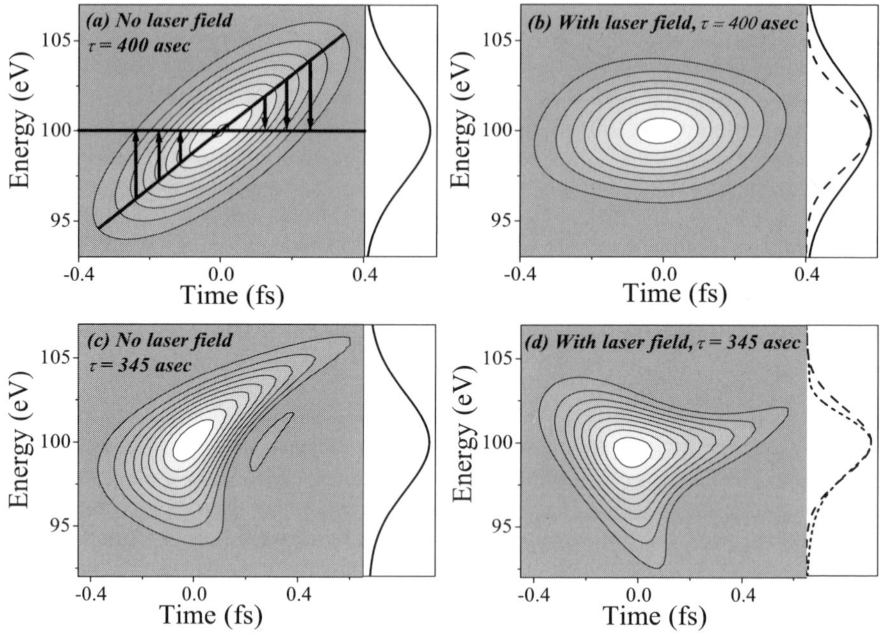

Fig. 4. Wigner distributions and energy spectra of the photoelectron wave packet at $\theta = 0$, for two different x-ray pulses, with and without the dressing laser field. In (a) and (b), the attosecond pulse is the same as in Fig. 3 (400 asec chirped pulse). In the right panel of (b), the photoelectron spectra with (dashed line) and without (full line) the laser field are compared. In (c) and (d), a small third-order spectral phase has been added to the previous x-ray pulse. The duration now is 345 asec. In the right panel of (d), the photoelectron spectra with (short-dashed line) and without (dashed line) this third-order phase are compared. Note that these two spectra slightly differ in width, but also in shape. The arrows in (a) sketch the streaking effect ($\partial\delta W/\partial t < 0$) of the laser field ($\lambda = 800$ nm, $I = 0.6$ TW/cm^2, phase of π at the maximum of the x-ray pulse) on the photoelectron energy.

Comparing Eq.(3-4) and Eq.(8) shows that the dressing laser field can be used as a temporal gate for FROG measurements on electron wave packets, and hence on x-ray pulses. If $|\mathbf{d}_\mathbf{v}|$ varies negligibly with $\mathbf{v}$, the laser field acts as a pure "phase" gate. This does not impede FROG measurements [31], although the measured spectrograms will not be intuitive. The technique used in [13] to characterize a single attosecond pulse may actually be considered as such a FROG measurement, although it was carried out in a slightly more complex configuration (large angular collection of the photoelectrons).

Figure 5 shows FROG traces calculated in the $\theta = 0$ direction, for three different pulses with the same spectral intensity. These spectrograms clearly contain information on the detailed shape of the x-ray pulse. For a given delay τ_0, the spectrum $S(\omega, \tau_0)$ can be considered as a streak camera spectrum, each delay corresponding to a different streaking speed. Thus, a more appropriate name for this method would be frequency-resolved optical streaking (FROST).

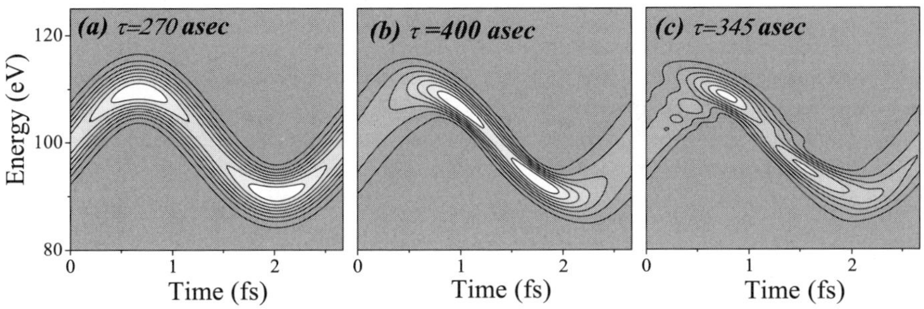

Fig. 5. FROG traces in the $\theta = 0$ direction, for the three pulses considered in Fig. 3 and 4: (a) 270 asec Fourier-limited pulse, (b) 400 asec positively-chirped pulse, (c) 345 asec pulse, identical to (b) but with an additional third-order spectral phase. The laser intensity is 2.2 TW/cm^2 (λ = 800 nm, U_p = 0.13 eV). The horizontal axis corresponds to the delay between the dressing field and the attosecond pulse.

While attosecond SPIDER and the attosecond streak camera are restricted to sub-cycle attosecond pulse, this method should be applicable to arbitrary attosecond fields, such as femtosecond trains of arbitrary attosecond pulses. For a single sub-cycle attosecond pulse, the delay τ has to span at least a full period of the phase modulation $\phi(t)$ [25], as shown in Fig. 5. In the case of a train of attosecond pulses, the delay τ has to be scanned until the train and the dressing laser pulse no longer overlap [25]. In other words, the whole femtosecond laser pulse -including its envelop- will be used as a gate. This can be considered as an extension of the technique used in [14] to characterize the average shape of attosecond pulses in a train. The maximal temporal extension of attosecond trains that can be characterized by this method will be limited by the spectral resolution of electron spectrometers.

4 Conclusion

Techniques used to characterize femtosecond light pulses can be transposed to attosecond x-ray pulses, using electron wave packets as an intermediate tool. A dressing low-frequency laser field acting as an ultrafast electron phase modulator enables to characterize these wave packets, and thus to gain information on the x-ray pulses. This modulator is naturally synchronized with the attosecond fields obtained by HHG, and its frequency is perfectly determined by the laser wavelength and the direction of observation of the photoelectrons.The amplitude of the modulation can also be controlled by the laser intensity. Since spatiotemporal coupling is expected to be strong for the pulses produced by HHG, the next challenge will be to find methods to simultaneously determine their spatial and temporal characteristics.

We emphasize that, as opposed to other techniques, attosecond FROG applies equally well to isolated attosecond pulses and to femtosecond trains of attosecond pulses. In the future, well-characterized trains will be useful : they are easier to generate than isolated pulses, and they can be used for attosecond pump-probe experiments with an increased repetition rate, provided the relaxation time of the process under study is smaller than the time separation between two pulses in the train.

These characterization techniques are likely to have an impact on attosecond science that goes beyond the temporal characterization of attosecond pulses: in analogy with the experiment presented in [1], they could be extended to study ultrafast phenomena occurring on an attosecond time-scale, such as inner-shell electronic relaxation processes, with both phase and amplitude resolution.

F. Quéré gratefully acknowledges C. Dorrer for very helpful discussions on characterization methods for visible pulses. This work is supported in part by Photonics Research Ontario.

References

1. M. Drescher et al. *Nature*, 439:803, 2002.
2. H. Niikura et al. *Nature*, 417:917, 2002.
3. H. Niikura et al. *Nature*, 421:826, 2003.
4. Y. Kobayashi et al. *Opt. Lett.*, 23:64, 1998.
5. J. M. Schins et al. *Phys. Rev. Lett.*, 73:2180, 1994.
6. T. E. Glover et al. *Phys. Rev. Lett.*, 76:2468, 1996.
7. A. Scrinzi, M. Geissler, and T. Brabec. *Phys. Rev. Lett.*, 86:412, 2001.
8. E. Hertz et al. *Phys. Rev. A*, 64:051801, 2001.
9. J. Itatani et al. *Phys. Rev. Lett.*, 88:173903, 2002.
10. M. Kitzler et al. *Phys. Rev. Lett.*, 88:173904, 2002.
11. H. G. Muller. *Appl. Phys. B*, 74:S17, 2002.
12. F. Quéré, J. Itatani, G.L. Yudin, and P.B. Corkum. *Phys. Rev. Lett.*, 90:073902, 2003.
13. M. Hentschel et al. *Nature*, 414:509, 2001.
14. P. M. Paul et al. *Science*, 292:1689, 2001.
15. T. F. Gallagher. *Phys. Rev. Lett.*, 61:2304, 1988.
16. P. B. Corkum, N. H. Burnett, and F. Brunel. *Phys. Rev. Lett.*, 62:1259, 1989.
17. M. Lewenstein et al. *Phys. Rev. A*, 49:2117, 1994.
18. A. M. Dykhne and G. L. Yudin. *Sov. Phys. Usp.*, 20:80, 1977.
19. A. M. Dykhne and G. L. Yudin. *Sov. Phys. Usp.*, 21:549, 1978.
20. E. S. Toma et al. *Phys. Rev. A*, 62:061801(R), 2000.
21. M. Drescher et al. *Science*, 291:1923, 2001.
22. L. Cohen. *Proc. of the IEEE*, 77:941, 1989.
23. C. Dorrer and I. Kang. *Opt. Lett.*, 28:477, 2003.
24. C. Dorrer and I. Kang. *Opt. Lett.*, to be published, 2003.
25. C. Dorrer and I. Kang. *Opt. Lett.*, 27:1315, 2002.
26. C. Iaconis and I. A. Walmsley. *Opt. Lett.*, 23:792, 1998.
27. E. Goulielmakis. *Appl. Phys. B*, B74:197, 2002.
28. R.Kienberger et al. *Science*, 297:1144, 2002.
29. R. Trebino et al. *Rev. Sci. Instrum.*, 68:1997, 1997.
30. D.J. Kane. *IEEE J. Quant. Elec.*, 35:1999, 1999.
31. M.D. Thomson, J.M. Dudley, L.P. Barry, and J.D. Harvey. *Opt. Lett.*, 23:1585, 1998.

Techniques for Temporal Characterization of Ultrafast XUV Radiation

Taro Sekikawa, Teruto Kanai, and Shuntaro Watanabe

Institute for Solid State Physics, Univesity of Tokyo, 5-1-5 Kashiwanoha, Kashiwa 277-8581, Japan, watanabe@issp.u-tokyo.ac.jp

1 Introduction

Since the invention of a camera with a strobe light, we have been able to observe a motion taking place at lighting speed. Now, the shortest pulse duration of a strobe light generated by a visible laser reaches to sub-4 fs, which is very close to a monocycle of the optical frequency, i.e. the ultimate limit in the visible region [1,2]. One way to obtain a shorter strobe light in the attosecond regime is using a higher carrier frequency. High harmonics generation is an attractive candidate for attosecond pulse generation. Recent progress in the research of high harmonics generation has broken into the attosecond time scale and opened the way to attosecond metrology [3]. However, a pulse shape and phase have not been characterized yet.

To date, several methods to estimate the pulse duration of high harmonics have been experimentally demonstrated [3–10]; for example, the cross-correlation between extreme ultraviolet (XUV) and infrared pulses yields strong signals and is applicable to XUV pulses. However, this method requires precise and careful analysis to deconvolute the pulse duration of short unknown pulses with longer reference pulses, and the phase cannot be measured in spite of its importance in pulse shaping, compression, coherent control, etc. Recently, an energy-resolved cross-correlation measurement of XUV pulses was proposed and both pulse durations and linear chirp were estimated by assuming a Gaussian pulse shape and a linear chirp [10]. However, the method requires a well-defined shorter pulse and cannot measure an exact pulse shape and higher order dispersion because of large ponderomotive shift.

Therefore, a method for full characterization of XUV high harmonic pulses has not yet been established. Frequency-resolved optical gating (FROG) is a method that can be used to completely characterize ultrashort pulses [11]. It involves measuring the time-dependent spectrum (spectrogram) of the autocorrelation or cross-correlation and solving the pulse retrieval problem mathematically without assuming a pulse shape and a phase function. In the previous work, we reported the FROG measurement of the fifth harmonic of a Ti:sapphire (TiS) laser in the vacuum ultraviolet (VUV) region by using two-photon ionization (TPI) as a nonlinear optical process in the autocorrelation scheme [9]. Nevertheless, it is difficult to apply the autocorrelation scheme to XUV high harmonic pulses higher than 13th harmonic of TiS laser because of low pulse energy and small TPI cross sections [6,12].

One way to circumvent this difficulty is to enhance TPI probability by using the cross-correlation scheme with strong well-defined fundamental pulses. A laser-induced free-free transition is a nonlinear process available for cross-correlation

FROG (XFROG) [13]. In the presence of the fundamental pulses, a primary photoelectron (PE) spectrum resulting from XUV pulses has sidebands at an interval of the photon energy of the fundamental pulses, corresponding to the absorption of a XUV photon together with the absorption and emission of a fundamental photon [4,5]. The spectra of the neighboring sidebands are identical to those of the sum or difference frequency generation. As a consequence, a spectrogram called a FROG trace can be created by measuring the PE spectra of a sideband as a function of the delay time between the fundamental and XUV pulses. One essential difference between the XFROG and the conventional cross-correlation measurements is that a FROG trace contains both temporal and frequency information. Thus, temporally longer pulses with a narrower bandwidth can be used as sharp knives to slice unknown shorter pulses with wider bandwidths in the frequency domain. Consequently, XFROG makes it possible to retrieve unknown shorter pulses accurately by using relatively longer reference pulses in contrast with traditional cross-correlation.

The final target of the pulse measurement is the full characterization of a few cycle femtoseconds or attosecond pulses. Up to the present, the attosecond pulse train was suggested by measuring the relative phase among several harmonics [14]. Attosecond pulses was demonstrated by the harmonic continuum of a few cycle TiS laser, where the half cycle near the peak of the pulse envelope contributed to select a single pulse [3]. There is another route toward attosecond pulses generation from a single isolated harmonic with a substantially large bandwidth. For this purpose, the wavelength of a driving laser should be short to avoid the spectral overlap of neighboring harmonics, and the short pulse width is required to ensure enough bandwidth for the attosecond pulses. Along this line, a sub-10 fs, sub-TW blue laser has been developed [15].

In this article, we illustrate the newly developed XUV-XFROG in Section 2. Toward attosecond pulse generation, the frequency-doubling scheme to generate sub-10 fs pulses at 400 nm is described in Section 3.

2 Cross-correlation frequency-resolved optical gating

In this section, the XFROG measurement of the ninth harmonic pulses of the second harmonic (SH) of a TiS laser is described. The amplitude and phase of the pulses are discussed taking into account the single atom response calculated by using a zero-range-potential (ZRP) model, including a binding state through a three-dimensional δ-function and offering a fairly completely explanation of high harmonic generation above the third harmonic [16]. We are especially interested in the phases of the XUV pulses, because the XUV pulses were generated under a peak intensity of 9×10^{14} W/cm^2, where the gas medium is ionized in contrast with the previous VUV-FROG measurements [9]. Under such conditions, high harmonics are generated not at the peak but at the leading edge of the driving pulse because of ionization. Thus, the contribution of the atomic dipole phase to high harmonic generation is expected to be different from that in the fifth harmonic generation. In this work, the ninth harmonic pulses generated by negatively chirped, nearly transform-limited, and positively chirped pulses were characterized to explore the possibility of coherent control of high harmonic generation for attosecond pulse generation.

The experimental setup for the XFROG measurement is shown in Fig. 1. A TiS laser system delivering 10-mJ pulses at a repetition rate of 1 kHz was used to generate 500-μJ SH pulses with a 500-μm-thick lithium triborate (LBO) crystal [17]. The ninth harmonic was generated by focusing SH pulses with an annular beam profile into a tube continuously filled with argon gas to a pressure of 26 mbar. The peak intensities of the SH pulses used for high harmonic generation were 6.8×10^{14} W/cm^2, 9×10^{14} W/cm^2, and 3×10^{14} W/cm^2 with negative, nearly zero, and positive chirp, respectively, changed by the pulse compression condition in the TiS laser system. High harmonic generation was observed only when the gas tube was placed at the focus of the laser beam because of the short confocal parameter of 2.5 mm. The fundamental pulses of the TiS laser used as reference pulses in XFROG were characterized in advance by second harmonic generation (SHG) FROG, taking into account the window of the vacuum chamber. The temporal direction of the fundamental pulses was determined so that the results of FROG were consistent with the compression condition. Time-delayed fundamental pulses propagated with the harmonic pulses coaxially and the SH pulses were eliminated spatially from high harmonics by an aperture. In this case, we used the fundamental pulses as a reference to separate the sidebands from the neighboring main peaks.

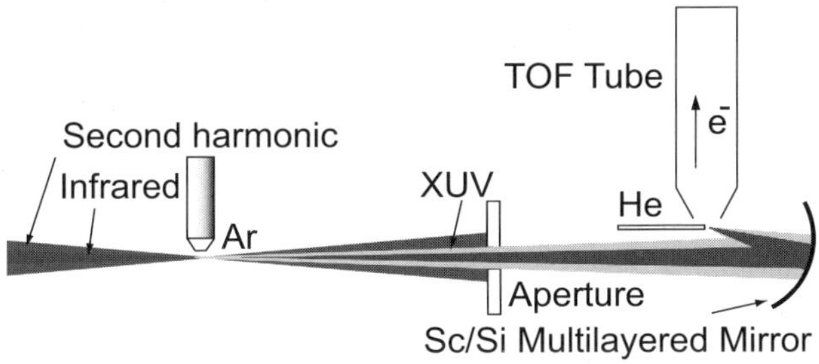

Fig. 1. Experimental setup for XUV-XFROG measurement. TOF is time-of-flight.

Both the ninth harmonic of the SH (photon energy, 27.9 eV) and the fundamental pulses were focused into helium gas from a metal tube by a concave scandium-silicon (Sc/Si) multilayer mirror with a curvature radius of 100 mm. A magnetic-bottle time-of-flight (TOF) PE spectrometer was used to detect PEs ejected by the TPI process of absorbing XUV light and emitting the fundamental light, corresponding to the difference frequency generation. The residual and scattered SH did not produce any effects on PE spectra. The polarizations of the XUV pulse and the reference pulse were parallel in the direction of the TOF drift tube. Because the two pulses were propagated coaxially, there was no broadening of the correlation widths due to geometrical factors. The beam diameters of the XUV and fundamental beams in the focus were estimated at 5.4 and 49 μm, respectively, by using a diffraction theory assuring a homogeneous spatial distribution of the dressing beam over the focus of the XUV beam. The peak intensity of the reference pulse in the ionization region was less than 5×10^{10} W/cm^2 and the ac Stark shift of the PE peak was undetectable. In the TOF tube, a retardation electric field was applied to improve spectral resolution. PE spectra were recorded by a digital oscilloscope at each optical delay of 2 fs. Figure 2a is a typical PE spectrum as a function of delay time showing the peaks of the ninth harmonic and its sidebands. A PE spectrum of a sideband was extracted to create a FROG trace. An XUV pulse was retrieved by the use of the iterative Fourier-transform algorithm with generalized projections [13].

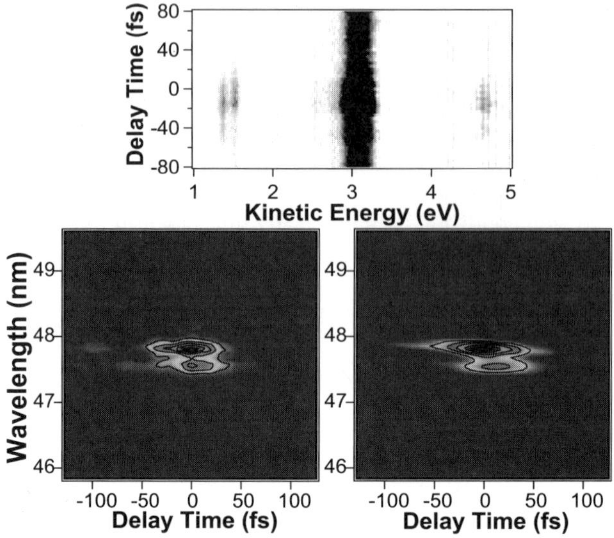

Fig. 2. (a) Typical photoelectron spectrum by the ninth harmonic with two sidebands as a function of delay time. (b) Measured and (c) reconstructed XFROG traces of the ninth harmonic. The peak intensity of the SH pulse was 6.8×10^{14} W/cm^2.

Figures 2b and 2c show the measured and retrieved 128 × 128 pixel XUV-XFROG traces, respectively, by using negatively chirped SH pulses with a peak intensity of 6.8×10^{14} W/cm^2. Figure 3 shows the retrieved fundamental pulses (left column) and the ninth harmonic of the SH pulses (right column), when the peak intensities of the SH pulses were 6.8×10^{14} W/cm^2 (top), 9×10^{14} W/cm^2 (center), and 3×10^{14} W/cm^2 (bottom). The pulse durations of the fundamental pulses were 59, 44, and 136 fs with negative chirp, no chirp, and positive chirp, respectively, from top to bottom. The corresponding pulse durations of the XUV pulses were 11, 10, and 21 fs with FROG errors of 0.023, 0.024, and 0.033, respectively.

An *a priori* temporal resolution of XFROG is the optical cycle of the reference pulse. In the present case, the temporal resolution is 2.7 fs and can be easily improved to be 0.9 fs by using the third harmonic of a TiS laser. Within this limit, XFROG by using a well-defined longer pulse can determine the possible shortest pulse duration accurately from the frequency bandwidth. However, it is not trivial how sensitive to the phase dispersion. Thus, a FROG trace of a transform-limited Gaussian 10-fs pulse was compared to those of a chirped 11-fs pulse with the bandwidth of a 10-fs pulse and a chirped 10-fs pulse with the bandwidth of a 9-fs pulse

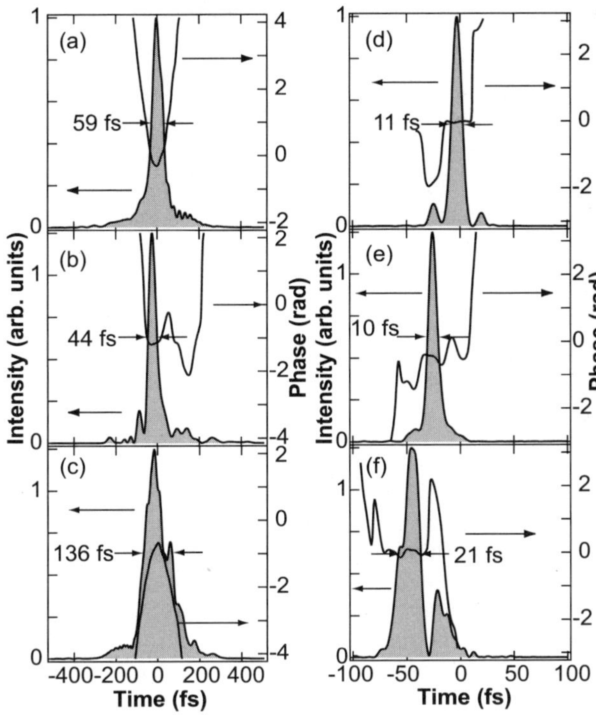

Fig. 3. Amplitudes and phases of the fundamental TiS laser pulses, when the peak intensities of the SH pulses were (a) 6.8×10^{14} W/cm^2, (b) 9×10^{14} W/cm^2, and (c) 3×10^{14} W/cm^2. The chirp signs of the TiS laser were negative, transform-limited, and positive, respectively, from top to bottom. Segments (d) - (e) show the amplitudes and phases of XUV pulses retrieved by using the pulses shown in (a)-(c), respectively.

by assuming a Gaussian 50-fs transform-limited reference pulse. Three XFROG traces are shown in Fig. 4. The differences among them are easily discernible under the present experimental condition, while it is difficult to find differences in the conventional cross-correlation. Therefore, we concluded that measurements errors were attributable to noise or fluctuation during the experiment rather than the longer pulse duration of the reference. We think the obtained intensity profiles and phases contain several percent errors [18].

One interesting feature of the observed phases of the XUV pulses is that the phase change within the main peak of the XUV pulse is less than 0.5 rad in spite of a large change in the chirp of the driving laser pulses. Thus, all XUV pulses generated by the different peak intensities are almost transform-limited, even though the temporal structure varies dramatically with the fundamental pulse. These results are in contrast with the chirped XUV pulse generation by chirped laser pulses with lower peak intensities [10]. Thus, the present results suggest that the generation

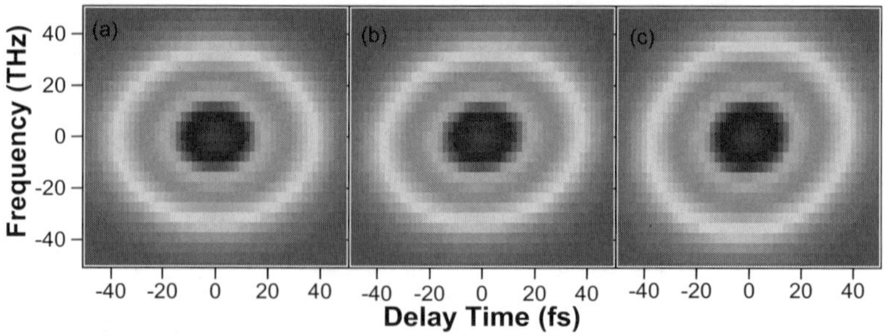

Fig. 4. Simulated XFROG traces of (a) a transform-limited 10-fs pulse, (b) a chirped 11-fs pulse with the bandwidth of a 10-fs pulse, and (c) a chirped 10-fs pulse with the bandwidth of a 9-fs pulse. The reference pulse is a transform-limited 50-fs pulse.

process of high harmonics by using a laser field above an ionization threshold is different from the process that occurs under the threshold.

In order to understand the weak dependence of the phase on the chirp of the driving laser pulse, we simulated the ninth harmonic generation by using a ZRP model to calculate the single atom response. Because the pulse duration of the SH pulse is much longer than the optical cycle, a model in the adiabatic regime was employed. The amplitude of an SH pulse is assumed to be that of the squared fundamental pulse shown in Fig. 3, validated from the pulse durations of the SH pulses estimated by cross-correlation. The peak intensity of the SH pulse is above the ionization threshold of Ar gas (2.5×10^{14} W/cm^2) and the dipole amplitude of an ion is much smaller than that of a neutral atom. Thus, the decrease in the conversion efficiency due to the depletion of neutral atoms was included using Ammosov-Delone-Krainov (ADK) ionization rates [19]. The total phase $\Phi_{\text{total}}(t)$ of an XUV pulse at a given time t is composed of the atomic dipole phase $\Phi_{\text{atom}}(t)$ and the phase of the driving laser pulse $\Phi_{\text{laser}}(t)$ and can be expressed as $\Phi_{\text{total}}(t) = \Phi_{\text{atom}}(t) + q \times \Phi_{\text{laser}}(t)$, where q is the harmonic order. $\Phi_{\text{atom}}(t)$ is the atomic dipole phase obtained from the dipole moment calculated by using the ZRP model. $\Phi_{\text{laser}}(t)$ divides into the following three phases: (1) the original laser chirp $\Phi_{\text{org}}(t)$ evaluated from twice the fundamental phase measured by FROG; (2) the phase induced by self-phase modulation (SPM) $\Phi_{\text{spm}}(t)$; and (3) the refractive index change stimulated by the ionization of the gas medium $\Phi_{\text{ion}}(t)$. The refractive index $n(t)$ of the ionizing medium used in $\Phi_{\text{ion}}(t)$ is expressed as $n(t) = \sqrt{1 - N_e(t)e^2/\varepsilon_0 m\omega^2}$, where ω, $N_e(t)$, e, ε_0, and m are the angular frequency of a pulse, the electron density, the electric charge, the dielectric constant, and the electron mass, respectively. $\Phi_{\text{spm}}(t)$ makes the pulse positively chirped and $\Phi_{\text{ion}}(t)$ shifts the carrier frequency higher [20–22]. In the simulation, the nonlinear refractive index of Ar gas was set at 9.8×10^{-20} cm^2/Watom [23] and $N_e(t)$ was calculated using ADK ionization theory. The spatial beam profile of the driving laser pulse was assumed to be Gaussian and the calculated results were averaged spatially. The interaction length was set at 1 mm, which is a typical interaction length for the ninth harmonic generation discussed below.

The simulated intensity profiles and the total phases of the XUV pulses generated by the SH pulses with peak intensities of 6.8×10^{14} W/cm^2, 9×10^{14} W/cm^2, and 3×10^{14} W/cm^2 are shown in Figs. 5a-c, respectively. Both the amplitudes and phases of the simulated and experimental results are consistent with each other: the simulated pulse durations are 16, 12, and 22 fs in Figs. 5a-c, respectively, and the calculated phases have small variations of less than 0.5 rad within the pulse durations. Based on the model calculation, we can qualitatively say that the steepness of the leading part of a pulse is determined by the nonlinear response of the dipole moment to the laser field, and that the trailing part of the pulse is switched off by the depletion of neutral atoms due to ionization. A qualitative explanation of the small change in the phase is not intuitive. The pulse envelope and four phase components are shown in Fig. 5d, when the peak intensity of the SH pulse was 6.8×10^{14} W/cm^2. Figure 5d illustrates two points: first, that phase change due to SPM is negligible, and second, that the other three phases are almost linear as a function of time within the pulse duration. As a result, second order phase variation is small, leading to small frequency chirp. The other two cases can also be explained in the same manner.

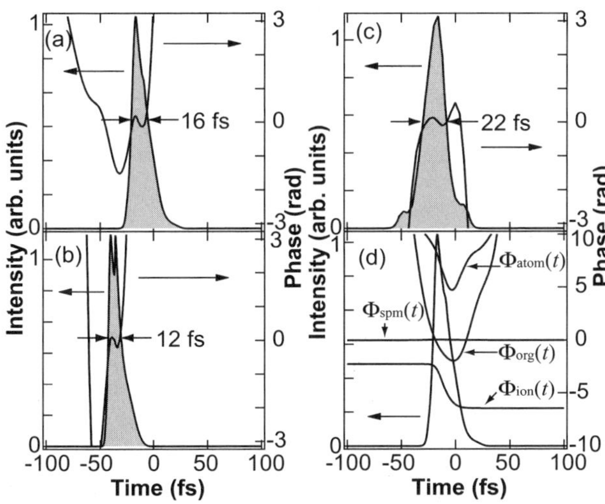

Fig. 5. Calculated amplitudes and phases of the XUV pulses when the peak intensities of the SH pulses were (a) 6.8×10^{14} W/cm^2, (b) 9×10^{14} W/cm^2, and (c) 3×10^{14} W/cm^2. Segment (d) shows the calculated amplitude and phases, $\Phi_{\mathrm{atom}}(t)$, $\Phi_{\mathrm{spm}}(t)$, $\Phi_{\mathrm{org}}(t)$, and $\Phi_{\mathrm{ion}}(t)$, when the peak intensity of the SH pulses was 6.8×10^{14}W/cm^2.

3 Toward attosecond pulse generation

In the previous section, the SH beam was used instead of the fundamental to avoid the overlap of photoelectron side bands. The wavelength of a driving laser should also be short to avoid the spectral overlap of neighboring harmonics for attosecond pulse generation from a single harmonic. The pulse duration of SH used in the previous section is too long (30-40 fs) for this purpose. Sub-10 fs pulses are necessary to ensure the bandwidth enough for attosecond generation. Sub-10 fs ultraviolet or blue pulses have been generated so far by TiS laser amplifiers followed by hollow fibers [24–27]. However, the pulse energy and average power were limited to 15 μJ and 15 mW, respectively. This power level is not high enough for high harmonic generation.

The bandwidth of SH is narrowed by group velocity mismatch, resulting in pulse broadening. To overcome the spectral narrowing, broadband frequency doubling (BFD) was proposed by two groups [28, 29], and was demonstrated in the sub-ps regime by a dye laser [29].

In this section, we applied this idea to sub-10 fs pulses for the first time, and have demonstrated sub-10 fs pulses with an average power of 1.9 W, a pulse energy of 1.9 mJ, and a peak power of 0.16 TW at 400 nm by use of broadband frequency doubling.

Two TiS laser systems with a repetition rate of 5 kHz and 1 kHz were employed in this experiment. The 5-kHz system can produce 21-fs pulses at an average power above 20 W [30]. The 1-kHz system has a capability of 22-fs pulses with an energy above 10 mJ/pulse [17]. The configuration of the BFD system is shown in Fig. 6. The fundamental beam is directed to the grating (G1, 150 lines/mm) by an incident angle of $\alpha = 8^\circ$, and is diffracted with an angular dispersion $\mathrm{d}\beta/\mathrm{d}\lambda$. The image on G1 is relayed by a concave mirror of R = 1 m on a 300-μm thick β-barium borate (BBO) crystal with a magnification M so that the angular dispersion is converted to the desired dispersion of the phase matching angle $\mathrm{d}\theta/\mathrm{d}\lambda$, i.e., $\mathrm{d}\theta/\mathrm{d}\lambda = (\mathrm{d}\beta/\mathrm{d}\lambda)/M$. In this experiment, M is 0.23 from $\mathrm{d}\theta/\mathrm{d}\lambda = 6.42 \times 10^{-4}$ rad/nm and $\mathrm{d}\beta/\mathrm{d}\lambda = 1.50 \times 10^{-4}$ rad/nm at 800 nm. The latter half is configured symmetrically to the former half except for the twice groove density of the grating (G2, 300 lines/mm), which compensates the angular dispersion. The first derivatives of angular dispersion and the phase matching angle are compensated in the above configuration. But each of the second derivatives is slightly different. Thus, the SH intensity shows the quadratic dependence to the spectral width and is expressed as

$$I_{SH} \propto \left[\sin\left(aL\Delta\lambda^2\right) / \left(aL\Delta\lambda^2\right)\right]^2 , \qquad (1)$$

where a is the constant and is π / 1.18 $\mu\mathrm{m}^{-3}$at 800 nm with a 300-μm thick BBO, L is the crystal length, and $\Delta\lambda$ is the wavelength difference with respect to the center wavelength (800 nm) [28, 29]. The acceptable spectral width in this condition is estimated to be 77 nm (the frequency width = 3.6×10^{13} Hz) at the full width at half maximum (FWHM). Therefore, 4.4 fs is possible for a squared hyperbolic secant pulse by considering that the frequency width is doubled by SHG. In this experiment, the spectral width of SH is limited by that of the driving laser, typically 55 nm (FWHM), depending on the spectral shape. The spectral width around 400 nm was 17 nm, in good agreement with 18 nm calculated from the square of the fundamental spectral shape. The spectral shape was controlled by changing an etalon distance in the regenerative amplifier so as to obtain the shortest pulse width [31].

Figure 7 shows the SH conversion efficiencies of the BBO crystal at 1 and 5 kHz. The efficiency increased with increasing fundamental intensity and saturated.

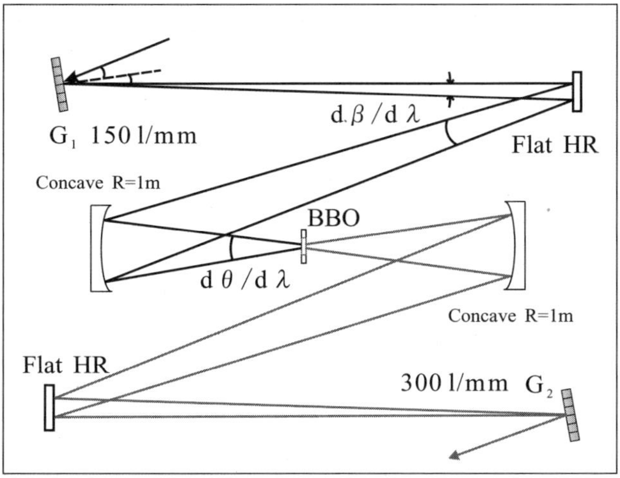

Fig. 6. Schematic of broadband frequency doubling system. α: incident angle; β: diffraction angle; G1: Au-coated grating (150 lines/mm); G2: Al-coated grating (300 lines/mm); HR: high reflector.

The intensity on the BBO crystal was changed by the input beam diameter, which was magnified by the combination of concave and convex mirrors. At an intensity of 3.4 TW/cm^2, the conversion efficiencies reached to 45 % at 1 kHz and 42 % at 5 kHz, respectively. The fundamental pulse width is 25 fs for both cases. The final SH outputs were 1.6 W at 5 kHz and 1.9 W at 1 kHz, with fundamental inputs of 7.3 W and 8 W, respectively, by considering 74 % and 70 % diffraction efficiencies of G1 and G2. After all, the total throughput of this BFD system is 22 % at 5 kHz and 23 % at 1 kHz. The final SH output energy was 1.9 mJ at 1 kHz with a fundamental input of 8 mJ.

The SH pulse shape and phase were determined by FROG with use of self-diffraction (SD) in a 100-μm thick sapphire plate [24]. Figures 8 and 9 show the retrieved pulse shapes, spectra, and phases. In Fig. 8a, the pulse width was 12 fs. In this condition, the SH output power was optimized by changing the intensity of the fundamental beam on BBO as described above. The 5-kHz system was employed as a driving laser for this measurement.

The design of both the 5 and 1 kHz systems are essentially the same. The group delay dispersion (GDD), third order dispersion (TOD), and the high order dispersion are not equal to zero at the center of the angular frequency, but they are balanced with each other so as to reduce the GDD in the wide range of the optical frequency. The calculated total phases of these systems were matched quite well to the FROG measurement of the fundamental pulses. In Fig. 8b, the calculated phase at SH is shown by the solid line by doubling the phase and angular frequency of the fundamental pulse. The calculated phase agrees quite well with the measured one in the wide range of frequency. Next, we tried to expand the spectrum of the fundamental by changing the etalon distance in the regenerative amplifier to generate the shorter SH pulse. A thinner (150 μm) crystal was used for this experiment. Figures 9a and 9b show the pulse shape and spectrum along with

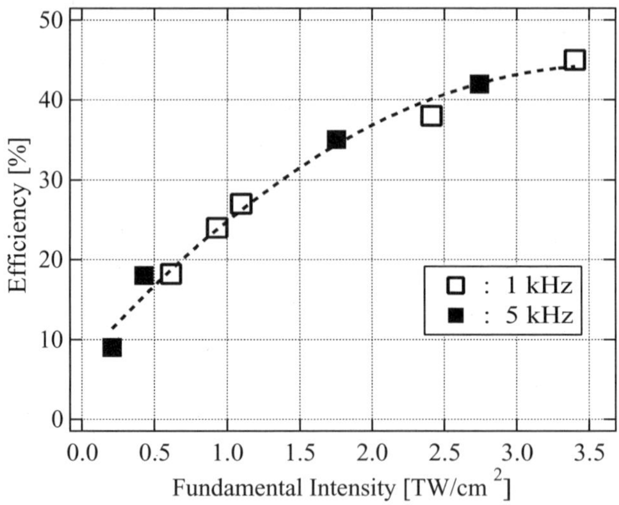

Fig. 7. SH conversion efficiencies versus intensity of the fundamental beam. The open and closed squares correspond to the 1 and 5 kHz repetition rate, respectively.

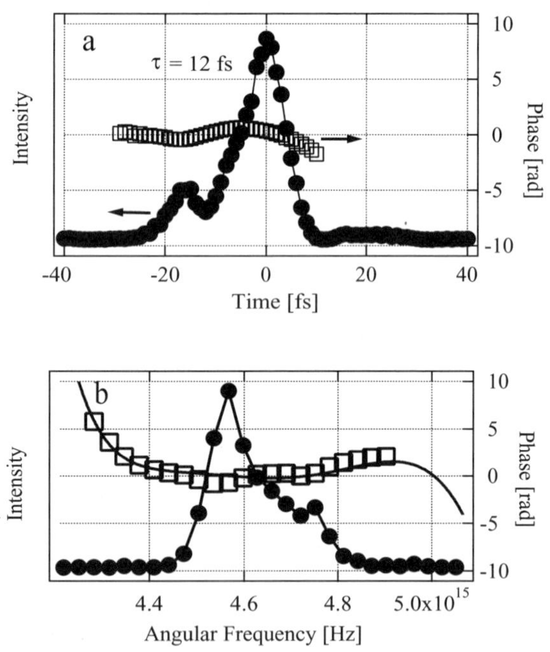

Fig. 8. Retrieved pulse shape (a), and spectrum (b) by SD-FROG. Related phases are shown in (a) and (b). A 300-μm thick BBO was used for BFD. The solid line in (b) is derived from the calculated phase of the fundamental beam.

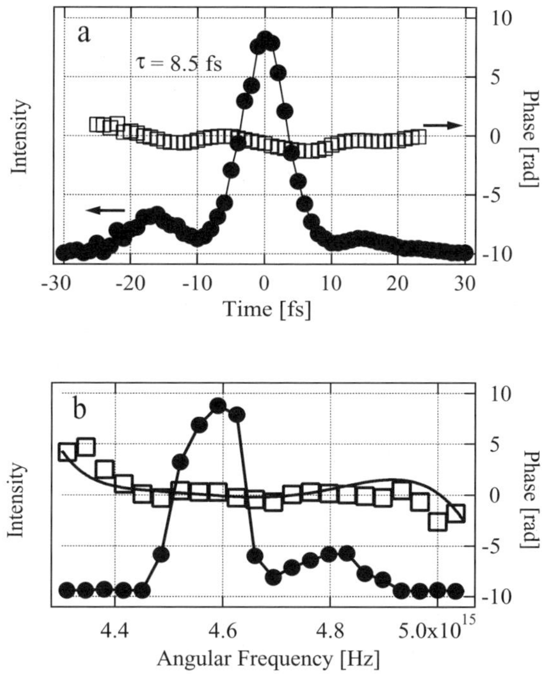

Fig. 9. Retrieved pulse shape (a), and spectrum (b) by SD-FROG. Related phases are shown in (a) and (b). A 150-μm thick BBO was used for BFD. The solid line in (b) is derived from the calculated phase of the fundamental beam.

the related phases. The pulse width was 8.5 fs with an average power of $\sim$1 W. The flat phase was realized over the entire spectrum. Throughout this experiment, the adjustment of dispersion was done by changing the separation and the angle of the grating pair of the compressor in the TiS laser system so as to obtain the highest SD signal.

The peak power obtained in this experiment is high enough for harmonic generation. It is not easy to anticipate the frequency width of a single harmonic because harmonic generation is highly nonlinear. However, we expect the developed light source will open another route to attosecond pulse generation.

Figure 10 shows the pulse shape of the ninth harmonic simulated by the procedure in Section 2 by assuming the 10-fs pulse duration of a second harmonic pulse with a peak intensity of 6×10^{15} W/cm^2. The generated pulse duration is 1.5 fs. The pulse shape is very sensitive to the peak intensity of because high nonlinearity in ionization. Attosecond pulses would be possible not only by reducing the pulse duration of the blue light source but by adjusting a peak intensity of a driving laser.

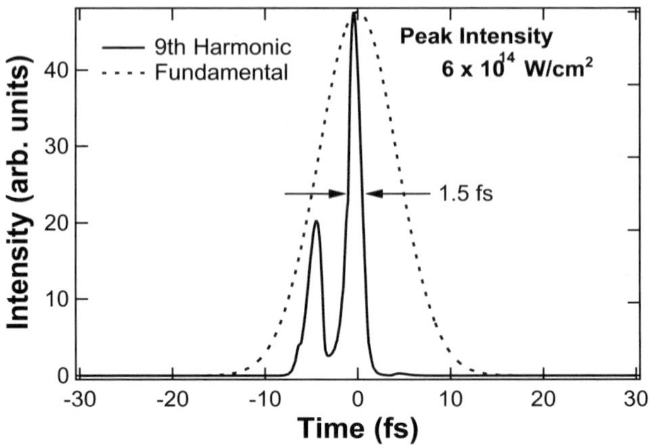

Fig. 10. Simulated ninth harmonic pulse by a 10-fs blue laser with a peak intensity of 6×10^{14} W/cm^2. The pulse duration is 1.5 fs.

4 Conclusions

We have characterized both the amplitudes and the phases of the ninth harmonic pulses of the SH pulses of the TiS laser by developing XUV-XFROG. XUV pulses were found to be nearly transform-limited, which can be explained well by the single atom response calculated by a ZRP model including the depletion of neutral gas.

Toward attosecond pulse generation, 8.5 fs pulses were generated at 400 nm by BFD. The average SH output powers were 1.6 W at 5 kHz and 1.9 W at 1 kHz with a 12-fs pulse width. The peak power and pulse energy were 0.16 TW and 1.9 mJ at a 1-kHz repetition rate, respectively. The throughput of the BFD system was 23 % at 1 kHz. This new blue light source with ultrashort pulses will be used for attosecond pulse production by high harmonics.

References

1. K. Yamane, Z. Zhang, A. Suguro, R. Morita, and M. Yamashita, in *Conference on Lasers and Electro-Optics*, Baltimore, (2003), CThPDA2.
2. B. Schenkel, J. Biegert, U. Keller, M. Nisoli, G. Sansone, S. Stagira, C. Vozzi, S. De Silvestri, and O. Svelto, in *Conference on Lasers and Electro-Optics*, Baltimore, (2003), CFI3.
3. M. Hentschel, R. Kienberger, Ch. Spielmann, G. A. Reider, N. Milosevic, T. Brabec, P. Corkum, U. Heinzmann, M. Drescher, and F. Krausz, Nature **414**, 509 (2001).
4. T. E. Glover, R. W. Schoenlein, A. H. Chin, and C. V. Shank, Phys. Rev. Lett. **76**, 2468 (1996).
5. J. M. Schins, P. Breger, P. Agostini, R. C. Constantinescu, H. G. Muller, A. Bouhal, G. Grillon, A. Antonetti, and A. Mysyrowicz, J. Opt. Soc. Am. B **13**, 197 (1996).

6. Y. Kobayashi, T. Sekikawa, Y. Nabekawa, and S. Watanabe, Opt. Lett. **23**, 64 (1998).
7. T. Sekikawa, T. Ohno, T. Yamazaki, Y. Nabekawa, and S. Watanabe, Phys. Rev. Lett. **83**, 2564 (1999).
8. E. S. Toma, H. G. Muller, P. M. Paul, P. Breger, M. Cheret, P. Agostini, C. L. Blanc, G. Mullot, and G. Cheriaux, Phys. Rev. A **62**, 061801 (2000).
9. T. Sekikawa, T. Katsura, S. Miura, and S. Watanabe, Phys. Rev. Lett. **88**, 193902 (2002).
10. J. Norin, J. Mauritsson, A. Johannson, M. K. Raarup, S. Buil, A. Persson, O. Duhr, M. B. Gaarde, K. J. Schafer, U. Keller, C.-G. Wahlstrom, and A. L'Huillier, Phys. Rev. Lett. **88**, 193901 (2002).
11. R. Trebino, K. W. DeLong, D. N. Fittinghoff, J. N. Sweetser, M. A. Krumbugel, and D. J. Kane, Rev. Sci. Instrum. **68**, 3277 (1997).
12. N. A. Papadogiannis, L. A. A. Nikolopoulos, D. Charalambidis, G. D. Tsakiris, P. Tzallas, and K. Witte, Phys. Rev. Lett. **90**, 133902 (2003).
13. S. Linden, H. Giessen, and J. Kuhl, Phys. Status Solidi B **206**, 119 (1998).
14. P. M. Paul, E. S. Toma, P. Breger, G. Mullot, F. Auge, P. Balcou, H. G. Muller, and P. Agostini, Science **292**, 1689 (2001).
15. T. Kanai, X. Zhou, T. Sekikawa, S. Watanabe, and T. Togashi, Opt. Lett., accepted for publication (2003).
16. W. Becker, S. Long, and J. K. McIver, Phys. Rev. A **50**, 1540 (1994).
17. Y. Nabekawa, Y. Kuramoto, T. Togashi, T. Sekikawa, and S. Watanabe, Opt. Lett. **23**, 1384 (1998).
18. D. N. Fittinghoff, K. W. DeLong, R. Trebino, and C. L. Ladera, J. Opt. Soc. Am. B **12**, 1955 (1995).
19. M. V. Ammsov, N. B. Delone, and V. P. Krainov, Sov. Phys. JETP **64**, 1191 (1986).
20. E. Yablonovitch, Phys. Rev. Lett. **60**, 795 (1988).
21. C.-G. Wahlstrom, J. Larsson, A. Persson, T. Starczewski, S. Svanberg, P. Salieres, P. Balcou, and A. L'Huillier, Phys. Rev. A **48**, 4709 (1993).
22. T. Sekikawa, T. Kumazaki, Y. Kobayashi, Y. Nabekawa, and S. Watanabe, J. Opt. Soc. Am. B **15**, 1406 (1998).
23. H. J. Lehmeier, W. Leupacher, and A. Penzkofer, Opt. Commun. **56**, 67 (1985).
24. S. Backus, J. Peatross, Z. Zeek, A. Rundquist, G. Taft, M. M. Murnane, and H. C. Kapteyn, Opt. Lett. **21**, 665 (1996).
25. E. T. J. Nibbering, O. Duhr, and G. Korn, Opt. Lett. **22**, 1335 (1997).
26. O. Duhr, E. T. J. Nibbering, G. Korn, G. Tempea, and F. Krausz, Opt. Lett. **24**, 34 (1999).
27. C. G. Durfee III, S. Backus, H. C. Kapteyn, and M. Murnane, Opt. Lett. **24**, 697 (1999).
28. O. E. Martinez, IEEE Journal of Qunatum electronics **25**, 2464 (1989).
29. G. Szabo and Z. Bor, Appl. Phys. B **50**, 51 (1990).
30. Y. Nabekawa, T. Togashi, T. Sekikawa, S. Watanabe, S. Konno, T. Kojima, S. Fujikawa, and K. Yasui, Appl. Phys. B **70**, S171 (2000).
31. C. P. J. Barty, G. Korn, F. Raksi, C. Rose-Petruck, J. Squier, A.-C. Tien, K. R. Wilson, V. V. Yakovlev, and K. Yamakawa, Opt. Lett. **21**, 219 (1996).

Characterization of Attosecond Electromagnetic Pulses

Ellen M. Kosik[1 2], Eric Cormier[3], Christophe Dorrer[4], Ian A. Walmsley[1], and Louis F. DiMauro[5]

[1] Clarendon Laboratory, University of Oxford, Parks Rd., Oxford OX1 3PU, UK
kosik@optics.rochester.edu
[2] The Institute of Optics, University of Rochester, Rochester, NY 14627, USA
[3] CELIA, Université Bordeaux I, 351, cours de la Libération, 33405 Talence CEDEX, France
[4] Bell Laboratories, Lucent Technologies, 101 Crawfords Corner Rd., Holmdel, NJ 07733, USA
[5] Brookhaven National Laboratory, Upton, NY 11973, USA

1 Introduction

Recent advances in laser science have enabled dynamics in physical systems to be observed on attosecond timescales [1]. Since several fundamental processes, such as electron motion in atoms and molecules, occur with this brevity, these advances herald a new era in time-resolved spectroscopy of matter. The demonstrations of attosecond duration pulses suggest that improved metrology will assist in the development of better source diagnostics and new experimental schemes. Thus, it will prove useful to have a technique to characterize X-ray ultraviolet (XUV) pulses with varying durations.

Work in attosecond metrology to date, has concentrated on methods for estimating the duration of the XUV pulses rather than determining their exact shape [2–8]. This requires a nonstationary element most often implemented using nonlinear interactions, primarily the mixing of the XUV pulse with an optical pulse in an atomic gas to produce a photoelectron.

More recently, there has been much interest in techniques aimed at providing a more complete (field amplitude and phase) characterization of XUV pulses. The electric-field of fifty femtosecond vacuum ultraviolet (VUV) pulses has been measured by a spectroscopic technique based on frequency-resolved optical gating (FROG) [9,10]. This method, which encodes the VUV field characteristics onto electrons liberated by two-photon ionization, may be translated to attosecond duration pulses with some modification. Two-color, two-photon ionization has been used as a nonstationary element for an interferometric technique derived from spectral phase interferometry for direct electric-field reconstruction (SPIDER) [11] which measures the average temporal shape of individual attosecond pulses in a pulse train [12].

In this paper, we outline a technique for the complete characterization of solitary XUV pulses based on SPIDER. We will also discuss the feasibility of implementing a tomographic technique, simplified chronocyclic tomography, in the XUV. Both techniques utilize the ionization of an electron by the XUV pulse in the presence of an optical field as the nonstationary element.

2 The Optical Field as a Phase Modulator

To better understand the effects of the optical and XUV fields on the ionized photoelectrons, we use the mathematical formalism set out in ref. [13] under the strong field approximation (SFA). From this theory, we can write the photoelectron spectrum, $b(W)$, in terms of the laser fields,

$$b(W) = -i \int_{-\infty}^{\infty} \mathrm{d}t \boldsymbol{d}_{\boldsymbol{p}(t)} \boldsymbol{E}_x(t) e^{i\Phi(t)} , \tag{1}$$

here W is the final electron energy, $\boldsymbol{E}_x(t)$ is the XUV field, and $\boldsymbol{d}_{\boldsymbol{p}(t)}$ is the associated dipole moment. The action of the optical field, $\boldsymbol{E}_l(t) = E_{lo}(t)\cos(\omega_l t)\boldsymbol{e}_L$, is completely contained in the phase term,

$$\Phi(t) = \frac{v^2}{2}t + I_p t + \frac{E_{lo}^2}{4\omega_l^2}t - \frac{E_{lo}^2}{8\omega_l^3}\sin(2\omega_l t) - \frac{vE_{lo}}{\omega_l^2}\cos(\omega_l t) \tag{2}$$

where v is the initial velocity of the ionized electron and I_p is the atomic ionization potential. Thus, the optical pulse plays the role of a phase modulator for the electron wave function and the techniques developed for femtosecond optical pulse metrology via phase modulation can be applied to attosecond XUV pulses where the electron energy plays the role of the photon energy in the conventional techniques. We note that in the absence of an optical field, $b(W)$ reduces to the XUV pulse energy spectrum, $\widetilde{E}_x(W) = |\widetilde{E}_x(W)|e^{i\phi_x(W)}$.

The effective phase modulation imparted by the optical field is dependent on the duration of the XUV pulse with respect to the optical frequency, ω_l. For XUV pulses which are very short with respect to ω_l, $\Phi(t)$ can be approximated by expanding the sine and cosine functions as polynomials around the mean arrival time of the XUV pulse (See Fig.1(**a**)). To first order, $\Phi(t)$, is a linear function of the absolute phase and intensity of the optical field, yielding a frequency shift proportional to E_{lo} and to v.

Recently, Quéré et al. have outlined a method for characterizing pulses in this regime based on SPIDER [14]. However, the assumption that the XUV pulse duration is much shorter than an optical cycle limits the maximum measurable pulse duration to approximately 100 as. There does not appear to be a fundamental limit to the shortest pulse measurable with this technique.

For XUV pulses which are long with respect to the optical period, the oscillations in $\Phi(t)$ result in energy shifts in multiples of the optical photon energy, $\hbar\omega_l$. The resulting photoelectron spectrum is a comb,

$$b(W) = \sum_{n=-\infty}^{n=\infty} c_n \widetilde{E}_x(W + n\hbar\omega_l) \tag{3}$$

where c_n is the amplitude of a given spectral peak and is dependent on the optical field intensity (Fig.1(**c**)). For relatively weak optical fields, only one sideband is significant, and the position of this sideband with respect to the "bare" photoelectron peak can be adjusted by varying ω_l. In this regime, the process is analogous to sum frequency generation in nonlinear crystals. This enables a direct XUV analog to the

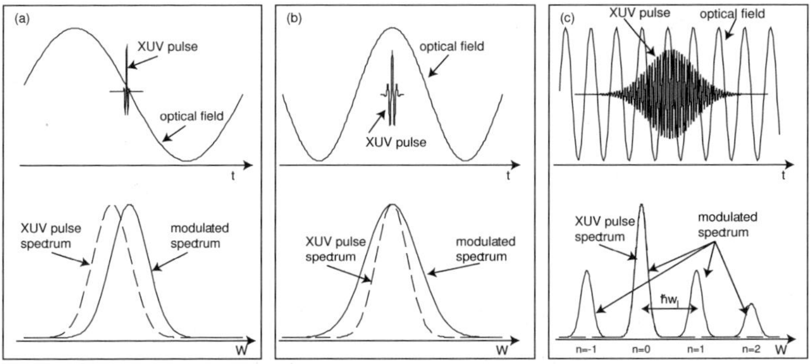

Fig. 1. Illustration of the different phase modulation regimes. (**a**)For very short XUV pulses, the XUV energy spectrum experiences a shift dependent on the absolute phase and intensity of the optical field at the time of ionization. (**b**)For XUV pulse durations comparable to an optical frequency, a time-dependent phase modulation is imparted on the energy spectrum. (**c**) For long XUV pulses, the phase modulation results in a comb of energy spectra spaced at one optical photon energy, $\hbar\omega_l$. In this figure, the n=0 spectral peak obscures the unmodulated XUV energy spectrum

femtosecond pulse characterization method, SPIDER, as will be discussed in Sect. 3.

For XUV pulses which are on the order of an optical frequency, $\Phi(t)$ imparts a time dependent phase modulation onto the XUV pulse. The degree of modulation depends on the absolute phase of the optical pulse with respect to the XUV pulse. For example, if the XUV pulse is aligned with a maxima of the optical field oscillation and is shorter than half the optical frequency, it would acquire a predominantly quadratic phase as shown in Fig.1(**b**). Quadratic temporal phase modulators of this type have been used in several pulse characterization techniques [15–17]. We show in Sect.4 that one of these methods can be used to characterize XUV pulses in this regime.

3 X-SPIDER

In this section, we outline a scheme for measuring pulses as short as 800 as by implementing the SPIDER method in the XUV. We make the assumption that the XUV pulses are long compared to the frequency of the optical field.

SPIDER is a variation on spectral shearing interferometry which extracts the pulse phase from the spectrally resolved interference of two time-delayed and spectrally sheared pulse replicas. In X-SPIDER, two replicas of the XUV pulse are delayed with respect to each other by a time, τ, and interact with a chirped optical pulse in a jet of helium gas. For a sufficiently chirped optical pulse, each of the XUV replicas interact with a different monochromatic frequency slice of the chirped pulse, ω_l and $\omega_l + \Omega$ respectively. The resultant photoelectron energy spectra will consist of a central peak due to the ionization of electrons from the XUV pulses alone and will have sidebands separated from the central peak by the optical photon energies, $\hbar\omega_l$ and $\hbar(\omega_l + \Omega)$. It is these sidebands which contain the phase information needed for SPIDER. The photoelectron spectrum of the first positive sideband as recorded by an integrating detector can be derived from Eq.3 and is given by:

$$S(W) = \left| \widetilde{E}_x(W + \hbar\omega_l) + \widetilde{E}_x(W + \hbar(\omega_l + \Omega))e^{iW\tau} \right|^2 . \tag{4}$$

This is a direct analog to the conventional SPIDER interferogram and the spectral phase of the XUV pulse, $\phi_x(\omega)$, can be extracted from the photoelectron interferogram by the same linear Fourier transform technique used in the optical case [11]. An exact calibration of the time delay between the pulses, τ, and an accurate measure of the pulse spectrum are needed to perform the data inversion. Because the central (n=0) spectral peak does not experience a spectral shift, the τ calibration and spectrum can be easily and directly extracted from this peak. This feature allows for single shot operation of X-SPIDER, although this may not be a practical reality due to low ionization efficiencies of the XUV pulse.

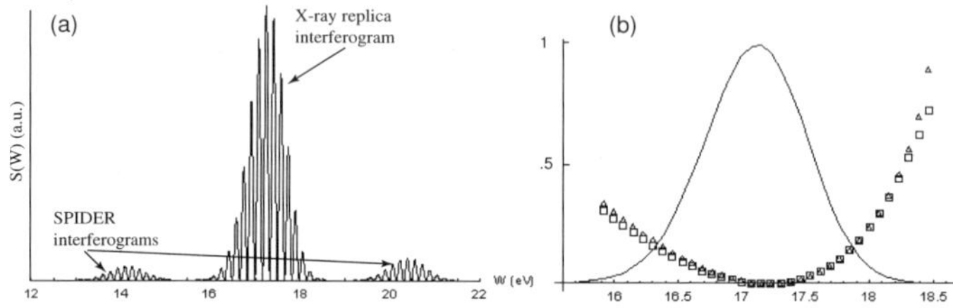

Fig. 2. (**a**)Photoelectron spectra created using SFA theory. Each of the sidebands (n=-1,0,1 peaks are shown) contain the information needed for an X-SPIDER reconstruction. (**b**)The phase reconstructed from the n=1 sideband (triangles) is compared to the actual phase (squares). The pulse spectrum (solid line) is shown as a reference

We have performed numerical simulations of the X-SPIDER technique using the formalism of Sect. 2. To confirm the validity of the assumptions used in that model, we compared these results to photoelectron spectra obtained by a numerical integration of the 3D Time-Dependent Schrödinger Equation (TDSE). This theory has given good agreement with experiment in the past [18, 19]. We found that the SFA treatment agrees well with the TDSE results for the parameters we present here. The simulated X-SPIDER results are shown in Fig.2. This data corresponds to a pair of 2 fs, 30 nm (27th harmonic of Ti:Sa) pulses delayed by 25 fs and a 400 nm (2nd harmonic of Ti:Sa) chirped pulse of intensity $10^{12}\,W/cm^2$. Figure 2(**a**) shows the simulated photoelectron interferogram while Fig. 2(**b**) is a comparison between the input XUV field amplitude and phase profiles and that reconstructed from the X-SPIDER trace. As a figure of merit, we use the RMS deviation, ε, between the original, $E_x(t)$, and reconstructed, $E_{rec}(t)$, fields; $\varepsilon = \left[\int_{-\infty}^{\infty} \mathrm{d}t |E_x(t) - E_{rec}(t)|^2 \right]^{\frac{1}{2}}$. The error for the reconstruction shown in Fig.2 is $\varepsilon = 2.5 \times 10^{-4}$.

These simulations use a second harmonic chirped pulse in order to resolve the X-SPIDER sideband from the central peak. It is possible to characterize even shorter pulses by using higher harmonics for the chirped pulse. We estimate that the short pulse limit is about 800 as for the fifth harmonic of Ti:Sa. This is set by the ability to create harmonics with sufficient intensity for use as the chirped pulse. In X-SPIDER the spectral shear is created by frequency mixing and not by the ponderomotive shift. Therefore the only requirement for the pulse intensity is that it is large enough to give a measurable signal at the sideband.

4 Simplified Chronocyclic Tomography for XUV pulses

Recently, Dorrer and Kang have demonstrated a technique for reconstructing the electric field of visible pulses using a temporal phase modulator [17]. This technique, Simplified Chronocyclic Tomography (SCT), can also be applied to XUV pulses provided the pulse duration is comparable to the period of the optical field used.

The projection of the Wigner function, rotated by an angle α in the chronocyclic space, onto the frequency axis is the fractional power spectrum, I_α. In chronocyclic tomography, the Wigner function is reconstructed from measurements of the fractional power spectrum taken at many different angles within the range $-\pi/2 < \alpha < \pi/2$ (Fig.3). A simplified version of chronocyclic tomography, SCT, requires only two projections, for very small rotations in the chronocyclic space, to reconstruct the electric field. This reconstruction is based on the fact that the group delay of the pulse, $\frac{\partial}{\partial\omega}\phi_x$, is related to the angular derivative of the fractional power spectrum, $\frac{\partial I_\alpha}{\partial\alpha} = -\frac{\partial}{\partial\omega}\left[I_0 \frac{\partial\phi_x}{\partial\omega}\right]$. This derivative can be obtained from measurements of I_α for two small rotations in the chronocyclic space.

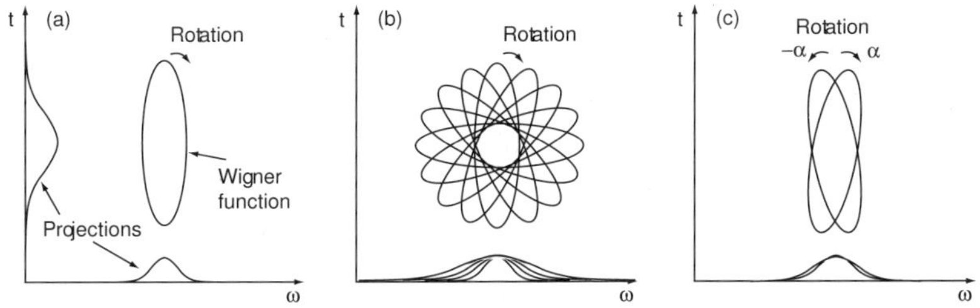

Fig. 3. (**a**) The Wigner function in chronocyclic space. With no rotation, the projections onto the frequency and time axes are the pulse spectral and temporal profiles, respectively. (**b**) In chronocyclic tomography, projections for many rotations allows for complete reconstruction of the Wigner function. (**c**)SCT uses the projections from only two small rotations

We have performed numerical simulations of the SCT technique for XUV pulses, again using the formalism discussed in Sect.2. The two rotations were achieved by aligning the XUV pulse, 30 nm (27th harmonic of Ti:Sa), with first a maximum and then a minimum of the optical field, 800 nm. The reconstructed and actual phase profiles are shown in Fig.4 for two XUV pulse durations. The reconstruction for the

100 as pulse agrees very well with the original field, giving a reconstruction error of $\varepsilon = .06$. The reconstruction is slightly worse in the case of the 500 as pulse, giving an error of $\varepsilon = .11$. This is due to the quadratic phase modulation approximation breaking down for longer pulses. In general, we believe it is possible to characterize pulses with durations from tens of attoseconds to approximately 500 attoseconds using SCT given an optical field at 800 nm.

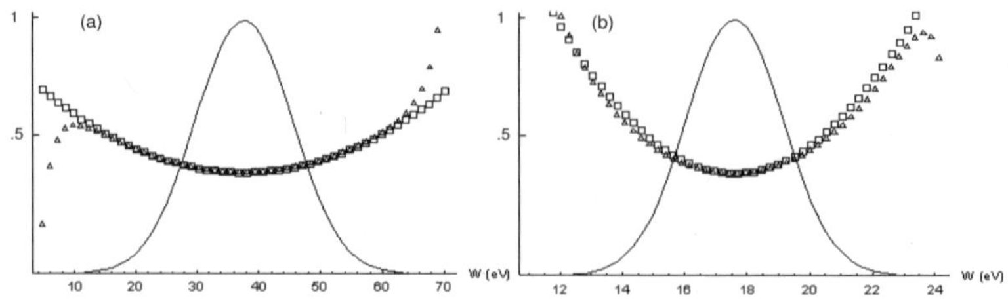

Fig. 4. Fields reconstructed using SCT. The reconstructed phase (triangles) is compared to the actual phase (squares) for two different input pulse widths. (**a**)100 as and (**b**) 500 as. The pulse spectrum (solid line) is shown as a reference

5 Conclusion

It is possible to fully characterize the amplitude and phase of attosecond XUV pulses using information in the energy spectrum of electrons ionized by the XUV field in the presence of an optical field. This nonlinear interaction can be understood as the optical field acting as a phase modulator on the electron wave function. The phase of the XUV pulse also affects this wave function, and the two effects combined can be used to measure the spectral phase of the XUV pulse. The effective phase modulation is a function of the XUV pulse duration with respect to the frequency of the optical field.

We outline a method for characterizing attosecond pulses using an XUV analog to the SPIDER method which we call X-SPIDER. We estimate this technique can measure XUV pulses as short as 800 as. For shorter XUV pulses, 500–50 as, it is possible to use Simplified Chronocyclic Tomography to fully characterize the pulse. Much shorter pulses, essentially any duration less than 100 as, can also be measured using derivatives of SPIDER.

6 Acknowledgments

This work was supported by the US Department of Energy and the UK Engineering and Physical Sciences Research Council under the Basic Technology Program.

The authors acknowledge useful conversations with Paul Corkum, Misha Ivanov, Ferenc Krausz, and Fabien Quéré.

References

1. M. Hentschel *et al.*, Nature **414**, 509 (2001).
2. M. Drescher *et al.*, Science **291**, 1923 (2001).
3. J. Itatani *et al.*, Phys. Rev. Lett. **88**, 173903 (2002).
4. P. M. Paul *et al.*, Science **292**, 1689 (2001).
5. T. E. Glover *et al.*, Phys. Rev. Lett. **76**, 2468 (1996).
6. J. M. Shins *et al.*, J. Opt. Soc. Am. B **13**, 197 (1996).
7. E. Constant *et al.*, Phys. Rev. A **56**, 3870 (1997).
8. T. Brabec and F. Krausz, Rev. Mod. Phys. **72**, 545 (2000).
9. T. Sekikawa *et al.*, Phys. Rev. Lett. **88**, 193902 (2002).
10. R. Trebino *et al.*, Rev. Sci. Inst. **68**, 3277 (1997).
11. C. Iaconis and I. A. Walmsley, Opt. Lett. **23**, 792 (1998).
12. H. G. Muller, Appl. Phys. B **74**, S17S21 (2002).
13. M. Lewenstein *et al.*, Phys. Rev. A **49**, 2117 (1994).
14. F. Quere *et al.*, Phys. Rev. Lett. **90**, 073902 (2003).
15. M. Beck *et al.*, Opt. Lett. **18**, 2041 (1993).
16. R. G. M. P. Koumans and A. Yariv, IEEE J. Quant. Electron. **36**, 137 (2000).
17. C. Dorrer and I. Kang, Opt. Lett. (to be published).
18. E. Cormier and P. Lambropoulos, Euro. Phys. J. D **2**, 15 (1998).
19. H. Bachau *et al.*, Rep. Prog. Phys. **64**, 1815 (2001).

Characterization of Attosecond Laser Pulses Using Angle-resolved Photoelectron Spectroscopy

S. Aseyev[1], Y. Ni[1], L. J. Frasinski[2], H. G. Muller[1], and M. J. J. Vrakking[1]

[1] FOM Institute for Atomic and Molecular Physics (AMOLF), Kruislaan 407, 1098 SJ Amsterdam, The Netherlands vrakking@amolf.nl

[2] J.J. Thomson Physical Laboratory, University of Reading, Whitenights, Reading RG6 6AF, UK

1 Introduction

In the last few decades the development of ultrashort laser pulses has led to significant advances in physics, chemistry and biology. With femtosecond lasers the elementary motion of atoms and molecules can be observed, leading - for example - to a better understanding of complex biological systems. Though electron dynamics often plays a crucial role in these systems, experimental studies have thus far been mostly limited to investigating the "slow" particles (i.e. the atoms) that move under the influence of a potential that results from a time-average over all interactions involving electrons. In order to investigate electron dynamics in real-time, attosecond laser pulses are needed.

Within the last few years substantial progress has been made in attosecond science. The generation of attosecond pulses has been made possible by the development of high harmonic generation (HHG). In HHG, a gaseous medium is exposed to an intense femtosecond laser. Electrons tunnel through the saddle point in the Coulomb+laser field potential and are accelerated by the laser field, before being driven back towards the ion core. Here recombination can take place accompanied by the emission of an extreme-ultraviolet (XUV) photon. The possibility for the generation of attosecond XUV pulses arises in this process since the electron recollision time is short compared to the duration of one half-cycle of the driving laser field (which itself lasts only 1.35 fs for a 800 nm Ti:Sapphire laser). Alternatively, the possibility for attosecond pulses can be regarded as arising from the fact that with many-cycle femtosecond driver pulses HHG leads to the formation of a comb of frequency components which are odd multiples of the fundamental driving laser frequency. Superposition of these harmonics with an appropriate relative phase can lead to bunching of the light in sub-femtosecond bursts, similar to the way that femtosecond laser pulses are formed in a mode-locked laser.

The first conclusive demonstration of the generation of attosecond laser pulses was obtained in an experiment by Paul et.al [1], where interferences in photoelectron spectra resulting from two-color two-photon ionization involving high harmonics and IR from the femtosecond driver laser were used to determine the relative phase of a series of harmonics. The conclusion of the experiment was that XUV bursts as short as 250 as were generated. Since the experiment was performed with a 40 fs driver laser, these bursts formed a pulse train that extended over an interval of several tens of fs (with a 1.35 fs time separation between the pulses, corresponding

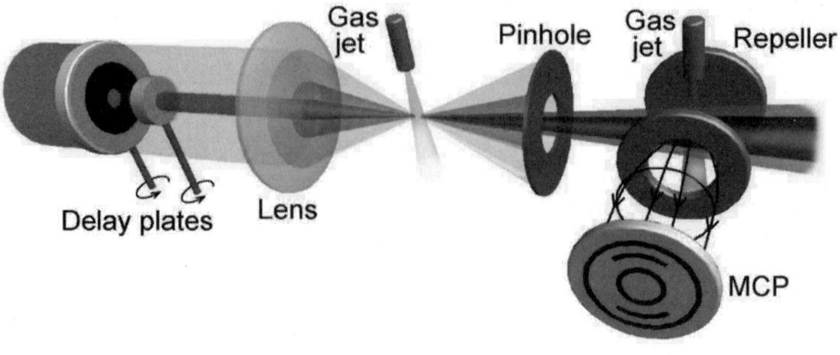

Fig. 1. Schematic view of the experiment: The output of a 50 Hz Ti:Sapphire laser system (70 fs, 10 mJ, 800nm) is split into an intense annular outer and a weak inner beam, which are delayed with respect to each other by passage through two pieces of glass cut from the same antireflection-coated plate. The annular beam is used to generate harmonics in a rare gas jet and is subsequently blocked by a masking plate with a ϕ=0.5mm pinhole. The high harmonics and the weak inner beam pass through the masking plate into a spectrometer chamber, where the lasers cross an atomic beam and where the resulting photoelectrons are detected using a velocity map imaging spectrometer, consisting of an electrostatic field geometry that projects the electrons onto a position-sensitive channel-plate detector. The 3D velocity distribution is recovered from the measured projection by applying an iterative procedure

to one half-cycle of the driving laser). Since then this work has been significantly extended by the group of Krausz and co-workers [2], who demonstrated the formation of *isolated* attosecond pulses using a 5 fs driver pulse, where the conditions for efficient harmonic generation were limited to a *single* half-cycle of the driver pulse. The interference technique pioneered by Paul et.al [1] makes use of the fact that for many-cycle driver pulses (leading to the generation of a train of attosecond pulses) the harmonics occur at odd multiples of the driving laser frequency and can easily be spectrally resolved. Therefore, electrons resulting from two-color XUV+IR two-photon ionization have different kinetic energies than electrons resulting from one-photon ionization by a single harmonic photon, and thus can be easily distinguished using photoelectron spectroscopy. Extension of the technique to shorter driver pulses that give rise to isolated attosecond pulses is not possible since in this limit the XUV spectrum is continuous and single-photon ionization by a harmonic produces electrons with a continuous kinetic energy distribution. In the experiments of Krausz and co-workers this problem was overcome by using a strong field XUV-IR cross-correlation technique. Here we propose an alternative approach based on extending the two-color interference technique of Paul et.al where we measure both the energy and the angular distribution of the ejected electrons, the latter being sensitive to the number of photons involved in the ionization pro-

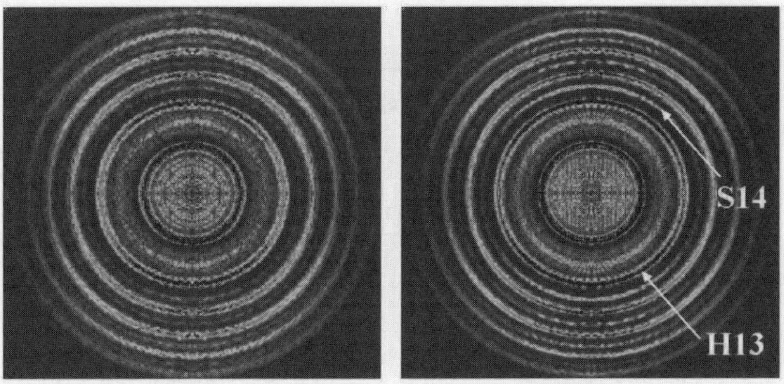

Fig. 2. Inverted photoelectron images recorded following two-color XUV+IR two-photon ionization of Ar atoms by high harmonics generated in Ar gas and the probe IR beam. The image, which represents a cut through the recovered 3D velocity distribution of the ejected electrons, shows a series of bright rings, corresponding to one-photon ionization by a single odd harmonic, and weaker rings at intermediate radii arising from two-color XUV+IR ionization. The strength and the angular distribution of the two-color sidebands depends on the time delay between the XUV and the IR beam. The image on the left was taken at a time delay where the interference was predominantly destructive (leading to weak sideband signals), and the one on the right at a time delay where the interference was predominantly constructive.

cess being considered. One- and two-photon ionization have opposite parity, which introduces odd Legendre polynomials in the angular distributions and allows unambiguous separation of the photoelectron signal at any energy into single- and two-photon contributions, thereby allowing the same kind of analysis as in the experiment of Paul et.al. Using angular information in the perturbative limit even single attosecond pulses can be fully characterized [3].

We have performed an experiment where we have characterized trains of attosecond pulses made by high harmonic generation in Ar, Kr and Xe gas, using angularly resolved photoelectron spectroscopy [4]. As shown in Figure 1, the front end of the experiment consisted of a setup similar to the one used in the experiments of Paul et.al [1] resulting in the formation of a high harmonic beam and - at a time delay that can be controlled with attosecond precision - an IR probe beam to be used in the interference experiment. On the back end of the experiment we have replaced the magnetic bottle photoelectron spectrometer used by Paul et.al (which measured angle-integrated energy spectra) by a velocity map photoelectron imaging detector (capable of measuring both the kinetic energy and angular distribution of the electrons) [5]. In a velocity map imaging detector charged particles produced at the crossing point of a laser and an atomic or molecular beam are accelerated towards a two-dimensional detector, which in our case consists of a dual microchannel plate (MCP) detector followed by a phosphorscreen and a CCD camera that records the positions of impact of the electrons on the MCP. As a result of the acceleration there is a well-defined time-of-flight to the detector for all electrons and

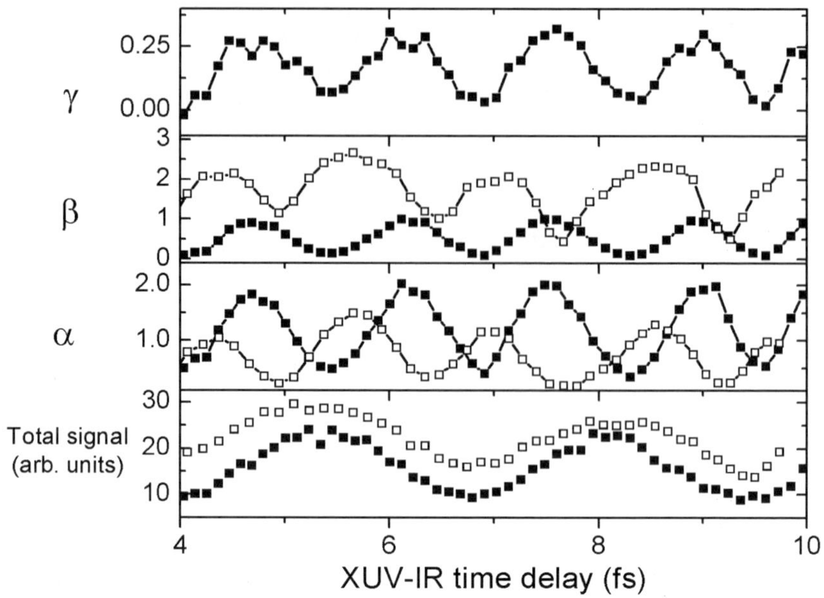

Fig. 3. Extracted angular parameters α, β and γ (see text for definitions) for sideband S16 that contains interfering contributions from ionization by the 15th and 17th harmonic, as a function of the time delay between the high harmonic beam and the IR dressing beam. Both α, which describes the intensity of the sideband, and parameters β and γ, describing the angular distribution of the ejected photoelectrons oscillate as a function of the delay with half the IR period, and allow extraction of the phase shift between the 15th and 17th harmonic. The total signal oscillates with the full IR period. The shift between the oscillations in α for experiments using Ar (solid squares) and Xe (open squares) indicates that for both harmonic generation gasses the attosecond burst appears at a different point during the field oscillation of the femtosecond IR driving laser

thus the position of an electron on the detector represents a measurement of two of its three initial velocity components. Recovery of the three-dimensional electron velocity distribution (and hence, both the kinetic energy and angular distribution) is achieved using an iterative inversion algorithm [6]. We note that a further difference with the experiment of Paul et.al is the fact that an in-line geometry is used for the experiment, which does not make use of a refocussing mirror.

At a given time delay between the harmonic beam and the IR probe beam an image is recorded that yields both the kinetic energy and the angular distribution of the photoelectrons, and hence both the strength of the two-color interference sidebands and the angular distribution (with respect to the laser polarization axis) of the photoelectrons. A typical example is shown in Figure 2. For a single sideband n, the measured signal can be expressed in terms of a superposition of Legendre

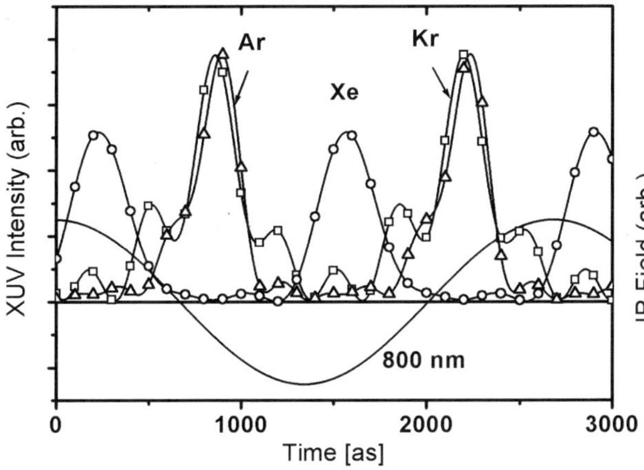

Fig. 4. XUV temporal shape for argon, krypton and xenon as harmonic generation gas, on the basis of the harmonic phase differences extracted from the XUV-IR time delay scans. The harmonic amplitudes needed for the reconstruction were obtained from photoelectron spectra recorded in the absence of the IR dressing beam.

polynomials as S(n,θ) = S(n) [α +$\beta P_2 \cos(\theta)$ + $\gamma P_4 \cos(\theta)$], where θ is the angle between the photoelectron ejection and the polarization axis of the laser. Here the parameter α determines the magnitude of the signal integrated over all ejection angles, and parameters β and γ describe the angular anisotropy. Measurements of α, β and γ as a function of the time delay between the harmonic beam and the IR probe beam can be used to determine the phase relationship between two adjacent odd harmonics that contribute to a particular sideband. In the experiment of Paul et.al [1] this analysis was performed purely on the basis of parameter α, however the analysis can similarly be performed on the basis of parameters β and γ.

In Figure 3 α, β and γ are shown as a function of the XUV-IR time delay, for sideband S16 (resulting from interfering ionization channels involving the 15th and the 17th harmonic) obtained using Ar and Xe gas in the harmonic generation and Ar in the two-color XUV+IR two-photon ionization. Also shown is a measurement of the total photoelectron yield as a function of the XUV-IR time delay. Oscillations in the total photoelectron yield reflect a modulation of the IR intensity in the harmonic generation source that results from an interference between the intense outer IR beam that is predominantly responsible for the harmonic generation and the weak inner IR beam that is later used in the interference experiment (see caption Figure 1). This modulation provides an absolute calibration of the time axis of the XUV-IR time delay experiment. In the experiment regular oscillations are observed in all three parameters α, β and γ. The oscillations in α can be described by a term $\cos(2\phi_{IR}+\phi_{q-1}-\phi_{q+1}+\Delta\phi_{atomic,\alpha})$, as can, to a good approximation, the oscillations in β and γ. Using calculated values of the atomic phases $\Delta\phi_{atomic}$ [7], the experiment thus enables the extraction of the phase difference $\phi_{\alpha,\beta or \gamma} = [\phi_{q-1}$-$\phi_{q+1}]$ (derived from the measurements of α, β and γ). Along with a measurement of the intensity of the individual harmonics in a measurement where the probe IR beam is blocked, this allows a reconstruction of the temporal profile of the

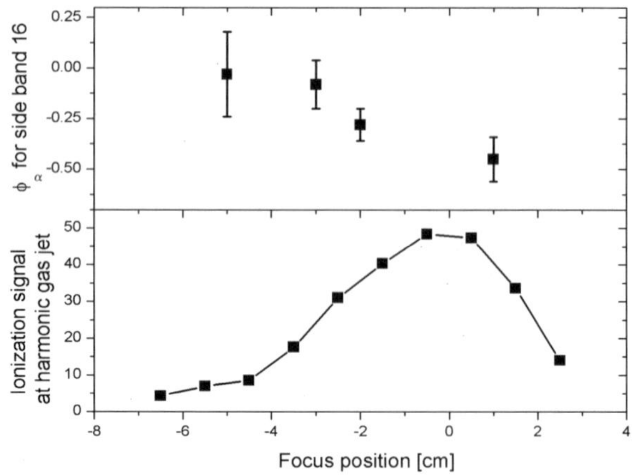

Fig. 5. Dependence on the phaseshift ϕ_α for sideband S16 on the focus position of the femtosecond IR driver laser, as determined by monitoring the total ionization signal in the harmonic generation jet

harmonic beam. This reconstruction is shown in Figure 4 for experiments using Ar, Kr and Xe as harmonic generation gas. Several aspects are noteworthy in this Figure. (1) For all three harmonic generation gases used a train of attosecond pulses is observed, with the shortest pulses (247 as FWHM) being obtained in the case of Ar. (2) The measurement shows that the attosecond pulses formed using Xe as harmonic generation gas appear at a remarkably different time with respect to the field oscillation of the IR driver pulses, as compared to the experiments using Ar and Kr as harmonic generation gas. This feature can also be directly inferred from the raw data shown in Figure 3, by realizing that the delay of the attosecond burst is related to the frequency-derivative of the spectral phase, and hence proportional to ϕ_α. Experimentally we have observed a remarkable robustness of the results shown in Figure 4 to the experimental conditions used. All experiments were repeated a number of times, with inevitably changes in the power and pulse compression of the femtosecond driver laser, the harmonic gas density and the position of the laser focus with respect to the pulsed jet used for the harmonic generation. However, in all cases very similar results were obtained, the typical variation in the measured phase-shifts being less than 0.1 rad. Systematic measurements were performed using Xe as harmonic generation gas and scanning the laser focus from 4.5 cm before to 1.5 cm after the harmonic generation gas jet (using a measurement of the yield of doubly-charged Ar atoms as a means to calibrate the position of the focus). As shown in Figure 5, the maximum variation in ϕ_α for sideband 16 was only 0.4 radian throughout these measurements. Similarly, the density of the harmonic generation gas jet was varied by a factor of 2, without noticable influence on the experimental results.

As discussed, an important motivation for the experiments reported above was an exploration of the way that measurements of the photoelectron angular distribution can be used for retrieval of the attosecond pulse shapes obtained by high harmonic generation. To analyze the success of this approach, the observed side-

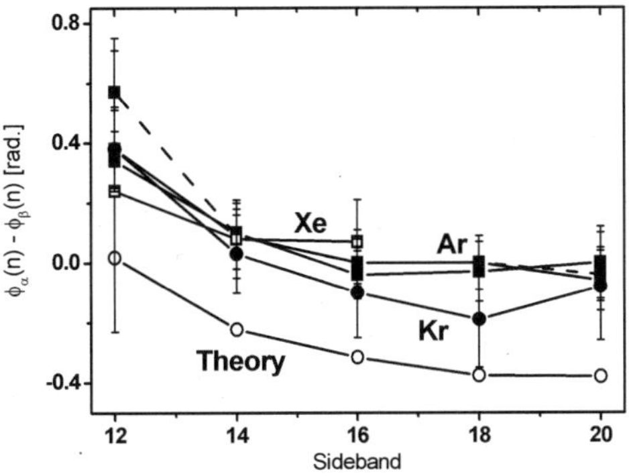

Fig. 6. Difference between the extracted phases ϕ_α and ϕ_β for three experiments using argon (solid squares), krypton (solid circles) and xenon (open squares) as harmonic generation gas, along with a theoretical prediction based on the mixed-color two-photon ionization model (open circles) [7]. Since the phase difference only weakly depends on the harmonic amplitudes and is independent of the harmonic phases, the experimental results lie on top of each other, and agree reasonably well with the theoretical prediction.

band angular distributions were compared to predictions following from the model that was used to calculate the atomic response [7]. In doing so, use was made of the fact that the corrections to ϕ_β and ϕ_γ resulting from the atomic response only weakly depend on the harmonic intensities. In Figure 6 the differences between the measured ϕ_α and ϕ_β are shown for five experiments using Ar (3), Kr and Xe as harmonic generation gas, and compared to a prediction from the model for the atomic response. The theoretical curve [7] reproduces the trend in the experimental data (which, within experimental uncertainties, are identical for all five measurements) quite well, albeit with a slight offset. This suggests that extraction of the relative phase of the harmonics can not only be performed on the basis of the intensities of the sidebands, but similarly, is encoded in the angular distributions of the ejected photoelectrons. A procedure for the extraction of the attosecond pulse shapes from kinetic energy and angularly resolved photoelectron emission for arbitrarily shaped driver pulses (including few-cycle driver pulses that lead to the formation of a single attosecond laser pulse) is currently under development.

This work is part of the research program of the Stichting voor Fundamenteel Onderzoek der Materie (FOM), which is financially supported by the Nederlandse organisatie voor Wetenschappelijk Onderzoek (NWO). It is further funded by the EC grant HPRN-2000-00133 (ATTO network).

References

1. Paul, P.M. et.al, Observation of a train of attosecond pulses from high harmonic generation, Science 292, 1689 (2001).
2. Hentschel, M. et.al, Attosecond metrology, Nature 414, 509 (2001).

3. Muller, H.G., Reconstruction of attosecond harmonic beating by interference of two-photon transitions, Appl. Phys. B 74, S17 (2002).
4. Aseyev, S.A., Ni, Y., Frasinski, L.J., Muller, H.G. and Vrakking, M.J.J., Attosecond angle-resolved photoelectron spectroscopy, submitted for publication.
5. Eppink, A.T.J.B. and Parker, D.H., Velocity map imaging of ions and electrons using electrostatic lenses: Application in photoelectron and photofragment ion imaging of molecular oxygen, Rev. Sci. Instrum., 68, 3477 (1997).
6. Vrakking, M.J.J., An iterative procedure for the inversion of two-dimension ion/photoelectron imaging experiments, Rev. Sci. Instrum. 72, 4084 (2001).
7. Toma, E.S. and Muller, H.G., Calculation of matrix elements for mixed extreme-ultraviolet infrared two-photon above-threshold ionization of argon, J. Phys. B: At. Mol. Opt. Phys. 35, 3435 (2002).

Part V

Ultraintense Pulses, Hard X-Ray Generation and Particle Acceleration

Isolated Attosecond Pulse Generation in the Relativistic λ^3 Regime by Reflection/Deflection/Compression

G. A. Mourou[1], N. M. Naumova[1], I. V. Sokolov[2], B. Hou[1], and J. A. Nees[1]

[1] Center for Ultrafast Optical Science, University of Michigan, Ann Arbor, MI 48109-2099 mourou@eecs.umich.edu
[2] Space Physics Research Laboratory, University of Michigan, Ann Arbor, MI 48109-2143

Summary. Lasers that provide a focal energy encompassed by a volume of a few cubic wavelengths (λ^3) can produce relativistic intensity with maximal gradients, using minimal energy. With particle-in-cell simulations we found, that single 200 attosecond pulses could be efficiently generated in a λ^3 laser pulse reflection, via deflection and compression from the relativistic plasma mirror driven by the pulse itself. An analytical model of coherent radiation from a charged layer confirms the pulse compression and is in good agreement with simulations. This novel technique is efficient (~10%) and can produce single attosecond pulses from the millijoule to the joule level.

1 Introduction

The future path of high field science will be paved in with investigations that bring the shortest pulses to the finest possible focus. This is the most efficient path to the highest intensity. We describe this as the lambda-cubed (λ^3) regime, where the energy of a laser pulse is confined to near the single wavelength Fourier transform limit in both transverse dimensions (λ^2), *as well as* in the longitudinal dimension ($c\tau \sim \lambda$).

Of great interest in high-field science, and essentially defining the field, is the physics of materials driven relativistically. This means that the intensity of the light is sufficient to produce a quiver motion greater than the rest-mass-energy of the particles. We have recently produced relativistic intensity in the kHz regime with short pulses of less than 10 fs [1], composed of only a few cycles, focused on a single wavelength spot size [2]. In this case, the entire laser pulse energy is contained within a focal volume of few λ^3.

Driven relativistically, electrons acquire a quiver energy exceeding their rest mass energy $m_e c^2$. When considering relativistic laser interactions, it is convenient to express intensity in terms of the normalized vector potential $a_0 = e|A|/m_e c^2$. With $a_0 > 1$ corresponding to intensities above $2 \cdot 10^{18}$ W/cm^2 for 0.8 μm light, the plasma dielectric constant must take relativistic effects into account. Accordingly, $\epsilon = 1 - \omega_{p0}^2/\gamma\omega^2$, where $\gamma = (1 + a_0^2)^{1/2}$ with $\omega_{p0} = (4\pi n_{e0} e^2/m_e)^{1/2}$.

Systems driven by such intense radiation often are described by a scaling factor of $I\lambda^2$. Here, the intensity term, I, represents the driving field strength and the λ^2 term indirectly defines the period of time in which charges may be accelerated by the incident fields, being inversely proportional to the square of the optical period. If we now consider the nature of this scaling parameter subject to a constraint of tight temporal and spatial focusing, we can use the limiting values for focal area and pulse duration along with the pulse energy, E, in defining the peak focal intensity:

$$I = \frac{E}{\tau \cdot A} \Rightarrow \frac{E}{\lambda/c \cdot \lambda^2} = \frac{Ec}{\lambda^3},$$

so,

$$I\lambda^2 \Rightarrow \frac{Ec}{\lambda}.$$

This does not indicate that all relativistic experiments should now be considered to scale as E/λ, but it does indicate that work in the relativistic λ^3 regime will favor shorter rather than longer wavelengths.

The relativistic λ^3 regime may be addressed at the lowest level by lasers with quite meager energies and consequently high repetition rates. As little as 35 μJ of energy is required to reach 10^{18} W/cm^2, provided light is concentrated exactly to a single wave in all three dimensions. Practically, available 20-fs laser systems operating at a kHz have the capability to approach this regime with well-corrected tight focusing. With the addition of gas-filled-capillary pulse shortening techniques and the use of dispersion controlling mirrors it is possible to achieve a focused pulse volume of less than $(2\lambda)^3$. Likewise, planned short-pulse petawatt lasers may be used in the relativistic λ^3 regime by maintaining high-numerical aperture, well-corrected focus, and several-cycle pulse duration. 30J in 30fs focused to 0.8 μm would yield a normalized vector potential of ~ 250.

2 Features of the relativistic λ^3 regime

As stated, the most direct advantage of working in the relativistic λ^3 regime is that the highest intensity is available for a given pulse energy. This in turn leads to a reduction in the requirements for pump lasers and aperture in the laser system producing the pulses. For a given intensity a laser operating in the relativistic λ^3 regime can have a higher repetition rate. As a reference, the first laser to produce relativistic intensity was the HELIOS laser at Los Alamos, commissioned *ca.* 1980. It produced kilojoule energies and fired once in 20 minutes. The first relativistic λ^3 laser, at the University of Michigan, produces a million times lower energy per pulse and has a million times the repetition rate.

The spatial coherence of an uncorrelated source, such as a laser driven plasma, scales as the inverse square of the source size. With a focal spot of a single wavelength the relativistic λ^3 laser tends to drive interactions within a very limited volume. Leading to a high degree of transverse spatial coherence for radiation and particles produced in the relativistic interaction. Additionally, the driving signal incident on targets in the relativistic λ^3 regime is naturally smoothed on the order of a spot, enhancing the uniformity of the source on that scale.

In the time domain there is also an increased coherence between the optical driving fields and the reaction of electrons in targets with density near that of solids. As the optical pulse is comprised of very few optical cycles, the number of cycles that can drive a coherent ensemble motion of electrons before the onset of turbulent motion is a greater portion of the entire pulse. This allows greater relative coherence between the temporal structure of the driving signal and the reacting material. This effect is clearly seen in the generation of attosecond pulses from near critical density plasma.

Gradients in the optical fields play a strong role in the activity of relativistically driven charged particles. In the relativistic λ^3 regime a maximal gradient exists in both forward and transverse directions. Most notably, the light has the capability to drive central electrons through a trajectory that displaces them from the focal axis by a significant portion of a spot size within half an optical cycle. This, we will show, has a profound effect on the behavior of the driving signal and its reflection.

High longitudinal gradients inherent the relativistic λ^3 regime provide the ponderomotive drive needed to directly accelerate particles. But the advantages of the high field strength and gradient might be limited to only a few wavelengths due to the extremely short focal depth associated with the tightly focused geometry. This can be addressed by selecting a self-trapping or a guided-wave structure for the interaction.

The complex behavior of plasma in the relativistic regime is open to understanding through physical models and computer simulation. When gas targets, long pulses and weak focusing are involved, one- and two-dimensional simulations can provide a great deal of predictive power. Solid-density simulation of tightly focused interactions is inherently 3-dimensional. They involve the longitudinal dimension, which is the primary direction of the ponderomotive force and the direction of the light propagation. The transverse direction that is the direction of motion for particle driven only by the transverse electric field is required. Finally, the second transverse direction is involved because of the asymmetry in the ponderomotive force parallel to and orthogonal to the driving electric field. Nevertheless, strong guidance can be taken from 2-dimensional simulations and final results compared with 3-dimensional simulations. The relativistic λ^3 regime is particularly interesting with solid density targets. Where particle-in-cell (PIC) simulations are to be used, the total interaction volume including the temporal dimension is significantly limited, allowing 3D simulation.

3 Attosecond pulse production through reflection/deflection/compression

The generation of sub-femtosecond pulses has been proposed [3] and demonstrated [4] using laser atom interaction in the non-perturbative regime in gases at intensities of the order of 10^{14} W/cm^2. In this domain attosecond pulses may be generated, but even with quasi-periodic phase matching [5] produce efficiency far below a percent. It may, however, be possible that higher intensity coupled with dense plasma and relativistic effects could generate attosecond pulses with very high efficiency.

The relativistic effect on ϵ has mainly been studied in sub-critical density plasma conditions, that is in *refraction* – commonly known as relativistic self-focusing – where the laser wavefront is modified due to the dielectric constant dependence on a_0. This effect was discussed by Litvak [6] and C. Max et al. [7]. It was observed with excimer lasers [8] and with chirped pulse amplified lasers [9]. Mourou et al. [10] proposed, that this effect could be studied in the λ^3 regime, provided that the numerical aperture of the relativistic filament is matched with that of the focusing optics. This regime has the advantage that it requires only millijoule energies which are easily produced at kiloherz repetition rates [1, 2, 4, 5].

We suggest that the concept of the relativistic self-focusing be extended from *refraction* to *reflection/deflection* for the case when the λ^3 laser pulse interacts with an over-critical density plasma. For $a_0 > 1$ the light pressure can significantly modify the shape and motion of the reflecting surface of the plasma which in turn will change the reflected wavefront. We discovered by PIC simulations that the plasma mirror produced by the laser pulse can be *deflective* enabling us to isolate single attosecond pulses. By using the appropriate optics this effect can be used to make a sub-period cut from the bulk pulse to obtain isolated sub-femtosecond pulses.

It is crucial to operate in the λ^3 regime for the following reasons: (1) The driving beam should consist of well corrected and tightly focused fundamental frequency light, that allows one to limit wave front distortions and work with the *coherently* reflected radiation. (2) Using only a few optical cycles makes possible a better discrimination between the initial collective action of electrons and their complex response to successive cycles. (3) The narrow focus impresses a strong slope, increasing the angular separation of the reflected electromagnetic energy.

Two more effects should be mentioned. First, deformation of the plasma profile results in significant changes of the local incidence angle (defined to be the angle between the original wavefront and the deformed reflecting surface). To enhance this effect the plasma density n_0 must be slightly over-critical, making the critical density surface more responsive to light pressure. Second, charge separation produces an electric field which is mostly normal to the plasma reflecting surface. Through these two effects we come to one more relativistic effect, which is the most important for us: along with the intense and coherent relativistic electron motion in the direction *perpendicular* to the wave vector of the reflected wave $\mathbf{k_r}$, which is responsible for the *magnitude* of the reflected wave, there must be also the coherent motion *parallel* to $\mathbf{k_r}$, which, according to the Doppler effect, should be responsible for dramatic *shortening/elongation* of the reflected pulse, depending on the sign of the parallel velocity.

Relativistic effects in super-critical plasmas have been discussed in their application to the generation of harmonics [11–13] and attosecond pulse trains [14] by weakly and tightly [13] focused long pulses. For the relativistic λ^3 regime discussed here, the Doppler phase *compression* becomes so significant that the reflected wave is disseminated in several impulses propagated in *different* directions, instead of being the quasi-periodic wave propagating in one direction.

4 Particle-in-cell simulation

To demonstrate this relativistic reflection/deflection/compression, we perform fully relativistic 2D PIC simulations, and study the highly nonlinear regime of the interaction. The computation box is $20\lambda \times 20\lambda$, with spatial resolution as high as 100 cells per λ. To resolve the density gradient we take 16 electrons and 16 ions per cell.

A linearly polarized laser pulse with its electric field along y direction has been initiated at the left boundary ($x = -14$) in vacuum and focused to a 1λ spot normal to the plasma layer. The laser pulse has a Gaussian profile and a duration equal to 5 fs ($\approx$ 2 cycles, full width at half maximum). The maximal intensity in the focus is $I = 2 \cdot 10^{19}$ W/cm^2. For $\lambda = 0.8$ μm this corresponds to the dimensionless amplitude $a_0 = 3$. The preionized plasma layer has a uniform profile, of thickness 2λ, and density $1.5n_{cr}$, corresponding to $n_{e0} = 2.6 \cdot 10^{21}\text{cm}^{-3}$. The electron-ion collision frequency is negligible: $\nu_{ei} \approx 10^8$ s$^{-1} \ll \omega$. We choose $t = 0$ to be the instant when the peak of the pulse envelope reaches the plasma boundary at $x = 0$. Space coordinates are measured in laser wavelengths, and time in cycles.

In Fig.1 the electromagnetic energy density is shown at $t = 11$ and $t = 4$ for $x < -1$. We see that the reflected radiation has been split into separate impulses, each moving in its own direction. At $t = 11$ the most intense impulse (3) is located toward the upper left-hand corner of the box, the next most intense impulse (1) is closer to the middle of the box, and the least intense impulse of the three (2) is located toward the lower left-hand corner.

We plot the reflected radiation at the half-intensity level (Fig.2) and find that only a sub-period pulse has been reflected in the upward direction. Tracing this impulse, we find that it has a divergence of $\sim 20^o$. The maximum intensity values of this impulse follow a $\sim 40^o$ direction with respect to normal incidence.

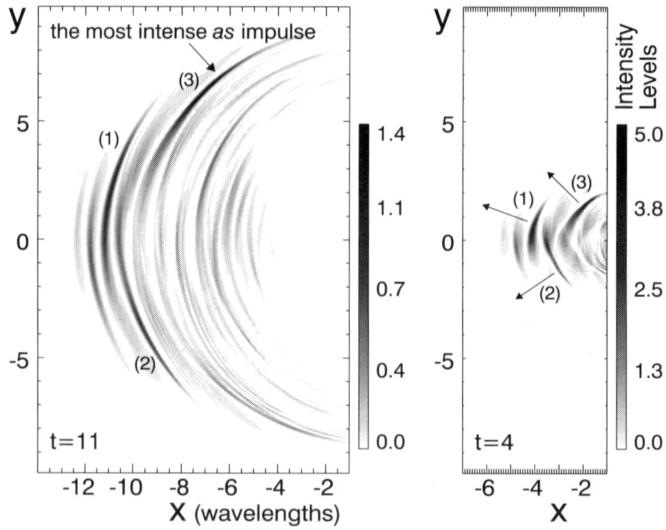

Fig. 1. The electromagnetic energy density of the reflected radiation $(E^2 + B^2)$ at instants $t = 11$ and $t = 4$. Numbers (1), (2), and (3) indicate the most intense impulses in the reflected radiation: (3) being the most intense. Arrows at $t = 4$ indicate the direction of pulse propagation. Parameters of the simulation: $a = 3$, $\tau = 5$fs, and $n_0 = 1.5n_{cr}$.

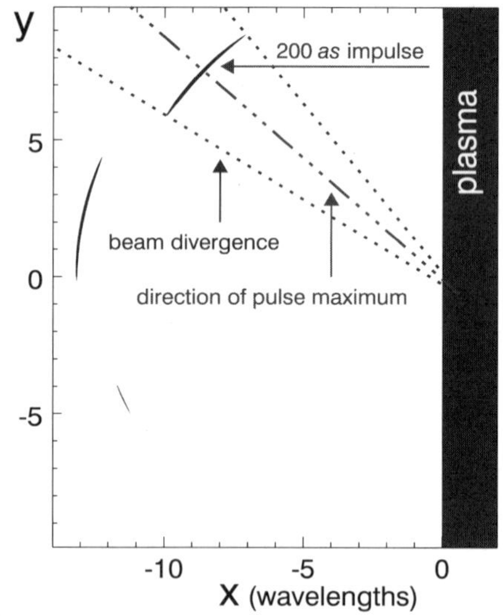

Fig. 2. Half-intensity level of the reflected radiation at the instant $t = 13$. The arrow indicates a single attosecond impulse. Dotted lines show the beam divergence. A dash-dotted line shows the direction of pulse maximum. The plasma layer is indicated schematically.

The evolution of the electron density (Fig.3) shows the motion of the reflecting surface driven by the laser pulse. Electron layers, pushed by the pulse, deflect each half-cycle into a new direction depending on the phase and amplitude of the particular half-cycle. In addition, the reflecting layer focuses these impulses in the backward direction with varying focal lengths. Thus, each half cycle has its own divergence and its own virtual source point within the $\sim\lambda^3$ volume. Due to the very tight focus and short duration of the incident pulse this discrimination in the direction and curvature between separate cycles is possible. The third frame in Fig.3, for $t = 0.9$, shows the surface from which the impulse (3) has been reflected/deflected. The surface, that redirects this impulse in the upward direction (shown by an arrow), has the minimum focal length.

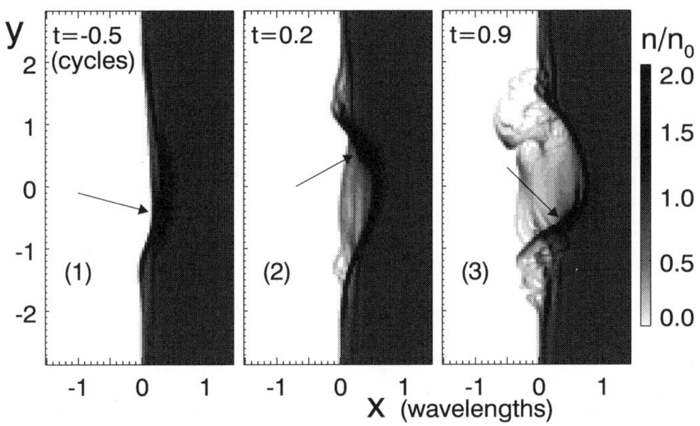

Fig. 3. Snapshots of the electron density at instants $t = -0.5$, 0.2, and 0.9. The arrows indicate the instantaneous orientation of the surfaces from which impulses (1), (2), and (3) were reflected at the approximate times of reflection.

Our simulations show that the laser-plasma interaction is phase sensitive due to the short incident pulse duration. Changing the phase of the pulse by π, we obtain a symmetrically mapped distribution for the reflected light, the most intense sub-period going to the lower left-hand corner.

We plot the time dependence of the electromagnetic energy density at the point $x = -3.5$, $y = 3$ in Fig.4 (along the maximal intensity path of impulse (3)). We find that the duration of this impulse is only 200 attoseconds, yet it contains 10% of the incident pulse energy. The pulse has been compressed under the motion of the reflective layer of electrons in the direction parallel to $\mathbf{k_r}$.

The process of attosecond pulse formation also scales to higher energies. With an intensity increase of two orders of magnitude and plasma density

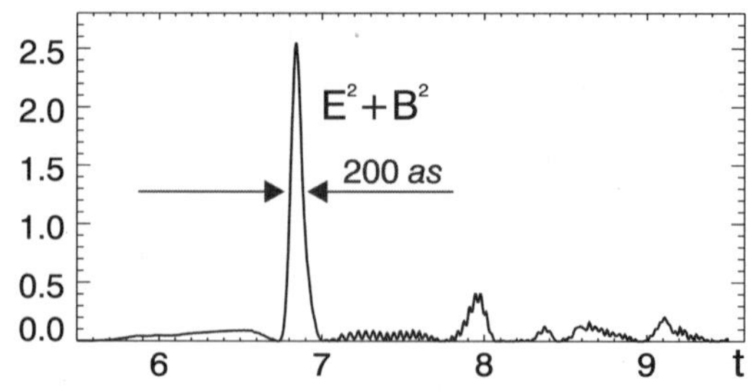

Fig. 4. The time evolution of $E^2 + B^2$ at the point $x = -3.5$, $y = 3$. The arrows indicate the half-intensity level of an isolated impulse that contains 10% of the incident pulse energy.

inscrease of one order, substantially the same effect occurs and attosecond pulses are formed efficiently.

We performed 3D simulations in addition to 2D runs. The 3D results agreed with those in 2D geometry, with some natural differences due to the anisotropy caused by the polarization effects [15]. On the other hand, in 3D simulations it is presently hard to achieve higher spatial resolution, so the 2D result is more reliable.

5 Analytical model of pulse compression

An analytic model, presented for the sake of better clarity, involves only one non-linear input to the component of the electron velocity, which is parallel to the plasma gradient, from the p-polarized electric field component of an incident plane wave. For short driving pulses the total electron dispacement can be *negative*, from plasma towards the laser, resulting in an extremely sharp reflected pulse.

Let us consider the reflection of a short relativistically strong obliquely incident p-polarized plane electromagnetic pulse arriving at a foil. The angle of incidence, θ_0, is defined with respect to the x direction, the x axis being normal to the foil, the wave comes from the $x < 0$ half of space. The foil has thickness $l \ll \lambda$ and its density profile can be described by the Dirac delta function $n_e = N_0\delta(x)$. While calculating the charge motion we assume the fields created by the currents and charges in the foil to be small as compared to the fields in the incident wave. This is allowed by the following condition: $\epsilon_0 = 2\pi e^2 N_0/m_e\lambda \ll 1$ (see [12]). It is known that the problem for non-zero angle of incidence θ_0 can be reduced to the problem for $\theta_0 = 0$ [16] by using the reference frame K' which moves with the velocity as: $V_y = c\sin\theta_0$, with respect to the original frame of reference K. In the K' frame the incident electric field has only one component E_y, the only component of the vector potential being A_y. Below our considerations are in the K' frame.

While in the original frame of reference K the plasma was at rest in the absence of the incident wave, at the same time all the particles move with the velocity $-V_y$ in the frame of reference K'. Particularly, the conserved integrals of motion are: $\kappa_y = p_y - eA_y/c = -m_e c \tan\theta_0$ and $\mathcal{E} = (m_e^2c^4 + p_x^2c^2 + p_y^2c^2)^{1/2} - p_x c = m_e c^2/\cos\theta_0$. These relations should be used while estimating the electron current $I_{y,e}$. The ion current in the frame of reference K' is as follows: $I_{y,i} = -eN_0V_y$.

In neglecting the radiation back reaction as well as the electron-ion interaction ($\epsilon_0 \ll 1$), the electron motion in the field of an arbitrary plane electromagnetic wave is given by well-known formulae, which describe this motion in a parametric form in terms of ξ, the parameter $2\pi\xi/\lambda$ being the phase of the wave (the incident wave in the present context) [17]:

$$x_e(\xi) = \int cp_x\mathcal{E}^{-1}d\xi, \qquad ct' = \xi + x_e(\xi), \tag{1}$$

$$p_x(\xi)/m_e c = \alpha(\alpha/2-1)\sin\theta_0 \tan\theta_0, \tag{2}$$

$$v_x(\xi)/c = 1 - (1+\alpha(\alpha/2-1)\sin^2\theta_0)^{-1}, \tag{3}$$

where $\alpha = a(\xi)/\tan\theta_0$, $a(\xi) = eA_y(\xi)/m_e c^2$. The longitudinal velocity v_x achieves the minimal value at $\alpha = 1$, if $a_0 = max|a| > \tan\theta_0$.

We introduce here the time t' related to the electron motion to distinguish it from the time t, which is used below to define the dimentional quantity associated with the phase of the reflected wave $\xi_r = ct + x$. The typical relation for the argument of the retarded potential is: $ct + x = ct' + x_e(t')$, which allows us to relate the instant t, at which the retarded potential is determined at the point x (see below Eq.(4), note that $x < 0$ and that the reflected wave direction of propagation is opposite to the x-axis direction), to the instant t', when this reflected wave was actually emitted by the electron layer.

We find the vector potential in the reflected wave, by integrating the current:

$$\frac{A_r}{2\pi} = \int_0^t d\tau \int_{-\infty}^{+\infty} d\eta \, [I_{y,e}\delta(\eta - x_e(\tau))\theta(\xi_r - c\tau - \eta) +$$
$$+ I_{y,i}\delta(\eta)\theta(\xi_r - c\tau)] = \int_0^{t'} d\tau I_{y,e}(\tau) + I_{y,i}\xi_r, \tag{4}$$

where $t'(\xi_r)$ is the solution of the equation as follows: $ct' = \xi_r - x_e(t')$, and $\theta(x)$ is the step function.

We use then, the relations: $\partial\xi/\partial t' = c - v_x(t')$, $\partial\xi_r/\partial t' = c + v_x(t')$, $\partial\xi_r/\partial\xi = 1 + 2cp_x\mathcal{E}^{-1}$, and finally find the electric field $E_{y,r}(\xi_r) = -\partial A_{y,r}/\partial\xi_r$ of the reflected wave in parametric form as follows (both $E_{y,r}$ and ξ_r are expressed in terms of ξ):

$$E_{y,r} = \frac{2\pi N_0 e(c\kappa_y + eA_y(\xi))\mathcal{E}}{m^2c^4 + (c\kappa_y + eA_y(\xi))^2} + \frac{2\pi N_0 e}{c}V_y, \tag{5}$$

$$\xi_r = \int (m^2c^4 + (c\kappa_y + eA_y(\xi))^2)\frac{d\xi}{\mathcal{E}^2}. \tag{6}$$

The illustration of this analytical model for the initial pulse profile $E(\xi) = -a_0 \exp(-2(\xi - 0.5)^2)\sin(2\pi\xi)$ and the parameters $a_0 = 2.5$, $\theta_0 = \pi/3$, $\epsilon_0 = 0.5$ is shown in Fig.5(a). We limit the incident electric field to one cycle, having in mind that the successive cycles create the reflection surfaces at different angles. Electron velocity v_x is negative, indicating that electrons move toward the pulse. That is why the reflected pulse becomes much shorter due to the phase compression. On changing the sign of $E(\xi)$ in Fig.5(b) the reflected pulse becomes longer due to the phase decompression.

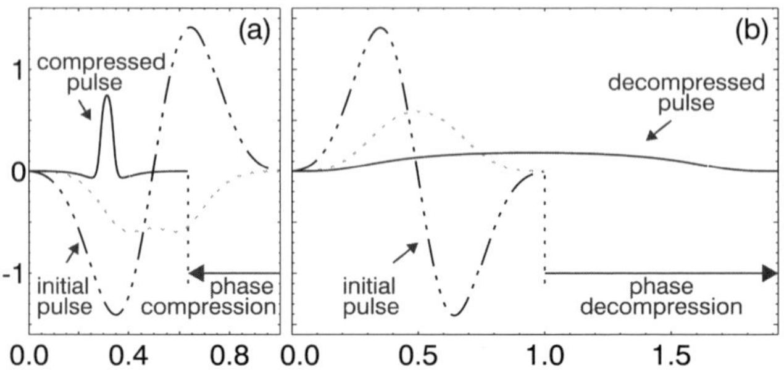

Fig. 5. The incident electric field $E(\xi)$ (dash-dotted line), electron velocity $v_x(\xi)$ (dotted line), and the reflected electric field $E_r(\xi_r)$ (solid line) for the parameters: $a_0 = 2.5$, $\theta_0 = \pi/3$, $\epsilon_0 = 0.5$ in the analytical model for the cases of pulse (a) compression and (b) decompression. The arrow shows the phase change for the reflected pulse: $-2\Delta x/\lambda$, where $\Delta x = x_e(\xi = 1)$.

Note that, just as in the numerical result, for short pulses, compression effects crucially depend on the relationship between the pulse phase and the incidence angle. On changing the sign of the electric field in the pulse at the fixed incidence angle, compression changes to decompression.

6 Conclusion

It is worthwhile to compare our technique and previous techniques [3–5]. The result is shown in Table 1. The suggested scheme is much more efficient than those involving gas targets. This is because in our scheme the signal encounters near-total reflection from a relativistic critical surface, in contrast with attosecond pulses generated from gaseous targets.

With PIC simulation and analytical modelling we have shown that reflections in the relativistic λ^3 regime are accompanied by deflection and compression leading to the generation of isolated attosecond pulses with 10%

Table 1. The comparison of the techniques for single attosecond pulse generation.

Techniques → Features ↓	Electron-Ion/Atom Interactions	Relativistic λ^3 Interactions
Focusing, NA	mild	~ 0.5
Efficiency	$10^{-7} \div 10^{-4}$	10^{-1}
Scalability	No	Up to 1J

efficiency. This technique can be scaled from the millijoule to the joule level. It also illustrates many of the advantages of working with tightly focused, ultrashort pulses, in the relativistic λ^3 regime.

Acknowledgement. We are thankful to S. V. Bulanov for useful discussions. The work was supported by the NSF grant 0114336.

References

1. D-F. Liu, J. Nees, H-W Wang, G. Mourou, Z. Chang, O. Albert, Conference on Lasers and Electro-Optics (CLEO 2000, San Francisco), (IEEE Cat. No.00CH37088), PDP 5-1/9.
2. O. Albert, H. Wang, D. Liu, Z. Chang, G. Mourou, Opt. Lett **25**, 1125 (2000).
3. P. B. Corkum, Phys. Rev. Lett., **71**, 1994 (1993).
4. A. Baltuška, Th. Udem, M. Uiberacker, M. Hentschel, E. Goulielmakis, Ch. Gohle, R. Holzwarth, V. S. Yakovlev, A. Scrinzi, T. W. Hansch, F. Krausz, Nature **421**, 611 (2003).
5. A. Paul, R. A. Bartels, R. Tobey, H. Green, S. Weiman, I. P. Christov, M. M. Murnane, H. C. Kapteyn, S. Backus, Nature **421**, 611 (2003).
6. A. G. Litvak, Zh. Eksp. Teor. Fiz. **57**, 629 (1969).
7. C. Max, J. Arons, A. B. Langdon, Phys. Rev. Lett. **33**, 209 (1974).
8. A. B. Borisov, A. V. Borovskiy, O. B. Shiryaev, V. V. Korobkin, A. M. Prokhorov, J. C. Solem, T. S. Luk, K. Boyer, C. K. Rhodes, Phys. Rev. A **45**, 5830 (1992).
9. P. Monot, T. Auguste, P. Gibbon, F. Jakober, G. Mainfray, Phys. Rev. Lett. **74**, 2953 (1995).
10. G. Mourou, Z. Chang, A. Maksimchuk, J. Nees, S. V. Bulanov, V. Yu. Bychenkov, T. Zh. Esirkepov, N. M. Naumova, F. Pegoraro, H. Ruhl, Plasma Physics Reports **28**, 12 (2002).
11. S. V. Bulanov, N. M. Naumova, F. Pegoraro, Phys. Plasmas **1**, 745 (1994); R. Lichters, J. Meyer-ter-Vehn, A. M. Pukhov, Phys. Plasmas **3**, 3425 (1996).
12. V. A. Vshivkov, N. M. Naumova, F. Pegoraro, S. V. Bulanov, Phys. Plasmas **5**, 2752 (1998).
13. S. V. Bulanov, T. Zh. Esirkepov, N. M. Naumova, I. V. Sokolov, Phys. Rev. E **67**, 016405 (2003).
14. L. Roso, L. Plaja, K. Rzazewski, D. von der Linde, Laser and Particle Beams **18**, 467 (2000).

15. N. M. Naumova, S. V. Bulanov, K. Nishihara, T. Zh. Esirkepov, F. Pegoraro, Phys. Rev. E **65**, 045402 (2002).
16. A. Bourdier, Phys. Fluids **26**, 1804 (1983).
17. L. D. Landau, E. M. Lifshits, *The Classical Theory of Fields*, (Pergamon, Oxford, 1983).

Diode-Pumped Regenerative Amplifier Front End for the Petawatt Laser Chain at LULI

N. Forget[1], C. Felix[1], E. Baynard[1], C. Le Blanc[1], V.Bagnoud[2], X. Ribeyre[2], and L. Videau[2]

[1] Laboratoire pour l'Utilisation des Lasers Intenses (LULI), UMR 7605 CNRS-CEA-Ecole Polytechnique-Université Pierre et Marie Curie, Ecole Polytechnique 91120 Palaiseau, France
nicolas.forget@polytechnique.fr

[2] Laboratoire de Physique du Laser, Commissariat à l'énergie atomique CEA-CESTA BP2, 33114 Le Barp Cedex, France

1 Introduction

In most high power laser chains running at 1 μm since the 90's, the amplification media is Ti:sapphire because of its wide emission spectrum. With a single Ti:sapphire regenerative amplifier the mJ can be easily reached at 800 nm but at 1053 nm the gain is very low and an additional stage of amplification is needed to reach tens of mJ [1].

The main power amplifiers of the Petawatt chain at LULI require from the front end at least a few mJ with a minimum bandwidth of 7 nm FWHM centred at 1054 nm for injection and several dozen of mJ at 1 Hz in narrow spectrum for alignment and diagnostic purposes. For the sake of compacity, stability but also extraction efficiency we have chosen to build a top-hat-profile diode-pumped Nd:glass regenerative amplifier. The use of Nd:glass as gain medium allows a much higher gain than Ti:sapphire and greatly simplifies the front-end scheme since only a single regenerative cavity is needed to reach more than 50 mJ [2].

In this paper we present the regenerative cavity setup, the diode-pumped laser head and the design the mode selecting element. We report numerical and experimental results on the beam profile quality and on the spectral shaping element needed to broaden the spectrum up to 7 nm.

2 Cavity setup

The regenerative cavity is a 4.9 m long cavity closed by a flat mirror and modified concave mirror (figure 1). The laser head is diode-pumped with 8 stacks of 6 diodes emitting at 800 nm (Thales LD). The total pumping energy is 2.5 J in 400 μs and the whole laser head, pumping included, fits in 10x10x10 cm^3 [5]. The low thermal production allows working at 1 Hz without any cooling system. The diodes are directly placed against the 4x4x80 mm^3 phosphate laser rod so that the pumping is transverse and highly uniform.

In order to get the greatest overlap between the gain medium and the laser beam we use a mode selecting element (MSE) included in the concave cavity mirror and propagate a supergaussian mode inside the regenerative cavity. Because of the very narrow gain bandwidth of our laser rod (16 nm for APG-1) we use an intracavity spectral filter made of a 5 mm quartz plate (Lyot filter [4]).

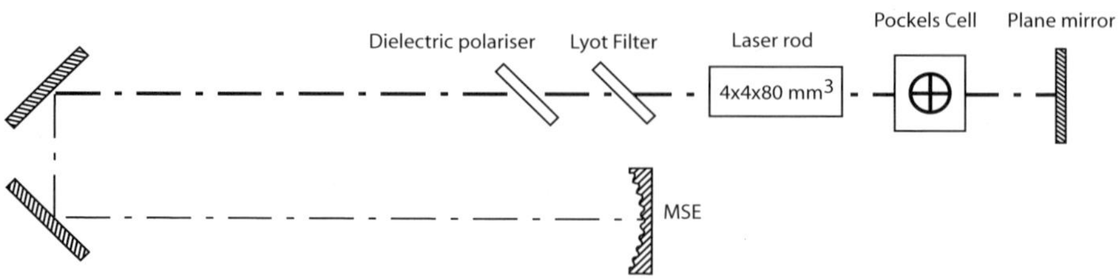

Fig. 1. Regenerative cavity setup

3 Design of the Mode Selecting Element

To shape the fundamental spatial mode of the cavity into a 2.7 mm supergaussian, a MSE was engraved into the concave mirror. As a supergaussian mode cannot propagate over a cavity round-trip without being diffracted, an amplitude or a phase modulation is needed to compensate for diffraction. We chose to use a phase mask directly engraved in the concave mirror. This phase mask is designed to conjugate the electric field incident on the concave mirror and thus to allow the mode to recover its initial shape when propagating backward. As a result the spatial mode remains the same after a round-trip and is therefore an eigenmode of the cavity.

We used a Fox-Li algorithm [3] to compute the eigenmodes of our cavity. This algorithm is based on modal selection : a random field is introduced in the cavity and propagated (2D Fresnel diffraction) until a single stable mode remains in the cavity. As the random field is propagated the high order eigenmodes contained in the random field [6] experience higher losses than the fundamental eigenmode. After a few hundred round-trips the high order modes disappear and the fundamental mode remains.

This algorithm shows that with a phase conjugating mirror the fundamental eigenmode is indeed the desired supergaussian mode and that the highest modal discrimination is reached for an optimal length of the cavity. This optimal length is found close to the Rayleigh length of the beam and thus depends on the mode size. For a 2.7 mm supergaussian of 10th order the optimal cavity length ranges from 4 m to 5 m depending on the aperture size of the optics. In figure 2, the losses of the fundamental mode and of the next higher order mode are plotted against the length of the cavity. For cavities longer than 3 m and shorter than 6 m the diffraction losses of the fundamental mode are less than 1% whereas the first order mode losses are about 50% per round-trip. Because of some technical constraints on the manufacture of the MSE (depth of the modulations and curvature radius of the available mirrors) the length of the cavity was set to 4.9 m.

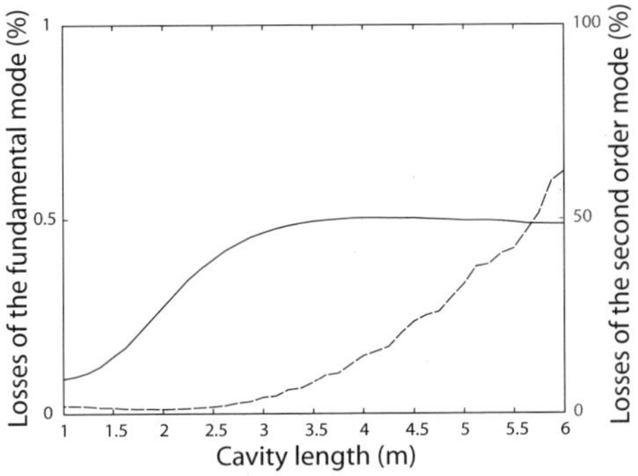

Fig. 2. Dashed line : round-trip losses of the fundamental (supergaussian) mode. Solid line : round-trip losses of the next eigenmode.

The radial phase modulation needed for phase conjugation of a 2.7 mm supergaussian and a 4.9 m cavity is given in figure 3 (left). The phase modulation can be divided into a quadratic phase part and a residual phase part. The quadratic phase gives the radius of the concave mirror (6 m) required and the residual part gives the phase mask to be engraved into this mirror.

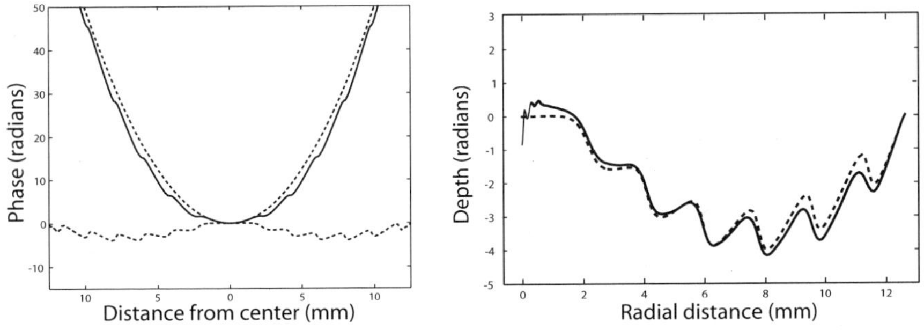

Fig. 3. Left : theoretical phase modulation of the phase mirror along a diameter of the mirror. Solid line : phase modulation needed for phase conjugation. Dots : quadratic phase (top) and residual phase (bottom). Right : comparison of the radial profiles of the theoretical (dots) and measured phase masks (solid line).

Figure 3 (right) shows the theoretical profile of the phase mask and the measured profile of the mirror used in our regenerative cavity. This mirror was engraved by ion-etching processing by R. Mercier and M. Mullot (IOTA, Orsay, France) and then antireflection coated. As seen on figure 3 the profile of the experimental profile is close to the theoretical profile. Nevertheless, deviation from the theoretical profile will lead to small ripples on the spatial profile as observed on the experimental beam profile.

4 Experimental results

For a cavity of 4.9 m (figure 1) the measured spatial mode is close to a 2.8 mm supergaussian of order 10. The calculated beam profile shown in figure 4 takes into account the measured profile of the MSE and also the $4 \times 4 \times 80\text{mm}^3$ laser rod placed at 40 cm from the plane mirror, and the finite diameters of the mirrors. The observed grid-shape on the mode come from diffraction on the edges of the laser rod.

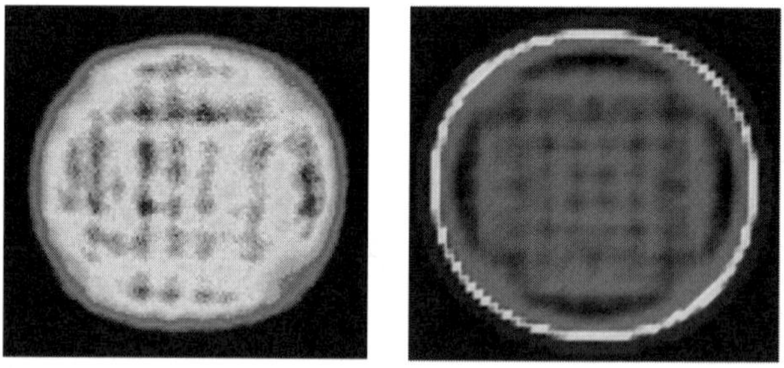

Fig. 4. Left : experimental mode. Right : calculated mode. The graylevels are not matched.

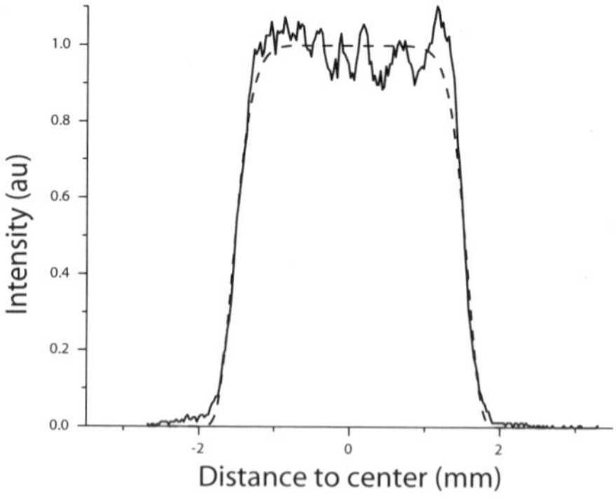

Fig. 5. Solid line : experimental profile. Dots : 2.8 mm supergaussian of order 10.

With the MSE the spatial mode is sensitive to the cavity length since the phase mask is designed for a fixed length of propagation. Figure 6 shows how the shape of the fundamental mode changes along with the cavity losses.

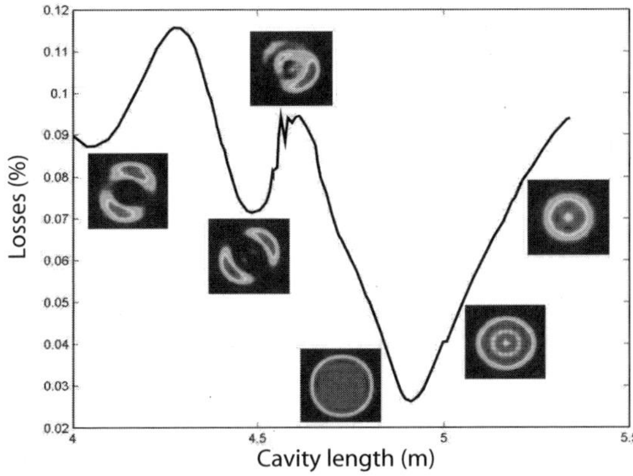

Fig. 6. Diffraction losses per round-trip as a function of the cavity length with the experimental phase mirror and the laser rod simulated by a square diaphragm.

As expected the mode is supergaussian and the losses are very low for cavity lengths around 4.90 m and remain so over 30 cm. For longer cavities the modes get closer to a Gaussian mode and for shorter cavities the mode is close to TEM_{10} mode. Contrary to Gaussian cavities, this kind of cavity is highly sensitive to a misplacement of the center of the phase mirror relative to the axis defined by the laser rod and the plane mirror. Simulation shows that if the phase mirror is misaligned by more than 500 μm the mode splits into two parts. These numerical results are in full agreement with observations and very helpful to optimize both cavity design and alignment.

5 Spectral shaping

Figure 7 presents the spectrum at the output of the femtosecond oscillator and after the stretcher. The spectrum of the oscillator is 18 nm wide FHWM and the bandwidth acceptance of the stretcher is 40 nm. The gain bandwidth of APG-1 is 16 nm FHWM and therefore gain narrowing is very strong as seen in the figure. At full pumping level we extracted 50 mJ with a spectrum of 1.5 nm FHWM. To broaden the gain bandwidth of the gain medium, a 5 mm quartz plate was placed inside the cavity near the Brewster angle. As quartz is birefringent it introduces wavelength-dependant losses and thus spectral modulation. When correctly tuned the gain bandwidth can be flattened. With the Lyot filter the first results show a spectrum of 7 nm for an energy of 35 mJ.

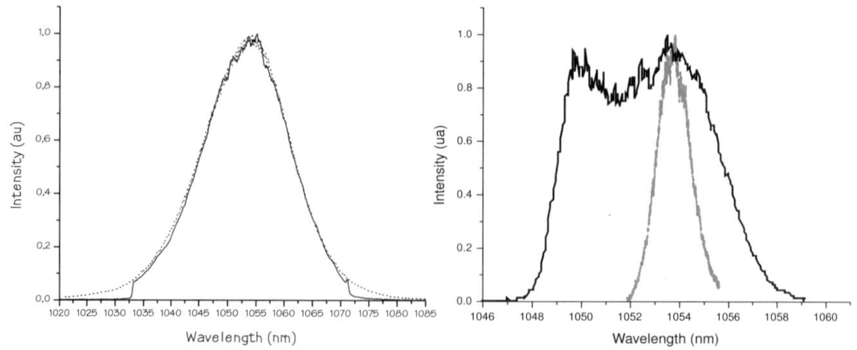

Fig. 7. Left : spectra of the 80 fs pulses before and after (clipped spectrum) the stretcher. Right : gain bandwidths with (black curve) and without (gray curve) the Lyot filter

6 Conclusion and prospects

We have demonstrated a new kind of front end for ultra-high intensity lasers such as the Petawatt project at LULI. We extracted 35 mJ from a single regenerative cavity with a 7 nm spectrum and a flat-top spatial profile.

Simulations show that the cavity could be shorten down to 3 m at the expense of a slightly lower modal discrimination and we intend to use some new engraved mirrors to improve performance. In addition we plan to use a programmable acousto-optic modulator (Dazzler^{TM} from FastLite) to shape the pulses before the regenerative cavity in order to broaden the spectrum and precompensate for dynamic saturation in the power amplifiers. Injection in the power amplifiers and recompression are expected in a few months.

This work was partly funded by the Conseil Regional Ile-de-France under agreements E.1127 and E.1258.

References

1. Perry, M.D., et al: Petawatt laser pulses. Opt.Lett., **24**, 160–162 (1999)
2. Bagnoud, V, et al: Diode pumped regenerative amplifier delivering 100 mJ single mode laser pulses. Opt.Lett., **26**, 337–339 (2001) (1999)
3. Fox, A.G., Li, T.: Resonant modes in a maser interferometer. The Bell technical journal. , **40**, 453–488 (1961) (1999)
4. Barty, C., et al: Opt.Lett., **21**, 219–221 (1996)
5. Patent CEA: #2768565 (2001)
6. Slater, J.C.: Microwave Electronics, Von Nostrand, Princeton, New Yersey, (1950)

The Petawatt Laser Glass Chain at LULI: Technical Issues and First Results

C. Le Blanc, C. Felix, N. Forget, J. C. Lagron, Ph. Hollander, A. M. Sautivet, C. Sauteret, and A. Migus

Laboratoire pour l'Utilisation des Lasers Intenses, Unité Mixte Ecole Polytechnique- CNRS-CEA-Université Paris VI n 7605, 91128 Palaiseau, France
catherine.leblanc@polytechnique.fr

Summary. We present the Petawatt Laser facility project at LULI, from the front end diode-pumped regenerative amplifier, delivering 60 mJ at 1 Hz, to the main power amplifiers and compact compressor based on high-damage-threshold dielectric gratings. This facility will offer a unique combination of a Petawatt laser synchronized with a 1 kJ long-pulse laser, a 100 TW laser, and three 100 J ns lasers.

1 Introduction

The first Petawatt laser was demonstrated by Livermore in 1998 [1]. Since the dismantling of Nova in Livermore (USA) and Phebus at CEA Limeil (France) in 1999, several Petawatt glass lasers are under development, with some of them already operational (e.g., Rutherford). The LULI 2000 installation is composed of two 1 kJ laser arms and four additional diagnostic beams at the 100 J level. One arm, working in the nanosecond regime, delivers 1 kJ pulses at 1.05 μm, while the other one is working in the nanosecond or femtosecond regime with a capability of delivering up to 1 Petawatt. The nanosecond version of the kJ laser is now operational and first experiments are starting summer 2003. We describe the progress of the Petawatt laser in terms of front end, power amplifiers and compressor.

2 General presentation

The PW laser is illustrated in figure 1.

3 Front end

The front end laser starts with a commercial oscillator and an Offner triplet using large optics (30 cm diameter, 2 m radius of curvature for the convex mirror) that stretch the 18 nm bandwidth pulse to 3.5 ns. The pulse is then sent to a side-pumped Nd:phosphate regenerative amplifier that delivers pulses of up to 60 mJ in a single TEM mode. The laser beam is mode matched

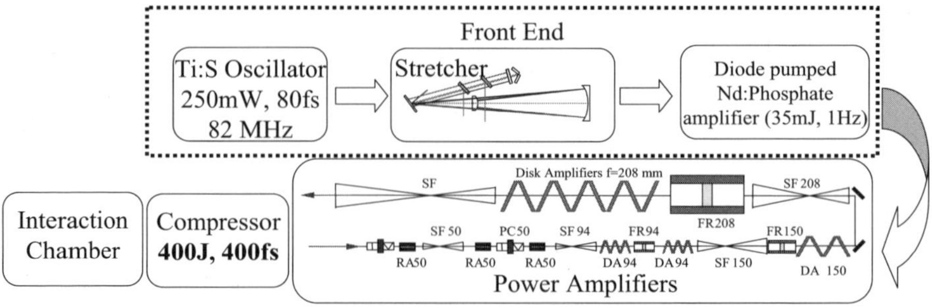

Fig. 1. Scheme of the LULI Petawatt Laser, consisting of a femtosecond front end which delivers 50 mJ at 1 μm, the main laser chain amplifiers and a vacuum compressor

to the amplification medium using an intra-cavity mirror where the phase is adapted to a tenth-order super-Gaussian beam. As a result of good spatial overlap and low loss, an optical-to-optical conversion efficiency of up to 10 % was measured for a pumping current of 80 A and output pulses of more than 60 mJ [2]. Very strong gain-narrowing requires the use of a birefringent filter in order to widen the spectrum from 1.5 nm to 7 nm, inducing a loss in extraction efficiency. The energy required to seed the main amplifiers is only 2 mJ. Figure 2 presents the experimental results concerning beam quality and spectral shape after the front end.

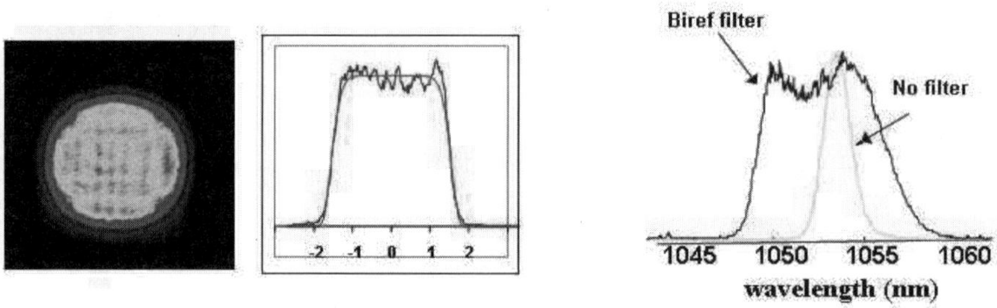

Fig. 2. Beam profile out of the regenerative amplifier (left). Output spectrum using a birefringent filter which gives 7 nm FWHM bandwidth (black), and without the filter (1.5 nm FWHM (grey)).

4 Power amplifiers

The main power amplifier is based on a series of Nd:Phosphate amplifiers from 50 mm diameter rods to 200 mm disc amplifiers capable of delivering energy of 1 kJ in the ns regime. First results in the nanosecond regime have shown for each laser chain an energy of more than 1 kJ (on one laser chain 1.5 kJ in 5 ns has been obtained). The beam profile (figure 3) shows a nice top-hat beam, meanwhile it is three times the diffraction limit.A 200 mm diameter deformable mirror will be implemented before the last 200 mm diameter disc amplifiers to correct this default. In the broad-band regime, expected energy is 600 J, for an input spectrum of 7 nm and 1 ns duration (MIRO calculation). The final calculated B integral won't exceed 1 rad.

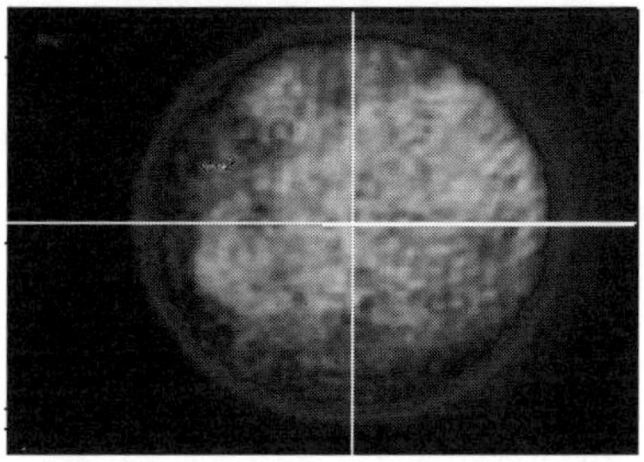

Fig. 3. Beam profile of the 1 kJ , ns pulse from the power amplifiers chain in the LULI 2000 laser.

5 Compressor

A key parameter in the design of the compressor is the damage threshold of the large aperture compression gratings. The Petawatt laser chains already demonstrated in Livermore [1] have used 90 cm diameter gold coated gratings in compressors with a distance between the two gratings of more than 10 meters. Our goal was to reduce the size of the gratings and the compressor chamber. For this reason we have designed a final compressor scheme based on four high damage threshold, 48 x 33 cm, dielectric gratings, presently under fabrication at Jobin-Yvon SA. Preliminary tests of same technology gratings almost as large (42 x 21 cm) are presently being performed on the LULI 100 TW facility [3]. The compact layout of our double-pass compressor will allow a chamber of only 3.2 meters diameter (figure 4), 3-4 times smaller than any compressor chamber developed for the same kind of laser. The distance between the gratings is 1.8 m and with a 60 degrees angle of incidence, this design will allowed to compress a 4 nm wide spectrum. To be able to increase the bandwidth, 2 pairs of segmented gratings will be used in order to get more than 10 nm bandwidth acceptance.

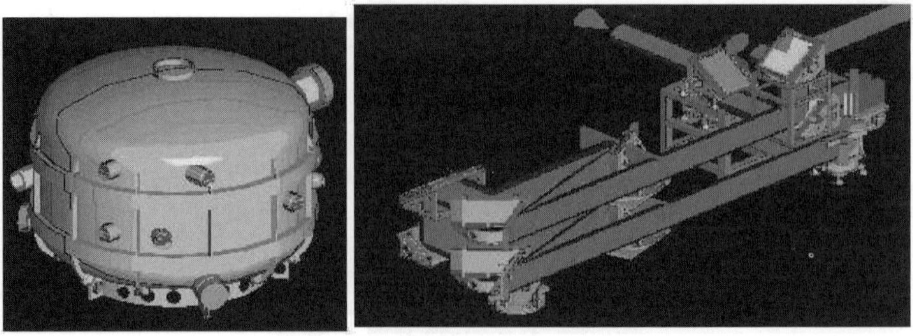

Fig. 4. Schematic of the 3.2 m diameter compressor chamber(left) and mechanical setup inside the compressor chamber with the beam trajectory (right).

6 Conclusion

Meanwhile the front end is ready to seed, the power amplifiers deliver their expected energy, we still waiting for our large dielectric gratings and the compressor chamber. We plan to get all the parts ready for the end of 2004.

References

1. Perry, M.D., et al: Petawatt laser pulses. Opt.Lett., **24**, 160–162 (1999)
2. Bagnoud, V, et al: Diode pumped regenerative amplifier delivering 100 mJ single mode laser pulses. Opt.Lett., **26**, 337–339 (2001)
3. Le Blanc, C., et al: Large size multilayer dielectric gratings for laser pulse compressor, performance review. Cleo Baltimore (2003).

High Repetition Rate Femtosecond Laser Driven Hard X-Ray Source and its Application for Diffraction Experiments

N. Zhavoronkov[1], Y. Gritsai[1], P. Mikheev[2], A. Savelev[2], R. Bernath[3], G. Korn[4], and T. Elsaesser[1]

[1] Max-Born Institute for Nonlinear Optics and Ultrafast Spectroscopy , Max-Born-Str. 2A, 12489-Berlin, zhavoron@mbi-berlin.de
[2] Moscow State University, Vorobievy Gory, Russia,
[3] School of Optics, Univ. Central Florida, USA
[4] Katana Technologies GmbH, Albert-Einstein-Ring 7, Kleinmachow, Germany

The capability of intense femtosecond pulses to create hard X-ray short-duration plasma-emitters has allowed time-resolved X-ray studies with sub-picosecond resolution in different branches of physic, material science and biology. Most of these studies were performed in a single-shot or 10 Hz regime with large multi-stage laser facilities and intensities above 10^{16} W/cm^2. For ultrashort X-ray sources to be more widely used they must operate in a more continuous regime. Most applications, that can be developed, will benefit from higher average power and high repetition rate, when low-intensity signal averaging is necessary. In the past few years there have been rapid improvements in high power Ti:sapphire femtosecond lasers systems operated at kiloherz repetition rates, some capable to generate peak intensities of 10^{18} W/cm^2 [1]. This systems are a good candidate to be used as a primary source for laser driven plasma X-ray sources.

In this paper, we present our approach to design a high repetition rate, compact hard X-ray source operated at 1 kHz. As a target we have used our recently developed liquid Ga-jet [2]. The typical flow velocities for such target are within 10-100 m/s that is enough to ensure the jet surface replenishment after each interaction with laser pulses up to 100 kHz repetition rate. The laser system consists of a chirp-mirrors controlled Ti:sapphire oscillator, generating 22 fs, 5 nJ pulses, a stretcher, a regenerative amplifier, two double-pass amplifiers and a compressor. The output of laser system can be arranged either after the first or after the second amplifiers. Pulses of 45 fs duration and energies of 1.5-2 mJ or 5-10 mJ respectively are generated. The test of output stability performed in real time the RMS was measured to be less than 1 %. The spatial beam profile observed in the focus of 5 m focal length lens was Gaussian with beam diameter not larger than 1.2 times of the diffraction limit. We directed the laser radiation after the first amplifier towards the vacuum chamber where the Ga-jet was streming vertically. The laser radiation was focused by a 7 cm focal length plano-convex lens at 3 mm out from the jet nozzle to 6.7 μm beam diameter at FWHM. The effective contrast ratio of the amplified pulse was 10^4 and the intensity on target was up to $2\text{x}10^{17}$ W/cm^2. The spectrum of the x-ray emission emanating from the plasma measured with a Si-photodiode-based energy-dispersive, x-ray detector (XR-100CR Amptek Inc.) is shown on Fig. 1. It consists of a broad continuum and two narrow features at 9.2 keV and 10.3 keV. These two peaks correspond to the characteristic K_α and K_β lines of Ga. The broad continuum is predominantly free-free collisional emission (Bremsstrahlung) from the plasma. The K_α and K_β lines originate from the collision of the high energy electrons with cold target material as they propagate forward through the jet, or from those electrons that stream outward from the plasma interaction, toward the laser, are captured by the Coulomb field, and stream back into the cold jet. The electron impact produces the K-shell holes in the cold material and as a result characteristic K-shell emission is observed.

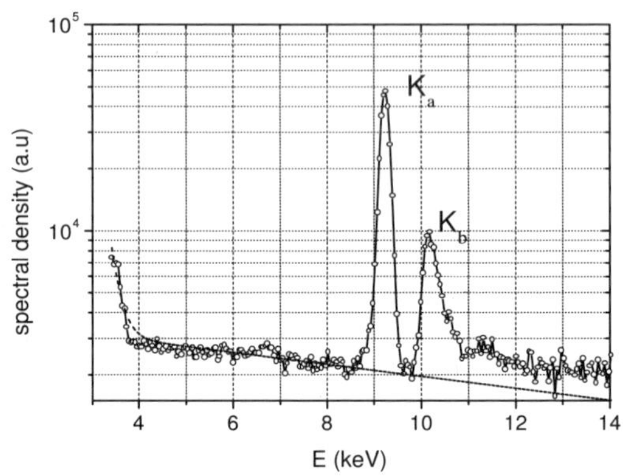

Fig. 1. X-ray spectrum from Ga-jet.

Due to the short pulse electron burst and the short stoping range generated K-lines are also short and allow subpicosecond time resolved optical pump and x-ray probe measurements [4, 5]. On the assumption of isotropic emission into 4π steradians a total emission flux of X-ray's is of $5.1\text{x}10^{10}$ photons /sec, with the K_a-line feature measured to be $2.6\text{x}10^{10}$ photons/sec. The experimental spectral density is assigned with a good accuracy by two-exponential function $F\ [A * \exp(-E/T_t) + B * \exp(-E/T_h)]$, where T_t and T_h are thermal and hot electron temperature respectively. The best agreement is obtained for the T_t=0.2 keV and T_h=12 keV. Using the modified band-pass filter method [3] we have measured the pulse-to-pulse temperature distribution as it is shown on Fig. 2.

The most probable electron temperature is lower than the optimal for the K-lines generation, but the quantity of the electrons with the suitable temperature large enough to create efficient K-lines radiation. The efficiency of laser energy to X-ray photon energy conversion is estimated to be $9.6\text{x}10^{-6}$. As an

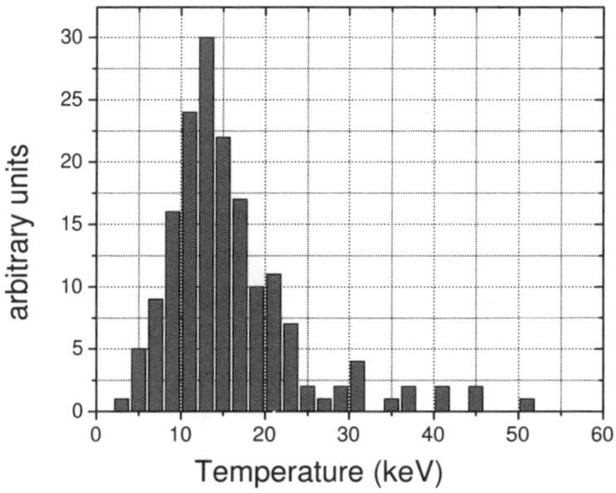

Fig. 2. Hot electron temperature distribution.

alternative to jet targets we have used Cu and Ti band targets (design from Prof. E.Foerster, Dr. I.Uschmann group, FSU Jena). K_α photon fluxes were measured to be 3.2x10^{10} phot/sec and 5.2x10^{10} phot/sec, with conversion efficiencies of 4.6x10^{-6}and 1.3x10^{-5} for Ti- and Cu-targets correspondingly.

The part of the generated X-ray radiation was separated by a led aperture to form the beam with necessary angular distribution for Bragg-diffraction experiments. A GaAs crystal was cut along [111] direction was used to observe Bragg diffraction. The diffracted beam was detected by X-ray CCD camera ANDOR DO-434 shown on Fig. 3. The integrated data from the CCD camera are shown on Fig. 4.

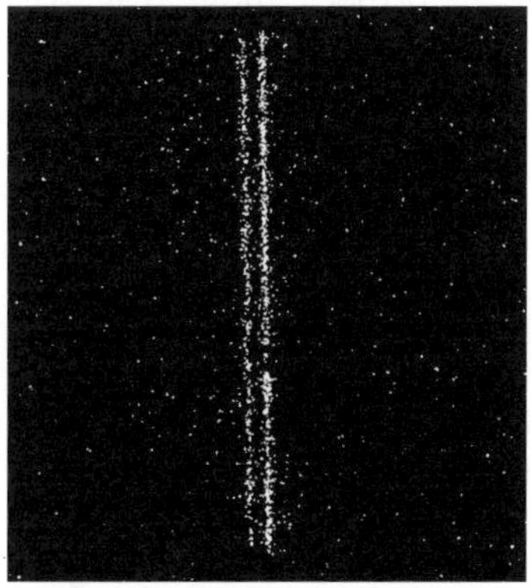

Fig. 3. Bragg-diffraction pattern.

The two K_{a1} and K_{a2} lines with energy of 9251 eV and 9225 eV from Ga are clearly resoled. The angular resolution of 0.12 arcmin/pixel with signal/bakground contrast 500:1. The diffraction pattern for GaAs [333] and

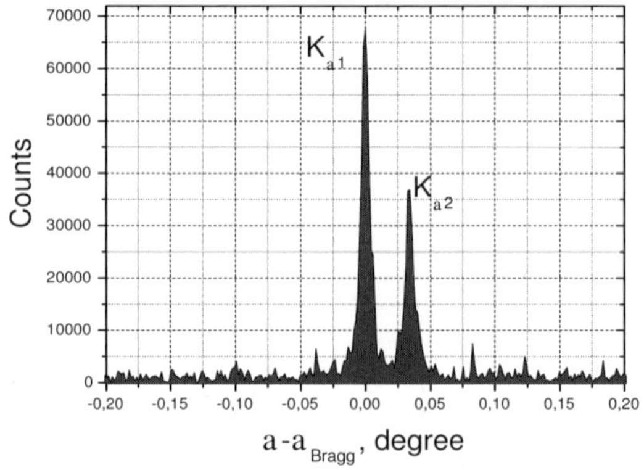

Fig. 4. Ga K_{a1} and K_{a2} lines.

GaAs/AlGaAs superlattice structure also was obtained. The average acquisition time for each pattern was only 5 sec.

In conclusion, we have demonstrated the principle of operation and the application of a new type of high repetition rate X-ray source. Most probably the conversion efficiency can be improved by optimization of the laser pulse and target irradiation conditions. Moreover other target materials will provide a broad spectral range of x-ray emission, either from plasma continuum emission, or from the characteristic K_α and K_β emission lines. These sources are also very attractive for dynamic EXASF studies of chemical bonds changes. Recently we also demonstrated the generation of X-ray continuum up to 5 keV using 10 W, 10 kHz repetition rate regenerative amplifier. R. Bernath participated in this project under NSF REU project EEC0244109

References

1. O. Albert, H. Wang, D. Liu, Z. Chang, G. Mourou, Opt. Lett., **25,** 1125 (2000).
2. G. Korn, A. Thoss, M. Faubel, T. Elsaesser, et al, Optics Letters, **27,** 866 (2002).
3. V.Gordienko, I. Lachko, P. Mikheev, et.al. Plasma Phys. Control Fusion, **44,** 2555 (2002).
4. C. Rose-Petruck et al., Nature **398,** 310 (1999).
5. C. Rissel, A. Rousse et al., Nature **390,** 490 (1997)

Generation, Amplitude and Phase Characterization of 10^{21}W/cm^2 Intensity

S.-W. Bahk, V. Chvykov, G. Kalintchenko, A. Maksimchuk, G. A. Mourou, N. Saleh, and V. Yanovsky

Center for Ultrafast Optical Science, University of Michigan, Ann Arbor, MI 48109-2099 swbahk@engin.umich.edu

Summary. The problem of intensity measurement at high-field regime is examined. We propose a method based on wave-front measurement that can evaluate ultra-high power density at the focus and compensate for the aberrations including the final focusing optics.

The advent of ultra-intense laser technology has been providing many experimental opportunities from producing high-energy electron beams to exploring the realm of relativistic optics and even electron-positron pair production [1]. In the ultra-intense regime, laser intensity is an important parameter both in categorizing laser-matter interaction and in analyzing data. In this paper, we point out problems in estimating the intensity of laser using conventional techniques and propose a better method.

One common way to estimate intensity at the focus is to use focal spot size and the encircled energy in the first Airy ring [2]. In this method, the small error in the spot size measurement could translate into large variation in the Strehl ratio. One theoretical example is shown in Fig. 1 with varying spherical aberrations. We can see that 0.7% FWHM error results in 50% Strehl ratio error. Spot size measurement errors always occur since camera CCD area only gives digitized information and the objective used to image the focal spot might introduce aberrations. Also, one cannot exclude errors in the calibration process that converts from real dimensions at the focus to CCD dimension. The often elliptical spot shape raises another question how one should specify focal size. When we apply this method to high-energy Chirped Pulse Amplification systems, the additional aberrations arising from thermal lensing/stress and intensity dependent phase are unaccounted for since we can only measure focal spot size at low energy. The other method, called the ionization method, is not documented in high-field regime and it only gives information about local hot spots. In contrast to these problems, the Strehl ratio is less sensitive to wave-front measurement error. If we represent wave-front with its r.m.s value(σ [in radians]), the Strehl ratio (S) is proportional to $\exp(-\sigma^2)$ [3]. For less than quarter-wave r.m.s. value ($\sigma < \pi/2$) , we can show that

$$(|\Delta S|/S) = 2\sigma^2(|\Delta\sigma|/\sigma) \leq 5(|\Delta\sigma|/\sigma),$$

from which, for example, only 5% error in Strehl ratio will be introduced by a 1% error in σ. We set the upper-limit to quarter-wave r.m.s. since the worse wave-front will give less than 0.08 Strehl ratio and it would usually not be good for any experiment. The error in σ mainly comes from the imaging optics. It is also notable that, if σ is very small, the sensitivity of Strehl ratio decreases.

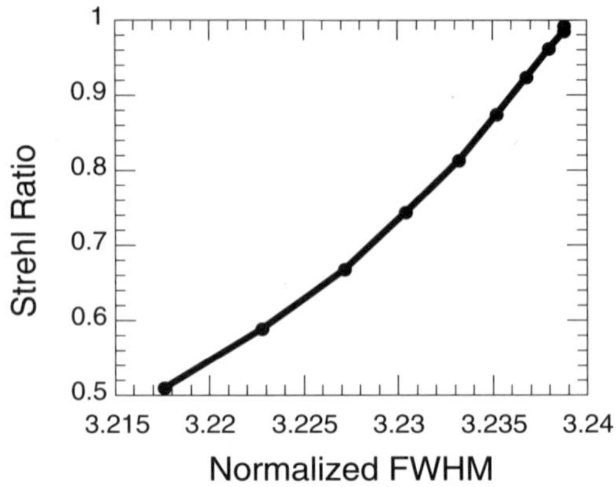

Fig. 1. Strehl ratio vs. spot size with spherical aberration. $\Phi = A(\rho^4 - \rho^2)$, where $A = 0 \ldots 12$ rad.

Our experimental schematic is shown in Fig. 2(a). The beam diameter is 36 mm and the actual focal length from the center of the off-axis paraboloid surface to focus is 67.7 mm. The deformable mirror plane is imaged through a paraboloid, an infinity-corrected objective and a 25 cm lens onto the wave-front sensor. For focal spot imaging, a flipper mirror is introduced between the objective and the 25 cm lens and the objective is moved by a measured distance along the optical axis. The deformable mirror used in this experiment is from Xinetics, Inc. The mirror surface is controlled by 37 actuators made of lead magnesium niobate. Maximum inter-actuator stroke is 2 μm and the clear aperture of the mirror is 42 mm. The wave-front sensor is a Shack-Hartmann wave-front sensor from Imagine Optic. The sensor area consists of a 32 by 32 miro-lens array and a CCD area of about 4 mm by 4 mm. The software for this sensor generates the first 63 Zernike coefficients and facilitates our calculation of the optimum voltage for the deformable mirror. Using a He-Ne laser, all 37 actuator response shapes are recorded as a form of Zernike coefficient representation and they are later used to reconstruct any given wave-front shape by linear superposition. The algorithm for the optimum voltage calculation is a simple matrix equation resulting from the least sqaure error method.

The laser system used in this experiment ("HERCULES") consists of one regenerative amplifier stage and two amplifier stages [4]. For convenience, we

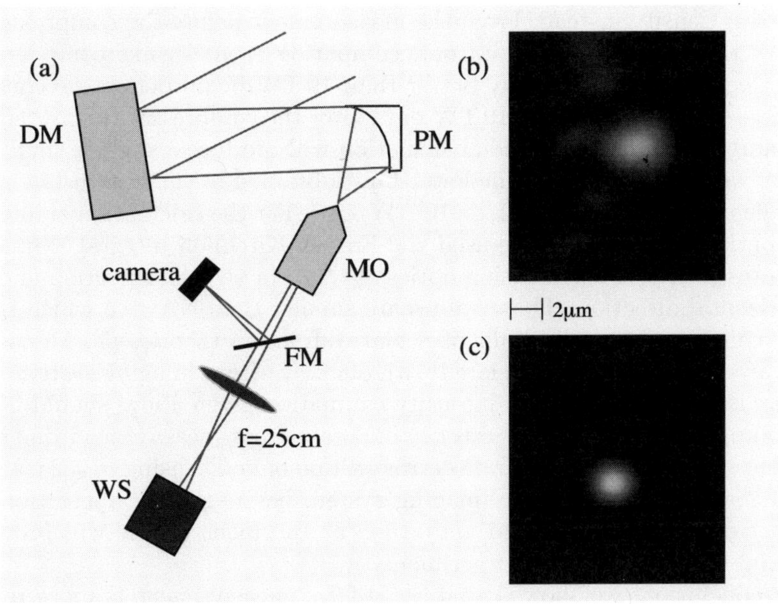

Fig. 2. (a) Experimental setup: DM(Deformable mirror), PM(2 inch paraboloid, F/1, 60°), WS(Wavefront sensor), FM(Flipper mirror), MO(Microscope objective, NA=0.66, ∞, 40×) (b) Regen focus before correction (c) Regen focus after correction. FWHM of spot (c) is 1.9 μm and energy in the first Airy ring is 75%. Both (b) and (c) have same spatial scales.

will designate 'Regen beam' for the regenerative amplifier beam only (~5 mJ) and '10 TW beam' when the first amplifier is activated and '30 TW' when the first and the second amplifiers are all activated; we also call these TW beam even though they may be measured before the compressor. For alignment and wave-front measurement after the paraboloid, we used the Regen beam with attenuation. Attenuation is done with a wave-plate and filters. Our first procedure is to measure the wave-front change from low-energy to high-energy operation and its standard deviation in each Zernike component. For this we measured 10 shots of wave-front with Regen, 10 TW beam, and 30 TW beam before the compressor and Regen and 10 TW beam after the compressor. The standard deviation of each Zernike component is less than $\lambda/15$ in all cases indicating the shot-to-shot stability of the system. The next procedure is to measure the wave-front at the Regen level as described in Fig. 2(a) and use a deformable mirror to compensate for any aberrations. We improved the wave-front of the Regen from 0.5λ r.m.s. value to 0.065λ. See Fig. 2(b)(c) for the improvement in the focal spot shape. The final step is to add the measured high-energy additional aberrations to low-energy compensated aberrations and calculate the focal intensity. Although 30 TW wavefront after compressor(near DM) was not measured, we can set a lower-limit for the power density instead. From the measurement before the compressor, we observe that the 30 TW wave-front, comparing each Zernike component, is almost equal to or marginally better than 10 TW beam before the compressor. Therefore, if we use the 10 TW data after the compressor (measured near the paraboloid), the diffraction calculation will produce a slightly smaller intensity value at the focus. The lower-limit obtained in this way using energy equivalent to 30 TW level is 1.1×10^{21} W/cm^2. For the calculation of intensity distribution at the focus, we used the Fresnel-Kirchhoff integral. Numerical apertures (NA) for our focusing optics are 0.26 in vertical direction and 0.22 in horizontal direction. These values are smaller than NA=0.5 which is generally considered as valid limit of Scalar diffraction theories [5]. Meanwhile, there are some aberrations related with our wave-front imaging system. Even if there is an aberration, our intensity estimation given above should be reliable. Suppose we have a 10% error in our wave-front measurement due to the imaging system. The Strehl ratio error will be only 3% using $\sigma = 2\pi \times 0.065$ in the stated formula. If the imaging system has negligible aberrations, we can better predict phase and amplitude at the focus and it will provide a powerful tool for high-field experiments.

In conclusion, we showed that wave-front measurement is more reliable than spot-size measurement in evaluating the focused intensity. By measuring the difference in the high-energy and low-energy wave-front, we were able to estimate ultra-high power density at the focus. It is also shown that we can use the wave-front information to compensate for the aberrations of a low-energy beam, including final focusing optics, using a deformable mirror.

Acknowledgement. This work was supported by the National Science Foundation through the Center for Ultrafast Optical Science at the University of Michigan under PHY-0114336.

References

1. G. Mourou, et al., Plasma Physics Reports, **28**, 12 (2002).
2. J. D. Bonlie, et al., Appl. Phys. B, **70[Suppl.]** , S155 (2000).
3. V. N. Mahajan, J. Opt. Soc. Am., **73**, 860 (1983).
4. V. Yanovsky, et al., CLEO 2002, CMK4.
5. H. H. Hopkins, Proc. Phys. Soc., **55**, 116 (1943).

Ultrafast Hard X-Rays from Electron Accelerators

P.H. Bucksbaum[1], D.A. Reis[1], and J. Hastings[2]

[1] FOCUS Center, Physics Department, University of Michigan, Ann Arbor, MI 48109-1120 phb@umich.edu
[2] Stanford Synchrotron Radiation Laboratory, Stanford Linear Accelerator Center, Menlo Park, CA

1 Introduction

Modern third-generation synchrotrons are many orders of magnitude brighter than laser-based laboratory sources of x-ray radiation (see figure 1 and reference [8]). Their importance in *ultrafast* x-ray science has been limited in the past, because the x-ray pulse duration is generally no shorter than the electron bunch lengths in synchrotrons, which are constrained by various technical considerations in electron storage rings to a range of tens of picoseconds or longer [15]. Some important dynamical phenomena can be studied in this sub-nanosecond time range (for example, [11]); however, the fundamental motion of interatomic bonds is in the range of 10's to 100's of femtoseconds. Therefore there is a strong research incentive to find new ways to shorten the pulse duration of x-rays from synchrotron radiation. (See, for example, the recent review [1].)

Another contribution in this volume describes an electron beam slicing method developed at the Berkeley Laboratory's Advance Light Source, which uses ultrafast lasers to displace the orbit of a small segment of the synchrotron electron bunch [10]. This sub-picosecond segment then produces ultrafast x-rays in an undulator or bending magnet beam line. There are also plans for a new generation x-ray source at Berkeley, which would disperse the electron momenta in a clever way so that the resulting dispersed x-rays could be compressed in a specially designed crystal spectrometer [14].

This paper presents two other approaches to this problem. Rather than slice the electrons at synchrotrons, we can slice the x-rays themselves [7]. This is the method we have adopted at the Advanced Photon source (APS) at the Argonne National Laboratory. Alternatively, we can compress highly relativistic electrons in a linear accelerator, where bunch lengths can be shortened to less than 100 fsec. When these ultrashort electron bunches travel through an undulator, they produce ultrafast pulses of x-rays [5]. This new approach has now been demonstrated at the Sub-Picosecond Pulse Source (SPPS) experiment at the Stanford Linear Accelerator Center (SLAC). This short review will describe the current progress and future prospects in each of these programs.

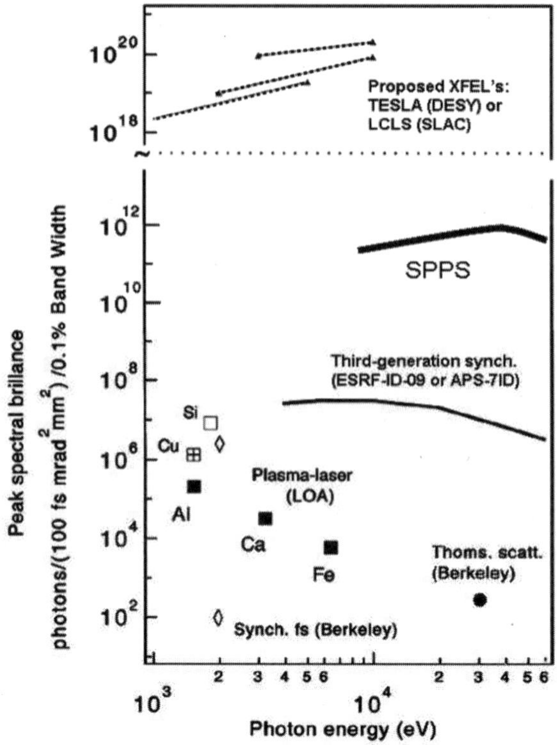

Fig. 1. Comparison of the peak brilliance of laser-based and accelerator-based sources of x-rays. This graph was adapted from [8], and modified to include the SPPS source.

2 Ultrafast x-rays at the Advanced Photon Source

The APS and its close cousins, the ESRF in Grenoble and and SPring-8 in Harima, Japan, are the world's most powerful research sources of sub-Angstrom x-rays. The APS is a 1km synchrotron ring with multiple bunches of electrons circulating at 7GeV. In its usual operating configuration, the APS delivers between one and ten million 10 keV photons in each sixty picosecond pulse; with a pulse separation of 152 nanoseconds. The x-rays are generated in undulators, and are collimated to about 10 $\mu r \times 30 \mu r$ and monochromitized to a bandwidth of about 1 eV.

X-ray pulse switching experiments are performed on the Sector 7 MHATT-CAT undulator beam line at the APS. The principle of the x-ray switch is laser-excitation of Bragg crystals [7]. The excitation produces a transient change in the scattering condition, leading to a switch action. The change can either be incoherent, such as simple laser heating, or coherent excitation of motion (coherent phonons) in the crystal. Several physical mechanisms have been demonstrated or proposed for this transient excitation, and the switch time scale can range from nanoseconds to less than 1 psec, depending on the details of the switch material and geometry.

2.1 Transmission geometry

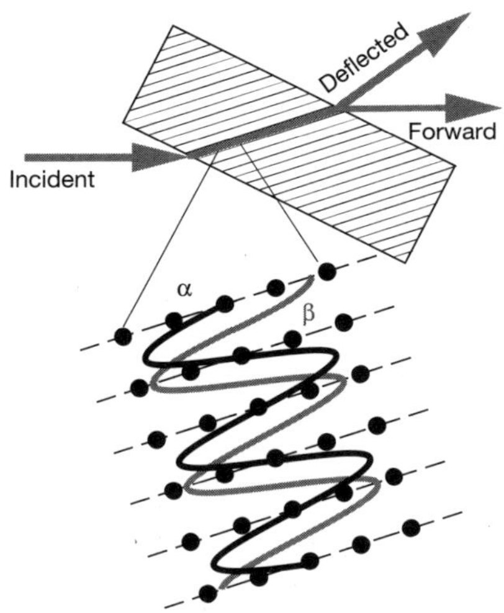

Fig. 2. Geometry for an ultrafast x-ray switch utilizing the Borrmann effect (from [2]). Anomalous transmission occurs for x-rays channelling through the Bragg planes in the α mode. The orthogonal β mode experiences anomalously strong absorption.

An x-ray switch using a transmission geometry is shown in figure 2 [2]. This switch uses the anomalous transmission (Borrmann effect), in which lattice planes form parallel plate waveguides for the x-rays. Since the x-ray electric field nodes are located at these atomic planes, absorption is minimized in this geometry. Nearly 50% transmission (the theoretical maximum) is possible for 10keV x-rays traversing a $300\mu m$ silicon wafer. When the x-rays exit the crystal, the transverse standing wave in the crystal waveguides becomes two traveling waves in free space, so that the crystal is a coherent beamsplitter for x-rays. The angle between the two exiting waves is the supplement of the Bragg angle.

The x-rays can be redistributed coherently between the two exiting beams by altering the boundary conditions at the exit face of the crystal. Alternatively, both beams can be switched off simultaneously by changing the input face boundary conditions. We have studied both of these effects in crystalline Ge, using laser absorption to launch acoustic waves in the crystal, as shown in figure 3. The acoustic wave created by an ultrafast laser travels into the crystal with very little dispersion or attenuation. When the x-rays intercept the acoustic disturbance, their field is redistributed from the transmission waveguide mode into an orthogonal mode where the antinodes are on the crystal planes. These two modes then coherently interfere at the exit face to change the distribution of energy into the two free space x-ray beams.

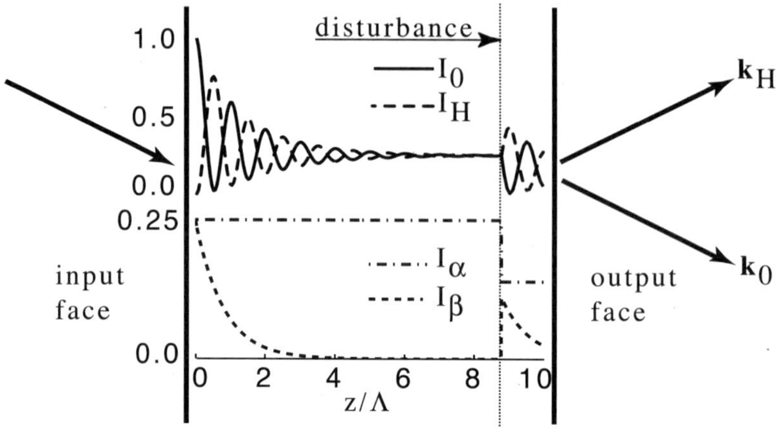

Fig. 3. In the Borrmann geometry, output x-ray beams can be switched on and off by coherent redistribution of x-rays from the α mode to the β mode at an internal discontinuity in the crystal [4].

The speed of the transient switch depends on several factors. One key factor is the frequency of the acoustic phonons, since the switch cannot be turned on and off faster than one-half of a wave cycle. We have been studying ways to employ high frequency optical phonons, where sub-picosecond switching will be possible [3].

2.2 X-ray streak cameras

X-ray streak cameras have been used to resolve ultrafast x-ray phenomena that are faster than the synchrotron pulse duration. Here scattered synchrotron radiation is modulated by ultrafast phenomena in the scatterer, and the modulation is recorded by the streak camera image. Melting, laser-induced crystal strain, and molecular rearrangement have been observed in this way. Recent developments have pushed the time resolution of x-ray streak cameras down to less than 1 ps, so these have become competitive with other ultrafast x-ray methods [6].

3 SPPS: Ultrafast x-rays at SLAC

Relativistic electron beams can be chirped and compressed using techniques that are analogous to the standard methods of ultrafast optics [13]. Since the velocity of a highly relativistic electron is nearly independent of its energy, space charge forces don't broaden the pulse. On the other hand, magnetic bends disperse the beam like a prism. It's possible to phase the rf accelerating cavities to impose a nearly linear energy chirp on the electrons. In this case, a four-magnet dog-leg chicane can act the same for the electrons as a four-prism compressor for a chirped optical beam.

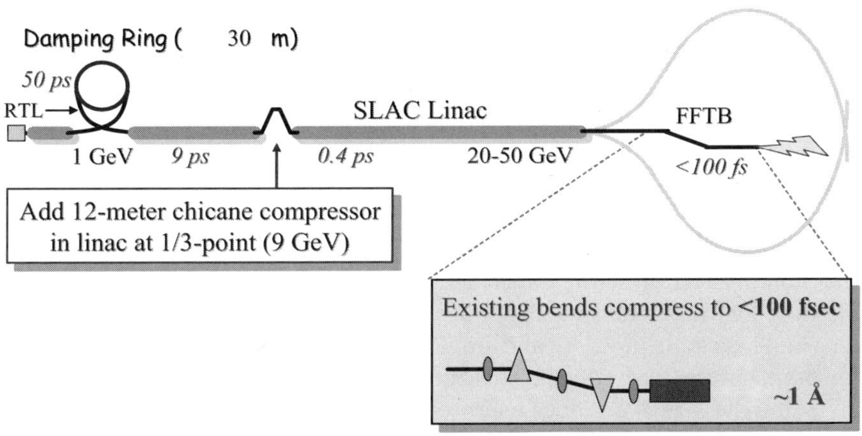

Fig. 4. Schematic layout of the SPPS experiment at SLAC, showing the placement of the compression stages along the length of the linac (from the SPPS collaboration).

The Sub-Picosecond Pulse Source (SPPS) is a new experiment at the Stanford Linear Accelerator Center (SLAC), which employs this relativistic chirped pulse compression on the beam from SLAC to produce sub-100 femtosecond bunches of electrons [5]. These in turn produce pulses of ultrafast x-ray radiation by passing through an undulator. SPPS is an international collaboration including SLAC/SSRL, DESY, Uppsala University (Sweden), University of Copenhagen (Denmark), the University of Michigan FOCUS Center, the Advanced Photon Source, BioCARS, the University of California at Berkeley, and the National Synchrotron Light Source. SPPS receives support from the U.S. DOE Office of Basic Energy Sciences, the NSF FOCUS Physics Frontier Center, and from the collaborating laboratories and universities.

SPPS can use the existing system that presently injects electrons into the PEP-II storage ring. The principal modification to the linac is the sequence of 4 bending magnets, which compress the electron bunch partway down the linac. Following acceleration, the electrons enter the SLAC Final Focus Test Beam (FFTB), where there is a final stage of compression, an undulator to produce x-rays, and a multilayer mirror monochrometer to deflect the x-rays into the experimental area. With 3.4 nC per bunch accelerated to 28 GeV and a 2.5 m long undulator, SPPS is designed to generate 80 fsec pulses of 10^7 8-10 keV x-rays. The peak brightness is $10^{25} photons/(s - mm^2 - mrad^2 - 0.1\% bandwidth)$.

The collaboration recently completed its first six-week operating period, when the x-ray and optical beams were installed, synchronized, and tested. The primary goals for the initial run were to test the operation of the new SPPS x-ray beam line, to measure x-ray pulse parameters, to evaluate the synchronization of the new ultrafast laser system with the x-ray pulses, and to measure the scattered x-ray intensity from a variety of samples. The measured x-ray intensity, 2×10^7 photons per pulse in 1% bandwidth, agrees well with predictions. Measurement of the electron beam pulse length after the first stage of compression demonstrated a pulse length of 300 fs, with beam parameters such that the final compression reduces this by a factor of 4. This means that the x-rays pulse duration in the undulator is inferred, but not yet directly measured, to meet the design of 80 fs. Direct measurements of this compressed electron bunch length and the final x-ray pulse width are planned for the next run. The x-ray scattering power and background rates for single crystal, powder, and liquid samples were measured and feasibility for pump-probe experiments was established.

Since the x-rays are not generated directly by the laser at SPPS, ultrafast timing is not automatic. Synchronization is maintained by feedback to the laser oscillator length, referenced to the rf that controls SLAC. The short-term synchronization jitter between laser and x-ray pulses was found to be less than 2 ps over 10-50 sec. Residual jitter on this order may be very difficult to reduce. In that case ordinary pump-probe spectroscopy would not be possible at the 100 fsec level, and so we are developing new sampling methods based on single-shot measurements of the laser-x-ray relative delay.

To measure the arrival time of the x-rays *without wasting any x-rays in the process*, a single shot electro-optic sampling apparatus is placed directly upstream of the undulater. This apparatus can also be used for direct measurements of the electron bunch length and shape. It uses the coherent terahertz half-cycle pulse which accompanies each relativistic electron bunch, to induce a transient electro-optic effect in a thin crystal [9]. The origin of the field pulse is the Coulomb field of the electrons, which undergoes extreme relativistic enhancement ($\gamma = (1 - v/c)^{-1} \simeq 60,000$). We have placed the crystal one centimeter from the electron beam axis, where the peak electric fields are expected to exceed $1MV/m$. An ultrafast laser pulse from the same oscillator that supplies light for the pump-probe experiments crosses the crystal at an angle, so that only part of the beam will be in the crystal at the same time as the terahertz field. We measure the spatial distribution of the polarization of the laser pulse, which is modified by the transient birefringence induced by the terahertz field. A number of difficulties are presented

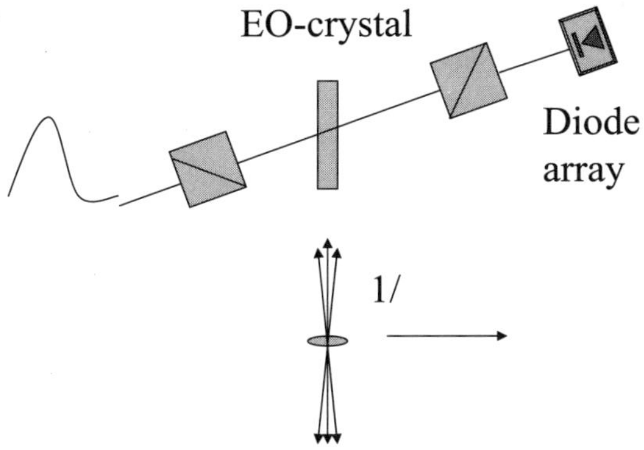

Fig. 5. The relative delay between the x-rays and the laser can be measured using the electro-optic effect induced by the electrons.

by the great distance (150 m) between the electron beam and the laser, which is near the x-ray experimental hutch. Optical delay lines simply cannot be used to compensate this distance, and so we use a previous pulse from the Kerr-lens modelocked Ti:Sapphire oscillator to perform the EO experiment. This pulse is transmitted through a polarization-preserving optical fiber, and we must compensate for dispersion with a combination grating compressor and liquid-crystal pulse shaper, to obtain an ultrafast pulse at the EO experiment. The optical path length must also be actively controlled over this distance. Details are in a separate publication [12].

The SPPS will share SLAC with other experiments in accelerator and high energy physics until 2006. Then it will be dimantled to make way for the construction of a much more powerful ultrafast x-ray source, the LCLS x-ray free-electron laser.

4 Acknowledgments

We acknowledge Paul Fuoss and the other members of the SPPS collaboration for their assistance in providing information for this brief review. This work is supported by the Department of Energy Basic Energy Sciences, Office of Energy Research under Contract No. W-31-109-Eng-38 and Contract No. DE-AC03-76SF00515; and by the National Science Foundation FOCUS Physics Frontier Center.

References

1. Barletta,W.A., and Winick, H.: Introduction to special section on future light sources, Nuclear Instruments and Methods in Physics Research A **500** 110 (2003).
2. DeCamp,M.F., Reis, D.A., Bucksbaum, P.H., Adams, B., Caraher, J.M., Clarke, R., Conover, C.W.S., Dufresne, E.M., Merlin, R., Stoica, V., and Wahlstrand, J.K.: Coherent Control of Pulsed X-ray Beams, Nature, **413**, 825-828(2001).
3. DeCamp, M.F., Reis, D.A., Bucksbaum, P.H., and Merlin, R.: Coherent Control and Dynamics of High Amplitude Optical Phonons in Bi, Phys. Rev. B. **64**, 2301 (2001).
4. DeCamp, M.F., Reis, D.A., Cavalieri, A., Bucksbaum, P.H., Clarke, R., Merlin, Dufresne, E.M., Arms, D.A., Lindenberg, A.M., MacPhee, A.G., Chang, Z., Lings, B., Wark, J.S., and Fahy, S.:Supersonic strain front driven by a dense electron-hole plasma, cond-mat/0301002 (2003).
5. Krejcik, P., Arthur, J., Carr, R. Cornacchia, M., Emma, P., Iverson, R., Safranek, J., Tatchyn, R.: The Sub-Picosecond Photon Source Design Study, SLAC-Pub-8950 (2000).
6. Liu J.Y., Wang J., Shan B., et al.: An accumulative x-ray streak camera with sub-600-fs temporal resolution and 50-fs timing jitter, Appl. Phys. Lett. **82** 3553-3555 (2003).
7. Bucksbaum, P.H., and Merlin, R.D.: The phonon Bragg switch: A proposal to generate sub-picosecond x-ray pulses, Solid State Comm. **111**, 535-539 (1999).
8. Rousse, A., Rischel, C., and Gauthier, J.-C.: Femtosecond x-ray crystallography. Rev. Mod. Phys., **73**, 17–32 (2001)
9. Srinivasan-Rao, T., Amin, M., Castillo, V., Lazarus, D.M., Nikas, D., Ozben, C., Semertzidis, Y.K., Stillman, A., and Kowalski, L.: Novel single shot scheme to measure submillimeter electron bunch lengths using electro-optic technique, Phys. Rev. ST Accel. Beams **5**, 042801 (2002).
10. Schoenlein, R.W., Chattopadhyay, S., Chong, H.H.W., Glover, T.E., Heimann, P.A., Shank, C.V., Zholents, A.A., Zolotorev, M.S.: Generation of femtosecond pulses of synchrotron radiation, Science **287** 2237-2240 (2000).
11. Schotte F., Lim M.H., Jackson T.A., Smirnov A.V., Soman J., Olson J.S., Phillips G.N., Wulff M., and Anfinrud P.A.: Watching a protein as it functions with 150-ps time-resolved X-ray crystallography, Science **300**, 1944-1947 (2003).
12. Lee, S., Fritz, D., Caveleri, A., and Reis, D., to be submitted (2003).
13. Uesaka, M., Tauchi, K., Kozawa, T., Kobayashi, T., Ueda, T., and Miya, K., Generation of a subpicosecond relativistic electron single bunch at the S-band linear accelerator, Phys. Rev. E 50, 3068-3076 (1994).
14. Zholents A., Heimann P., Zolotorev M., and Byrd J.: Generation of subpicosecond X-ray pulses using RF orbit deflection. Nucl. Inst. and Meth. A, **425**, 385-389 (1999).
15. Wiedemann, H.: Particle accelerator physics : basic principles and linear beam dynamics. Springer-Verlag, New York (1993)

Amplification of Multikilovolt Xe(L) Hollow Atom Transitions with Xe Clusters in Confined Plasma Channels

Alex B. Borisov[1], Xiangyang Song[1], Yevgeniya Koshman[1,2], Jack Davis[3], Yang Dai[2], Keith Boyer[1], and Charles K. Rhodes[1,2,4,5]

[1] Laboratory for X-Ray Microimaging and Bioinformatics, Department of Physics, University of Illinois at Chicago, Chicago, IL 60607-7059, USA
[2] Department of Bioengineering, University of Illinois at Chicago, Chicago, IL 60607-7052, USA
[3] Plasma Physics Division, Naval Research Laboratory, Washington, DC 20375, USA
[4] Department of Electrical and Computer Engineering, University of Illinois at Chicago, Chicago, IL 60607-7053, USA
[5] Department of Computer Science, University of Illinois at Chicago, Chicago, IL 60607-7042, USA

Summary. The recent observation of strong amplification on multikilovolt Xe(L) hollow atom transitions in the ∼2.8 Å spectral region can be seen as a consequence of the combination of (1) a new concept for amplification that involves the creation of a highly ordered state combining ionic, plasma, and coherent radiative components and (2) the use of two recently discovered (c. ∼1990) forms of radially symmetric energetic matter, namely, hollow atoms and self-trapped plasma channels. This approach enables the demanding power densities necessary for x-ray amplification ($\sim 10^{19}$ W/cm^3) to be reached under conditions for which (α) the effective phase space volume of the interaction is profoundly limited and (β) the energy transfer is radiation dominated.

1 Historical Introduction

In their seminal paper "Infrared and Optical Masers" published in *The Physical Review* in 1958, Arthur L. Schawlow and Charles H. Townes concluded with a section in which they discussed the high-frequency limits of amplification [1]. They wrote "unless some radically new approach is found," the x-ray zone is out of reach. Furthermore, the last sentence of the same article stated "... continuous tuning over larger ranges of frequency will require materials with very special properties." In relation to the work described below, the former requirement is satisfied by the development of a new concept that involves radiation dominated energy flow under conditions for which the phase space of the interaction is precisely controlled while the latter is met by the use of hollow atoms. These hollow atom states, which characteristically yield a broad spontaneous emission profile [2] of the type shown for Xe in Fig. 1, and which were completely unknown at the time of the Schawlow and Townes piece, constitute an important new form of excited matter that was discovered decades later [3]. "Hollow atoms" are atoms (ions) that intrinsically possess an "inverted" electronic configuration consisting of deeply bound inner-shell vacancies, perhaps multiple, with the simultaneous retention of several electrons in relatively weakly bound outer orbitals [4]. Accordingly, these states are optimally suited for the prompt emission and amplification of x-rays. In this way, the new results on x-ray amplification reported briefly herein fulfill the Schawlow/Townes prognosis and confirm the astonishing foresight and accuracy of their conclusions.

2 Power Density Requirement and Amplification Mechanism

In order to achieve significant amplification in the multikilovolt spectral region, the ability to produce controllably high power compression ($\sim 10^{19}-10^{20}$ W/cm^3) in appropriately configured material is both difficult and key. Experimental evidence gathered over the last several years together with corresponding theoretical analyses led to the conclusion that the system composed of Xe hollow atom states, that produce the characteristic Xe(L) spontaneous emission spectrum at $\lambda \cong 2.9$ Å illustrated in Fig. 1 and arise from the excitation of Xe clusters [5, 6] with an intense pulse of ultraviolet (248 nm) radiation [7], closely represents the ideal situation sought.

The solution to the fundamental power compression issue was found in the alliance of two recently studied phenomena. They are (α) the direct multiphoton excitation of hollow atoms from clusters with ultraviolet radiation [5] and (β) a nonlinear mode of confined propagation in plasmas resulting from a relativistic/charge-displacement mechanism of self-channeling [8, 9]. Basically, the concept for amplification is founded on the proper matching [10, 11] of the cluster excitation with the channeling process and the characteristics of the confined electromagnetic propagation. Consequently, the amplification proceeds through the creation of an excited and highly ordered state that is comprised of four mutually coupled components; namely, atomic (ionic) matter, plasma electrons, and two coherent radiative fields. Since this dynamical mechanism enables the energy transfer (1) to be confined to a small fraction of the available phase space and (2) to occur through a radiation dominated interaction, both (α) the potential for high efficiency and (β) the ability to control the spectrum of effective x-ray amplification are conferred on the system. It is this combination of factors that leads to the high amplifications observed [11].

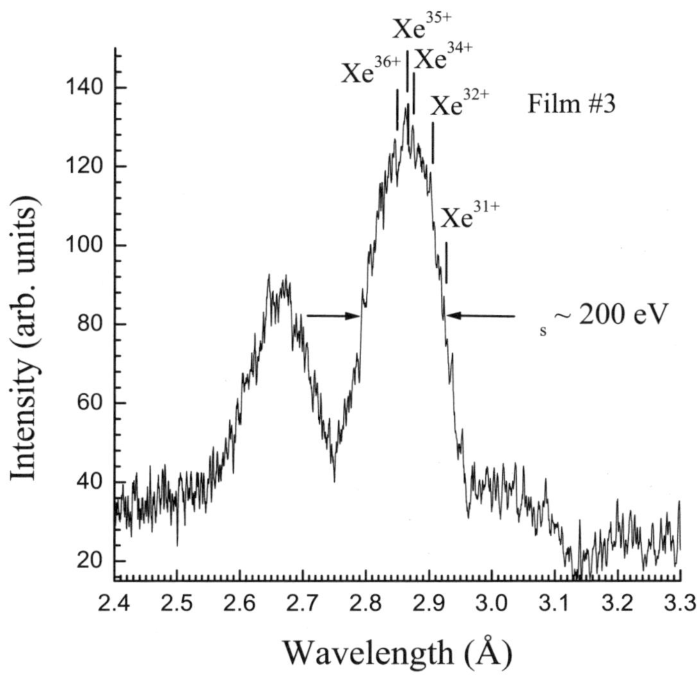

Fig. 1. Characteristic unamplified spontaneous emission profile of the Xe(L) 3d−2p hollow atom spectrum (film #3) produced from Xe clusters with intense ($\sim 10^{19}$ W/cm^2) femtosecond 248 nm excitation *without* plasma channel formation. The splitting between the major and minor lobes arises from the spin-orbit interaction of the 2p vacancy. The full width of the main feature is $\delta_\sigma \sim 200$ eV. The positions of selected charge state transition arrays (Xe^{31+}, Xe^{32+}, Xe^{34+}, Xe^{35+}, and Xe^{36+}) are indicated. The spectral resolution of these film data is ~ 4 eV.

3 Précis of Experimental Findings

Amplification has been demonstrated by the existence of a quantitative interlocking concordance of physical evidence [11]. Specifically, the data include (a) the observation of strong enhancement of selected spectral components of several Xe^{q+} hollow atom transition arrays (q = 30, 31, 32, 34, 35, 36, 37) radiated axially from confined plasma channels, (b) the measurement of line narrowing that is spectrally correlated with the amplified transitions, (c) evidence for spectral hole-burning in the spontaneous emission, a manifestation of saturated amplification, that corresponds spectrally with the amplified lines, and (d) the detection of an intense narrow ($\delta\theta_x \sim 0.2$ mr) directed beam of radiation in the far field of the source. An estimate of the peak spectral brightness achieved in the initial experiments [11] gave a value of $\sim 10^{31}$–10^{32} photons·s^{-1}·mm^{-2}·mr^{-2} (0.1% Bandwidth)$^{-1}$.

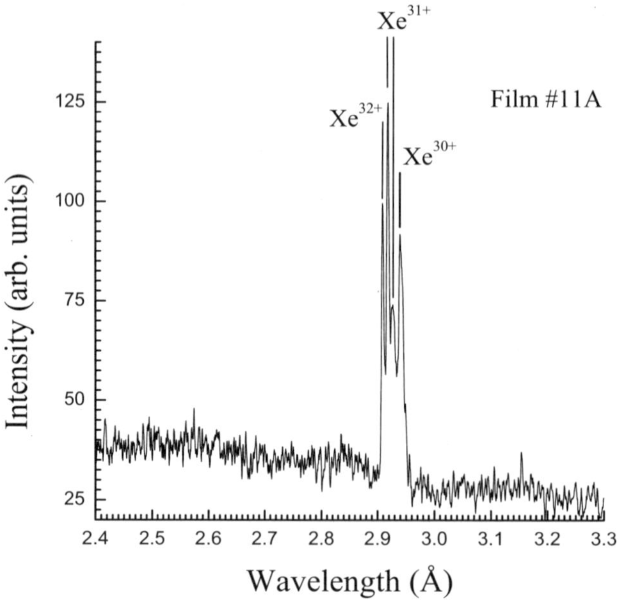

Fig. 2. Axially amplified lines corresponding to the Xe^{30+}, Xe^{31+}, and Xe^{32+} 3d→2p transition arrays observed (film #11A) with a von Hámos spectrograph equipped with Kodak RAR 2492 film. The emission originates from a self-trapped plasma channel generated by the intense ($10^{19}-10^{20}$ W/cm^2) femtosecond 248 nm pulses that provide the excitation of the Xe(L) hollow atom states from the xenon clusters.

A spectrum typical of the x-ray beam axially amplified in the narrow plasma channel and recorded on film with a von Hámos spectrograph is illustrated in Fig. 2. In this case, the strongly enhanced lines observed correspond to the Xe^{30+}, Xe^{31+}, and Xe^{32+} charge state arrays of the major lobe shown in Fig. 1. This observation, along with additional spectra showing comparable results of several other transition arrays in the $\lambda \sim 2.71-2.93$ Å region, indicate that the hollow atom states are strongly inverted and that the amplification can be tuned across a substantial fraction of the spectral profile illustrated in Fig. 1. Additional single-pulse observations [12] of deep spectral hole-burning that showed the formation of a spectral gap with a width $\Delta h\omega_x \cong 60$ eV support this conclusion. These single-pulse data were obtained with a modified von Hámos instrument equipped with a multi-element linear CCD array located in the film plane.

Direct measurements of the divergence of the amplified x-ray beam emitted from the channel have been made with the experimental configuration shown in Fig. 3. The pattern of damage typically produced by a single x-ray pulse on the Ti foil protecting the film pack at a distance of 12.5 cm from the source is illustrated in Fig. 4. At a separation of 12.5 cm from the exit of

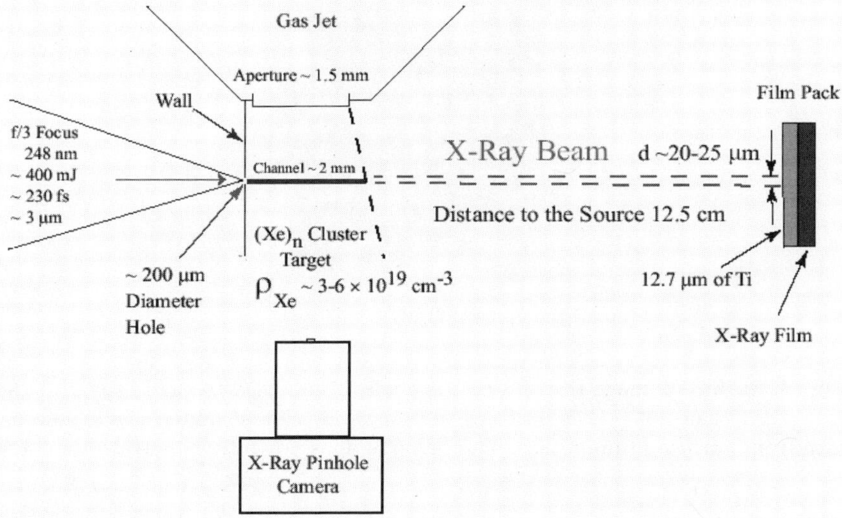

Fig. 3. Experimental configuration used for the observation of the low divergence x-ray beam produced by amplification of Xe(L) radiation in self-trapped channels inside an evacuated chamber. The x-ray pinhole camera was equipped with a $\sim$10 μm thick Be foil enabling the morphology of the channel to be visualized by the Xe(M) emission ($\sim$1 keV). The observed channel length typically is $l \approx$1.5–2.5 mm. The wall defining the entrance plane having the 200 μm aperture was fabricated from $\sim$100 μm thick steel and the incident 248 nm pulse was focused with an $f/3$ off-axis parabolic optic to a spot size of $\sim$3 μm. The average Xe density in the target was estimated to be $\rho_{\mathrm{Xe}} \sim$3–6$\times 10^{19}$ cm^{-3}. The location of a film pack used for measurement of the amplified x-ray beam composed of a 2 cm square 12.7 μm thick foil backed by a matching piece of x-ray film is shown. This detector was placed on the channel axis in a perpendicular orientation at a distance of 12.5 cm from the cluster target. For a channel diameter of $\sim$3 μm, the location of the film pack corresponds to the far field of the source. Typical single-pulse damage caused on the Ti foil by an incident x-ray pulse had a diameter $d \approx$ 20–30 μm, a range that corresponds to a divergence $\delta\theta_X \sim$ 0.2 mr.

the amplifying channel, the Ti foil shown in Fig. 4 is clearly damaged over a region having a characteristic diameter $d \cong$25 μm. Microscopic examination of the Ti surface revealed that material had been removed by the exposure to a significant depth, but the foil had not been fully penetrated. The corresponding divergence of the x-ray beam given by the diameter of the damaged region is $\delta\theta_X \sim$ 0.2 0.2 mr.

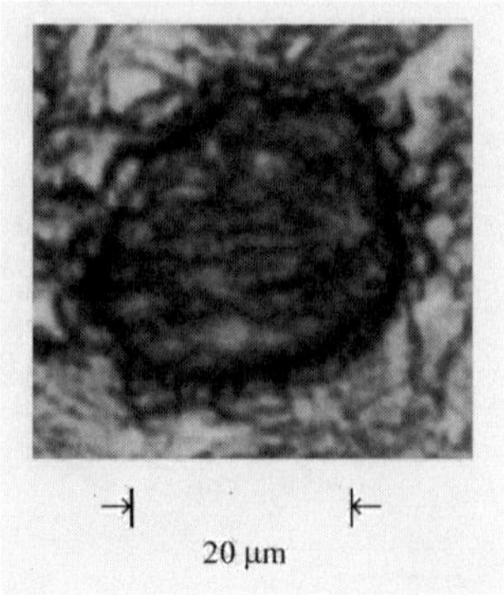

Fig. 4. Surface damage to the 12.7 μm thick Ti foil used to shield the film pack at an axial distance of $\sim$12.5 cm from the source as viewed under a light microscope with front surface illumination. The damage does not fully penetrate the Ti foil.

4 Conclusions

Amplification on multikilovolt Xe(L) hollow atom transitions produced from the nonlinear excitation of Xe clusters has been observed in self-trapped plasma channels produced by intense femtosecond 248 nm pulses. The cluster excitation and the channeling mechanism do not act independently. Basically, the concept for amplification is founded on a proper matching of the cluster excitation with the channeling process and the characteristics of the confined electromagnetic propagation. Consequently, the amplification proceeds through the creation of an excited and highly ordered state [11–13] that is comprised of four mutually coupled components; namely, atomic (ionic) matter, plasma electrons, and two coherent radiative fields, specifically, the ultraviolet and x-ray waves. The chief consequence is energy flow conducted through a radiation dominated interaction that is confined to a small phase space volume.

Acknowledgements

Support for this research was partially provided under contracts with the Army Research Office (DAAG55-97-1-0310) and the Department of Energy at the Sandia National Laboratories. Sandia is a multiprogram laboratory operated by the Sandia Corporation, a Lockheed Martin Company, for the United States Department of Energy under contract no. DE-AC04-94AL85000. One author (J.D.) would like to thank ONR for support through the NRL 6.1 program. Correspondences should be addressed to C.K.R. (e-mail: rhodes@uic.edu)

References

1. A.L. Schawlow and C.H. Townes: Phys. Rev. **112,** 1940 (1958).
2. M.W. Clark, D. Schneider, D. DeWitt, J.W. McDonald, R. Bruch, U.I. Safronova, I.Yu. Tolstikhina, and R. Schuch: Phys. Rev. B **47,** 3983 (1993).
3. H. Winter and F. Aumayr: J. Phys. B **32,** R39 (1999).
4. N. Vaeck, J.E. Hansen, P. Palmeri, P. Quinet, N. Zitane, M. Godefroid, S. Fritzsche, and N. Klystra: Physica Scripta **T95,** 68 (2001).
5. A. McPherson, B.D. Thompson, A.B. Borisov, K. Boyer, and C.K. Rhodes: Nature **370,** 631 (1994).
6. A.B. Borisov, A. McPherson, K. Boyer, and C.K. Rhodes: J. Phys. B **29,** L43 (1996).
7. K. Kondo, A.B. Borisov, C. Jordan, A. McPherson, W.A. Schroeder, K. Boyer, and C.K. Rhodes J. Phys. B **30,** 2707 (1997).
8. A.B. Borisov, A.V. Borovskiy, V.V. Korobkin, A.M. Prokhorov, O.B. Shiryaev, X.M. Shi, T.S. Luk, A. McPherson, J.C. Solem, K. Boyer, and C.K. Rhodes: Phys. Rev. Lett. **68,** 2309 (1992).
9. A.B. Borisov, J.W. Longworth, K. Boyer, and C.K. Rhodes: Proc. Natl. Acad. Sci. USA **95,** 7854 (1998).
10. A.B. Borisov, A. McPherson, B.D. Thompson, K. Boyer, and C.K. Rhodes: J. Phys. B **28,** 2143 (1995).
11. A.B. Borisov, X. Song, F. Frigeni, Y. Koshman, Y. Dai, K. Boyer, and C.K. Rhodes: *Ultrabright multikilovolt coherent tunable x-ray source at* $\lambda \sim 2.71-2.93$ *Å*, J. Phys. B, submitted.
12. A.B. Borisov, X. Song, Y. Koshman, Y. Dai, J. Davis, K. Boyer, and C.K. Rhodes: *Saturated multikilovolt x-ray amplification with Xe clusters: single-pulse observation of Xe(L) spectral hole-burning,* J. Phys. B, submitted.
13. G. Marowsky and C.K. Rhodes: Neue Zürcher Zeitung, Nr. 254, 1. November 1995, S. 42.

Asymmetric Pulse Shapes in Grating-Based CPA Compressors and Optimal Electron Acceleration in Plasmas

Csaba Tóth, Jerome Faure, Jeroen van Tilborg, Cameron G. R. Geddes, Carl B. Schroeder, Eric Esarey, and Wim P. Leemans

L'OASIS Group, Center for Beam Physics, Lawrence Berkeley National Laboratory, BLDG 71R0259, 1 Cyclotron Rd., Berkeley, CA 94720, USA
ctoth@lbl.gov

Summary. The temporal shape (skew) of the envelope function of 50-200 femtosecond Ti:sapphire laser pulses has been controlled by fine-tuning the higher-order spectral phase coefficients to optimize the electron yield of a self-modulated plasma wake-field accelerator.

1 Introduction

The details of the time-envelope function of ultrashort light pulses ('pulse shape') plays a crucial role in many nonlinear optical phenomena. Two light pulses with the same full-width-half-maximum (FWHM), but having different rise- and fall-time characteristics, or amplitude modulation, can lead to vastly different results in processes where the excitation in the medium depends strongly on the instantaneous light intensity. An example is the yield and energy spectrum of accelerated electrons in a laser wake-field accelerator [1,2], where a driving pulse with a sharp leading edge will result in faster growth of the plasma wake-field, and consequently, higher energy electrons after acceleration. There are several methods of actively controlling the shape of ultrashort light pulses, such as frequency-domain filtering and spectral-to-time domain envelope transfer by using liquid crystal phase modulators [3] and acousto-optic phase and amplitude control [4]. Here we investigate the *intrinsic* pulse shaping behavior of the most widely used optical pulse compressor – the grating pair – with special emphasis on the pulse *skewness.*

The conventional description of the propagation and phase evolution of ultrashort light pulses is based on the Taylor series expansion of the optical phase around the center laser frequency, ω_0. The properties of ultrashort (sub-100 fs duration) optical pulsesstrongly depend on the higher order phase terms beyond the second order; the presence of higher order components (such as 'cubic' third order, or 'quartic' fourth order) modifies the shape of the pulse, resulting in modulation, and may even lead to the appearance of pre- or post-pulses [5–9].

2 Skew of optical pulses

In order to characterize quantitatively the *shape* of the pulse, it is worthwhile to introduce the skewness parameter, S. The simplest, self-explanatory definition of S could be based on the ratio of the 'tail-width-half-max' (TWHM) and the 'head-width-half-max' (HWHM) of the laser intensity, as, for example, S=(TWHM/HWHM)-1. In addition to this heuristic definition of the skewness, a variety of more rigorous analytical expressions involving more features of the envelope function, $I(t)$, can be used. In the particular case of laser-plasma accelerators, the following definition of skew based on second and third order moments [10] proved to be useful: $S = m_3/m_2^{3/2}$, where

$$m_k = \int_{t_a}^{t_b} (t - t_0)^k \, I(t)dt \Big/ \int_{t_a}^{t_b} I(t)dt \tag{1}$$

In practical cases the limits of integration, t_a and t_b, can be set according to the meaningful amplitude level of the retrieved pulse shape just above the noise level of the measurement. Figure 1 shows the main types of skewed pulses: S=0 corresponds to symmetric pulse shape, S<0 describes a pulse with a slowly increasing front part and suddenly dropping tail, while S>0 represents a fast increase and long tail.

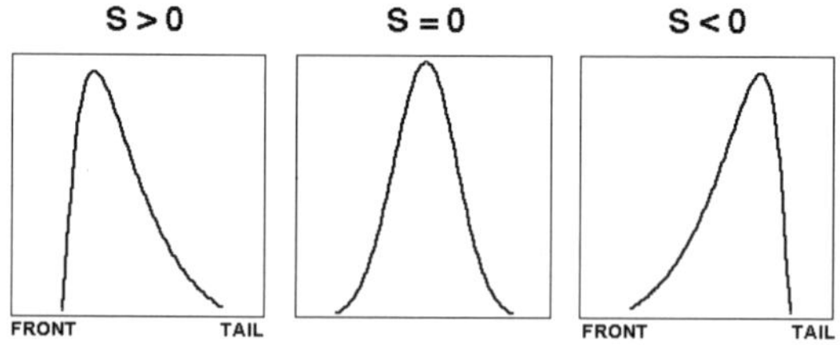

Fig. 1. Skewed pulse shapes characterized by the skewness parameter, S.

In a typical chirped pulse amplification (CPA) system, the final pulse-forming device is the grating pair pulse compressor. In contrast to the most simplistic view, by scanning the grating separation the experimenter changes not only the pulse length and the linear chirp, but also modifies all the higher order terms [7–9], and consequently the pulse shapes. The effects on the pulse shape become negligible at settings far from the shortest compression. On the other hand, when the pulse duration is in the range of a few times the minimum and the compressor angle is set slightly off from the optimum, or when the pulse contains not fully compensated higher-order terms, then a typical scan around the shortest pulse will produce significant changes to the pulse shape. For example, by compressing stretched pulses with a positive bias third-order phase component, the pulse shape in the course of a compressor separation scan initially will be skewed toward the head of the pulse ($S>0$), then flips to become a skewed pulse at the tail ($S<0$), and finally flips back again to the original asymmetric one ($S>0$) as the solid curve shows in Fig. 2a. It must be emphasized, that these changes of the pulse envelope are fundamentally different from the well-known and widely used mapping of the spectral amplitude to the time domain for strongly chirped pulses, where only the dominant, second-order phase coefficient plays any role [3]. Further details of this 'skew-flip' behavior can be examined in the three panels of Fig. 2, which shows the simulated pulse shape dependence for ~20 nm spectral bandwidth during a compressor scan around the shortest pulse for three specific third order bias values. The grating parameters are: groove density = 1480 mm^{-1}, angle of incidence = 50.0 deg, grating separation at the shortest pulse = 323 mm, and λ = 795 nm.

In a real laser system, however, the situation is complicated by the non-negligible presence of even higher order, generally uncompensated, phase terms beyond the third, associated again with the material dispersion in the components of the laser system. The solid curve in Fig. 3 shows the results of a full simulation of the pulse shape evolution in our Ti:sapphire CPA laser. The squares correspond to measured data, averaged for 5 laser shots. The pulses were fully characterized by retrieving the amplitude and phase functions from Polarization-Gate Frequency Resolved Optical Gating (PG-FROG) images [11, 12]. The two inserts in Fig. 3 show example retrievals at the positive chirp side (a) and at the negative chirp side (b), respectively. While both pulses have the same FWHM pulse duration (76 fs), the pulse at the positive chirp side has sharper rising edge than the pulse with negative chirp – in good agreement with the simulation results and the skew definitions above. The modulations at the rising edge of the negatively chirped pulse indicate the presence of unbalanced third and higher order phase terms. These unbalanced terms are eventually responsible for the seemingly general 'chirp/shape' relationship of 'positive chirp = faster rise time and slower fall time; negative chirp = slower rise time and faster fall time' observed in our laser setup. It should be noted, however, that for other experimental parameters (e.g., different mismatch of the stretcher/compressor or different stretcher/amplifier chain designs), balanced compression could manifests itself differently at 'off-shortest' pulse durations (off-optimal compressor settings), leading to different 'chirp/shape' combinations. The above analysis shows the possibility of laser pulse shape control by the simple method of grating alignment and change of separation without the insertion of sophisticated pulse shapers. For example, by decreasing the angle of incidence on the gratings by 0.6 degree, the effect of the added $\phi_3 = +30000\ fs^3$ third order phase component makes possible to generate positively- or negatively-skewed pulses at different positions along the grating scan.

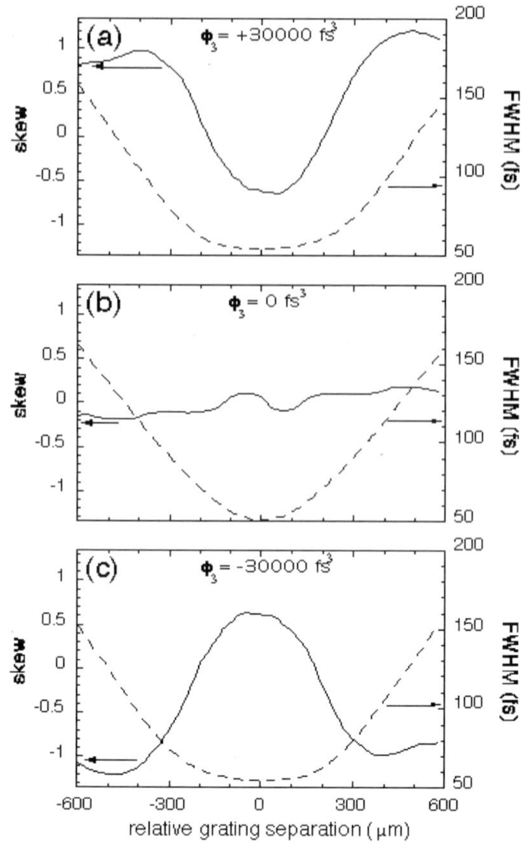

Fig. 2. Conventionally defined pulse width, FWHM (dashed curve) and skew, S, (solid curve) of ultrashort pulses in a CPA system plotted around the best compression in the function of the grating separation for three cases of third order bias phase components. (a): +30,000 fs^3, (b): 0 fs^3, (c): -30,000 fs^3.

3 Laser wake-field acceleration of electrons

To illustrate the practical importance of the skewness of ultrashort laser pulses, we describe a strongly nonlinear experiment: the acceleration of electrons by laser produced plasma wakes [13]. In laser plasma experiments performed in the self-modulated wake-field acceleration regime (Ti:sapphire CPA laser system, ~0.5 J compressed energy in >45 fs pulses, He gas target and ~2-4×10^{19} cm^{-3} electron density excited by ~1.5×10^{19} W/cm^2 peak light intensity), we observed strong asymmetry both in the yield of electrons and in the number of neutrons produced by the accelerated electrons of multiple-

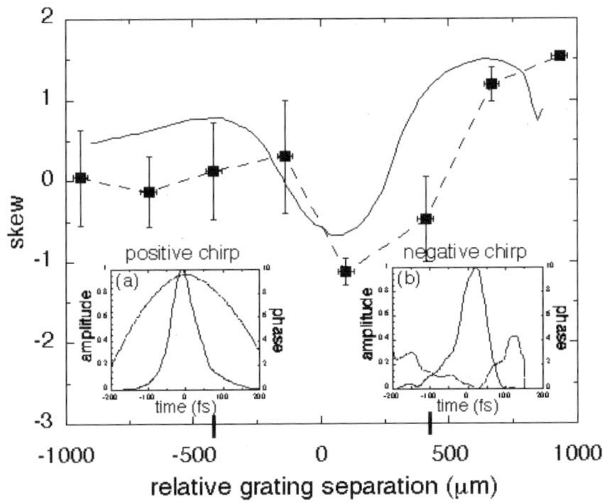

Fig. 3. Simulated (solid curve) and measured (squares) skew of pulses in a CPA system around the best compression plotted in the function of the grating separation. Inserts show retrieved amplitude and phase functions in the time domain at the marked relative grating separations. (a): -420 μm, positive chirp side, (b): +410 μm, negative chirp side

tens of MeV energy as a function of the compressor scans [1, 2]. The total charge of the emitted electrons has been found to depend strongly, and asymmetrically, on the position of the grating separation relative to the optimum pulse compression setting. Analytic modeling and simulation indicate that the effect of the intrinsic frequency chirp is less important and itself can not explain the observed anomaly [14]. On the other hand, taking into account the shape dependence, it is possible to explain the observed asymmetric enhancement; according to Raman instability analysis, faster rise times generate larger plasma wakes underneath the laser pulse envelope, resulting in larger plasma waves and eventually larger electron yield.

In summary, we have shown that appropriate knowledge and control of the pulse shapes in grating-based CPA systems can be used for optimization of a laser wake-field accelerator operating in the self-modulated regime. The interplay of pulse shapes, skewness, and sign of chirp can lead to an enhanced growth of the plasma wake responsible for electron acceleration. In the experiments, pulse shaping is achieved with a conventional grating compressor scan by intentionally offsetting the higher order phase components. This method of higher order phase control is a simple alternative to the recently developed acousto-optic phase modulation devices such as the DAZZLER [4, 15], and can conveniently be used at arbitrarily high power and energy levels limited only by the damage threshold of the compressor gratings.

References

1. Leemans, W.P., Rodgers, D., Catravas, P.E., Geddes, C.G.R., Fubiani, G., Esarey, E., Shadwick, B.A., Donahue, R., Smith, A.: Gamma-neutron activation experiments using laser wakefield accelerators. Phys. Plasmas, **8**, 2510–2516 (2001)
2. Leemans, W.P., Catravas, P., Esarey, E., Geddes, C.G.R., Tóth, C., Trines, R., Schroeder, C.B., Shadwick, B.A., van Tilborg, J., Faure, J.: Electron-yield enhancement in a laser-wakefield accelerator driven by asymmetric laser pulses. Phys. Rev. Lett., **89**, 174802–4805 (2002)
3. Weiner, A.M., Leaird, D.E., Patel, J.S., Wullert, J.R.: Programmable shaping of femtosecond optical pulses by use of 128-element liquid crystal phase modulator. IEEE J. Quantum Electron., **28**, 908–920 (1992)
4. Verluise, F., Laude, V., Cheng, Z., Spielmann, C., Tournois, P.: Amplitude and phase control of ultrashort pulses by use of an acousto-optic programmable dispersive filter: pulse compression and shaping. Opt. Lett., **25**, 575–577 (2000)
5. Fittinghoff, D.N., Walker, B.C., Squier, J.A., Tóth, C., Rose-Petruck, C., Barty, C.P.J.: Dispersion considerations in ultrafast CPA systems. IEEE J. Sel. Top. Quantum Electron., **4**, 430–440 (1998)
6. Backus, S., Durfee III, C.G., Murnane, M.M., Kapteyn, H.C.: High power ultrafast lasers. Rev. Sci. Inst., **69**, 1207–1223 (1998)
7. Sullivan, A., White, W.E.: Phase control for production of high-fidelity optical pulses for chirped-pulse amplification. Opt. Lett., **20**, 192–194 (1995)
8. Squier, J., Barty, C.P.J., Salin, F., Le Blanc, C., Kane, S.: Use of mismatched grating pairs in chirped-pulse amplification systems. Appl. Opt., **37**, 1638–1641 (1998)
9. Tóth, Cs., Fittinghoff, D.N., Walker, B.C., Squier, J.A., Barty, C.P.J.: Pulse characterization techniques for sub-30 femtosecond terawatt lasers. In: Elsaesser, T., Fujimoto, J.G., Wiersma, D.A., Zinth, W. (eds) Ultrafast Phenomena XI, Springer, Berlin (1998) pp.109–111
10. CRC Standard Mathematical Tables, Beyer, W.H. (ed) CRC Press, Inc., Boca Raton, FL, (1981) p. 508
11. Kane D.J., Trebino, R.: Single-shot measurement of the intensity and phase of an arbitrary ultrashort pulse by using frequency-resolved optical gating. Opt. Lett., **18**, 823–825 (1993)
12. Trebino, R.: Frequency-Resolved Optical Gating: The Measurement of Ultrashort Laser Pulses. Kluwer Academic, Boston, (2002), p. 101
13. Esarey, E., Sprangle, P., Krall, J., Ting, A.: Overview of plasma-based accelerator concepts. IEEE Trans. Plasma Sci., **24**, 252–288 (1996)
14. Schroeder, C.B., Esarey, E., Shadwick, B.A, Leemans, W.P.: Raman forward scattering of chirped pulses. Phys. Plasmas, **10**, 285–295 (2003)
15. Ohno, K., Tanabe, T., Kannari, F.: Adaptive pulse shaping of phase and amplitude of an amplified femtosecond pulse laser by direct reference to frequency-resolved optical gating traces. J. Opt. Soc. Am. B., **19**, 2781–2790 (2002)

Part VI

Ultrafast Nonlinear Optics

Progress in Chirped Pulse Optical Parametric Amplifiers

R. Butkus, R. Danielius, A. Dubietis, and A. Piskarskas

Laser Research Center, Vilnius University, Sauletekio Ave. 9, bldg.3, LT-2040, Vilnius , Lithuania algis.piskarskas@ff.vu.lt

Summary. We discuss the main issues of the optical parametric chirped pulse amplification and overview recent progress in the field.

Key words: Nonlinear optics, Optical parametric amplification, Optical parametric chirped pulse amplification, Non-collinear optical parametric amplification, Ultrashort pulses

Introduction

The first experimental demonstrations of optical parametric amplifier (OPA) and optical parametric generator (OPG) have been carried out more than three decades ago [1, 2]. Very fast progress of ultrashort pulse lasers and advent of effective nonlinear crystals in nineties boosted OPA/OPG's on the top of extensive research in numerous scientific laboratories demonstrating unique possibilities of OPA/OPG's to produce pico- and femtosecond pulses broadly tunable from 0.2 up to 20 micrometers (see [3,4] for extensive review).

Optical parametric amplification offers several advantages as compared to amplification in laser amplifiers, including reduced amplified spontaneous emission, low thermal effects, and great wavelength flexibility. Although OPA has to be operated with precise pump and signal timing, the spatial and temporal characteristics of the seed signal are not altered during the amplification process. Broad amplification bandwidth of the OPA and high gain achievable per single pass makes these devices attractive for the optical parametric amplification of chirped pulses (OPCPA).

The OPCPA principle for the first time has been proposed and demonstrated by Dubietis *et al.* [5]. The idea to use energetic nanosecond pump to amplify chirped and stretched pulse and to boost ultrashort pulse power to TW level has been clearly formulated by the authors. Later on, an extensive development of the OPCPA technique towards the generation of ultrahigh peak power pulses has been initiated by Ross *et al.* [6]. Another important work has been done by Wilhelm *et al.* [7], exploiting extremely broad amplification bandwidth of non-collinear optical parametric amplifiers (NOPA's). Five years later this technique provided optical pulses shorter than 4 fs, as demonstrated by Baltuška *et al.* [8]. Eventually, most of the NOPA's share the basic aspects with the OPCPA technique, since they involve chirped pulse amplification and compression. Nevertheless, there are main distinctive features of NOPA technique: i) a white-light-continuum seed and a pump pulses are derived from the same femtosecond laser; ii) output pulse can be broadly tunable or compressed to a few cycle limit. Practical implementation of a NOPA system is much simpler than that of typical OPCPA setup, however its application is mostly limited to pump-probe spectroscopy.

1 Chirped pulse OPA concept and pioneering experiment

Chirped pulse parametric amplification technique has originated from the chirped-pulse-amplification (CPA) concept proposed for the high energy laser amplifiers [9], which nowadays is exploited by any ultrashort pulse medium-to-high power laser system. Chirped pulse OPA concept maintains all the basic requisites, just the laser amplifier is replaced by the OPA (Fig. 1).

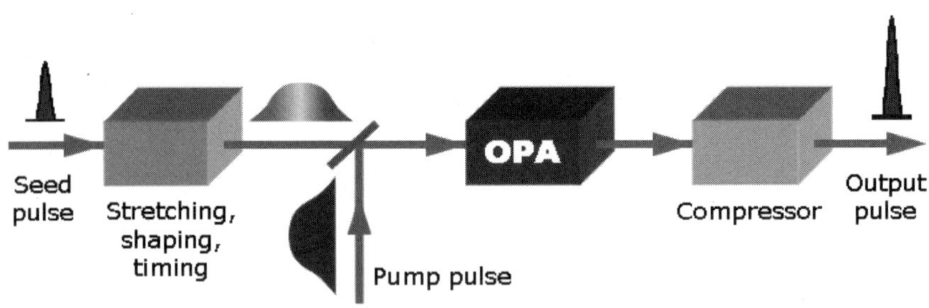

Fig. 1. Basic layout of chirped pulse optical parametric amplification.

The pioneering experiment, which lies on the basis of the OPCPA concept has been demonstrated a decade ago in [5]. Spectral broadening due to self-phase-modulation along with temporal stretching of input 1.7 ps pulses centered at 1055 nm was performed in a single mode fiber, yielding linearly chirped 155 cm^{-1}, 5 ps seed pulses at the input of the OPA. The pump pulse was regeneratively amplified and frequency doubled, and temporally shaped to match the signal pulse width. Seed pulses were amplified in 8 mm long type I phase matching BBO crystal by factor 2×10^4 with no bandwidth reduction and no impact on the chirp character, and compressed down to 70 fs, to the time-bandwidth limit (Fig. 2).

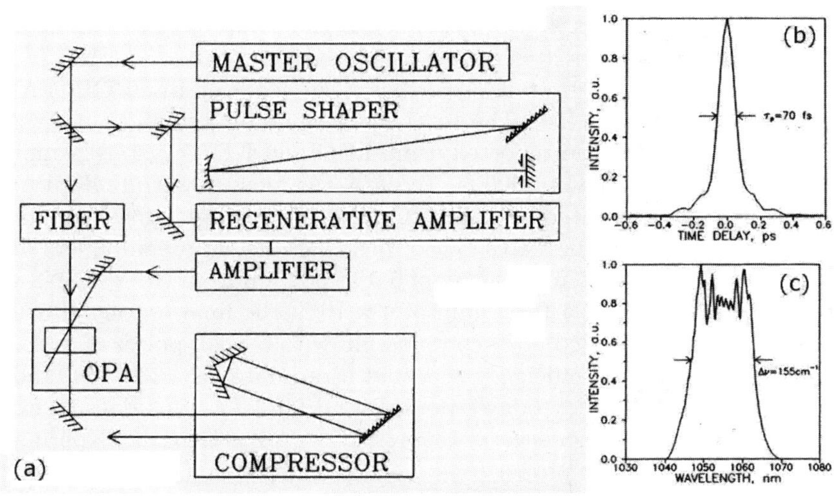

Fig. 2. Schematic (a) and experimental results (b,c) of the first OPCPA demonstration [5].

2 OPCPA: towards extremely high power pulses

With growing interest in the development of ultrahigh power (TW to PW) ultrashort-pulse laser systems, OPCPA has been identified as a feasible amplification technique [6, 10] and became an attractive alternative to existing broadband Ti:sapphire laser systems [11]. In recent years there was a rapid progress in design of OPA based systems towards more stable, more powerful and shorter pulses in TW peak power range [12–15].

There are several peculiarities in high output power OPCPA systems, which distinguish them from low-energy systems described in Section 3. Usually such powerful systems are pumped with Nd:glass or Nd:YAG lasers with pulse durations in the range of nanoseconds or sub-nanoseconds. The seed pulse is mostly taken from a separate properly synchronized femtosecond Ti:sapphire laser. The OPA is operated at the degeneracy or very close to it, several amplification stages can be used. Typically the last amplifier stage employs a crystal, that can be grown in large size to withstand high pump energies. Thus, high total gain and high output pulse energies can be obtained, but usually with a pulse durations not less than a few hundreds of femtoseconds. The repetition rate is rather low and limited by the pump laser.

Table 1 summarizes recent achievements in high energy OPCPA based systems. In 1997, a detailed project was presented, offering several practical designs of OPCPA configurations promising output powers above 10 PW [6]. Since then, no one has achieved anticipated powers, however there has been significant progress. Major challenge is the compression of nanosecond OPCPA output into sub-20 fs range.

More than 1 TW of output power has been achieved in an OPCPA system with two amplifiers [10]. The total energy gain of 10^{10} and output pulse energies of 0.5 J has been reported using LBO and KDP crystals. Amplified pulses were recompressed to 300 fs. To date, the most powerful OPCPA system has been demonstrated by Yang *et al.* [12]. Chirped pulses were amplified from 20 pJ to 900 mJ by two LBO optical parametric preamplifiers and a final KDP OPA with a pump energy of 5 J at 532 nm from Nd:YAG-Nd:glass hybrid amplifiers. With a total gain of 4×10^{10}, the final output of 579 mJ 155 fs pulses was obtained after compression with a peak power of 3.67 TW. The highest conversion efficiency reported up to date is $\sim$ 23% [14] and the highest repetition rate in today's powerful OPCPA systems doesn't exceed 10Hz [11]. There are also several reports, concerning OPCPA experiments, which remained without proper temporal characterization using BBO [16,17] and KD*P crystals [18].

Another approach is based on a novel hybrid chirped-pulse amplification scheme that uses a single pump pulse and combines OPA (two BBO crystals) and laser amplification (10 cm long Ti:sapphire crystal with a four-pass configuration) [13].

The OPCPA approach has been used also for amplification of pulses generated by Er-doped fiber lasers. With periodically poled $LiNbO_3$ crystal, pumped by Alexandrite amplifier 1 mJ pulses were delivered at 10 Hz repetition rate [19]. A repetition rate of 1-kHz was achieved using periodically poled $KTiPO_4$, pumped by all-solid-state laser system [20].

OPA permit the use of multiple pump beams if each of them satisfies the phase-matching condition with a single signal beam [21, 22]. The basic property of the parametric amplification process is that the phase difference between the pump and the signal pulses is transferred to the idler pulse, thus compensating for random differences between these two. The major benefit emerging from the use of the multiple-beam pump is that a number of low-to-medium power pump lasers (even with different wavelengths) can be combined into single signal beam. This appears quite promising for increasing pulse repetition rate and yielding high average output power.

Experimentally, multiple-beam pumped OPA was demonstrated by some of the present authors; the amplified signal energy exceeded that of the single pump, almost 80% conversion efficiency has been reported with no impact on the pulse temporal profile [23]. Another relevant feature of the multiple-beam pumped OPA has been demonstrated recently, showing that multiple pump beams used in a proper geometry could notably extend the amplification bandwidth in the case of the chirped broadband seed [24].

Table 1. Progress in (sub)ns-pulse-pumped chirped pulse OPA.

Reference	Year	λ_p, nm	λ_s, nm	τ_p, ns	τ_s, ns	E_p, J	G	η, %	E_{out}, J (E_{comp}, J)	τ_{out}, fs	P_{out}, TW	Crystal; L_c, mm	p.r.r., Hz
A.Dubietis *et al.*, [5]	**1992**	527	1055	0.008	0.005	0.003	2×10^4	3	10^{-4} (0.65×10^{-5})	70	$\mathbf{9\times 10^{-4}}$*)	BBO ; 8	1.5
I.N.Ross *et al.*, [6]	**1997**	527	1053	0.5	0.5	1490	2×10^{11}	30	400 (224)	21	**11000****)	LBO ; 1.5 LBO ; 6.8 KDP ; 28.6	n.a.
I.N.Ross *et al.*, [10]	**2000**	527	1054	0.6	0.3	2.5	10^{10}	20	0.5 (0.4)	300	**1.3**	LBO ; 20 KDP ; 30	n.a.
X.Yang *et al.*, [12]	**2002**	532	1064	0.5	0.15	5	4×10^{10}	18	0.9 (0.57)	155	**3.67**	LBO ; 15 LBO ; 18 KDP ; 30	n.a.
I.Jovanovic *et al.*, [13]	**2002**	532	1054	8.5	3	0.51	6×10^7	6	0.031 (0.015)	310	**0.05**	BBO ; 15 BBO ; 15 BBO ; 10	10
H.Yoshida *et al.*, [14]	**2003**	532	1053	9.5	3	0.29	10^8	23	0.065 (n.a.)	350	<**0.18**	BBO ; 15 BBO ; 15	6
L.J.Waxer *et al.*, [15]	**2003**	527	1054	1	0.7	0.025	6×10^6	14	0.005 (n.a.)	470	<**0.01**	LBO ; 25 LBO ; 23	5

λ_p - pump wavelength, λ_s - seed wavelength, τ_p - pump pulse duration, τ_s - seeded pulse duration, E_p - pump pulse energy, G - gain, η - pump-to-signal energy conversion efficiency, E_{out} and E_{comp} - output pulse energy before and after compression respectively, τ_{out} - output pulse duration, P_{out} - output power, L_c - crystal length, p.r.r. - pulse repetition rate.

*) a pioneering OPCPA experiment; **) project in prospect.

3 NOPA: towards ultimately short and tunable light pulses

Although collinear OPA's, based upon Type I phase matching crystals exhibit very broad bandwidth at the degeneracy as well as at some other special wavelengths [25,26]. application of non-collinear geometry enables one to create degeneracy-like situation all over the tuning range. This feature of the non-collinear interaction has been considered both theoretically and experimentally by a number of research groups [27–30], and combination of this approach with seeding by femtosecond white light continuum and prism compressor resulted in a versatile device capable of delivering sub-20 fs pulses tunable in the visible that became known as NOPA [7]. Later on, several groups contributed to its further development, mainly in the direction of pulse shortening. The main issue addressed was adjusting the negative GVD of the compressor to compensate for the continuum chirp as well as for the positive GVD of used optics. Minimization of the amount of dispersive material along with application of small apex angle prism compressor and chirped mirrors have lead to generation of sub-8 fs pulses [33,35]. Pre-compression of the seed pulse prior to amplification and compression using prism/grating or prism/chirped mirror combinations produced sub-5 fs pulses [36, 38]. And finally, spatially dispersed pump beam and complex adaptive compressor helped to reduce the pulse duration to < 4 fs [8]. Along with an obvious advantage of broad amplification bandwidth, application of non-collinear geometry creates problems as well. The most important one is the limitation on the applicable pump energy due to small spot size within which the pulses are overlapped in time. This limits the output pulse energy to ~ 10 μJ and even less if the spectral content is to be preserved within entire beam. Another shortcoming is that the use of the idler wave is not as straightforward due to spatial dispersion of the idler beam. At least two solutions for the removal of dispersion and spatial chirp have been suggested so far, employing either diffraction grating [36] or two prism sequence [41].

Due to rapid development of knowledge and technology, NOPA systems became commercially available. A commercial tunable sub-30 fs and up to 10 μJ energy per pulse **NOPA** system is available from Clark-MXR, Inc. [42]. **Topas-white**, developed by Light Conversion Ltd. [43], offers even shorter and the most powerful ultrashort broadly tunable pulses, to date. The latter device has some distinctive features. The pump pulses of the two-stage amplifier are tilted to match the seed pulse front, which allows one to significantly increase the output. The seed pulse is shaped by a grating compressor with fixed phase mask in such a way that signal pulse compression becomes feasible using positive GVD. Besides compactness and low loss of the compressor this also provides simple means for pulse pre-compression, which in turn allows one to control the bandwidth easily. Fig. 3 depicts some of the performance data of **Topas-white**. Table 2 represents the development of low energy OPCPA based systems (NOPA) during the last decade.

Table 2. Progress in ultrashort-pulse-pumped chirped pulse OPA.

Reference	Year	λ_p, nm	τ_p, fs	E_p, μJ	η, %	E_{out}, μJ (E_{comp}, μJ)	τ_{out}, fs	P_{out}, GW	Crystal; L_c, mm	Tunability range, nm
A.Dubietis *et al.*, [5]	**1992**	527	8000	3000	3	100 (65)	**70**	0.9	BBO ; 8	n.t.
V.Yakovlev *et al.*, [32]	**1994**	800	100	250	20	50 (n.a.)	**30**	< 1.6	BBO ; 7	1100-1300
T.Wilhelm *et al.*, [7]	**1997**	405	65	40	25	11 (n.a.)	**16**	< 0.7	BBO ; 2	475-750*)
G.Cerullo *et al.*, [33]	**1997**	390	180	7	15	1 (n.a.)	**11**	< 0.1	BBO ; 1	545-650
A.Shirakawa *et al.*, [34]	**1998**	395	150	100	3	3 (n.a.)	**14**	< 0.2	BBO ; 1	550-690
G.Cerullo *et al.*, [35]	**1998**	390	180	10	20	2 (1.8)	**7.2**	0.25	BBO ; 1	n.t.
A.Shirakawa *et al.*, [36]	**1998**	395	150	130	4	5 (0.5)	**6.1**	0.08	BBO ; 1	550-700*)
G.Cerullo *et al.*, [37]	**1999**	390	180	10	20	2 (n.a.)	**7.5**	< 0.26	BBO ; 1	n.t.
A.Shirakawa *et al.*, [38]	**1999**	395	150	100	6	6 (5)	**4.7**	1	BBO ; 1	n.t.
M.R.Armstrong *et al.*, [39]	**2001**	387	120	75	7	5 (2)	**7**	0.29	BBO ; 1	n.t.
M.Zavelani-Rossi *et al.*, [40]	**2001**	390	180	10	20	2 (n.a.)	**5.7**	< 0.35	BBO ; 1	n.t.
A.Baltuška *et al.*, [8]	**2002**	395	> 150	50	10	5 (0.5)	**4**	0.125	BBO ; 1	n.t.
Clark-MXR, Inc., [42]	**1999**	387	150	200	5	n.a. (10)	**10**	1	n.a.	470-650
Light Conversion Ltd., [43]	**2003**	400	150	500	10	n.a. (50)	**7.9**	6	BBO ; n.a.	490-750

λ_p - pump wavelength, τ_p - pump pulse duration, E_p - pump pulse energy, η - pump-to-signal energy conversion efficiency, E_{out} and E_{comp} - output pulse energy before and after compression respectively, τ_{out} - output pulse duration, P_{out} - output power, L_c - crystal length, n.t. - not tunable.

*) denotes a tunable idler wave also.

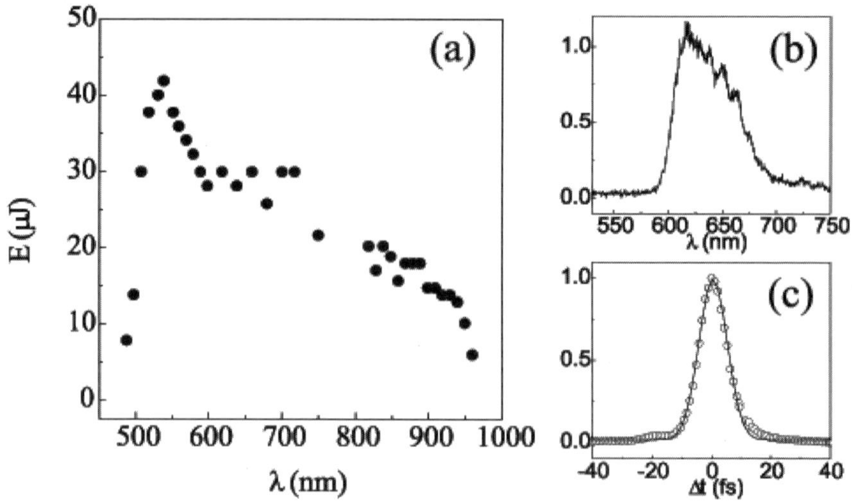

Fig. 3. Operational characteristics of TOPAS-white: (a) tuning curve, (b) spectrum and (c) autocorrelation trace of 7.9-fs pulse.

4 Conclusions

In this review we briefly discussed the progress in the field of chirped pulse parametric amplification, since both the idea and pioneering experiment have been suggested and demonstrated by some of the present authors a decade ago. Although we distinguish between the two operating modes of modern chirped pulse parametric amplifiers, OPCPA and NOPA, we reveal that both represent the same technique and share a common concept. Despite some disadvantages (lack of pump energy accumulation, limited aperture of the nonlinear crystals, tough requirements of pump and signal synchronization and losses introduced by the idler wave), chirped pulse OPA is very promising due to extremely broad gain bandwidth, high gain per single pass, energy combining of uncorrelated and incoherent pump beams, phase conjugated output signal and idler pulses, high output beam quality, high contrast ratio and low thermal distortions. The most outstanding experimental achievements reported up-to-date are 155 fs, $\sim$ 600 mJ pulses for the OPCPA, and 4 fs, 0.5 μJ pulses for the NOPA, respectively. We foresee that in the nearest future these results will be improved, bringing the OPCPA based systems on the top of the list of high-peak-power and ultimately short pulse lasers for applications in various fields of research, for instance, attosecond science [44].

References

1. S.A.Akhmanov, A.I.Kovrigin, A.S.Piskarskas, V.V.Fadeev, R.V.Khokhlov: JETP Lett. **2**, 191 (1965).
2. K.P.Burneika, M.V.Ignatavichus, V.I.Kabelka, A.S.Piskarskas, A.Yu.Stabinis: JETP Lett. **16**, 257 (1972).
3. R.Danielius, A. Piskarskas, A. Stabinis, G. P. Banfi, P. Di Trapani, R. Righini: JOSA B **10**, 2222 (1993); see also the "Feature Issue on Optical Parametric Oscillation and Amplification" (R. Byer and A. Piskarskas, eds.): JOSA B **9**, 1656-1791 (1993) and **10**, 2148-2243 (1993).
4. G.Cerullo, S.De Silvestri: Rev.Sci.Instr. **74**, 1 (2003).
5. A. Dubietis, G. Jonušauskas, A. Piskarskas: Optics Commun. **88**, 437 (1992).
6. I.N.Ross, P.Matousek, M.Towrie, A.J.Langley, J.L.Collier: Opt.Commun., **144**, 125 (1997).
7. T.Wilhelm, J.Piel, E.Riedle: Opt.Lett. **22**, 1494 (1997).
8. A.Baltuška, T.Fuji , T.Kobayashi: Opt.Lett. **27**, 306 (2002).
9. D.Strickland, G.Mourou: Opt.Commun. **56**, 219 (1985).
10. I.N.Ross, J.L.Collier, P.Matousek, C.N.Danson, D.Neely, R.M.Allot, D.A.Pepler, C.Hernandez-Gomez, K.Osvay: Appl.Opt. **39**, 2422 (2000).
11. I.Jovanovic, B.J.Comaskey, C.A.Ebbers, R.A.Bonner, D.M.Pennington, E.C.Morse: Appl.Opt. **41**, 2923 (2002).
12. X.Yang, Z.Xu, Y.Leng, H.Lu, L.Lin, Z.Zhang, R.Li, W.Zhang, D.Yin, B.Tang: Opt.Lett. **27**, 1135 (2002).
13. I.Jovanovic, C.A.Ebbers, C.P.J.Barty: Opt.Lett. **27**, 1622 (2002).
14. H.Yoshida, E.Ishii, R.Kodama, H.Fujita, Y.Kitagawa, Y.Izawa, T.Yamanaka: Opt.Lett. **28**, 257 (2003).
15. L.J.Waxer, V.Bagnoud, I.A.Begishev, M.J.Guardalben, J.Puth, and J.D.Zuegel: Opt.Lett. **28**, 1245 (2003).
16. X.Yang, Z.Xu, Z.Zhang, Y.Leng, J.Peng, J.Wang, S.Jin, W.Zhang, R.Li: Appl.Phys.B **73**, 219 (2001).
17. S.K.Zhang, M.Fujita, H.Yoshida, R.Kodama, M.Yamanaka, Y.Izawa, C.Yamanaka: Jpn.J.Appl.Phys. **40**, 3188 (2001).
18. G.Freidman, N.Andreev, V.Bespalov, V.Bredikhin, V.Ginzburg, E.Katin, E.Khazanov, A.Korytin, V.Lozhkarev, O.Palashov, A.Poteomkin, A.Sergeev, I.Yakovlev: Proc. SPIE, 4972A-13, (2003).
19. A.Galvanauskas, A.Hariharan, D,Harter, M.A.Arbore, M.M.Fejer: Opt.Lett. **23**, 210 (1998).
20. F.Rotermund, V.Petrov, F.Noack, V.Pasiskevicius, J.Hellstrom, F.Laurell, H.Hundertmark, P.Adel, C.Fallnich: Electron.Lett. **38**, 561 (2002).
21. A.Marcinkevičius, A.Piskarskas, V.Smilgevičius, A.Stabinis: Opt.Commun. **158**, 101 (1998).
22. A.Piskarskas, V.Smilgevičius, A.Stabinis, V.Vaičaitis: J.Opt.Soc.Am.B **16**, 1566(1999).
23. A.Dubietis, R.Danielius, G.Tamošauskas, A.Piskarskas: J.Opt.Soc.Am.B **15**, 1135(1998).
24. E.Žeromskis, A.Dubietis, G.Tamošauskas, A.Piskarskas: Opt.Commun. **203**, 435 (2002).
25. A.J.Campillo, R.C.Hyer, S.Shapiro, Opt. Lett. **4**, 357 (1979).

26. B.Bareika, A.Birmontas, A.Piskarskas, V.Sirutkaitis, A. Stabinis: Sov.J.Quantum Electron. **12**, 1654 (1982).
27. V.Krylov, A.Kalintsev, A.Rebane, D. Erni, U.P.Wild: Opt.Lett. **20**, 151 (1995).
28. P.Di Trapani, A.Andreoni, G.P.Banfi, C.Solcia, R.Danielius, A.Piskarskas, P.Foggi, M.Monguzzi, C.Sozzi: Phys.Rev.A **51**, 3164 (1995).
29. P.Di Trapani, A.Andreoni, C.Solcia, R.Danielius, A.Dubietis, A.Piskarskas: J.Opt.Soc.Am.B **12**, 2237 (1995).
30. G.M.Gale, M.Cavaliari, T.J.Driscoll, F.Hache: Opt.Lett. **20**, 1562 (1995).
31. R.Danielius, A.Piskarskas, P.Di Trapani, A.Andreoni, C.Solcia, P.Foggi: Opt.Lett. **21**, 973(1996).
32. V.V.Yakovlev, B.Kohler, K.R.Wilson: Opt.Lett. **19**, 2000 (1994).
33. G.Cerullo, M.Nisoli, S.De Silvestri: Appl.Phys.Lett. **71**, 3616 (1997).
34. A.Shirakawa, T.Kobayashi: Appl.Phys.Lett. **72**, 147 (1998).
35. G.Cerullo, M.Nisoli, S.Stagira, S.De Silvestri: Opt.Lett. **23**, 1283 (1998).
36. A.Shirakawa, I.Sakane, T.Kobayashi: Opt.Lett. **23**, 1292 (1998).
37. G.Cerullo, M.Nisoli, S.Stagira, S.De Silvestri, G.Tempea, F.Krausz, K.Ferencz: Opt.Lett. **24**, 1529 (1999).
38. A.Shirakawa, I.Sakane, M.Takasaka, T.Kobayashi: Appl.Phys.Lett. **74**, 2268 (1999).
39. M.R.Armstrong, P.Plachta, E.A.Ponomarev, R.J.D.Miller: Opt.Lett. **26**, 1152 (2001).
40. M.Zavelani-Rossi, G.Cerullo, S.De Silvestri, L.Gallmann, N.Matuschek, G.Steinmeyer, U.Keller, G.Angelow, V.Scheuer, T.Tschudi: Opt.Lett. **26**, 1155 (2001).
41. H.S.Tan, E.Schreiber, W.S.Warren: Opt.Lett. **27**, 439 (2002).
42. http://www.cmxr.com/Scientific/NOPA.htm
43. http://www.lightcon.com/lc/scientific/topas-white.htm
44. A.Baltuška, Th.Udem, M.Ulberacker, M.Hentschel, E.Goullelmakis, Ch.Gohle, R.Holzwarth, V.Yakovlev, A.Scrinzi, T.W.Hänsch, F.Krausz: Nature **421**, 611 (2003).

Optical Parametric Chirped-Pulse Amplification in Periodically-Poled $KTiOPO_4$ at 1053 nm

Igor Jovanovic, Jason R. Schmidt, and Christoper A. Ebbers

Lawrence Livermore National Laboratory, Mail Code L-490, 7000 East Avenue, Livermore, California 94550, USA jovanovic1@llnl.gov

1 Introduction

Generation of high peak power [1] and high focal intensity from Nd:glass short-pulse lasers is receiving much attention, particularly in large laser facilities where fast ignition [2] may be used for high-gain inertial confinement fusion. Optical parametric chirped pulse amplification (OPCPA) [3] has been recently studied as an alternative techology for the front end of a short-pulse Nd:glass laser. This is the result of superior characteristics of OPCPA when it is used for broadband amplification at wavelengths near 1 μm. It offers increased bandwidth over amplification in Nd:glass, while its gain is several orders of magnitude higher than that of Ti:sapphire at 1 μm, eliminating the need for regenerative amplification. Additionally, OPCPA exhibits high prepulse contrast as a result of cavityless amplification, which is of great importance for high-fidelity target experiments. Other favorable characteristics of OPCPA include high peak power scalability through large-aperture potassium dihydrogen phosphate (KDP) crystals [4] and potential high average power scalability due to negligible thermal deposition.

A double-crystal beta-barium borate (BBO) high-gain preamplifier for front end of a Nd:glass high peak power laser has been previously demonstrated [5], producing broad bandwidth ~1-mJ pulses when pumped by a commercial Q-switched pump laser. The low conversion efficiency (~1%) and beam ellipticity obtained in this configuration were primarily a consequence of relatively large birefringent walk-off compared to the beam diameter in crystal. Walk-off sets the minimum peak power limit for efficient OPCPA in bulk nonlinear crystals, resulting in poor prospects for efficient high average power OPCPA with low pump pulse energy.

Noncritical phase matching eliminates the limitation imposed by walk-off, but in bulk nonlinear crystals such as lithium triborate (LBO) operating at elevated temperatures, it is limited to specific wavelengths. OPCPA used in conjunction with quasi-phase-matching (QPM) is not limited by birefringent walk-off and utilizes the large nonlinear coupling coefficients of materials such as lithium niobate and potassium-titanyl phosphate (KTP). Nondegenerate OPCPA in bulk crystals exhibits broad bandwidth only at certain combinations of pump wavelength, signal wavelength, and noncollinear angle [7] in a particular nonlinear crystal, as determined by the crystal dispersion. In contrast, QPM OPCPA offers the possibility of achieving broad bandwidth by virtue of the poling design, such as chirped poling [8], fanned poling [9], and aperiodic poling. [10] Nondegenerate OPCPA in periodically-poled KTP (PPKTP) has been previously demonstrated at a wavelength of 1.573 μm, pumped by a Nd:YAG laser. [11] Here we present, to our knowledge, the first demonstration of broadband, efficient, nearly degenerate OPCPA in PPKTP at a wavelength of 1053 nm, with a potential to replace bulk-crystal OPCPA at millijoule-level energies.

2 Experimental Setup

Our experimental setup is shown in Fig. 1. Our seed pulse at a center wavelength of 1053 nm with a spectral bandwidth of 7 nm FWHM is generated by a Time-Bandwidth-Product GLX-200 Nd:glass oscillator. A single pulse is selected from the oscillator pulse train and injected into the pulse stretcher. The resulting 800-pJ, 1.2-ns seed pulse is resized to 1-mm diameter, with a gaussian mode. The 10-Hz pump pulse is provided by a commercial, Q-switched, frequency-doubled pump laser (Continuum Powerlite Plus) operating in a single longitudinal mode. Pumping using single longitudinal mode pump is necessary to avoid temporal (spectral) modulation of the amplified signal.The pump pulses are 6 ns FWHM long and exhibit a spatio-temporal buildup characteristic for an unstable resonator. The supergaussian pump beam is resized to 0.6-mm diameter and imaged onto the PPKTP optical parametric amplifier crystal.

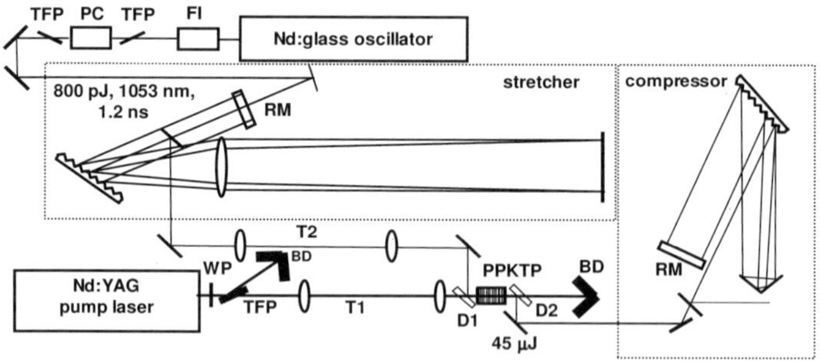

Fig. 1. Experimental setup. λ/2-waveplate, TFP-thin film polarizer, D - dichroic mirror, PC-Pockels cell, FI-Faraday isolator, RM-roof mirror, T-telescope, BD-beam dump.

Our optical parametric amplifier is a 5x1x15 mm^3-PPKTP crystal, used in ZZZ-coupling configuration with an effective nonlinear coefficient d_{eff}=8.7

pm/V and a QPM period of 9.04 μm. The temperature of the crystal is controlled using a crystal oven, with a phase matching temperature of 36.0 °C. Potential crystal degradation may be alleviated by running at higher temperatures, but we chose 36 °C for convenience. Seed and pump beams are coupled into the PPKTP crystal using a dichroic beamsplitter, and the seed pulse overfills the pump pulse, so that ~50% of the seed energy is contained in the area defined by the pump pulse.

3 Results

Up to 700 μJ of 532-nm pump energy was incident on the PPKTP crystal, with a peak intensity of ~30 MW/cm^2. Our selected pump intensity was such that the generated parametric fluorescence was < 1% of the amplified signal. We measured a maximum output signal energy of 45 μJ, for a conversion efficiency of 13% to signal and idler, or 65% of the theoretically maximum conversion efficiency based on the spatiotemporal overlap of seed and pump pulse (the seed and pump pulse have an estimated spatio-temporal overlap of 20%).

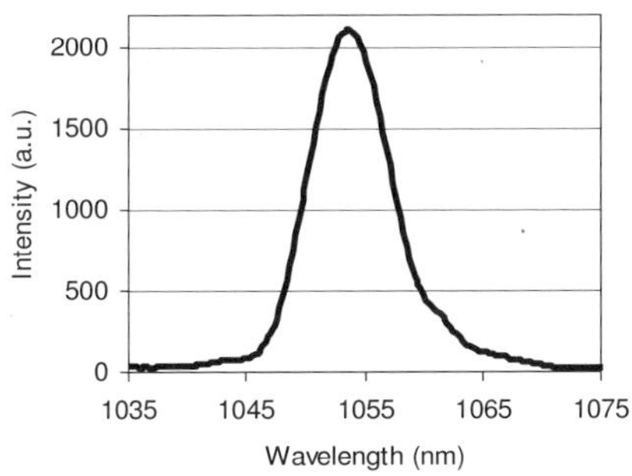

Fig. 2. Amplified signal spectrum. Spectral bandwidth is 8 nm FWHM.

The amplified spectrum (Fig. 2) exhibits spectral broadening to 8 nm FWHM as a result of operation in strong pump depletion regime. The calculated spectral bandwidth of 15-mm long PPKTP is 50 nm, with a potential of supporting sub-20-fs pulses. Spectral broadening has been previously observed in OPCPA [5] and is the result of nonuniform conversion over the signal spectrum.

We measured the pulse-to-pulse stability of our system over 3000 shots to be 2.7% rms (Fig. 3(a)). Several ~20% variations in output energy are the result of large temporal drift of the pump laser of ~3 ns. Stability of the system is greatly enhanced by operation in pump-depletion regime. The output signal beam profile (Fig. 3(b)) is circular and gaussian-like, which is consistent with previous experimental results. [12]

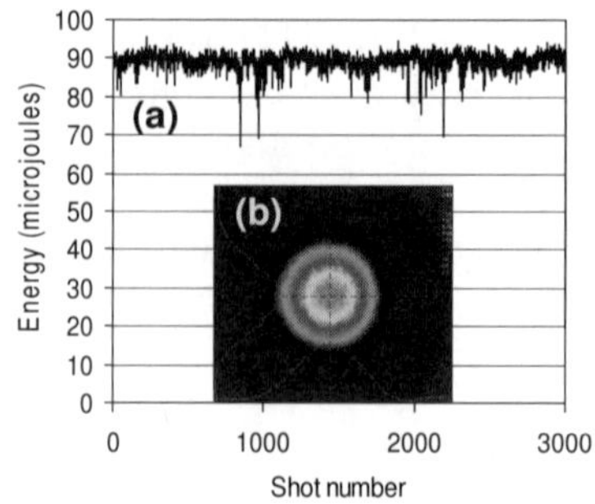

Fig. 3. (a) Shot-to-shot stability of OPCPA in PPKTP (several large fluctuations correlate with the temporal drift of the pump laser; (b) transverse intensity profile of the amplified signal.

We measured the angular sensitivity of OPCPA in PPKTP (Fig. 4) and determined it to be $\pm 2\,^{\circ}$, which is an improvement over type-I OPA in BBO of over 100. Large angular acceptance opens up a possibility of pumping with low beam quality or poorly collimated beams.

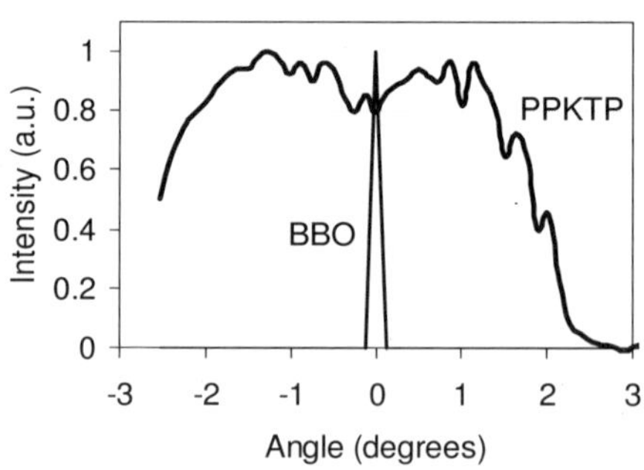

Fig. 4. Comparison of measured angular acceptance of PPKTP with the angular acceptance of BBO. Crystal length is 15 mm.

We compressed the output signal pulses using a gold-grating pulse compressor with 50% efficiency. The autocorrelation trace is shown in Fig. 5. Deconvolution of the autocorrelation trace using our measured spectrum yields the pulse width of 390 fs, which is nearly twice the transform limit of the signal spectrum. Imperfect pulse compression is the result of uncompensated phase between the compressor and the expander. In our configuration the expander includes a refractive telescope.

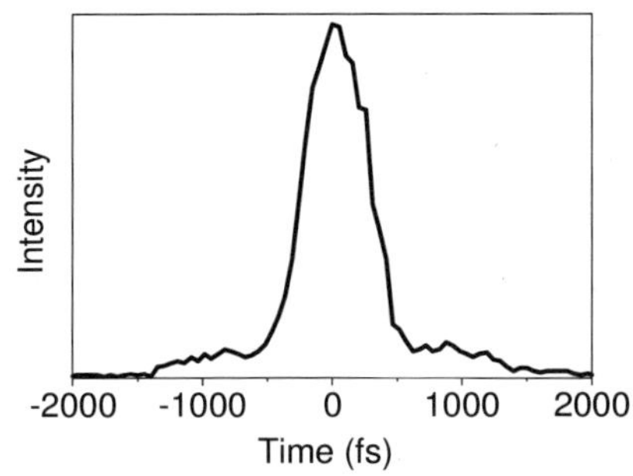

Fig. 5. Intensity autocorrelation of the recompressed amplified pulse. Deconvolved pulse width is 390 fs FWHM using the measured amplified spectrum.

4 Conclusion

In conclusion, we have demonstrated, to our knowledge, the first OPCPA high-gain preamplifier at 1053 nm in PPKTP. OPCPA in one PPKTP crystal produced a gain of 1.1×10^5, with a conversion efficiency of 13% to signal and idler, or 65% of the theoretical maximum of this configuration. The obtained beam profile is closest to the diffraction limit of all OPCPA outputs reported to date. We have measured the angular acceptance bandwidth of our crystal and determined it to be two orders of magnitude broader than the acceptance bandwidth of a bulk crystal such as BBO. We have not observed any grey-tracking degradation of the PPKTP crystal during ~24 hours of operation, but longer lifetime experiments are in progress.

Several improvements are suggested to increase the output energy of the PPKTP preamplifier. The use of a broad bandwidth (16 nm) oscillator should allow greater stretching ratio. A shorter PPKTP crystal would allow us to operate at an intensity of 100 MW/cm^2 without significant parametric fluorescence, increasing the output pulse energy by a factor of ~3. This intensity level has been previously reported as the limit for safe operation without crystal damage or degradation. Finally, periodic poling of KTP with larger crystal apertures (~3 mm) would allow scaling of this device to millijoule energies. Beam quality, spectral characteristics, and simplicity make OPCPA in PPKTP a desirable choice as a part of the front end of short-pulse Nd:glass laser. High conversion efficiency opens up a possibility of use of OPCPA in PPKTP for scaling of short pulses to high average power.

This work was performed under the auspices of the U.S. Department of Energy by Lawrence Livermore National Laboratory under Contract No. W-7405-Eng-48.

References

1. Perry, M.D., Pennington, D., Stuart, B.C., Tietbohl, G., Britten, J.A., Brown, C., Herman, S., Golick, B., Kartz, M., Miller, J., Powell, H.T., Vergino, M., Yanovsky, V.: Petawatt Laser Pulses. Opt. Lett. **24**, 160–162 (1999)
2. Tabak, M., Hammer, J., Glinsky, M.E., Kruer, W.L.,Wilks, S.C., Woodworth, J., Campbell, E.M., Perry, M.D., Mason, R.J.: Ignition and high gain with ultrapowerful lasers. Phys. Plasmas **1**, 1626–1634 (1994)
3. Dubietis, A., Jonasauskas, G., Piskarskas, A.: Powerful femtosecond pulse generation by chirped and stretched parametric amplification in BBO crystal. Opt. Commun. **88**, 437–440 (1992)
4. Ross, I.N., Matousek, P., Towrie, M., Langley, A.J., Collier, J.L.: The prospects for ultrashort pulse duration and ultrahigh intensity using optical parametric chirped pulse amplifiers. Opt. Commun. **144**, 125–133 (1997)
5. Jovanovic, I., Comaskey, B.J., Ebbers, C.A., Bonner, R.A., Pennington, D.M., Morse, E.C.: Replacing Ti:sapphire regenerative amplifiers with optical parametric chirped pulse amplifiers. Appl. Opt. **41**, 2923–2929 (2002)
6. Galvanauskas, A., Hariharan, A., Harter, D., Arbore, M.A., Fejer, M.M.: High-energy femtosecond pulse amplification in quasi-phase-matched parametric amplifier. Opt. Lett. **23**, 210–212 (1998)
7. Jovanovic, I., Ebbers, C.A., Stuart, B.C., Hermann, M.R., Morse, E.C.: Nondegenerate optical parametric chirped pulse amplification. In: Technical Digest, Summaries of papers presented at the Conference on Lasers and Electro-Optics **1**, 387–388, Opt. Soc. America, Washington, DC, (2002)
8. Arbore, M.A., Galvanauskas, A., Harter, D., Chou, M.H., Fejer, M.M.: Engineerable compression of ultrashort pulses by use of second-harmonic generation in chirped-period-poled lithium niobate. Opt. Lett. **22**, 1341–1343 (1997)
9. Powers, P.A., Kulp, T.J., Bisson, S.E.: Continuus tuning of a continuus-wave periodically poled lithium niobate optical parametric oscillator by use of a fan-out grating design. Opt. Lett. **23**, 159–161 (1998)
10. Arbore, M.A., Marco, O., Fejer, M.M.: Pulse compression during second-harmonic generation in aperiodic quasi-phase-matching gratings. Opt. Lett. **22**, 865–867 (1997)
11. Rotermund, F., Petrov, V., Noack, F., Pasiskevicius, V., Hellström, J., Laurell, F., Hundertmark, H., Adel, P., Fallnich, C.: Compact all-diode-pumped femtosecond laser source based on chirped pulse optical parametric amplification in periodically poled $KTiOPO_4$. Electron. Lett. **38**, 561–562 (2002)
12. Galvanauskas, A., Hariharan, A., Raksi, F., Wong, K.K., Harter, D., Imeshev, G., Fejer, M.M.: Generation of diffraction-limited femtosecond beams using spatially- multimode nanosecond pump sources in parametric chirped pulse amplification systems. Conference on Lasers and Electro-Optics, Technical Digest, TOPS **39**, Opt. Soc. America, San Francisco, CA (2000)

Nonlinear Phase-Locking of Parametric Optical Processes: the Sliding Resonance Effect

J.T. Mendonça[1] and H. Crespo[2]

[1] GoLP/Instituto Superior Técnico, Av. Rovisco Pais 1
1049-001 Lisboa, Portugal titomend@alfa.ist.utl.pt
[2] CLOQ/Dep. Física, Fac. Ciências, Univ. Porto, R. do Campo Alegre 687
4169-007 Porto, Portugal hcrespo@fc.up.pt

Summary. We describe a new basic type of nonlinear interaction between optical pulses which considers parametric processes occurring in the presence of cross-phase modulation. The resulting sliding resonance effect strongly influences the phase-matching conditions, giving rise to a nonlinear phase-locking mechanism. The explicit example of the resulting modified parametric amplification is considered in detail using both analytical and numerical approaches. Preliminary experimental results which evidence the extension of this effect to four-wave mixing processes will also be presented.

1 Introduction

There has been a growing interest in the study of new phase-matching mechanisms based on propagation effects as a means of drastically increasing the efficiency of specific nonlinear optical processes. It has been shown theoretically that optical ionisation caused by an ultrashort laser pulse propagating in a gas can result in self-phase matching of high order nonlinear processes such as high harmonic generation [1,2]. Geometrical phase-matching of two-colour pulses propagating in glass resulted in the efficient generation of high-order cascaded four-wave mixing processes up to the eleventh order [3]. Self-induced phase-matching of four-wave mixing processes due to a combination of self- and cross-phase modulation induced by a pump laser pulse has been experimentally observed in optical fibres [4].

Here we describe the possibility of occurrence of parametric processes where the necessary frequency and phase-matching conditions are provided by phase modulation. We will show that this situation can occur for short laser pulses propagating in a nonlinear medium.

Consider three optical beams, with central frequencies ω_0, ω_1 and ω_2, propagating in a nonlinear medium such that, initially, the frequency and/or phase-matching conditions are not satisfied for any wave-mixing process. They behave quite independently from each other, unless their short durations and high intensities allow for self and cross-modulation to take place. As a result, their frequency spectrum will change along propagation in such a way that, at some given position and time $t = \tau$ they eventually become resonant and phase-matched. If we consider the simple three-wave mixing process where the pulse ω_0 originates pulses ω_1 and ω_2, we can write:

$$\begin{aligned} \Delta\omega(\tau) &= \omega_0(\tau) - \omega_1(\tau) - \omega_2(\tau) = 0 \\ \Delta\boldsymbol{k}(\tau) &= \boldsymbol{k}_0(\tau) - \boldsymbol{k}_1(\tau) - \boldsymbol{k}_2(\tau) = 0 \end{aligned} \tag{1}$$

We will determine the time τ (for given initial values of the central frequencies $\omega_j(0)$ where $\Delta\omega(0) \neq 0$ and/or $\Delta\boldsymbol{k}(\tau) \neq 0$) and also the growth rates and the saturation levels of the resulting nonlinear interaction. In order to illustrate the concept of the resulting sliding resonance process, the particular case of parametric amplification will be considered in detail.

2 Basic amplitude equations

Let us start with the propagation equation for the transverse electric field $\boldsymbol{E}$ in a dielectric medium. For simplicity, we assume one-dimensional propagation along some direction Oz in an isotropic medium. These assumptions can easily be relaxed. We can then write:

$$\left(\frac{\partial^2}{\partial z^2} - \frac{1}{c^2}\frac{\partial^2}{\partial t^2} \right) E = \mu_0 \frac{\partial^2}{\partial t^2}(P_L + P_{NL}) \tag{2}$$

The linear polarization P_L characterizes the linear response of the medium, according to:

$$P_L(z,t) = \epsilon_0 \int_0^\infty \chi^{(1)}(\tau) E(z, t-\tau) d\tau \tag{3}$$

where $\chi^{(1)}$ is the linear response function, and the nonlinear polarization P_{NL} is, neglecting dispersion corrections, determined by:

$$P_{NL}(z,t) = \epsilon_0 \left[\chi^{(2)} + \chi^{(3)} E(z,t) \right] E(z,t) E(z,t) \tag{4}$$

Here, $\chi^{(2)}$ and $\chi^{(3)}$ are the second and third order nonlinear susceptibilities. We assume that the total electric field can be described as a superposition of three distinct optical pulses, $E(z,t) = \sum_j E_j(z,t)$, for $j = (0,1,2)$, and that each of the pulses can be described by:

$$E_j(z,t) = A_j(z,t) \exp(ik_j z - i\omega_j t) \tag{5}$$

where $A_j(z,t)$ are varying amplitudes, and k_j and ω_j are the wavenumbers and frequencies. We assume that $A_j(z,t)$ are slowly varying in both space and time, as compared with the space and time scales $1/k_j$ and $1/\omega_j$. For a weakly dispersive medium, the linear polarization associated with each of these modes can be determined by:

$$P_{Lj}(z,t) \simeq \epsilon_0 \left[\chi^{(1)}(\omega_j) A(z,t) + i \frac{\partial \chi^{(1)}(\omega_j)}{\partial \omega} \frac{\partial A(z,t)}{\partial t} \right] \exp(ik_j z - i\omega_j t) \quad (6)$$

Replacing this in the wave equation (2), and noting that the linear dispersion relation is satisfied:

$$k_j^2 c^2 = \omega_j^2 [1 + \chi^{(1)}(\omega_j)] \quad (7)$$

we can write the amplitude equation for each of the three modes in the form:

$$\left(\frac{\partial}{\partial z} + \frac{1}{v_j} \frac{\partial}{\partial t} \right) A_j = i \frac{\mu_0 \omega_j^2}{2k_j} P_{NL}(\omega_j) \exp(-ik_j z + i\omega_j t) \quad (8)$$

In this equation, the following expression for the group velocity v_j was used:

$$\frac{1}{v_j} = \frac{\omega_j}{k_j c^2} + \frac{\omega_j^2}{2k_j c^2} \frac{\partial \chi^{(1)}(\omega_j)}{\partial \omega} \quad (9)$$

The quantity $P_{NL}(\omega_j)$ appearing in the amplitude equation (8) contains the nonlinear polarization terms oscillating with frequencies close to ω_j, or in other words, it is the dominant nonlinear polarization contribution to the mode j. Let us first consider the mode $j = 0$. From equation (4), and adding the complex conjugate to equation (5), we can write this nonlinear polarization as:

$$P_{NL}(\omega_0) = \epsilon_0 \Big\{ \chi^{(2)}(\omega_1, \omega_2) A_1 A_2 \exp[i(k_1 + k_2)z - i(\omega_1 + \omega_2)t] + \left[\sum_{n=0,1,2} \chi^{(3)}(\omega_j, -\omega_j, \omega_0) |A_n|^2 \right] \exp(ik_0 z - i\omega_0 t) \Big\} \quad (10)$$

Replacing in equation (8), and assuming that the nonlinear susceptibilities are weakly dependent on the frequency, we obtain, for the mode $j = 0$:

$$\left(\frac{\partial}{\partial z} + \frac{1}{v_0} \frac{\partial}{\partial t} \right) A_0 = i \frac{\omega_0^2}{2k_0 c^2} \left[\chi^{(2)} A_1 A_2 \exp(i\Delta k z - i\Delta\omega t) + \sum_{n=0,1,2} |A_n|^2 A_0 \right] \quad (11)$$

Here, we have used the wavenumber mismatch: $\Delta k = k_1 + k_2 - k_0$, and the frequency mismatch: $\Delta\omega = \omega_1 + \omega_2 - \omega_0$. Comparing with equation (1) we can identify these quantities with $\Delta\boldsymbol{k}(0)$ and $\Delta\omega(0)$.

For the other two modes, with $(j, l) = (1, 2)$, similar amplitude equations can be established as:

$$\left(\frac{\partial}{\partial z} + \frac{1}{v_j}\frac{\partial}{\partial t}\right) A_j = i\frac{\omega_j^2}{2k_j c^2}\left[\chi^{(2)} A_0 A_{l\neq j}^* \exp(-i\Delta k z + i\Delta\omega t) + \sum_{n=0,1,2} |A_n|^2 A_j\right] \tag{12}$$

Equations (11) and (12) describe the coupling of the three distinct optical pulses by nonlinear resonant interaction, under the influence of phase modulation. However, because we have assumed that, initially, there is a mismatch: $\Delta\omega \neq 0$ and/or $\Delta k \neq 0$, this resonant interaction will only take place if the phase modulation compensates for this initial mismatch, as described bellow.

Notice that both the self- and the cross-phase modulation are retained in the equations. However, it is possible to find experimental conditions such that self-phase modulation can be considered negligible as compared with cross-phase modulation [3]. For this reason, and to keep the formalism simple, the self-phase modulation terms will be neglected in the following. In the more general case where self-phase modulation is not negligible, it could result in a similar contribution to phase-matching, and in particular a self phase-matching effect could be expected [1, 2].

3 Sliding resonance equations

At this point it is convenient to make a transformation of variables, from (z, t) to (η, τ), such that:

$$\eta = z - v_0 t \quad , \quad \tau = t \tag{13}$$

This leads to:

$$\left(\frac{\partial}{\partial z} + \frac{1}{v_j}\frac{\partial}{\partial t}\right) = \frac{1}{v_j}\frac{\partial}{\partial \tau} + \left(1 - \frac{v_0}{v_j}\right)\frac{\partial}{\partial \eta} \tag{14}$$

We can now assume envelope solutions for the amplitudes A_j, for $j = (0, 1, 2)$, of the form:

$$A_j(\eta, \tau) = a_j(\eta, \tau) \exp[i\phi(\eta, \tau)] \tag{15}$$

Replacing in equation (11), we obtain:

$$\frac{1}{v_0}\left(\frac{\partial a_0}{\partial \tau}+ia_0\frac{\partial \phi_0}{\partial \tau}\right)=i\frac{\omega_0^2}{2k_0c^2}\left[\chi^{(2)}a_1a_2e^{is(\eta,\tau)}+\chi^{(3)}(|a_1|^2+|a_2|^2)a_0\right] \tag{16}$$

where the total phase mismatch $s(\eta,\tau)$ is given by:

$$s(\eta,\tau)=\Delta k(\eta+v_0\tau)-\Delta\omega\tau+[\phi_1(\eta,\tau)+\phi_2(\eta,\tau)-\phi_0(\eta,\tau)] \tag{17}$$

We can now equate to zero the terms proportional to the envelope amplitude a_0, and obtain two separate equations for the amplitude and phase of the mode $j=0$, of the form:

$$\frac{\partial \phi_0}{\partial \tau}=\alpha_0\omega_0(|a_1|^2+|a_2|^2) \tag{18}$$

and:

$$\frac{\partial a_0}{\partial \tau}=iC_0a_1a_2e^{is(\eta,\tau)} \tag{19}$$

where the new coefficients α_0 and C_0 are defined below.

It is obvious from here that equation (18) describes the cross-phase modulation of the optical pulse $j=0$ due to the presence of the other two $j=(1,2)$. We can also recognize that equation (19) describes the nonlinear three-wave coupling. This nonlinear coupling can only become significant if, for some range of values η and τ, the total phase mismatch is reduced nearly to zero: $s(\eta,\tau)\simeq 0$. When that occurs, the variation of the wave amplitudes a_j will have a feedback influence on the phase modulation. The two effects, phase modulation and three-wave coupling, are therefore coupled to each other, as stated by the above equations (18) and (19).

Of course, these two equations have to be complemented by those associated with the other two modes. Replacing equations (14) and (15) in (12), we obtain for the phases:

$$\frac{\partial \phi_j}{\partial \tau}+\left(1-\frac{v_j}{v_0}\right)\frac{\partial \phi_j}{\partial \eta}=\alpha_j\omega_j(|a_0|^2+|a_{k\neq j}|^2) \tag{20}$$

and, for the amplitudes:

$$\frac{\partial a_j}{\partial \tau}+\left(1-\frac{v_j}{v_0}\right)\frac{\partial a_j}{\partial \eta}=iC_ja_0a_{k\neq j}^*e^{-is(\eta,\tau)} \tag{21}$$

where j and k can take the values 1 and 2.

The cross-phase modulation coefficients α_j, and the wave mixing coefficients C_j for $j=(0,1,2)$, are determined by:

$$\alpha_j=\frac{\omega_j v_j}{2k_jc^2}\chi^{(3)} \quad , \quad C_j=\frac{\omega_j^2 v_j}{2k_jc^2}\chi^{(2)} \tag{22}$$

This completes the derivation of the equations describing the sliding resonance effect, which is nothing but the three-wave mixing assisted by phase modulation. Similar equations could also be derived for sliding resonances of four-wave mixing processes, which are the most relevant parametric processes in centrosymmetric optical media where the second order susceptibility is absent ($\chi^{(2)} = 0$). In order to understand the physical relevance of the sliding resonance, and to illustrate its main features, we will consider the simple but important case of parametric amplification of a small optical signal a_1 by a strong wave a_0.

4 Parametric amplification

Let us then consider a situation where, initially, we have $a_2 = 0$ and $|a_1| \ll |a_0|$. Neglecting group velocity dispersion ($v_j \simeq v_0$), we can assume that $|a_0|$ is nearly time independent and we can write, for the amplitudes of the modes with $j = (1, 2)$:

$$\frac{\partial a_j}{\partial \tau} = iC_j a_0 a^*_{k\neq j} \exp[-is(\eta, \tau)] \tag{23}$$

The phase function $s(\eta, \tau)$ depends on the intensity and profile of the pump mode $j = 0$, as determined by:

$$\phi_j(\eta, \tau) = \phi_j(0) + \alpha_j \omega_j |a_0(\eta, \tau)|^2 \tau \tag{24}$$

¿From equations (23) we can easily obtain (using $t = \tau$):

$$\frac{\partial^2 a_j}{\partial t^2} = \kappa^2 a_j - i \frac{\partial s}{\partial t} \frac{\partial a_j}{\partial t} \tag{25}$$

where:

$$\kappa^2 = C_2 C_1^* |a_0|^2 \tag{26}$$

Let us write the expression of the phase function $s(\eta, \tau)$ in the form:

$$s(\eta, t) = \Delta k \eta + [\Delta k v_0 - \Delta\omega(\eta, t)]t \tag{27}$$

Let us consider for simplicity that $\Delta k = 0$, and that, for some values of η we can obtain a perfect matching condition on the frequencies: $\Delta\omega(\eta, t) = 0$, for $t = t_0$. We can then write: $\Delta\omega(\eta, t_0) = \Delta\omega(0) - \alpha t_0 = 0$, where we have used:

$$\alpha = \sum_{j=1,2} \alpha_j \omega_j(0) \frac{\partial}{\partial \eta} |a_0(\eta)|^2 \tag{28}$$

These expressions determine the value of t_0, for a given initial frequency mismatch $\Delta\omega(0) \neq 0$, and for a given pump beam profile $a_0(\eta)$. Around this value t_0, we can also write:

$$\Delta\omega(t) = \Delta\omega(0) - \alpha t = \alpha(t - t_0) \tag{29}$$

Let us first consider the case where the phase modulation effects are negligible, $\alpha = 0$. We can then replace $(\partial s/\partial t) = -\Delta\omega(0)$ in equations (25). The general solution of these equations, valid for the initial conditions $a_2(0) = 0$ and $(\partial a_1(0)/\partial t) = 0$, is:

$$a_1(t) = a_1(0) \cosh\left(\sqrt{\kappa^2 - \Delta\omega(0)/4t}\right) \exp[i\Delta\omega(0)t/2] \tag{30}$$

and:

$$a_2(t) = i a_1(0) \sinh\left(\sqrt{\kappa^2 - \Delta\omega(0)/4t}\right) \exp[i\Delta\omega(0)t/2] \tag{31}$$

For a perfect frequency matching $\Delta\omega(0) = 0$, this reduces to the well known results for parametric amplification: $a_1(t) = a_1(0)\cosh(\kappa t)$ and $a_2(t) = i a_1(0)\sinh(\kappa t)$. However, when the frequency mismatch is non negligible, equations (30) and (31) simply describe oscillations of the small signal $a_1(0)$ around its initial value. Now, let us consider that the existence of phase modulation induces a sliding resonance at $t = t_0$. Then, equation (25) takes the form:

$$\frac{\partial^2 a_j}{\partial t^2} = \kappa^2 a_j + i\alpha(2t - t_0)\frac{\partial a_j}{\partial t} \tag{32}$$

for the modes with $j = (1, 2)$. This is a Hermite equation, which can be solved in terms of confluent hypergeometric functions:

$$a_1(t) = a_1(0)\,{}_1F_1\left(-i\frac{\kappa^2}{2\alpha}, \frac{1}{2}, i\alpha(t - t_0/2)^2\right) \tag{33}$$

with a similar solution for $a_2(t)$.

In order to understand the physical relevance of this result, it is more appropriate to complete our discussion with a numerical illustration of the sliding resonance around $t = t_0$. This is shown in Figure 1, where the usual parametric amplification of phase and frequency matched optical beams is represented in Figure 1a . The existence of a frequency mismatch of the order of κ^2 prevents the optical amplification to occur, and reduces the coupling to low amplitude oscillations, as shown in Figure 1b. Inclusion of a sliding resonance will reinstall the parametric amplification, at the cost of a slight time delay, as shown in Figure 1c. This same figure also shows that the sliding amplification will stop at some level near $t = t_0$, after which some oscillations can take place, but now around the amplified values of a_1 and a_2. Of course, this sliding effect is dependent on the time delay between the interacting optical pulses, because it is determined by the derivative of the intensity profiles. Obviously, in the extreme cases where there is no such delay, or if the delay is larger than the pulse duration, the sliding resonance can never be observed (due to group velocity dispersion, such delays can also arise during the interaction process, but this effect is not important for short interacting distances and was neglected here.)

Similar results can also be obtained if, instead of the parametric approximation which considers that the mode a_0 is an infinite source of energy, we use the full three wave equations, where the amplitude of the mode $j = 0$ is allowed to change. In that case, the results can even be more favorable because the energy $|a_0(\eta)|^2$ starts to decrease near $t = t_0$ and the evolution of $\Delta\omega(t)$ is in some sense blocked, with the mismatch remaining close to zero for a longer time (or propagation distance), because the cross-phase modulation of the waves $j = (1, 2)$ also decreases due to the decrease in pump intensity. This means that the full three wave description of the parametric amplification leads to a decreasing value of the parameter α once the resonance region around $t = t_0$ is attained, which corresponds to a nonlinear phase-locking mechanism. On the other hand, the pump mode can also be affected by cross-phase modulation on the intensity profiles of modes 1 and 2. The sliding resonance of three wave mixing in the case of pump depletion is illustrated in Figure 2, whereas in Figure 1, the initial phase-mismatch prevented an efficient energy transfer.

5 Preliminary experiments

In the line of the experimental work already performed for highly nondegenerate cascaded four-wave mixing processes in dielectric media [3], a series of preliminary experiments based on a standard noncollinear pump-probe configuration was performed, this time using a single-colour Ti:sapphire laser-amplifier system (20 μJ, 35 fs - LOA/ENSTA), as shown in Fig. 3. The interaction angle θ was chosen in order to maximize frequency up-convertion processes for which the corresponding wavevectors are approximately given by $(m+1)k_1 - mk_2$ $(m = 1, 2, 3, ...)$. The nonlinear media was a 200 μm thick glass slide, pumped at intensities of around $10^{11} W/cm^2$, where the delay-dependent frequency shift due to cross-phase modulation allowed for the occurrence of highly nondegenerate third-order difference-frequency mixing processes. Spectra of the resulting multicoloured pattern where obtained for different delays between pulses, in steps of 33 fs (Fig. 4). Maximum efficiency occurred for a relative delay of 66 fs, where frequency up-converted beams where directly observed up to the 22nd order. For large delays, we see that the spectrum of the emerging pulses is practically unaffected by self-phase modulation.

6 Conclusions

In this work the concept of sliding resonance of nonlinear wave mixing processes was proposed, which extends the well known phenomenology of the parametric optical processes to the case of pulses with very short durations. Coupled equations for three wave mixing modified by cross-phase modulation were derived. Analytical solutions for the particular case of parametric amplification in the absence of pump depletion were given in terms of confluent hypergeometric functions. The physical relevance of the sliding resonance were illustrated with the help of numerical examples, where pump depletion was also included. It was shown that, when the initial phase-mismatch prevents energy exchange between the interacting pulses, phase modulation can lead to the occurrence of local phase-matching where energy transfer (in particular parametric amplification) can take place. This sliding resonance can only occur when the time delay between the pump and the secondary pulses is smaller than their duration, and cannot take place in the absence of time delay between the pulses. Finally, it should be added that the concept of sliding resonance still applies when we replace the three-wave mixing process considered in the present work by four-wave mixing or higher order nonlinear processes, as evidenced by preliminary experiments performed by our group at LOA [5], and by other experiments where a nonlinear phase-locking mechanism is believed to be responsible for efficient third-harmonic generation and propagation in a long filament produced by an ultrashort laser pulse in air [6]. In view of these results, we believe that the concept of sliding resonance of parametric optical processes presented here can find application in novel and more efficient methods of frequency conversion and harmonic generation of ultrashort laser pulses in nonlinear media.

References

1. G. Tempea, M. Geissler, M. Schnürer, and T. Brabec, *Phys. Rev. Lett* **84**, 4329 (2000).
2. M. Geissler, G. Tempea, and T. Brabec, *Phys. Rev. A* **62**, 033817 (2000).
3. H. Crespo, A. Dos Santos and J.T. Mendonça, *Opt. Lett.* **25**, 829 (2000).
4. J. Zhang, Q. Li, W. Pan, S. Y. Luo, and Y. L. Chen, *Opt. Lett.* **26**, 214 (2001).
5. H. Crespo et al. (to be published).
6. N. Aközbek, A. Iwasaki, A. Becker, M. Scalora, S. L. Chin, and C.M. Bowden, *Phys. Rev. Lett.* **89**, 143901 (2002).

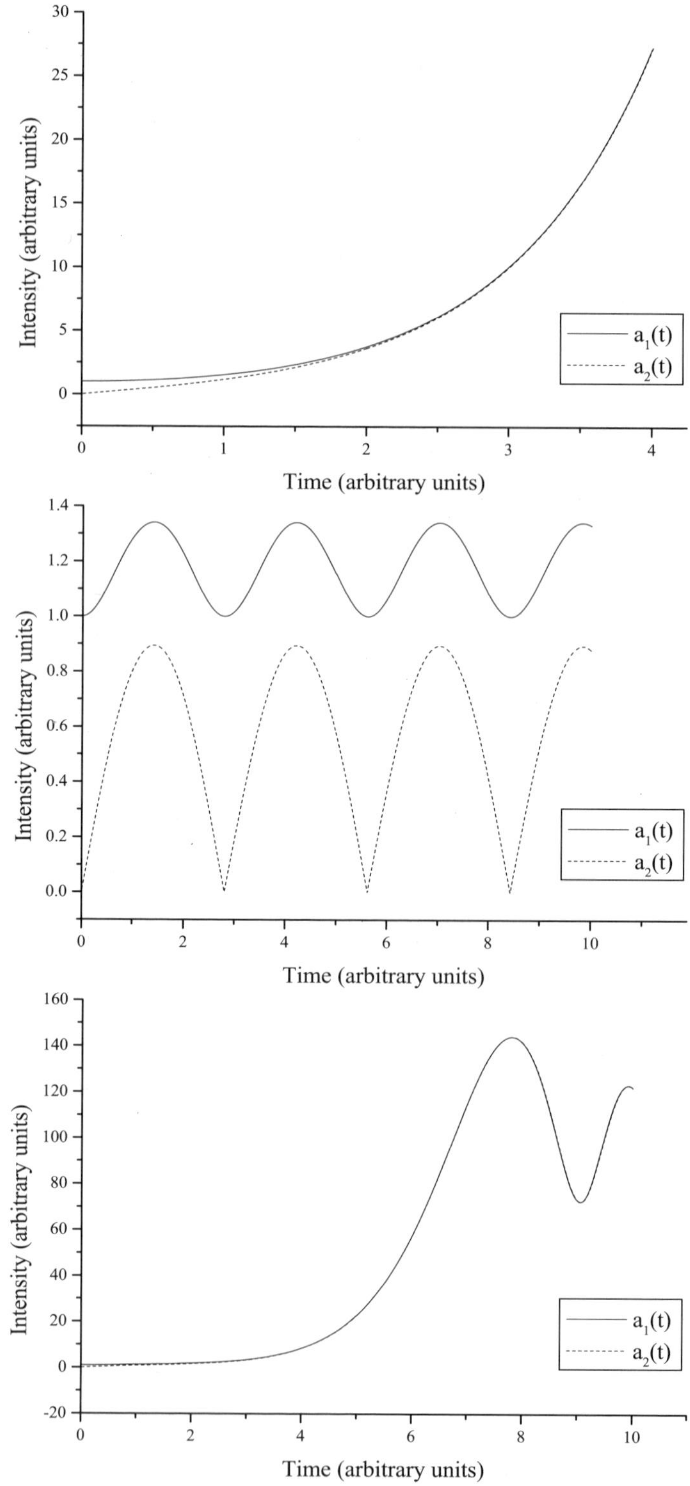

Fig. 1. Parametric amplification of a small optical signal a_2 by a strong wave a_1: a) ideal phase-matching; b) total phase-mismatch; c) sliding resonance (nonlinear phase-matching) - note that the difference between a_1(t) and a_2(t) is not noticeable at this scale.

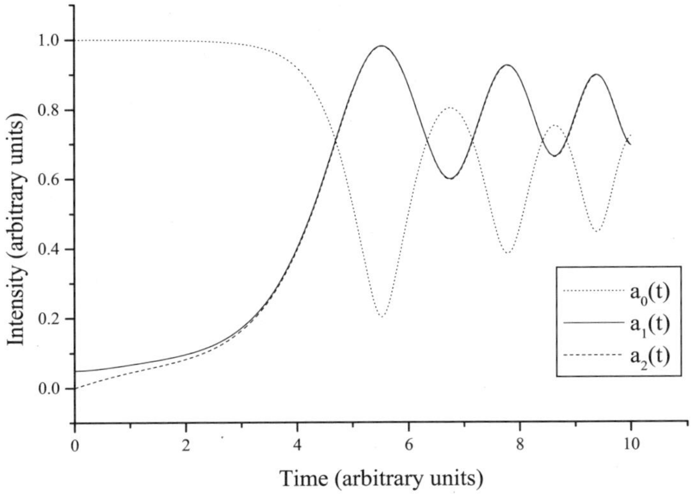

Fig. 2. Sliding resonance in three wave mixing with depletion of pump wave a_0.

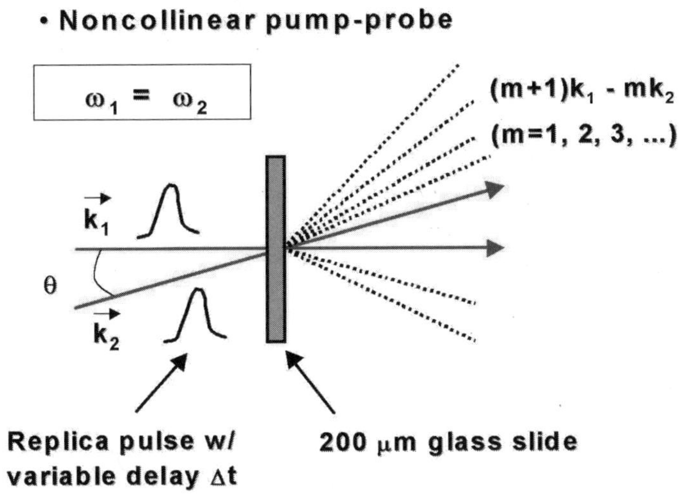

Fig. 3. Basic interaction setup.

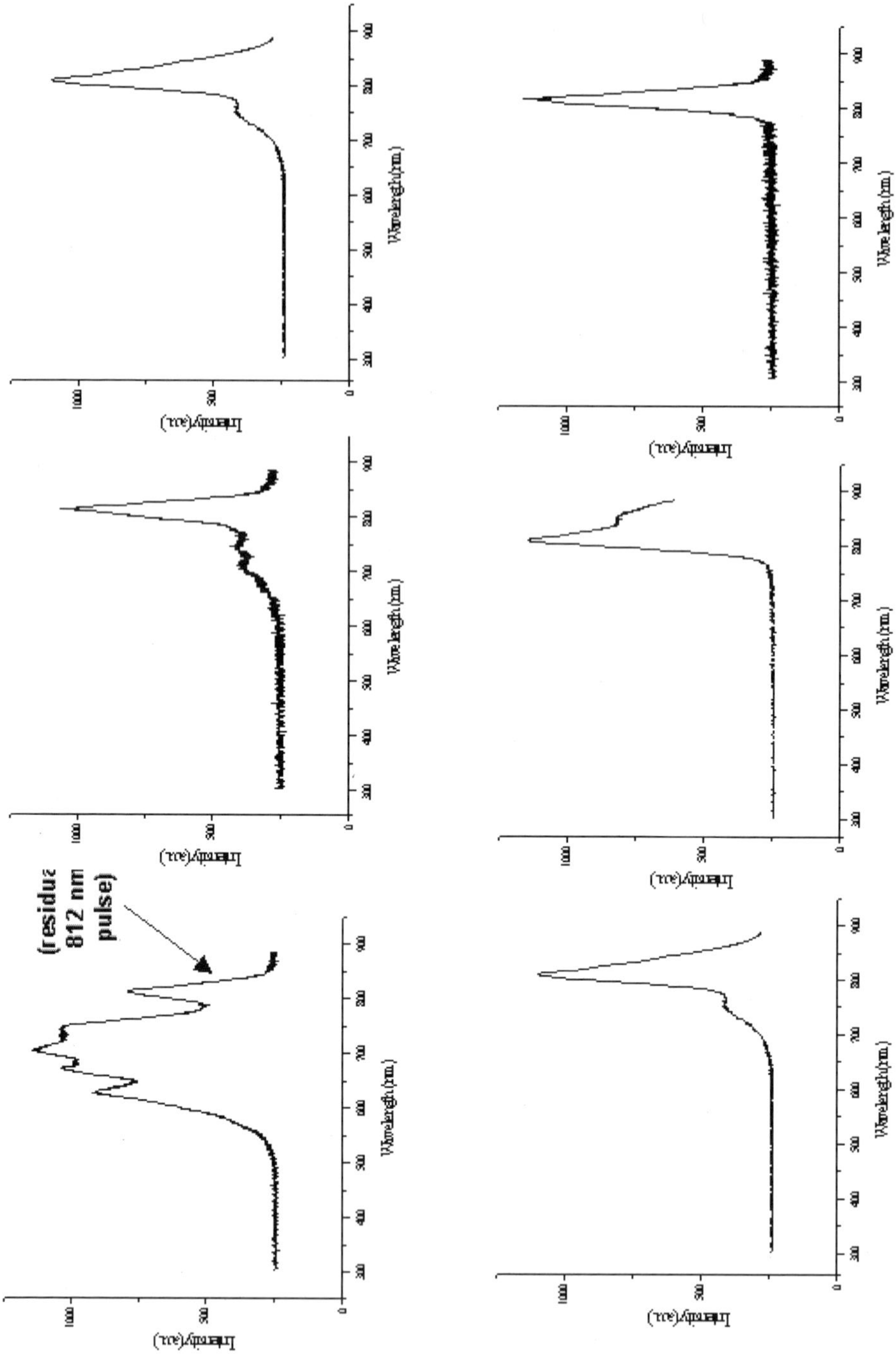

Fig. 4. Evolution of the total spectrum (Intensity vs. wavelength in the range 300-900 nm) as a function of relative pulse delay, in steps of 33 fs. Broadband, quasi-continuous spectra can be obtained

Low Threshold, High Repetition Frequency, Femtosecond Optical Parametric Oscillator Based on Chirped-Pulse Frequency Conversion

Karl A. Tillman[1], Derryck T. Reid[1], D. Artigas[2], J. Hellström[3], V. Pasiskevicius[3], and F. Laurell[3]

[1] Ultrafast Optics Group, David Brewster Building, School of engineering and Physical Sciences, Heriot-Watt University, Riccarton, Edinburgh EH14 4AS, UK e-mail:K.A.Tillman@hw.ac.uk

[2] Department of Signal Theory and Communications, Universitat Politecnica de Catalunya, Gran Capitan D3, 08034 Barcelona, Spain

[3] Laser Physics and Quantum Optics, Roslagstullsbacken 21, 106 91 Stockholm, Sweden

Summary. We report the first demonstration of a femtosecond optical parametric oscillator based on chirped-pulse frequency conversion in a long crystal of aperiodically poled potassium titanyl phosphate. The minimum pump threshold power was 14.4 mW and a signal slope efficiency of 35% was achieved. Continuous tuning from 1194–1455 nm was obtained for an average pump power of 750 mW.

In the context of laser physics the need for ultrafast tunable near to mid IR sources has always been of significant importance due to their research applications in fields such as time resolved spectroscopy and the telecommunications industry. However, due to a lack of suitable gain media, meeting this demand has always been problematic and as a result it is necessary to turn to processes like parametric frequency conversion as a means of addressing this problem. Frequency conversion using femtosecond laser sources requires that the nonlinear medium used in the generation process must possess sufficient acceptance bandwidth to support broadband pulse generation. This bandwidth constraint reduces the choice of nonlinear crystals with favourable phase-matching properties and sufficiently high nonlinear gain to a handful of materials.

Synchronously pumped femtosecond optical parametric oscillators (OPO) are now recognized as an effective means of creating tunable infrared ultrafast pulses. The first femtosecond OPO was demonstrated at Cornell University, [1] and was based on the birefringently phase matched (BPM) nonlinear crystal, potassium titanyl phosphate (KTP). KTP, due to its low nonlinear gain coefficient requires high pump powers for lengths short enough to provide sufficient phase-matching bandwidth and, as a consequence, the first demonstration required an intracavity-pumping scheme to reach threshold. This situation changed in the early 1990s with the introduction of Ti:sapphire lasers which were capable of producing femtosecond pulses of substantially higher peak powers than previously seen [2]. Shortly after this the first extracavity

Ti:sapphire pumped femtosecond OPO was demonstrated and the average output powers from these OPOs were in the hundreds of mW range [3]. This increase in output power meant that for the first time femtosecond OPOs were a practical option for research in the near to mid-IR wavelength range.

Despite this progress in femtosecond OPOs in the last decade a fundamental limit still remained due to two conflicting requirements regarding the crystal length. Firstly for broadband conversion a short crystal is necessary because the phase-matching bandwidth varies as the reciprocal of the crystal length. This dependence means that for long crystals excessive phase mismatch occurs resulting in back conversion of the generated pulses and a reduction in the conversion efficiency and effective gain available from the extra length. Conversely, long crystals produce more parametric gain, resulting in improved conversion efficiency due to the increased interaction length, but limits the conversion bandwidth. The need for phase-matching therefore introduces an apparent restriction in the crystal length requiring a compromise to be made between the best conversion efficiency and the highest available gain. In an attempt to improve this situation the technique of quasi-phasematching (QPM) [4], was first applied to femtosecond OPOs by groups at Stanford and Cornell Universities [5]. QPM made it possible to access much higher effective nonlinearities in ferroelectric crystals leading to higher gain and the work of Ref. [5] used lithium niobate in the periodically poled geometry known as PPLN. Access to the higher nonlinearity of periodically poled crystals allowed substantial reductions in the threshold value of femtosecond OPOs because QPM exploits the largest coefficient of the nonlinear susceptibility tensor, d_{33}. Despite this advantage, one drawback of using periodically poled crystals in OPOs is that the group velocity mismatch between the pump and the converted pulses is typically larger for QPM than for BPM. This meant that maintaining the necessary temporal coherence between the pump pulse and the resonant signal pulse was in fact harder in periodically poled crystals than for birefringently phase-matched crystals of the same length. So, although QPM techniques produced an overall improvement in OPO performance, they did not remove the need for a compromise between the mutually contradictory requirements of high crystal gain and broad crystal bandwidth [6].

The biggest practical restriction in today's ultrafast infrared OPO sources is still the need to use large, expensive pump lasers such as the self-modelocked Ti: sapphire for their very high average powers. Ti: sapphire lasers are still one of the few pump sources available with output powers high enough to easily exceed the oscillation thresholds of typical femtosecond OPOs. To change this situation it is necessary to develop new strategies for reducing OPO pumping thresholds to allow the use of lower power less expensive pump sources such as the diode pumped Cr:LiSAF systems.

The solution adopted by our work was to lengthen the crystal to increase the gain available for the conversion process but at the same time maintain the necessary conversion bandwidth by using the principle of chirped frequency conversion [7–11]. In order to achieve parametric conversion using a chirped or aperiodically-poled crystal, the pump pulses must have a chirp that matches that of the crystal. Figure 1 illustrates the optimum situation in which an unchirped signal pulse walks through a negatively chirped pump pulse as the two pulses co-propagate through the aperiodic grating. The pump pulse enters the crystal first and the longer wavelengths travelling on its trailing end are phasematched to the signal by the larger local grating period at the entrance of the crystal. By the time the pump pulse reaches the end of the crystal the situation has been reversed as the signal pulse travels with a faster group velocity and now leads. At this point phasematching occurs between the signal and the shorter pump wavelengths found on the leading edge of the pulse where they match the smaller local grating period near the exit face of the crystal. The wavelength and bandwidth of the pump pulses automatically define the mean crystal period and aperiodicity, and for a particular crystal length the optimum pump chirp is determined by the group-velocity dispersion properties of the crystal.

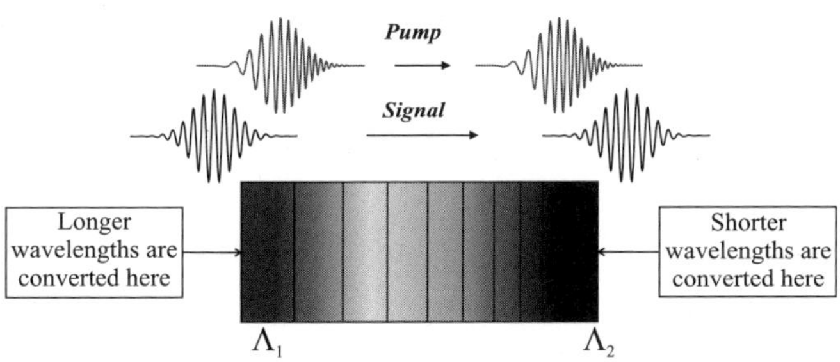

Fig. 1. A schematic illustrating the concept of chirped-pulse pumping in an aperiodically poled crystal.

In [6] we have shown that achieving the lowest threshold operation means using the longest crystal that can still permit a good overlap between the pump and signal pulses. In practice the maximum length is limited by material restrictions and the desire to maintain a tightly focused mode along the entire length of the crystal. The APPKTP crystal used in the experiment was 20 mm long with a grating period which varied linearly from 29.5 – 30.0μm and was designed to provide phasematching over a 10nm pulse bandwidth centred at a wavelength of 850 nm. Although small, this chirp provides a dramatic increase in the phasematching bandwidth when compared to the equivalent value (1.2 nm) for an unchirped crystal of the same length, which is only sufficient only to phasematch a transform-limited 1ps pump pulse.

Our system (Fig. 2) comprised of a 104 MHz self-modelocked Ti:sapphire oscillator operating at a wavelength close to 850nm with a power of 750mW and producing 190fs pulses. The OPO mirrors had a dual-band coating with high reflectivity at the signal wavelength of 1.25μm and high transmission at the pump wavelength of 850nm. The OPO cavity itself used a three mirror design comprising two concave reflectors (radii of 100mm and 75mm) and one plane output mirror that was either a high reflector or an 8% output coupler. The pump pulses were stretched and negatively chirped by a double pass of a four-prism pulse-stretcher before being mode-matched into the OPO cavity using an 80mm focal-length lens. After stretching, the pump pulses had been reduced in power to 92% of the original value and the pulses had been stretched to durations, inferred from autocorrelation measurements, of 900 fs.

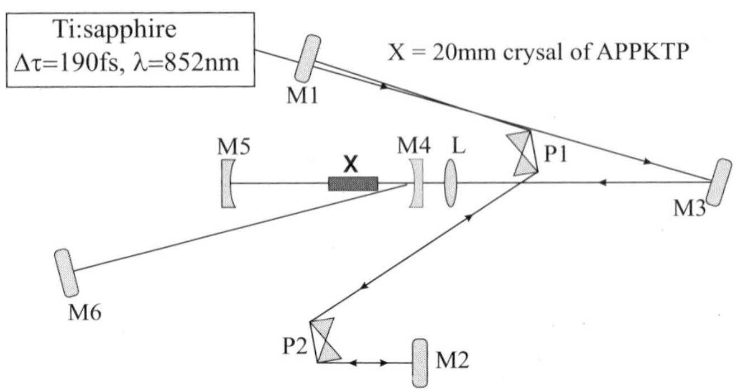

Fig. 2. Schematic showing the stretching prism arm sequence and the overall cavity configuration of the APPKTP OPO. M1 and M3 are steering mirrors to send the stretched pump pulse into the OPO, M2 is the end mirror of the Prism sequence, mirrors M4 and M5 provide OPO cavity focusing and M6 is the OPO end mirror/output coupler.

The OPO was initially characterized with all three of the cavity mirrors being high reflectors and under this condition the pump threshold was measured to be 14.4 mW before the focusing lens. We believe that this value represents the lowest threshold measured for a femtosecond OPO based on KTP and so these results confirm that chirped-pulse frequency conversion with an aperiodically poled crystal can offer advantages over conventional phase-matching approaches. Tuning of the OPO was accomplished in the usual way by adjusting the cavity length and the resulting signal pulses could be generated over a 260nm range from 1.194–1.455 μm. The cavity length movement required for this degree of tuning was around 130 μm and consequently the OPO exhibited very stable operation with little sensitivity to environmental perturbations. A selection of the signal spectra measured during cavity length tuning is presented in Figure 3. One noteworthy point to make is that when using the OPO for nonlinear microscopy involving slow ($\sim$100 second) frame rates, in contrast with results obtained using a conventionally phase-matched femtosecond OPO, we observed no image artifacts associated with power fluctuations.

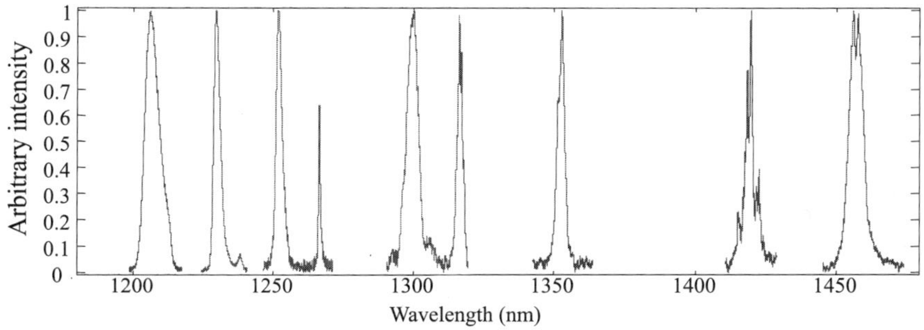

Fig. 3. Representative signal spectra recorded from the OPO as the length of the cavity was tuned over the total output range. The spectra have been normalized for clarity.

In the maximum output power configuration we used an 8% output coupler and with an average pump power of 678mW we were able to extract as much as 257mW of average signal power in a single beam at a wavelength of around 1.3μm. The measured slope efficiency was 35% (figure 4) with a pump depletion of greater than 80% (figure 5) and a comparison between this value and the measured signal extraction efficiency suggests that the oscillator would support a significantly higher output coupling.

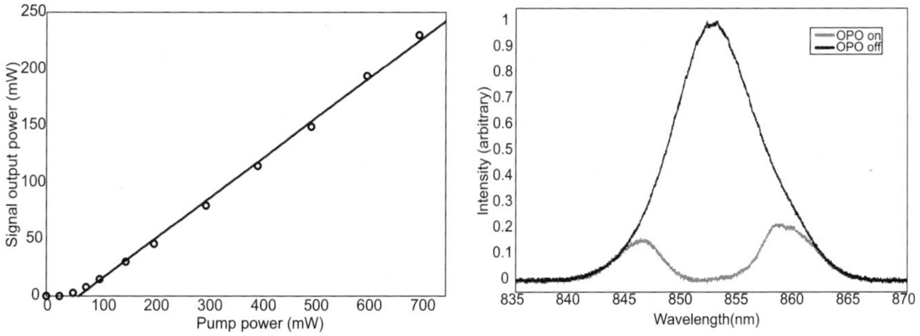

Fig. 4. Left: Slope efficiency of the 8% output coupler setup with a linear data fit. Right:Depletion of the pump spectrum comparing the cases of the OPO on in grey and off in black.

Autocorrelation measurements of the signal pulses from the OPO implied sub-picosecond pulse durations throughout the tuning range of the oscillator and the autocorrelation profile shown in Figure 6, implied a signal duration of 810fs assuming a sech^2 profile, for full details of this work refer to [12].

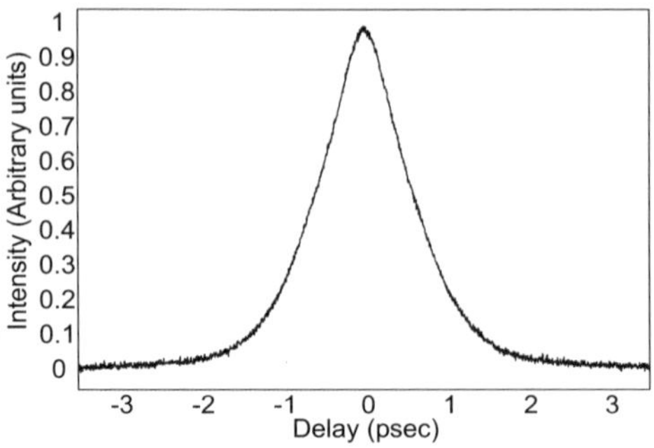

Fig. 5. Intensity autocorrelation indicating a pulse duration of 810fs assuming a sech^2 intensity profile.

In summary we have demonstrated the first femtosecond OPO using chirped-pulse frequency conversion in a long aperiodically poled crystal of KTP. The OPO showed superior performance in terms of the conversion efficiency and the pump oscillation threshold, was robust and widely tunable and our approach can be generally applied to other materials and wavelengths. The low thresholds obtainable using the chirped-pulse technique could for the first time make femtosecond OPOs compatible with the low average output power available from a simple ultrafast oscillator pumped directly by semiconductor laser diodes.

References

1. D.C. Edelstein, E.S. Wachman, C.L. Tang: Appl. Phys. Lett., **54**, 1728 (1989).
2. D.E. Spence, P.N. Kean, W. Sibbett: Opt. Lett., **16**, 42 (1991).
3. W.S. Pelouch, P.E. Powers, C.L. Tang: Opt. Lett., **17**), 1070 (1992).
4. J.A.Armstrong, N.Bloembergen, J.Ducuing, P.S.Pershan: Phys. Rev., **127**, 1918 (1962).
5. K.C. Burr, C.L. Tang, M.A. Arbore, M. M. Fejer: Appl. Phys. Lett., **70**, 3341 (1997).
6. D. Artigas, D.T. Reid, Opt. Lett., **27**, 851 (2002).
7. M. A. Arbore, O. Marco, M. M. Fejer: Opt. Lett., **22**, 865 (1997).
8. M. A. Arbore, A. Galvanauskus, D. Harter, M. H. Chou, M. M. Fejer: Postdeadline paper CPD6, Conf. on Lasers and Electro-Optics, Baltimore (1997).
9. A. Arbore, A Galvanauskas, D. Harter, M. H. Chou, M. M. Fejer: Opt. Lett., **22**, 1341 (1997)
10. T. Beddard, E. Ebrahimzadeh, D. T. Reid, W. Sibbett, Opt. Lett., **25**, 1052 (2000).

11. K. A. Tillman, D. T. Reid, D. Artigas, J. Hellström, V. Pasiskevicius, F. Laurell: Opt. Lett., **28**, 543 (2003).
12. K. A. Tillman, D. T. Reid, J. Hellström, V. Pasiskevicius, F. Laurell: J. Opt. Soc. Am. B, **20**, 1309 (2003).

Generation of sub-20 fs Tunable Visible Pulses from a 100 kHz NOPA For Measuring Ultrafast Heterogeneous Electron Transfer

R. Ernstorfer[1], L. Gundlach[1], C. Zimmermann[1], F. Willig[1], R. Eichberger[1], and E. Riedle[2]

[1] Hahn-Meitner-Institute Berlin, SE4 - Dynamics of Interfacial Reactions, Glienicker Straße 100, 14109 Berlin, Germany, eichberger@hmi.de
[2] LS für BioMolekulare Optik, Ludwig-Maximilians-Universität, Oettingenstr. 67, 80538 München, Germany, eberhard.riedle@physik.uni-muenchen.de

Abstract We describe the design and performance of a noncollinear optical parametric amplifier that can be pumped by the output of a commercial 100 kHz Ti:sapphire regenerative amplifier. The output of the NOPA is compressed to sub-20 fs length and tunable from 460 nm to beyond 1 μm. The maximum pulse energy is 750 nJ. The system is used to measure photoinduced electron transfer in a perylene/TiO_2 system with unprecedented temporal resolution.

1 Introduction

For many spectroscopic applications in physics, chemistry and biology widely tunable pulses with a duration of less than 20 fs are needed to resolve the ultrafast dynamics on the time scale of nuclear motions within the constituents. Many of these samples are seriously damaged when absorbing strong laser pulses. Therefore measurements are required with weak laser pulses at a high repetition rate. Photoinduced heterogeneous electron transfer is expected to occur on such a short timescale. Tunable sub-20 fs pulses are extremely hard to generate at the 80 MHz repetition rate available from Ti:sapphire oscillators [1]. Nonlinear optical conversion in the one kHz range is now well established [2]. This relatively low repetition rate combined with the necessary weak laser pulses for the above mentioned samples results in extremely long averaging times for weak spectroscopic signals. An intermediate repetition rate on the order of 100 kHz is therefore of great interest.

2 Experimental and Results

It is shown here that a single stage noncollinearly phase matched optical parametric amplifier (NOPA) can generate sub-20 fs tunable visible pulses at a repetition rate of 100 kHz. The NOPA is pumped by 800 nm pulses with 7 μJ energy and a pulse length of 50 fs (Coherent RegA 9050). An output pulse energy as high as 750 nJ is demonstrated. This output energy is considerably higher than previously obtained with a double stage collinear amplifier [3] and the pulse length is dramatically shorter. So far NOPAs have only been used with kHz Ti:sapphire pump systems and pump energies of above 100 μJ. The low energy pump requires a careful optimization of the setup. In a preliminary experiment we pumped the setup with pulses from a kHz system and slowly lowered the energy of the 775 nm pulses. In this way we were eventually able to reach stable operation of the NOPA at the input level available at 100 kHz. With fixed optical geometry we measured the green output as a function of the blue pump pulses (see inset). It can be seen that scalability can even be obtained at blue pump energies well below 1 μJ, i.e. at red input of 3 μJ or less (Fig. 1).

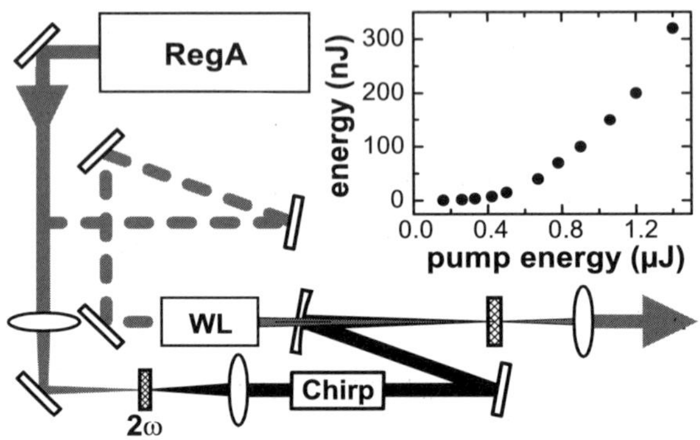

Fig. 1. 100 kHz NOPA system design and dependence of the output on the blue input pulses.

Only 10% of the red pump light is used to generate a single filament continuum in a 3 mm sapphire plate. The main part of the light is frequency doubled in a 300 μm BBO crystal placed slightly behind the tight focus of lens L_1 (Fig. 2). To ensure a good temporal overlap of the 2.7 μJ blue light with the chirped seed light a chirper made from two fused silica plates is introduced. The stretched 400 nm light is then used to amplify a broad spectral slice of the continuum seed light in a 2 mm BBO crystal. It has been shown before that an appropriate noncollinear geometry leads to simultaneous phase matching throughout most of the visible and consequently very broadband visible pulses [2]. For the 100 kHz pump system we found it best to place the amplifier crystal within the Rayleigh range of the focused blue light (Fig. 2). The intensity was adjusted to about 150 - 200 GW/cm^2 by matching the size of the pump beam on the focusing mirror M. After the doubling crystal the pump light is recollimated with the fused silica lens L_2. As can be seen from the lower part of Fig. 2, variation of the spacing d_T between the lenses L_1 and L_2 allows the adjustment of the proper intensity at the amplifier crystal while simultaneously avoiding damage on mirror M. The necessary change of the positions of lens L_1 and the BBO crystal is small and easily adjusted with sufficient precision. The NOPA is tunable from 460 nm to well beyond 1 μm (for sample spectra see Fig. 3). The NIR wavelengths are also generated on axis and utilize the NIR part of the seed light. At the maximum of the tuning curve a pulse energy of 750 nJ is obtained corresponding to a quantum efficiency of more than 30% from the blue and an overall energy conversion efficiency (800 nm to green) of about 10%. The average power of

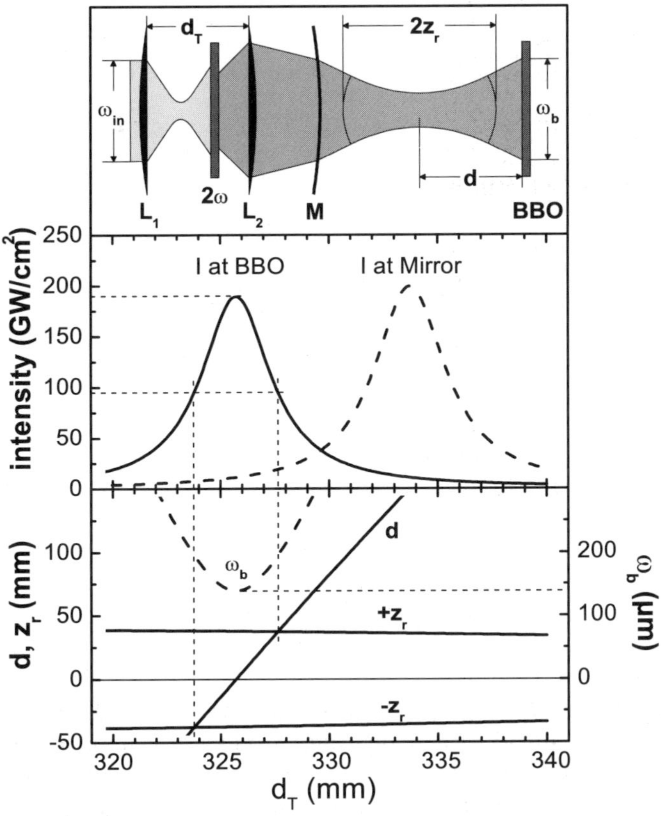

Fig. 2. Beam geometry of the red and blue pump pulses (upper part). Pump intensity and geometry of the Rayleigh range (lower part).

75 mW is considerably higher than obtained from comparable 1 kHz systems with sub-20 fs pulse length. The spectral bandwidth is sufficient for sub-20 fs pulses through most of the tuning range. With proper compression in

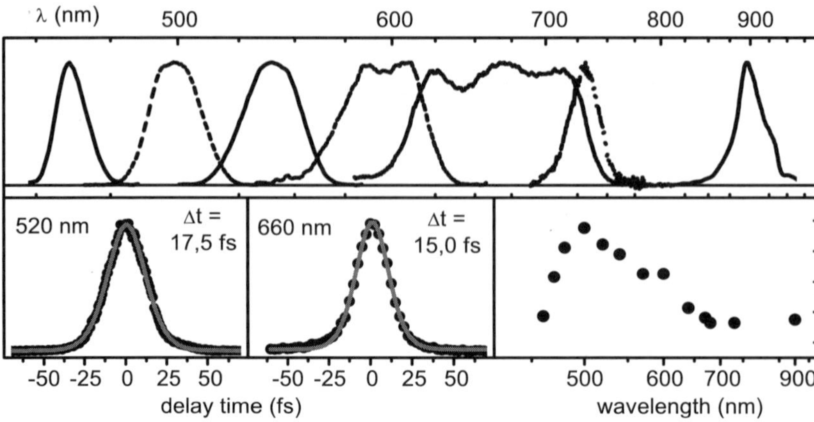

Fig. 3. Output spectra, autocorrelation traces and pulse energy for the 100 kHz NOPA.

a fused silica prism compressor we routinely measure durations well below 20 fs (see Fig.3). The spatial mode of the NOPA output was very close to a Gaussian fundamental. The efficiency falls off toward longer wavelengths because of signal/pump group velocity mismatch (GVM). The GVM is zero for signal pulses around 490 nm and increases to 150 fs/mm for 700 nm. The effective overlap of a given spectral component with the 400 nm pump pulses decreases correspondingly to only 1 mm out of the 2 mm amplifier crystal. The noncollinear phase matching only adjusts the signal/idler GVM to zero [2].

The 100 kHz NOPA was used for the measurement of photoinduced femtosecond electron transfer from the excited singlet state of the perylene chromophore into nanostructured anatase TiO_2 [4]. The experimental system requires low energy pump pulses in the range of 400–450 nm. Therefore we run the NOPA in the NIR and frequency double the output tightly focused in a 100 μm BBO crystal (Fig. 4), resulting in pulse energies of 20 nJ. Compression of the blue pulses below 15 fs was achieved by downchirping the NOPA output with a standard prism compressor setup. Weak probe pulses around 575 nm with a bandwidth of 35 nm were selected from a separately generated white light continuum and compressed with a prism compressor. The measurements were carried out in ultra-high-vacuum to ensure a very high long time stability and reproducibility of the system.

The perylene chromophore is an ideal model compound for spectroscopic studies of photoinduced electron transfer because the absorption spectra for the perylene ground, first excited singlet, and cationic ground state are well separated in the visible spectral range. Perylene was modified with covalently attached side groups (tertiary-butyl) to prevent dimerization and anchored on the semiconductor surface via a carboxylic acid group (Fig. 5). The molecular electronic ground state is located in the band gap of TiO_2 about 1.9 eV

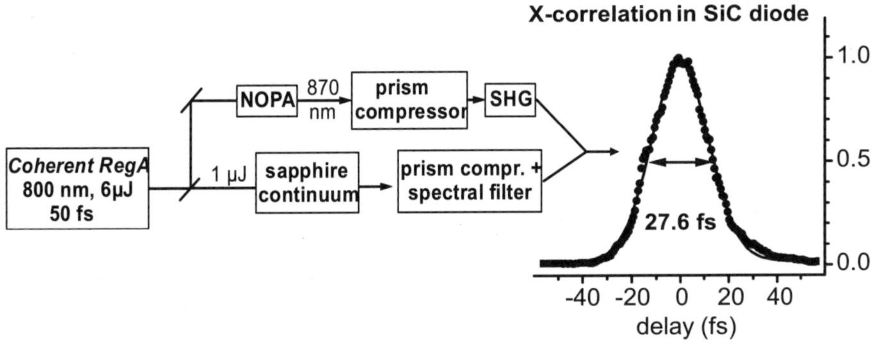

Fig. 4. Experimental setup of the pump probe experiment. The achieved time resolution was below 30 fs.

below the lower edge of the conduction band. The energetic position of the donor orbital of the chromophore is located about 800 meV above the conduction band edge of TiO_2 [5]. In this case all the Franck-Condon factors can be realized in parallel due to the wide energetic spread of electronic levels in the conduction band of the semiconductor and the injection time is essentially controlled by the electronic overlap between the donor orbital and the electronic acceptor states [6]. The electronic coupling between the large perylene chromophore and the surface of TiO_2 can be changed by inserting a molecular spacer group between the anchor group and the chromophore (Fig. 5) [7].

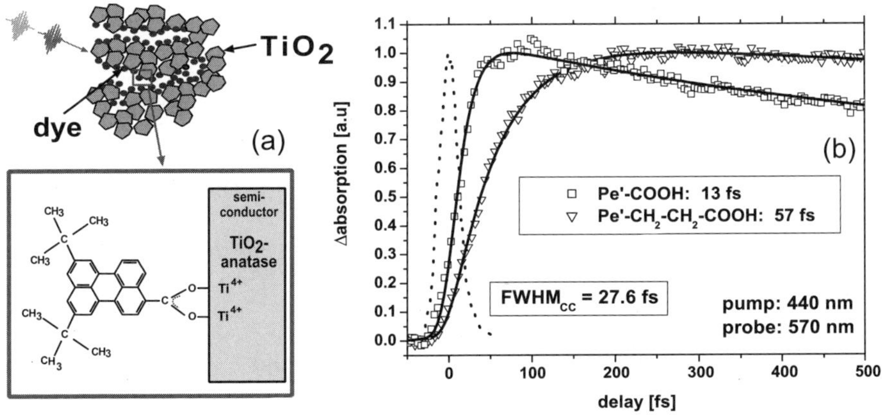

Fig. 5. Sketch of the experimental system. Chromophores were attached to a film of nanostructured TiO_2 of sponge-like structure (a). The electron injection dynamics were probed by the rise of the perylene cation absorption. The electronic coupling between chromophore and surface was changed by inserting a spacer group (b).

Injection of electrons from the photoexcited perylene chromophore into the transparent 2-3 μm thick nanostructured anatase TiO_2 film was investigated by measuring transient absorption signals for the rise in the population of ionized perylene (radical cation) at 575 nm (Fig. 5). Electronic cross phase modulation effects were minimized by using 40 μm thin display glass as substrate for the TiO_2 film. The shortest electron transfer time constant of 13 fs was measured for perylene directly attached to the TiO_2 surface via the carboxylic acid group. This value could be verified in a complementary measurement of the decay of the reactant state (e.g. excited state absorption at 710 nm). Inserting two methylene groups between the chromophore and the surface reduces the electronic overlap between the molecular donor state and the semiconductor acceptor states and thus slows down the injection dynamics by a factor of 4.

3 Summary

In conclusion the single stage NOPA pumped by 100 kHz 50 fs pulses is an extremely simple, compact and stable device that delivers tunable visible pulses with unprecedented pulse energy and in particular much shorter duration than previously available. Sub-20 fs pulses were employed for measuring 13 fs heterogeneous electron transfer dynamics. A decreased chirp of the seed continuum together with an improved compression scheme should even lead to sub-10 fs pulse durations.

References

1. G.M. Gale, M. Cavallari, T.J. Driscoll, and F. Hache, Opt. Lett. **20**, 1562 (1995).
2. E. Riedle, M. Beutter, S. Lochbrunner, J. Piel, S. Schenkl, S. Spörlein, and W. Zinth, Appl. Phys. B, **71**, 457 (2000).
3. L.D. Ziegler, J. Morais, Y. Zhou, S. Constantine, M. K. Reed, M. K. Steiner-Shepard, and D. Lommel: IEEE J. Quantum Electron., **QE-34**, 1758 (1998) and references therein.
4. C. Zimmermann, F. Willig, S. Ramakrishna, B. Burfeindt, B. Pettinger, R. Eichberger, and W. Storck, J. Phys. Chem. B **105**, 9245 (2001).
5. L. Gundlach, R. Ernstorfer, C. Zimmermann, S. Felber, R. Eichberger, and F. Willig, to be published.
6. D. Miller, G. McLendon, A. Nozik, W. Schmickler, and F. Willig, *Surface Electron Transfer Processes*, chapter 5 (VCH Publishers, New York 1995).
7. R. Ernstorfer, L. Gundlach, S. Felber, C. Zimmermann, R. Eichberger, and F. Willig, to be published.

Broadband Sum-Frequency Mixing for Indirect Phase Control of UV Pulses With a Sub-20-Fs TW-Class Ti:Sapphire Laser System

Yasuo Nabekawa and Katsumi Midorikawa

RIKEN (The Institute of Physical and Chemical Research), 2-1 Hirosawa, Wako-shi, Saitama 351-0198, JAPAN nabekawa@postman.riken.go.jp

1 Introduction

Optical phase control techniques have been widely used for the generation of shaped femtosecond pulses.

Although many of the results for shaped femtosecond pulses are now presented in the fields of physics and chemistry, the spectral range of these pulses has been limited to the near infrared (NIR) or visible (VIS) wavelength because phase control instruments, such as liquid crystal light modulators(LC-SLMs) and acousto-optic programmable dispersive filters (AOPDFs), are not transparent in other wavelength regions. Thus, arbitrary control of the phase of UV femtosecond pulses has not yet been realized although modulations of amplitude have been demonstrated with frequency doubling of phase-modulated pulses [1] and also with sum-frequency mixing of amplitude-modulated pulses [2].

A method of UV phase control is indirect phase control through NIR pulses. If the phase of NIR femtosecond pulses can be transferred to UV pulses with frequency mixing in a nonlinear optical medium, we can control the UV phase using the techniques in the NIR range.

Tan and coworkers demonstrated indirect phase control with a noncollinear optical parametric amplifier (NOPA) in an infrared (IR) region [3], and Witte and coworkers showed relative phase differences between two pulses in the mid-infrared (MIR) region generated by different-frequency mixing of a phase-controlled IR pulse with another bandwidth-limited pulse [4]. This method can be extended to sum-frequency mixing (SFM) of a phase-controlled NIR femtosecond pulse with a quasi-monochromatic VIS$\sim$violet pulse for the generation of UV pulses with indirect phase control.

The group delay walk off in a nonlinear crystal which limits the bandwidth of UV pulses, however, remains as a problem for the generation of pulses shorter than 100 fs in the UV region. In this paper, we show a novel scheme for broadband wave-vector matching between a monochromatic visible (VIS) pulse and a NIR (Ti:sapphire laser) pulse of sufficient bandwidth for generating a sub-20-fs UV pulse in a nonlinear crystal with conventional thickness ($\sim$1 mm).

2 Sub-20-fs, TW-class Ti:sapphire laser system with a phase controller

In this section, we show a sub-20-fs, TW-class laser system developed for the phase control of sub-20-fs UV pulses [5].

The laser is a typical chirped pulse amplification (CPA) system consisting of a mode-locked oscillator, an Offner stretcher, a regenerative amplifier, a four-pass amplifier, and a compressor. A phase controller with a LC-SLM [6] is placed between the stretcher and the regenerative amplifier [5]. The configuration of the phase controller is that of a typical 4-f telescope consisting of two concave mirrors with two gratings and an LC-SLM (Cambridge Research & Instrumentation, SLM-128) which is placed on the Fourier plane of the concave mirrors. The period of pixels in the LC-SLM corresponds to the 0.79 THz optical bandwidth.

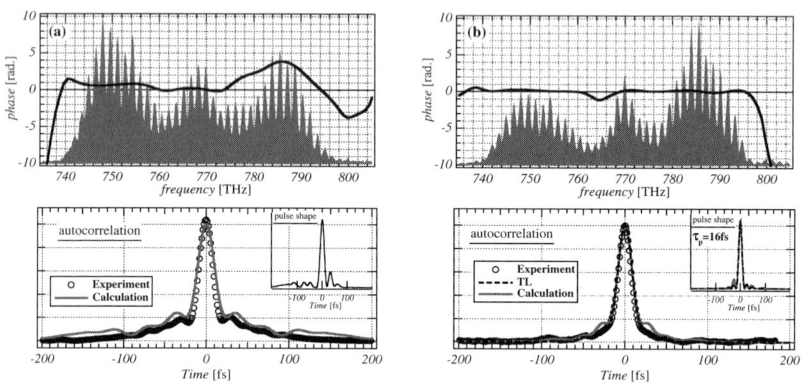

Fig. 1. Measured phases with spectral interferometry of direct electric field reconstruction (SPIDER) and measured single-shot autocorrelation traces (circles) of the TW-class Ti:sapphire laser system. Pulse shapes calculated from measured phase and spectrum are shown in insets as solid gray curves: (a) without compensation of unexpected phase errors. (b) with compensation. The dashed curve corresponds to the transform-limited pulse

The regenerative amplifier with a ring cavity, which consists of Pellin-Broca prisms instead of dielectric coated mirrors, eliminates the limitation of the spectral width on the dielectric coat. A spectral width exceeding 100 nm for 30 mJ pulses amplified by a system consisting of a four-pass amplifier following the regenerative amplifier is obtained and the pulses are compressed to a 16fs pulse width with an energy of 13 mJ.

We show the measured phases obtained by spectral interferometry for direct electric-field reconstruction (SPIDER) [7] and autocorrelation traces of the output pulses of the system in Fig. 1 (a), Fig. 1 (b), Fig. 2 (a), and Fig. 2 (b).

The phase controller successfully compensated for the unexpected phase error (Fig. 1 (a)), which might have been caused by distortions on the surfaces of the curved mirrors in the stretcher, with the result of a 16 fs pulse width (Fig. 1 (b)) in the terrawatt regime. An applied extra phase of sinusoidal modulation, $\phi_{\rm mdu}=\cos[2\pi(\nu-\nu_0)/\nu_{\rm prd}]$, which is shown as a dotted curve in Fig. 2 (a), is in good agreement with the measured phase shown as a gray solid curve. Positions of the subpeaks in the calculated (dashed curve) and measured (gray solid curve) autocorrelation traces reflect the inverse of the period ($\nu_{\rm prd}$=10[THz]) of the modulation. By applying a phase jump ($\phi_{\rm jmp}=\pi\theta(\nu-\nu_0)$, where θ is Heviside's step function) shown as a dashed curve Fig. 2 (b), we could obtain the phases and the pulse shapes that we had expected (gray solid curves in Fig. 2 (b)), although the measured slope of the phase jump is unsharpened by the band-pass filter of the SPIDER calculation in the time domain. To our knowledge, the results in Figs. 2 (a) and (b) are the first demonstration of shaped sub-20 fs pulses with an energy of more than 10 mJ.

3 Group-delay-matched sum-frequency mixing

The mismatch of the pulse widths between the quasi-monochromatic and the femtosecond pluses can be solved by stretching the NIR femtosecond pulse with a constant chirp, as is performed in optical parametric chirped-pulse

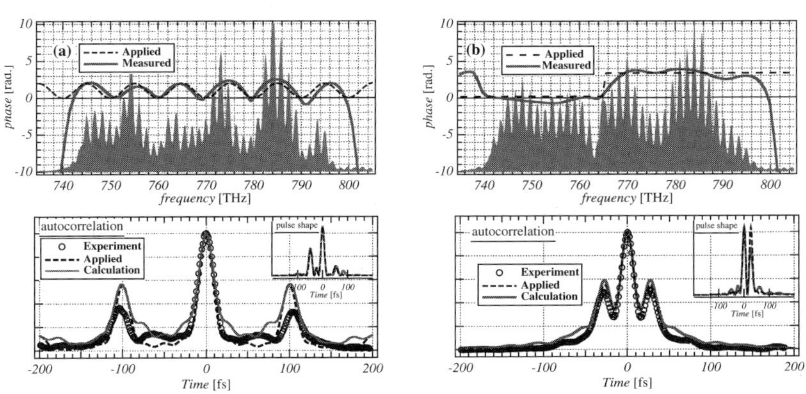

Fig. 2. Measured phases with the SPIDER apparatus and measured single-shot autocorrelation traces (circles) with phase modulations. Pulse shapes calculated from measured phase and spectrum are shown in insets as solid gray curves. Applied phases to the LC-SLM are also shown as dashed curves: (a)with periodic modulation, (b)with a phase jump .

amplification. The group delay walk off in a nonlinear crystal which limits the bandwidth of UV pulses, however, is the remaining problem for sub-100-fs pulses in the UV region [2]. In this section, we show a novel scheme for

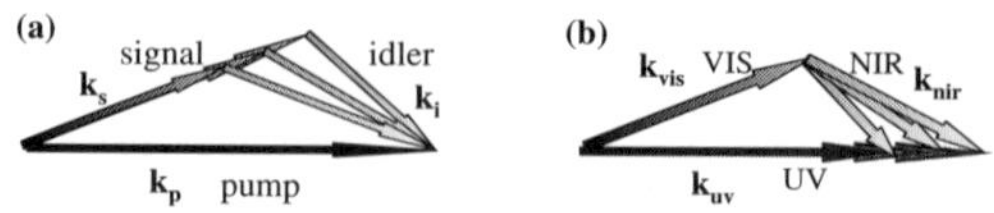

Fig. 3. (a) Phase matching (wave-vector conservation) in NOPA. (b) Noncollinear angularly dispersed phase matching in sum-frequency mixing.

the broadband phase matching between a monocromatic, VIS pulse and a NIR pulse of sufficient bandwidth for generating a sub-20-fs UV pulse [8]. The basic idea of this scheme is similar to that NOPA as shown in Fig. 3. Wave-vector conservation, $\boldsymbol{\Delta k} \equiv \boldsymbol{k}_{\rm uv}(\nu_{\rm uv}) - \{\boldsymbol{k}_{\rm vis}(\nu_{\rm vis_0}) + \boldsymbol{k}_{\rm nir}(\nu_{\rm nir})\} = 0$, can be fulfilled for a wide range of UV optical frequencies, $\nu_{\rm uv}$, while maintaining the direction of $\boldsymbol{k}_{\rm uv}(\nu_{\rm uv})$, only if we choose an appropriate noncollinear angle, $\alpha_{\rm vn_0}$, between the monochromatic visible wave vector, $\boldsymbol{k}_{\rm vis}(\nu_{\rm vis_0})$, and the wave vector, $\boldsymbol{k}_{\rm nir}(\nu_{nir_0})$, of the center frequency, ν_{nir_0}, of a broad NIR bandwidth. The condition for this broadband wave-vector matching can be expressed as a general formula [2];

$$0 = \left.\frac{d\boldsymbol{k}_{\rm uv}}{d\omega_{\rm uv}}\right|_0 - \left.\frac{d\boldsymbol{k}_{\rm nir}}{d\omega_{\rm nir}}\right|_0 , \tag{1}$$

where $\omega_{\rm uv}$ and $\omega_{\rm nir}$ are the optical angular frequencies of the UV and NIR waves, respectively, and subscript $_0$ denotes substitutions of the variables to the center values after differentiations. By imposing the condition that there is no angular dispersion in the UV output on Eq. (1), we can obtain the equation which determines the noncollinear angle, $\alpha_{\rm vn_0}$;

$$v_{go}^{\rm nir} \cos\left(\alpha_{\rm vn_0} - \alpha_{\rm vu_0}\right) = v_{go}^{\rm uv} . \tag{2}$$

We define $\alpha_{\rm vu_0}$ as a noncollinear angle between the visible and UV pulses at the center angular frequencies. Group velocities of the NIR and the UV pulses at the center angular frequencies are denoted $v_{go}^{\rm nir}$ and $v_{go}^{\rm uv}$, respectively. The angular dispersion of the input beam which satisfies Eq (1), $d\alpha_{\rm vn}/d\omega_{\rm nir}|_0$, is given by

$$\left.\frac{d\alpha_{\rm vn}}{d\omega_{\rm nir}}\right|_0 = -\tan\left(\alpha_{\rm vn_0} - \alpha_{\rm vu_0}\right) \frac{1}{k_{\rm nir_0}} \left.\frac{dk_{\rm nir}}{d\omega_{\rm nir}}\right|_0 . \tag{3}$$

Calculated results of the phase mismatch, $\boldsymbol{\Delta k} \cdot \boldsymbol{L}$, and the acceptance curve [2] for Type-I SFM of 532 nm monochromatic light and NIR light around 800

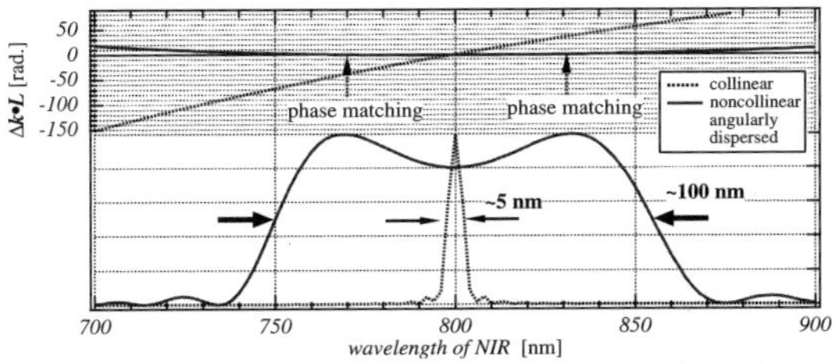

Fig. 4. Phase mismatches, $\boldsymbol{\Delta k}\cdot\boldsymbol{L}$, in collinear (dotted curve) and noncollinear angularly dispersed (solid curve) geometries for Type-I SFM of the second harmonic of the Nd:YAG laser and broad spectrum of the Ti:sapphire laser in a BBO crystal. Squares of the absolute value of the acceptance functions, $\eta(\lambda_{NIR})\equiv\{e^{i\boldsymbol{\Delta k}(\lambda_{NIR})\cdot\boldsymbol{L}} - 1\}/\boldsymbol{\Delta k}(\lambda_{NIR})\cdot\boldsymbol{L}$, are also shown below the phase mismatches.

nm in a 1-mm-thick BBO crystal are shown in Fig. 4 as solid curves, where $\boldsymbol{L}$ is directed to $\boldsymbol{k}_{UV}$ with a length of BBO.

We set $\alpha_{\mathrm{vn_0}}$=26.1° and the angle of the optical axis of the BBO crystal to be 54.8°. We can broaden the acceptable bandwidth to ∼100nm with slight tilt of the BBO from a perfect-phase-matching condition in spite of a central dip in the acceptance curve.

We conducted an experiment on this new scheme with the sub-20-fs Ti:sapphire laser system mentioned above [5] and a frequency-doubled Nd:YAG laser with energy of 25 mJ and a pulse duration of 4.5 ns. The amplified pulses, which bypassed the compressor in the CPA system, with an energy of 28 mJ and a pulse duration of ∼500 ps were led to a spatial disperser in which the image of the beam on the surface of a dispersion prism was relayed in the BBO crystal.

Measured spectra of the NIR and the UV pulses for wavelengths are shown in the inset of Fig. 5. The wavelength of the UV pulse is centered around 320 nm and the spectral width is ∼17 nm. By converting the wavelengths to optical frequencies, we can see that the power spectra of these NIR and UV pulses are almost the same except for the central dip which may be filtered by the acceptance function, as shown in Fig. 5.

The output beam was diverged with diffraction. However, the ratio of the beam diameter in the direction of angular dispersion to that in the perpendicular direction remained almost unity. Thus, we conclude that there is almost no angular dispersion. These results support the concept of broadband phase matching using this novel scheme.

The energy of the UV pulse is only 1.3 μJ at most. The low efficiency ($<10^{-4}$) of the conversion will be improved by increasing the intensities of the input pulses.

This work was supported by Grant-in-Aid for Scientific Research on Priority Areas No.14077222 and Grant from Research Foundation for Opto-Science and Technology.

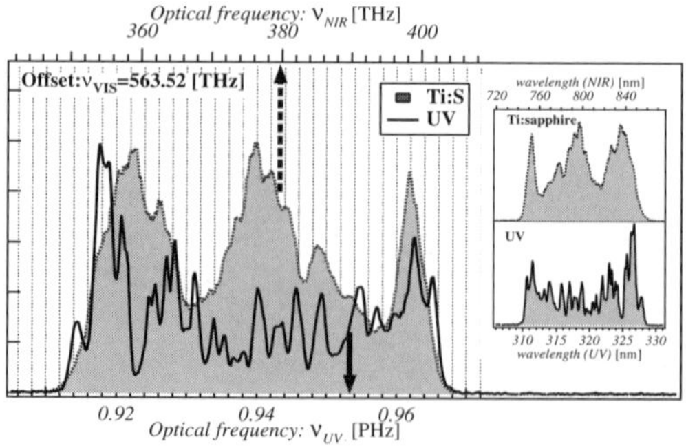

Fig. 5. Spectra of the input NIR pulse (dotted curve with hatched area, top axis) and the UV pulse (solid curve, bottom axis) for optical frequencies. The top axis is shifted with an offset of the mixed visible frequency.

References

1. Hacker, M., Netz, R., Roth, M., Stobrawa, G., Feurer, T., Sauerbrey, R.: Frequency doubling of phase-modulated, ultrashort laser pulses. Appl. Phys. B **73** (2001) 273– 277
2. Hacker, M., Feurer, T., Sauerbrey, R., Lucza, T., Szabó, G.: Programmable femtosecond laser pulses in the ultraviolet. J. Opt. Soc. Am. B **18** (2001) 866–871
3. Tan, H.S., Schreiber, E., Warren, W.S.: High-resolution indirect pulse shaping by parametric transfer. Opt. Lett. **27** (2002) 439– 441
4. Witte, T., Kompa, K.L., Motzkus, M.: Femtosecond pulse shaping in the mid infrared by difference frequency mixing. Appl. Phys. B **76** (2003) 467–471
5. Nabekawa, Y., Shimizu, Y., Midorikawa, K.: Sub-20-fs terawatt-class laser system with a mirrorless regenerative amplifier and an adaptive phase controller. Opt. Lett. **27** (2002) 1265–1267
6. Weiner, A.M.: Femtosecond pulse shaping using spatial light modulators. Rev. Sci. Instrum. **71** (2000) 1929–1960
7. Iaconis, C., Walmsley, I.A.: Self-Referencing Spectral Phase Interferometery for Measuring Ultrashort Optical Pulses. IEEE J. Quantum Electron. **35** (1999) 501–509
8. Nabekawa, Y., Midorikawa, K.: Broadband sum frequency mixing using noncollinear angularly dispersed geometry for indirect phase control of sub-20-femtosecond UV pulses. Opt. Express **11** (2003) 324–338

New Generation of Photoconductive Few-Cycle Terahertz Emitters

J. Darmo[1], G. Strasser[1], T. Roch[1], T. Müller[1], K. Unterrainer[1], G. Tempea[2], T. Le[3], and A. Stingl[3]

[1] Institute for Solid-State Electronics, Vienna University of Technology, Floragasse 7, A-1040 Vienna, Austria juraj.darmo@tuwien.ac.at
[2] Institute of Photonics, Vienna University of Technology, Gusshausstrasse 27, A-1040, Vienna, Austria
[3] Femtolasers Produktions GmbH, Fernkorngasse 10, A-1100 Vienna, Austria

1 Introduction

The generation of few-cycle Terahertz radiation into the free space from a biased photoconductive gap illuminated by an ultrashort laser pulse is known more than a decade [1]. Since then, many attempts were presented to increase the generated THz power without compromising the bandwidth of the output radiation [2,3]. In this contribution, we report on a low-temperature-grown GaAs (LT GaAs) based photoconductive THz emitter integrated with a semiconductor based Bragg mirror. This design improves the generators THz output power by about one order of magnitude. The optical resonance and the confinement of the photogenerated carriers in the high electric field region of the LT GaAs layer are responsible for the observed enhancement of the THz emission. In addition, we have focused on an optimization of the growth temperature of the LT GaAs layer with respect to a maximum photoresponse of the material and a maximum breakdown field. It is known that annealed LT GaAs changes its properties (resistivity, carriers' lifetime) with the growth temperature [4]. Therefore, we designed and tested a multilayer LT GaAs structure grown at an optimized temperature to increase the performance of the THz emitter. Finally, we have shown an application of the Bragg mirror based photoconductive THz antenna for the THz generation in the cavity of femtosecond Ti:Sapphire laser.

2 Optical model of emitter

The schematic layout of the structure of the photoconductive THz emitter is shown in Fig.1. The structure consists of a multilayered Bragg mirror with a semi-insulating GaAs substrate below and low-temperature-grown GaAs on the top. The Bragg mirror layers consist of AlGaAs and AlAs. The optical parameters of all semiconductor materials were taken from [5]. We have modelled the total absorption and reflection for the absorbing LT GaAs layer using the transfer matrix method. The calculated absorption is not a monotonic

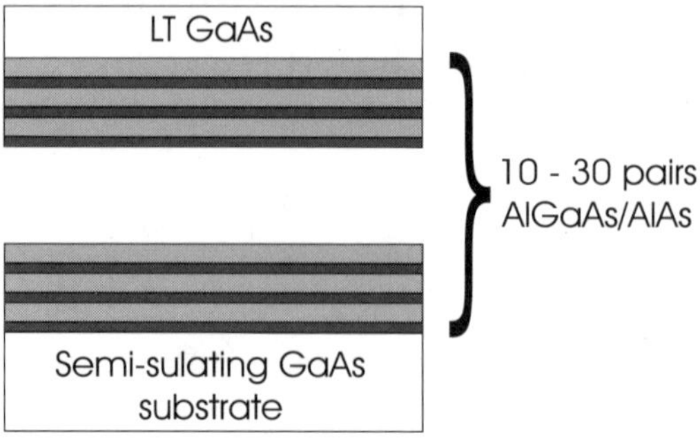

Fig. 1. Schematics of the layer structure of photoconductive THz emitter.

function of the layer thickness but shows resonance maxima and off-resonance minima (see Fig.2). The position of the maxima is given by the optical thickness of the layers equal to $1/2 * (1/2 + n) * \lambda_o$, where l_o is the wavelength of the incident light. An absolute maximum is achieved for a layer with an optical thickness of $5/4 * \lambda_o$. At this thickness the absorption reaches more than 90% for an absorption coefficient of $\alpha = 1.335 \times 10^4$ cm^{-1}, in GaAs at 800 nm [5]. This value is larger than that achievable for GaAs layer without Bragg mirror due to intrinsic 30% reflection at the air/GaAs interface. Here, the resonance in the optical cavity effectively cancels the reflection of the exciting light at the interface. The calculation of the absorption also suggests that nearly 50% of the energy can be absorbed in a layer as thin as 65 nm. This is relevant in applications where the total thickness of the absorbing layer plays a role, e.g. in the intra-cavity generation of the THz radiation [9].

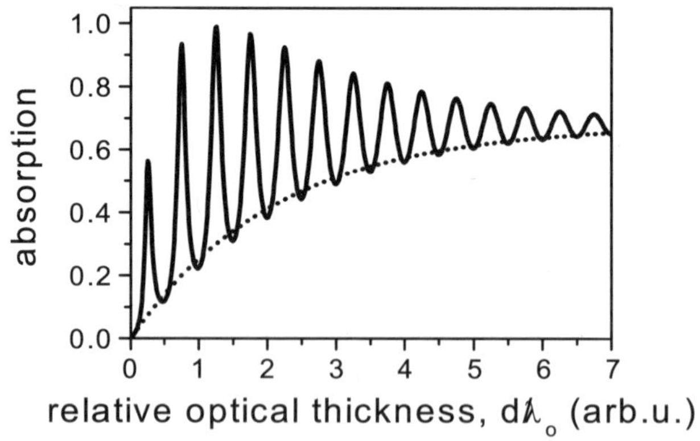

Fig. 2. Absorption of the 800 nm light in GaAs layer without (broken line) and with (solid line) the underlying Bragg mirror.

3 Experiment

A low-temperature MBE GaAs layer grown at temperatures 220 - 350 ^{o}C and annealed in-situ at 600 ^{o}C (10 minutes) was used as photoconductive material. The modified THz emitter is made of a 50 – 350 nm thick LT GaAs grown directly on a Bragg mirror (Fig.1). The Bragg mirror itself consisted of 10 – 30 pairs of $Al_{0.2}Ga_{0.8}As/AlAs$ layers designed for a center wavelength of 800 nm. As reference emitter a 2 μm thick LT GaAs layer grown directly on high resistive GaAs (100) substrate was used. Moreover, a bare semi-insulating GaAs was also used as reference. Electrical contacts to the photoconductive layer were made of Ti/Au metal and had the shape of coplanar 20 μm wide striplines separated by a gap of 300 μm.

The emitter chips of the size 2 x 3 mm were mounted onto a highly resistive silicon aplanar hemisphere lens to efficiently couple the generated THz radiation into free space. The emitters were excited using ultrashort optical pulses (pulse width $\sim$ 10 fs, center wavelength $\sim$ 800 nm, spectral width $\sim$ 120 nm, repetition rate $\sim$ 76 MHz, pulse energy $\sim$ 2.5 nJ) from a solid-state pumped mode-locked Ti:sapphire laser. The laser beam was focused onto the 300 μm gap between the polarized electrodes of the emitter. The shape of THz pulse and average output power were measured by electro-optic sampling [7] and by an autocorrelation technique with a calibrated Si bolometer [8], respectively.

4 Results

4.1 Role of Bragg mirror

First we have studied the contribution of the Bragg mirror to the performance of the THz emitter. We have compared emitter structures with and without Bragg mirror. The spectra obtained by fast Fourier transformation of the time domain THz signals from both emitter types are shown in Fig. 3. The THz electric field emitted from the reference emitter with a thin LT GaAs layer but with Bragg mirror is about two times stronger than that for the emitter without the Bragg mirror but with a 2 μm thick LT GaAs layer. In terms of power, the emitter with the Bragg mirror emits more than 4 times higher THz power at the peak frequency (Fig. 3).

Let us now quantify the change in the efficiency of the THz emission due to the Bragg mirror in the emitter structure. The calculation of the absorption of light with a wavelength of 800 nm yields 21% and nearly 69% for 326 nm and 2 μm thick GaAs layers, respectively. Thus with respect to the measured 4 times larger THz output power, an increase in the THz emission efficiency by more than 10 was achieved for the THz emitter with Bragg mirror [6]. The bias dependence of the emitted THz field shows for both types of emitters a linear increase up to 240–260 V. At biases above 260 V heating related device failure was observed for long time operation. The observed improved efficiency of the THz generation for the emitter with the Bragg mirror is explained by a larger electric field and a higher density of photogenerated carriers in the active region of modified THz emitter [6].

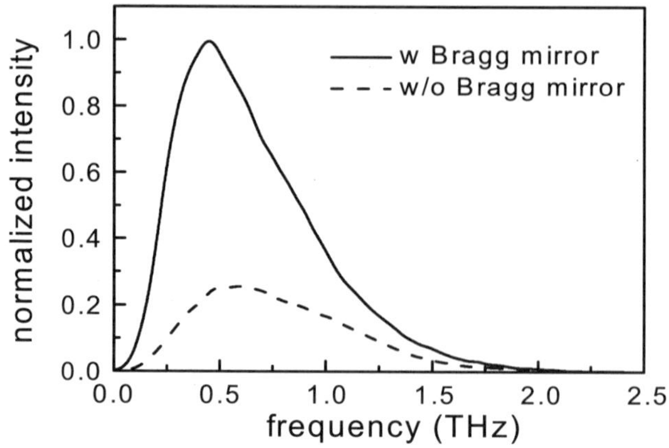

Fig. 3. Normalized power spectrum of the emission from a THz emitter with (solid line) and without (broken line) Bragg mirror.

4.2 Role of growth temperature

To access the photoresponse of LT GaAs layers grown at various temperatures, we used an attenuated unfocused laser beam from the Ti:sapphire laser to homogeneously illuminate the area between the electrodes of the emitter structures. In this configuration the measured DC photocurrent depends only on the lifetime and mobility of the photogenerated carriers. As expected, the photocurrent of the structures exhibited a clear monotonic decrease with a decreasing growth temperature of LT GaAs indicating that carriers' mobility and lifetime are reduced.

Fig. 4 shows the bias dependence of the amplitude of the THz transients for emitters made of 350 nm thick LT GaAs grown at temperatures between 220 and 350°C on the semiconductor Bragg mirror, when excited by a 10 *fs*-laser pulse. The steepest rise of the THz emission with emitter bias was observed for the structure made of LT GaAs grown at 320°C. On the other hand, the emitter structure made of LT GaAs grown at the lowest temperature of 220 ^{o}C exhibits the weakest increase of the THz radiation with bias. The lowest THz generation efficiency correlates well with the weakest photoresponse observed for this type of THz emitter and is due to the short lifetime of photogenerated carriers and to the low mobility. In all cases, the THz amplitude increases linearly with the applied bias up to about 200–260

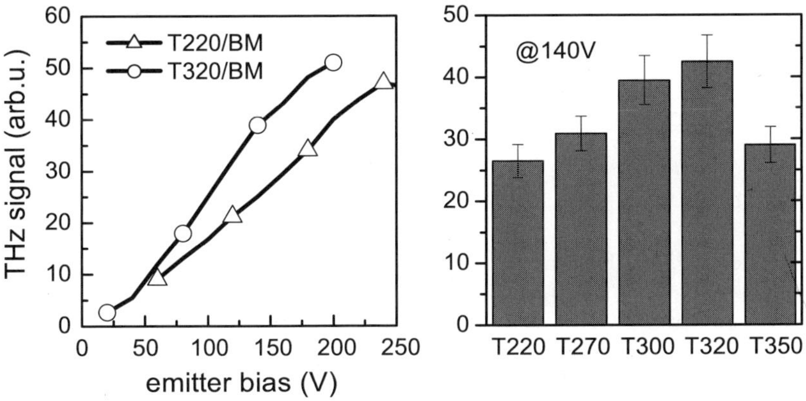

Fig. 4. THz emission from the different emitters as a function of bias (left panel), and at fixed bias of 140 V (right panel).

V depending on the growth temperature of LT GaAs. At higher biases the THz emitter exhibit a heating related device failure.

4.3 Intra-cavity emitter

A photoconductive THz emitter with the above discussed design was recently successfully used as an intra-cavity THz generator in a femtosecond Ti:Sapphire oscillator [9]. The number of layers of Bragg mirror for the emitter structure was chosen with respect to the high reflectivity needed for the intra-cavity mirror. The thickness of the absorbing LT GaAs layer was chosen to achieve about 20% absorption in the structure. Further details about the THz emitter design can be found elsewhere [9]. The emitter was placed into the cavity of a commercial *fs* Ti:sapphire laser (FemtoCompact of Femtolasers Produktions). For 4 W pump power the oscillator operated at about 900 mW of the intra-cavity power and the emitter delivered $\sim$7 μW of average THz power. This THz output power is voltage-controlled without disturbing the lasing and mode-locking, and can be electrically modulated up to 100 kHz [9].

5 Conclusions

The performance of photoconductive THz antenna emitters based on the epitaxial GaAs grown at different temperatures on a AlGaAs/AlAs Bragg mirror is presented. A maximum output THz power was observed for a growth at a temperature of about 320°C. The resonant cavity effect, the effective optical and electrical isolation of the photoconductive layer from the substrate by Bragg mirror, as well as optimized mobility-lifetime product in the LT GaAs leads to an enhancement of the optical-to-THz conversion efficiency.

The work was done within the framework of the European Community project IST-1999-10154 (Teravision). The authors acknowledge also support from the Austrian Fond zur Förderung des Wissenschaftlichen Forschung (SFB-ADLIS).

References

1. Grischkowsky, D., Keiding, S., van Exter, M., and Fattinger, C.: Far-infrared Time-domain Spectroscopy of Dielectrics and Semiconductors with Terahertz Beams, J. Opt. Soc. Am. B **7**, 2006–2015 (1990)
2. Darrow, J.T., Zhang, X.-C., Auston, D.H., and Morse, J.D.: Saturation Properties of Large-aperture Photoconducting Antennas, IEEE J. Quantum Electron. **28**, 1607–1616 (1992)
3. Cai,Y., Brener, I., Lopata, J., Wynn, J., Pfeiffer, L., and Federici, J.: Design and Performance of Singular Electric Field Terahertz Photoconducting Antennas, Appl. Phys. Lett. **71**, 2076–2078 (1997)
4. Look, D.C.: Molecular Beam Epitaxial GaAs Grown at Low Temperatures, Thin Solid Films **241**, 61–73 (1993).
5. Palik E.D.:Handbook of Optical Constants of Solid, Academic Press, London (1985)
6. Darmo, J., Müller, T., Strasser, G., Unterrainer, K., and Tempea G.: Terahertz Emitter with Intergrated Semiconductor Bragg Mirror, Electron. Lett. **39**, 460–462 (2003)
7. Wu, Q. and Zhang, X.-C.: Ultrafast Electro-optic Field Sensor, Appl. Phys. Lett. **68**, 1604–1606 (1996)
8. Kersting, R., Unterrainer, K., Strasser, G., Kaufmann, H.F., and Gornik, E.: Coherent Plasmon inn-doped GaAs, Phys. Rev. B **58**, 4553–4559 (1998)
9. Darmo, J., Müller, T., Strasser, G., Unterrainer, K., Le, T., Stingl, A., and Tempea, G.: Voltage-controlled Intracavity Terahertz Generator for Self-starting Ti:sapphire Lasers, Optics Lett. **27**, 1941–1943 (2002)

Magnetic-Field Induced Enhancement of THz-Radaiation from InAs Irradiated with 0.8 μm Laser up to 27 T and InSb Irradiated with 1.56 μm Laser up to 5 T

H. Takahashi[1], A. Quema[2], M. Goto[2], S. Ono[2], and N. Sarukura[2]

[1] Department of Photo Science, The Graduate University for Advanced Stdies, Shonan Village, Hayama 240-0193, Japan htakahas@ims.ac.jp
[2] Institute for Molecular Science Myodaiji, Okazaki 444-8585, Japan

Summary. Magnetic-field dependences of THz-radiation from InAs and InSb were investigated. For InAs irradiated with 0.8 μm laser , it is found that THz-radiation power exhibits oscillation-like behavior with increasing magnetic field, and reaches maximum at around 3 T. Furthermore, significant enhancement of THz-radiation power is successfully achieved for InSb by using a 1.56 μm exciation and an external magnetic field of 1.2 T.

1 Introduction

Various THz-radiation emitters including photoconductive switches,[1] semiconductor surface,[2] and nonlinear optical process [3] have been reported by utilizing ultrafast optical pulses. From the application viewpoint, the development of compact and intense THz-radiation emitter is strongly required. In this sense, semiconductor surfaces have attracted a great deal of attention. [4-6] Previously, we achieved significant enhancement of THz-radiation power by using InAs surface and reported the quadratic magnetic-field and excitation-intensity dependence of THz-radiation power. [7,8] Furthermore, anomalous magnetic-field dependence of THz-radiation power was observed, including saturation, decrease, and recovery. [9] In this report, the magnetic-field dependence of THz-radiation power from InAs surface was investigated up to 27 T. It is found that THz-radiation power saturates at around 3 T and 13 T, and THz-radiation power at 3 T is much higher than that at 13-T. Additionally, the magnetic-field dependence of THz-radiation power from InSb was also investigated up to 5 T by using a 1.56 μm laser as an excitation source.

2 Experiment and Results

The experimental setup is shown in Fig.1. A Ti:sapphire laser and a fiber laser (IMRA Model A50) are used as excitation sources. A mode-locked

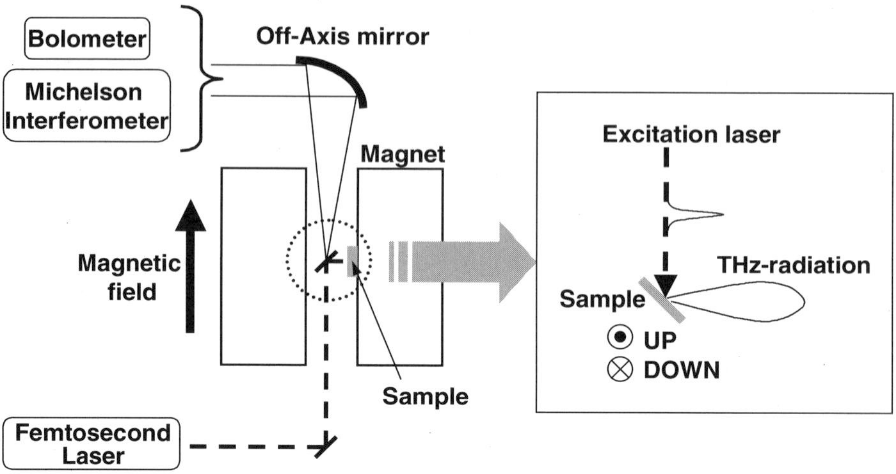

Fig. 1. Experimental setup for the THz-radiation emitter in a magnetic field. The magnetic field was applied parallel to the semiconductor surface.

Ti:Sapphire laser delivered 100-fs pulses at a wavelength of 800 nm with a 82-MHz repetition rate. A mode-locked fiber laser delivered 100-fs pulses at a wavelength of 1560 nm with a 50-MHz repetition rate. The sample used was undoped bulk InAs(100) and InSb(100), which have carrier concentrations of $3 \times 10^{16}\ cm^{-3}$ and $4 \times 10^{14}\ cm^{-3}$, respectively. Magnetic field is provided by the specially designed hybrid magnet for 28 T experiment [10], and the supperconducting magnet for 5 T experiment. A liquid-cooled Ge bolometer was provided for detecting the power of the total radiation. The fourier spectrum of THz-radiation was measured by a Michelson-interferometer.

Figure 2(a) illustrates the THz-radiation power measured in various magnetic fields. THz-radiation power exhibits quadratic dependence on the magnetic field, and increases until it saturates at around 3 T. Then, THz-radiation power decreases, and reaches a minimum value. Further increasing the magnetic field further, THz-radiation power recovers and saturates again at around 13 T, in which slight difference is observed depending on the magnetic-field direction. Figure 2(b) presents the THz-radiation spectra at selected magnetic fields. THz-radiation spectrum at 3 T is intense and exhibits broad-band structure. By applying higher magnetic fields above 12 T, THz-radiation spectrum is drastically changed and low-frequency component appears. Additionally, clear periodic structure is observed at magnetic-field above 12 T. From the comparison of interferograms for InAs with different thicknesses, the origin of this periodic structure is considered to be the interference of two THz pulses from front surface and backside of the substrate.

THz-radiation power measured in the presence of various magnetic fields for InSb is shown in Fig.3. Samples are irradiated with femtosecond laser un-

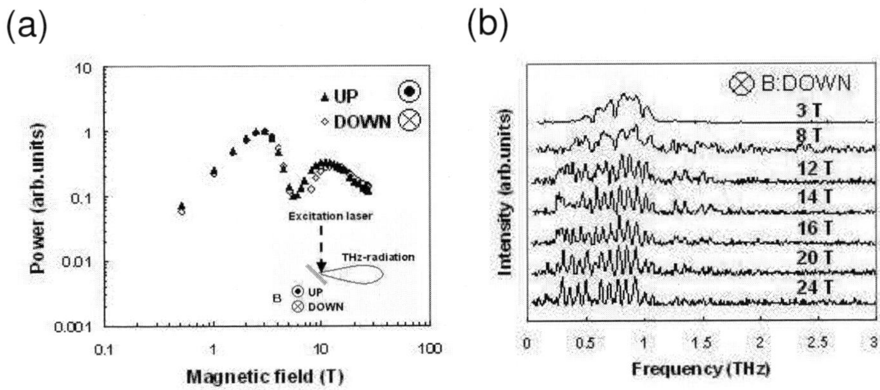

Fig. 2. (a) THz-radiation power measured in various magnetic fields. (b) THz-radiation spectrum measured at selected magnetic fields. InAs was irradiated with a femtosecond laser under excitation with 300-mW at 800-nm.

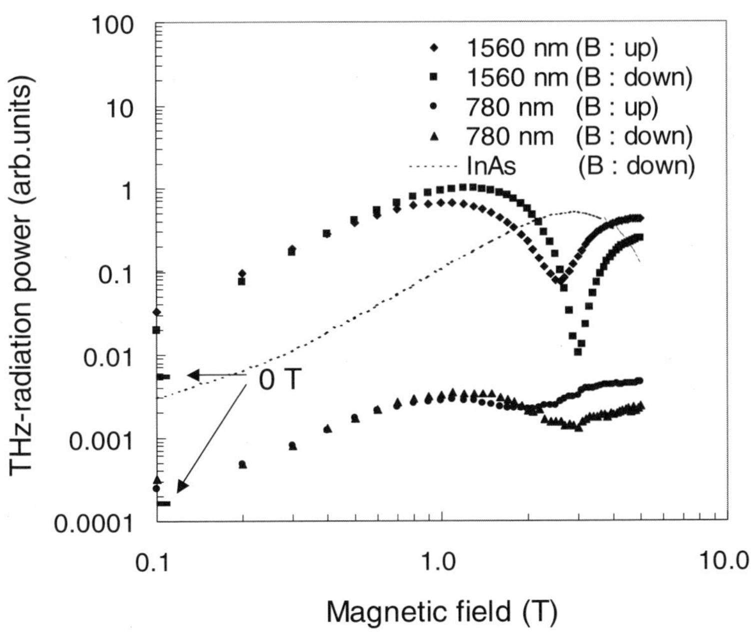

Fig. 3. Magnetic field dependence of THz-radiation power from InSb. InSb surface was irradiated with a mode-locked fiber laser under excitation condition of 160 mW at 1560 nm, 60 mW at 780 nm.

der excitation of 160 mW at 1560 nm, 60 mW at 780 nm. For 1560 nm optical excitation, THz-radiation power exhibits quadratic dependence on magnetic field, and increases until it saturates at around 1.0 T and 1.2 T depending on the magnetic-field direction. Then, THz-radiation power decreases, and reaches a minimum value. By increasing the magnetic field further, THz-radiation power recovers up to 5 T. For 780 nm optical excitation, magnetic-field dependence is very similar to that of 1560 nm optical excitation case, except that its power is lower by an order of magnitude. Part of the drastic enhancement of THz-radiation power due to excitation wavelength arises from the large number of carriers generated by the longer wavelength laser. However, this accounts for only a factor of at most 6, when comparing the average power and photon energy in both cases. Therefore, another effect should be considered to explain the enhancement of THz-radiation power. One of the possible explanations is given by the velocity overshoot. Under 780 nm optical excitation, photo-excited carriers gain high kinetic energy, and most of them are scattered into L valley due to electron-phonon scattering. The effective electron mass in L valley is much heavier than that of Γ valley. This effect leads to the less efficient THz-radiation. In contrast, under 1560 nm optical excitation, photo-excited carriers cannot gain sufficient energy to scatter into L valley, and can contribute to generate THz-radiation.

3 Summary

In summary, magnetic-field dependence of THz-radiation power from InAs surface was investigated up to 27 T by using specially designed hybrid magnet. It is found that THz-radiation power saturates at around 3 T and 13 T, and THz-radiation power at 3 T is much higher than that at 13-T. Furthermore, magnetic-field dependence of THz-radiation power from InSb is also presented. For 1.560 μm optical excitation, THz-radiation power from InSb is significantly enhanced by the external magnetic field, and its enhancement factor exceeds over 100. The maximum radiation power is obtained at a magnetic field of 1.2 T, which is easily achieved by conventional magnet.

References

1. Auston, D.H., Cheung, K.P., Smith, P.R.: Appl. Phys. Lett. **45**, 284–286 (1984).
2. Zhang, X.C., Hu, B.B., Darrow, J.T., Auston, D.H.: Appl. Phys. Lett. **56**, 1011–1013 (1990).
3. Kawase, K., Sato, M., Taniuchi, T., Itoh, H.: Appl. Phys. Lett. **68**, 2483–2485 (1996).
4. Zhang, X.C., Liu, Y., Hewitt, T.D., Sangsiri, T., Kingsley, L.E., Weiner, M.: Appl. Phys. Lett. **62**, 2003–2005 (1993).
5. Weiss, C., Wallenstein, R., Beigang, R.: Appl. Phys. Lett. **77**, 4160–4162 (2000).

6. Takahashi, H., Suzuki, Y., Sakai, M., Ono, S., Sarukura, N., Sugiura, T., Hirosumi, T., Yoshida, M.: Appl. Phys. Lett. **82**, 2005–2007 (2003).
7. Sarukura, N., Ohtake, H., Izumida, S., Liu, Z.: J. Appl. Phys. **84**, 654–656 (1998).
8. Ohtake, H., Ono, S., Sakai, M., Liu, Z., Tsukamoto, T., Sarukura, N.: Appl. Phys. Lett. **76**, 1398–1400 (2000).
9. Ohtake, H., Murakami, H., Yano, T., Ono, S., Sarukura, N., Takahashi, H., Suzuki, Y., Nishijima, G., Watanabe, K.: Appl. Phys. Lett. **82**, 1164–1166 (2003).
10. Muto, Y., Nakagawa, Y., Noto, K., Miura, S., Hoshi, A., Watanabe, K., Kido, G., Ichikawa, H., Fujioka, T., Sato, Y., Osaki, O., Takano, H.: Sci. Rep. RITU, A. **33**, 221–237 (1986)

Part VII

Novel Applications of Ultrashort Light Pulses

Ultrafast Electron Diffraction

From the Gas Phase to the Condensed Phase with Picosecond and Femtosecond Resolution

Vladimir A. Lobastov, Ramesh Srinivasan, Franco Vigliotti, Chong–Yu Ruan, Jonathan S. Feenstra, Songye Chen, Sang T. Park, Shoujun Xu, and Ahmed H. Zewail*

Laboratory for Molecular Sciences, Arthur Amos Noyes Laboratory of Chemical Physics, California Institute of Technology, Pasadena, CA 91125

1 Introduction

Diffraction techniques have allowed determination of three-dimensional *equilibrium* structures with atomic resolution in systems ranging from diatoms (NaCl) to DNA, proteins and complex assemblies such as viruses. For dynamics, the time resolution has similarly reached the fundamental atomic scale of motion. With the advent of femtosecond time resolution nearly two decades ago, it has become possible to study the dynamics of *non-equilibrium* molecular systems, also from the very small (NaI) to the very large (DNA, proteins and their complexes) [1]. A tantalizing goal is the potential to map out, in real time, the coordinates of all individual atoms in complex chemical and biological reactions.

X-ray and electron diffraction, if endowed with ultrafast temporal resolution, can provide the unique ability of revealing all internuclear coordinates of *transient* structures with very high spatial resolution. In our laboratory, the method of choice has been ultrafast electron diffraction (UED), for the following reasons: First, the cross-section for electron scattering is six orders of magnitude larger compared to x-ray scattering. Second, UED experiments are 'tabletop' scale and can be implemented with ultrafast (femtosecond and picosecond) laser sources. Third, electrons are less damaging to specimens per useful elastic scattering event. Fourth, electrons, because of their short penetration depth arising from strong interaction with matter, can reveal transient structures in gases, surfaces and (thin) crystals.

The development of UED has evolved through different phases. The "dreaming phase" was conceptual [2, 3] and was followed by technical and theoretical advances in order to develop the methodology and obtain our first UED images — the "exploration phase" [4–12]. The two phases culminated in the breakthrough developments of the third-generation apparatus (UED-3) for the gas phase in 2001 [13] and the fourth-generation apparatus (UED-4) for the condensed phase and biological systems in 2003 [14]. These two generations launched the "explosion phase", where we have been able to

---

* Based on the opening keynote lecture at UFO IV, Vienna (June 30, 2003)

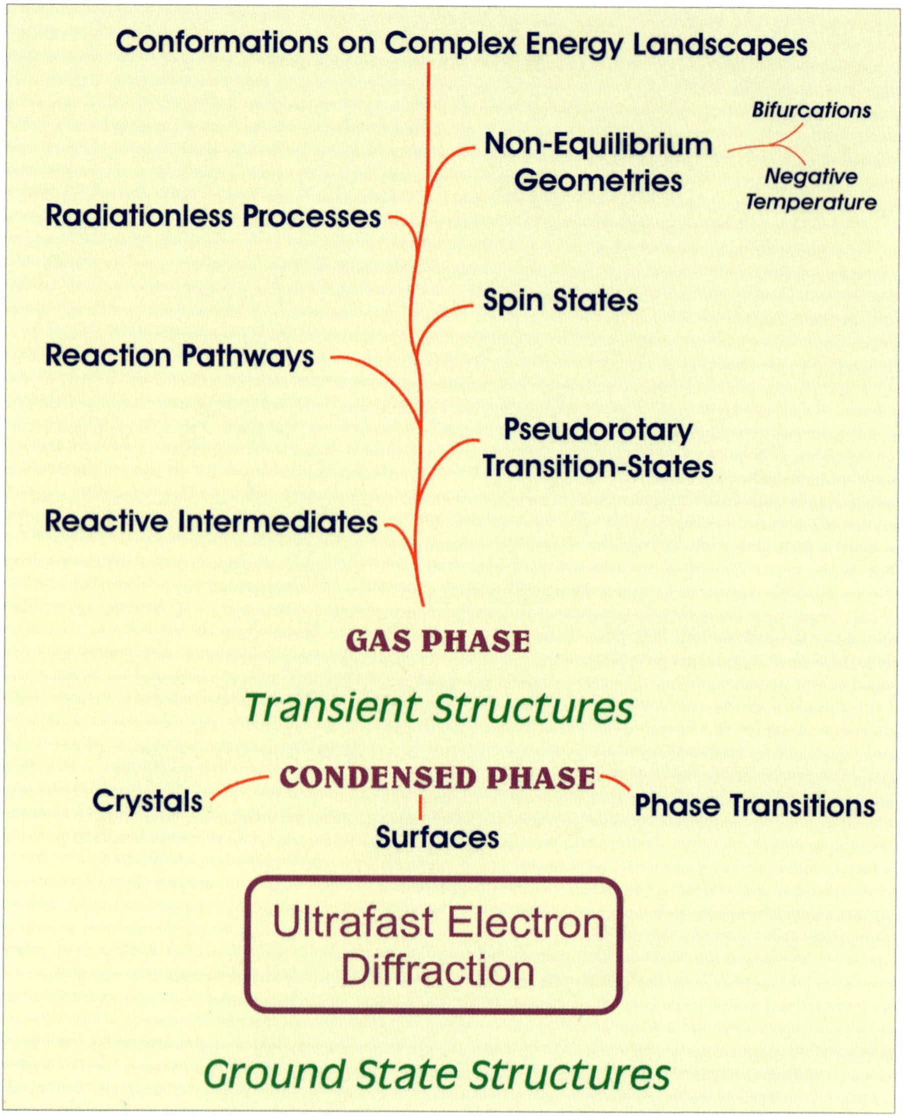

Fig. 1. Scope of UED applications

image complex molecular structures [15–22] in the four dimensions of space and time with atomic-scale resolutions. The detection sensitivity of structural change is as low as 1% in the gas phase, and less than a monolayer in the condensed phase. A variety of phenomena have been studied (Fig. 1), and a comprehensive review of the principles and some applications has recently been published [22]. Here we will highlight some of the key concepts and challenges in UED, followed by illustrative examples in the gas phase and the condensed phase. Due to space constraints, we will limit references to the work from this laboratory — original citations are in [22].

2 UED Principles

The UED technique employs properly timed sequences of ultrafast pulses — a femtosecond laser pulse to initiate the reaction and ultrashort (femtosecond or picosecond) electron pulses to probe the ensuing structural change in the sample. The resulting electron diffraction patterns are then recorded on a CCD camera. This sequence of pulses is repeated, timing the electron pulse to arrive before or after the laser pulse; in effect, a series of snapshots of the evolving molecular structure are taken. Each time–resolved diffraction pattern can then, *in principle*, be inverted to reveal the three-dimensional transient structure that gave rise to the pattern at that specific time delay. However, *in practice*, a key challenge lies in recovering the structural information that is embedded in the as-acquired diffraction images.

A key advance in extracting the small population of changing structures from the large background signal has been the development of the Diffraction Image Referencing (Diffraction–Difference) Method; see [22]. The method consists of timing the electron pulses so as to establish an in situ reference signal (usually the ground state structure obtained at negative time or a precursor transient structure). The digital nature of our processing methodology then allows us to obtain the difference of each time-resolved diffraction pattern from this reference image, thus revealing the change from the reference structure in the form of difference rings (gases and liquids) or difference spots (crystals).

One of the most important features of the diffraction–difference method is the control over the sequence of electron pulses — t_{ref}. The choice of t_{ref} allows us to isolate structures of different transient species:

(1) By choosing t_{ref} to be at negative time, we can obtain the ground–state diffraction pattern. Also, by recording diffraction images at two different negative times (probing the same reactant structure at each of these times), we can obtain a control diffraction–difference image to verify the absence of rings or Bragg spots.

(2) By choosing t_{ref} to be at a specific positive time, we can isolate different transient species in, say, non-concerted reactions based on the relevant timescales of the non-concerted bond breaking (Sect. 4.1).

(3) Finally, we can also extract the molecular diffraction image resulting only from the transient species via the 'transient–only' or the 'transient–isolated' method.

This methodology, combined with the large electron scattering cross-section inherent in the UED experiment, provides the impetus for investigating diverse molecular phenomena. However, in order to realize this potential of UED, several other conceptual challenges had to be surmounted. First, there had not previously been a way to determine in situ the zero–of–time in UED. Second, for an ultrafast electron pulse, electron–electron repulsion takes place, and these space–charge effects broaden the pulse duration over time. Third, the orders–of–magnitude lower density of gas–phase samples relative to solids and surfaces results in much weaker scattering intensities. Last, an unprecedented sensitivity in the change in the number density of molecules detected had to be reached. These challenges in UED — the clocking of the change, the space–charge effect, the low densities in the gas phase or condensed phase monolayers — were major hurdles that had to be circumvented in order to reach the current state–of–the–art. Also, the theoretical apparatus for the structural analysis of complex systems had to be developed.

3 UED Generations

Figures 2 and 3 illustrate the different components of the apparatus for UED-3 and UED-4, respectively. Several disparate fields of study are involved: femtosecond pulse generation, electron beam optics, CCD detection systems, ultrahigh vacuum (UHV) technology and advanced computation. For UED-3 of isolated reactions, output from a femtosecond laser is split into a pump path and an electron–generation path. The pump laser proceeds directly into the vacuum chamber and excites a beam of molecules. The probe laser is directed toward a back-illuminated photocathode where the laser generates electron pulses via the photoelectric effect; the electrons are accelerated, collimated, focused, and then scattered by the isolated molecules; see [22] for original citations to the evolution of diffraction and detection methods on different time scales. The time delay between the arrival of the pump laser pulse and the probe electron pulse is controlled with a computer-driven translation stage. The resulting diffraction patterns are detected with a CCD camera, and the images are stored on a computer for later analysis. The UED-3 apparatus (Fig. 2) is also equipped with a time–of–flight mass spectrometer (MS–TOF) to aid in the identification of species generated during the course of chemical reactions.

In UED-4 (Fig. 3), the new features include three interconnected UHV chambers — the sample preparation, load-lock and scattering chambers. The crystal is mounted on a computer-controlled goniometer for high-precision (0.005°) angular rotation; it can be cooled to a temperature of 100 K. The preparation chamber has sputtering and cleaning tools, and is also equipped with low-energy electron diffraction (LEED) and Auger spectroscopy (AS) for characterization of the crystal surface.

As in UED-3, we have characterized the electron and light pulses used in UED-4. For the 30-keV electrons, we used in situ streaking techniques and, for light, the now standard autocorrelation method. We have obtained the fastest streaking speed of 140 ± 2 fs/pixel in UED-4, thus approaching the state–of–the–art in streak cameras. For the extraction field of 10 kV/mm, the

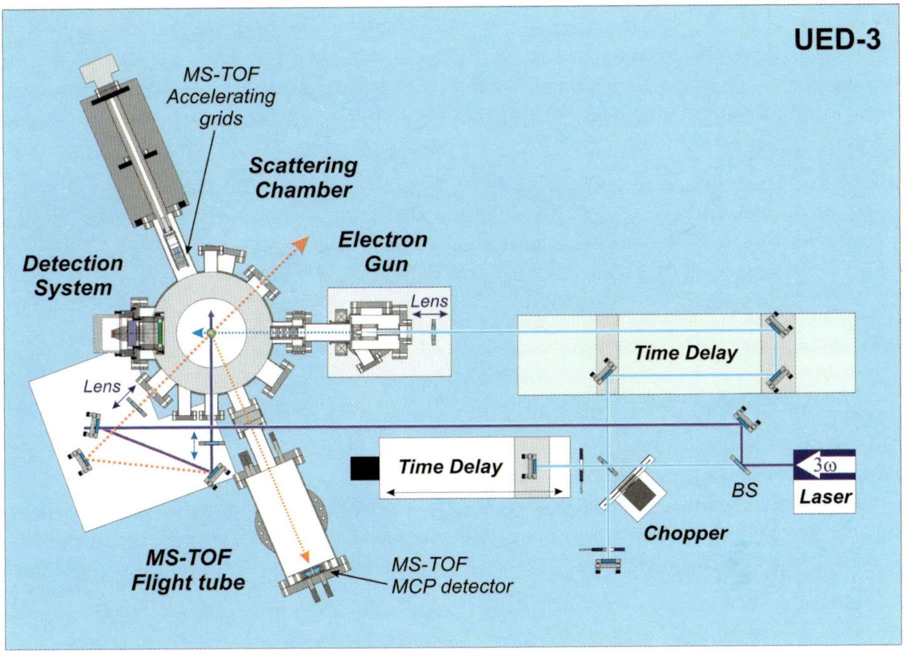

Fig. 2. UED-3 apparatus

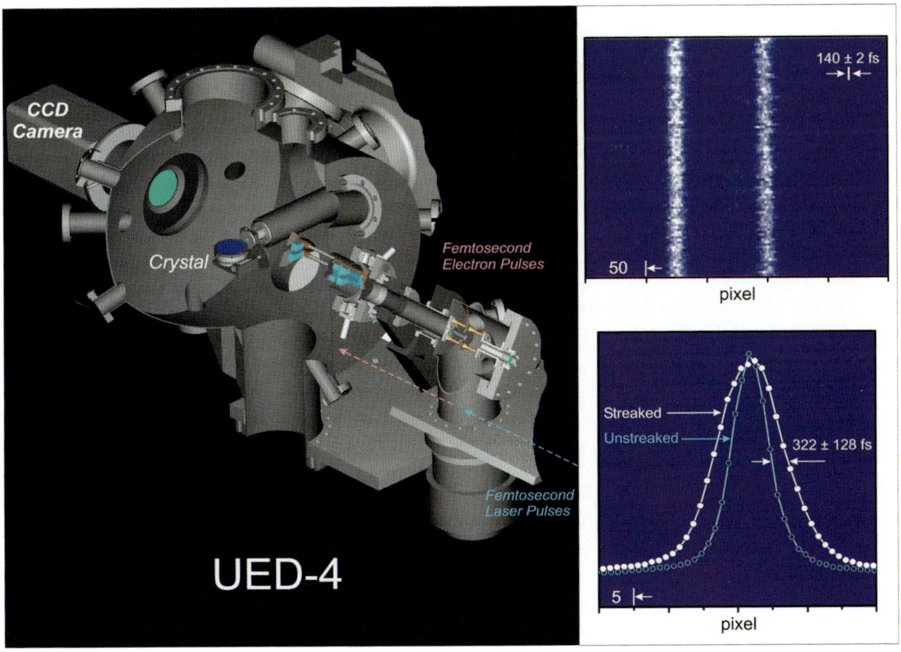

Fig. 3. UED-4 apparatus

spreading time is ~20 fs. The measured streaking broadening is 322 ± 128 fs (Fig. 3). Using the same electron gun design, Jianming Cao, a former member of this group and now at Florida State University, has achieved ~300 fs pulses, albeit with a streaking speed of 250 fs/pixel. The coherence time of electrons is 10 fs or longer giving a coherence length on the order of a micron. The initiation laser pulse duration is typically 120 fs, and the overall temporal resolution is determined by the geometry of the experiment and the relative widths (shapes) and speeds of the two pulses (see [22]).

4 UED Applications

The temporally and spatially resolved transient structures elucidated by UED include structures in radiationless transitions, structures in non-concerted organic reactions, structures in non-concerted organometallic reactions, structures of carbene intermediates, dynamic pseudorotary structures, non-equilibrium structures and conformational structures on complex energy landscapes, transient structures of surfaces and bulk crystals, and solid–to–liquid phase transitions. These studies were reported in the references cited above. Here we give a few select examples.

4.1 Transient Structures in Reactions

We have directly observed the structural dynamics of isolated chemical reactions via changes in diffraction patterns, exemplified by the 2D difference rings in Fig. 4. A prototypical case is that of the non-concerted elimination reaction of dihaloethanes. It demonstrates the UED methodology of using different electron pulse sequences — different t_{ref} — to isolate the *reactant*, *intermediates in transition*, and *product* structures. The reaction involves the elimination of two iodine atoms from the reactant $C_2F_4I_2$ to give the product C_2F_4. The structures of all intermediates were unknown, and the challenge lay in determining the structural dynamics of the entire reaction.

Structural Dynamics

Figure 5 shows the time-dependent difference radial distribution curves referenced to −95 ps, $\Delta f^{\mathrm{E}}(t;\ \textit{–95 ps};\ r)$, indicating the structural change occurring over the course of the reaction. These radial distribution functions were derived from the molecular scattering functions, which in turn were obtained from 2D CCD patterns. It is significant to note that the negative peak intensity at ~5.1 Å in the $\Delta f(r)$ curves remains constant after 5 ps, whereas the peak intensities around 2–3 Å continue to increase over a longer time scale. The negative peak at ~5.1 Å results from the loss of the I··I internuclear separation of the *anti* conformer of the parent $C_2F_4I_2$ molecules, while those at 2–3 Å result primarily from the depletion of C–I, F··I, and C··I distances.

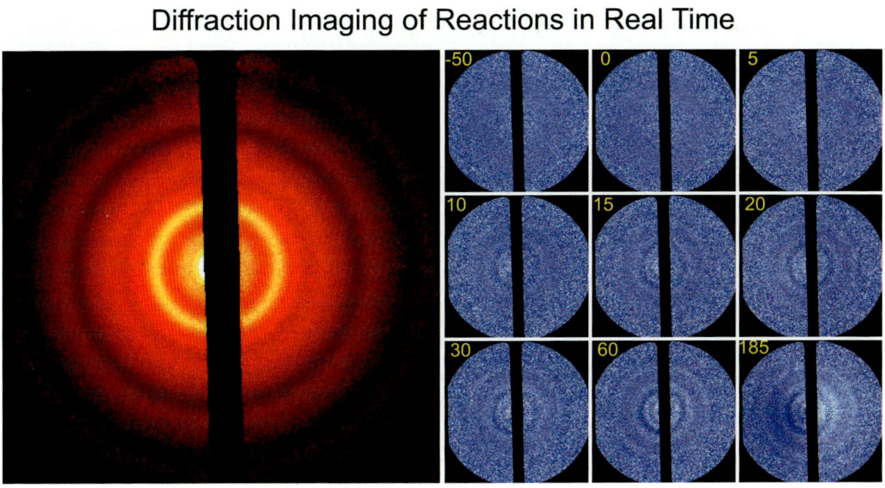

Fig. 4. 2D diffraction rings in the gas phase (pyridine)

These observations demonstrate the non-concerted nature of the structural changes in the reaction: The first step ($C_2F_4I_2 \longrightarrow C_2F_4I + I$) occurs on the femtosecond time scale (in Fig. 5, note the growth in peak intensity between 0 and 5 ps, and the lack of growth thereafter), whereas the second step ($C_2F_4I \longrightarrow C_2F_4 + I$) is considerably slower, taking place in 26 ± 7 ps, which we obtained from the change in structure population with time.

Since the reaction involving the reactant $C_2F_4I_2$ is complete within the first 5 ps, we generated a set of additional diffraction–difference curves with $t_{\text{ref}} = +5$ ps — in order to highlight the structural changes of the reaction intermediate and product only — with no contribution from any other species present. Fig. 5 shows $\Delta f^{\text{E}}(t;\ \mathit{5\ ps};\ r)$, which arises only from the transient C_2F_4I and final product C_2F_4 species, with the depletion of the C_2F_4I radical being evident at C–I, C··I, and F··I separations; note that the populations of other internuclear separations (e.g. C–F, C–C, and F··F) are essentially unchanged. The absence of an I··I component (~5.1 Å) in the $\Delta f^{\text{E}}(t;\ \mathit{5\ ps};\ r)$ curves clearly shows that we are observing solely the population change of the transient C_2F_4I structure forming C_2F_4. Theoretical $\Delta f^{\text{T}}(t;\ \mathit{5\ ps};\ r)$ curves (red curves in Fig. 5) agree well with the experimental results. The time-dependent population of C_2F_4 formed after 5 ps yields a time constant of 25 ± 7 ps, in complete agreement with the results of the $\Delta f^{\text{E}}(t;\ \mathit{-95\ ps};\ r)$ and reconfirms the time scale for the structural dynamics of the reaction.

Freezing the Intermediate and Determining the Structure

The molecular structure of the C_2F_4I intermediate was determined from the diffraction–difference curves $\Delta sM(t;\ \mathit{5\ ps};\ s)$. Both the bridged and classical C_2F_4I structures were considered in the analysis of the diffraction data. The

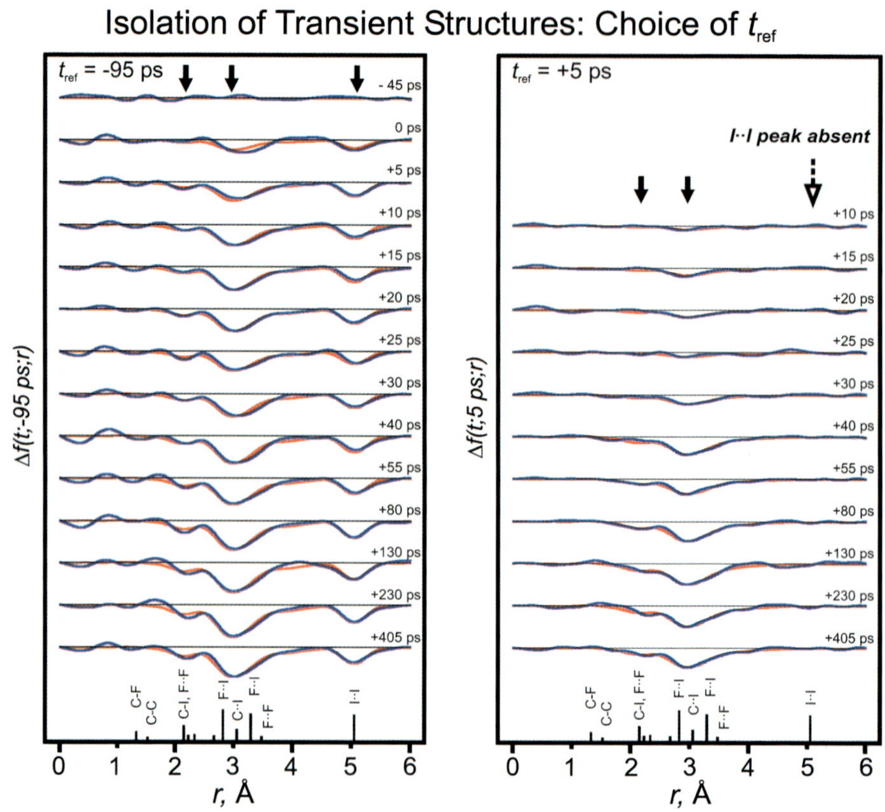

Fig. 5. Diffraction Image Referencing and the sequencing of pulses (t_{ref}) to isolate transient structures of a chemical reaction

theoretical curves for the mixture of classical structures (shown in [22]; Fig. 19) reproduce the experimental data very well, whereas the fit provided by the theoretical bridged structure is vastly inferior. Thus, we conclude that the structure of the C_2F_4I radical intermediate is, in fact, classical in nature — the iodine atom does not bridge the two carbons.

The structural parameters of the C_2F_4I intermediate are given in Fig. 6. The C–I and C–C distances of the C_2F_4I intermediate are, respectively, longer and shorter than those of the reactant, while the C–F′ internuclear distance in the radical site (–CF'_2) is shorter than that of the –CF_2I site. These results elucidate the increased C–C and decreased C–I bond order resulting from the formation of the transient C_2F_4I structure. Moreover, the ∠CCF′ and ∠F′CF′ angles become larger than the corresponding angles of the reactant (by $\sim 9°$ and $\sim 12°$, respectively), suggesting that the radical center (–CF'_2) of the C_2F_4I intermediate relaxes following loss of the first I atom. The structures (Fig. 6) and dynamics reported for this reaction are vital in describing the retention of stereochemistry in such class of reactions, as detailed in [22].

4.2 Light and Heavy Atom Effects

Non-equilibrium Conformations of Complex Landscapes

Complex transient structures with no heavy atoms are ubiquitous in chemical and biological reactions. To elucidate such structures, we performed UED in a homologous series of molecules — cyclohexadiene, cycloheptatriene and cyclooctatriene (COT3) — comprising of only carbons and hydrogens. These molecules also provide the opportunity for studying conformational changes on complex energy landscapes. Figure 7 shows the refined structure of the COT3 transient intermediate. Instead of *anti* character (which would be expected if the molecule were at equilibrium), the various conformations of the ring-opened structure show considerable *gauche* character indicating an inverted (non-Boltzmann) torsional distribution. The non-Boltzmann behavior is also revealed in the disparities of the torsional motions and in the corresponding carbon–carbon single bond distances. The two terminal torsions have broad distributions ($\pm 40°$); in contrast, the torsion at the central carbon–carbon bond is rather confined ($\pm 6°$), which establishes the *gauche* character of octatetraene (OT). Correspondingly, the central C–C bond distance was found to be lengthened to $\sim$1.487 Å while all the other carbon–carbon bonds remain at (near) equilibrium distances.

These far-from-equilibrium OT structures reflect large amplitude torsional motions which persist for long times and which are not in thermal equilibrium with the rest of molecular motions, indicating a bottleneck in energy redistribution. The longer times involved in these torsions and their localization and isolation may be a key for biological function in macromolecules; see [22].

Non-radiative Transitions

In Fig. 8, we show the refined ground–state structure obtained for chlorobenzene. We recently studied this system — benzene with a heavy atom (chlorine) — for several reasons, including the effect of a heavy atom on the local change in structure and the chemistry that follows non-radiative transitions. The ground–state structure was determined and the observed changes following the C–Cl bond breakage were dramatic. Elsewhere, we will detail the structural dynamics.

4.3 The Condensed Phase

Diffraction of Crystals and Surfaces

In UED-4, we studied the temporal evolution of bulk and surface structures with atomic scale spatial resolution. Here, we briefly discuss our studies on

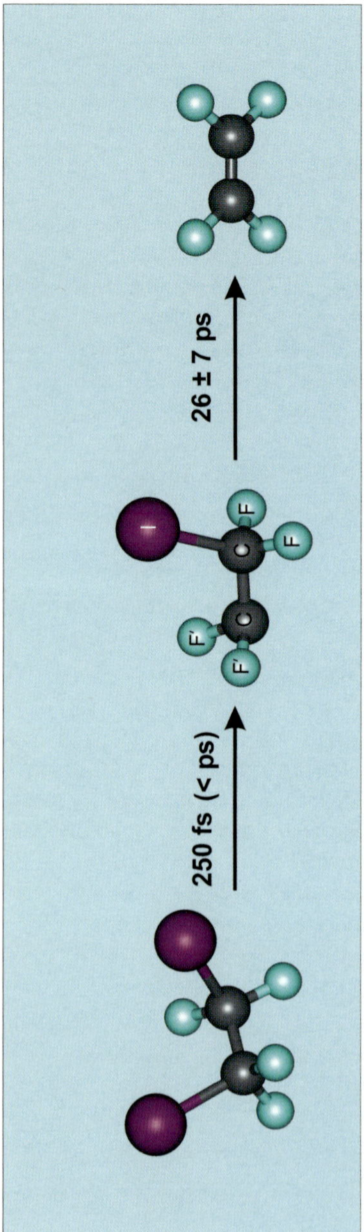

Reactant

Distances

	experiment	ab initio anti	ab initio gauche
r (C-C)	1.534 ± 0.013	1.532	1.540
r (C-F)	1.328 ± 0.003	1.320	1.323
r (C-I)	2.136 ± 0.007	2.159	2.147

Angles

	experiment	ab initio anti	ab initio gauche
α (C-C-F)	109.4 ± 1.0	109.0	107.6
α (C-C-I)	111.6 ± 1.0	111.9	114.8
α (F-C-F)	107.8 ± 1.0	108.7	107.9
ϕ^{anti}(IC-CI)	180 (fixed)	180.0	
ϕ^{gauche}(IC-CI)	70 ± 3		67.8

Intermediate

Distances

	Experiment	ab initio anti	ab initio gauche
r (C-C)	1.478 ± 0.049	1.503	1.508
r (C-F)	1.340 ± 0.037	1.322	1.327,1.323
r (C-I)	2.153 ± 0.013	2.164	2.149
r (C-F')	1.277 ± 0.027	1.304	1.309,1.307

Angles

	Experiment	ab initio anti	ab initio gauche
α (C-C-F)	108.6 ± 6.0	108.6	109.8,108.1
α (C-C-I)	115.0 ± 3.1	112.7	111.8
α (F-C-F)	108.0 ± 11.2	108.8	108.0
α (C-C-F')	117.9 ± 3.1	114.0	112.3,113.8
α (F'-C-F')	119.8 ± 7.8	111.8	111.2

Product

Distances

	Experiment	DFT
r (C-C)	1.311 ± 0.021	1.306
r (C-F)	1.319 ± 0.006	1.312

Angles

	Experiment	DFT
α (C-C-F)	123.8 ± 0.6	123.8

Fig. 6. Structural dynamics of elimination reactions: from reactant through intermediate to final product

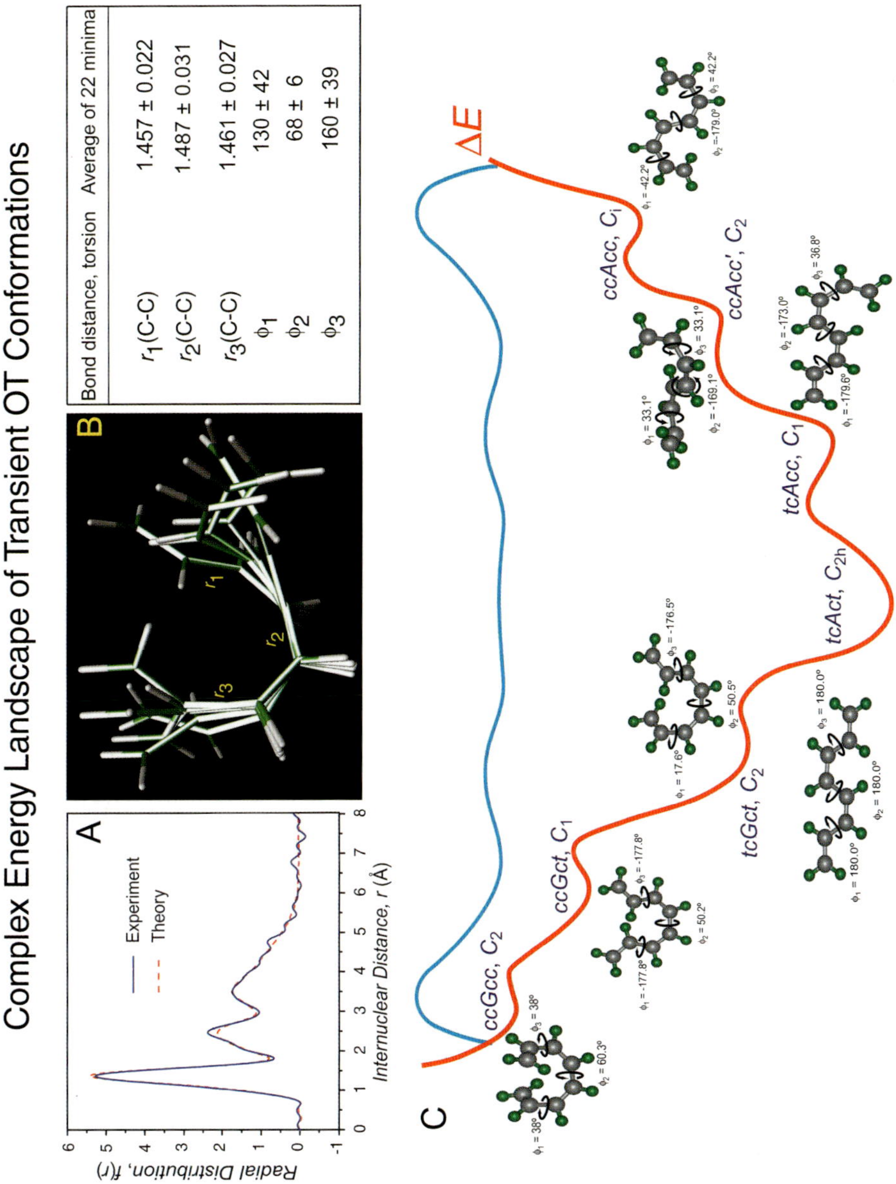

Fig. 7. Complex energy landscapes involving non-equilibrium conformational dynamics — structures determined by UED

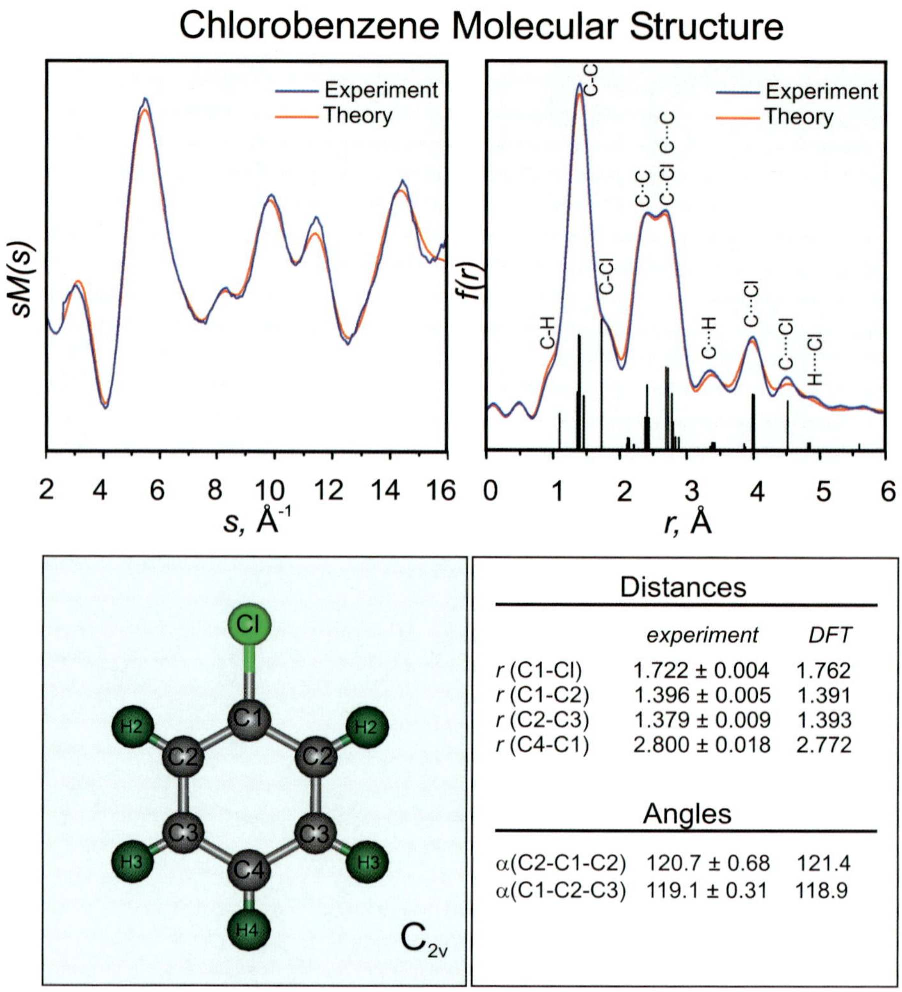

Distances

	experiment	DFT
r (C1-Cl)	1.722 ± 0.004	1.762
r (C1-C2)	1.396 ± 0.005	1.391
r (C2-C3)	1.379 ± 0.009	1.393
r (C4-C1)	2.800 ± 0.018	2.772

Angles

α(C2-C1-C2)	120.7 ± 0.68	121.4
α(C1-C2-C3)	119.1 ± 0.31	118.9

Fig. 8. Heavy atom effect in UED (chlorobenzene)

crystals of silicon (with and without adsorbates). Referencing the diffraction–difference to the ground–state structure, Fig. 9 shows the changes in the structure caused by the initiating pulse, from ground state pattern at negative time to the observed change at positive time. The structural change is evident in the shift with time of the in-phase Bragg peak of the rocking curve, while the increase in vibrational amplitude is reflected in the broadening (Fig. 10). The evolution takes place as a rise to a maximum shift and then a decay to the coordinates of the original structure. By gating on a Bragg spot, we can follow the changes with time. These results show the "instantaneous" structural change (1 ps in Fig. 10), a homogeneous expansion of the lattice 2.35-Å Si–Si bond by up to 0.04 Å at 50% level of damaging fluence. This is followed by lattice relaxation from highly non-equilibrium structure to the final ground state structure with Si–Si distances of 2.35 Å. The multiple time constants of restructuring describe the expansion in real space and vibrational temperature as reported elsewhere [14].

For the surface structures (and with hydrogen or chlorine), the results are striking. Gating the streak at the out-of-phase condition, we directly resolve the surface structural interferences and their evolution with time. Remarkably, the two spots change in time, and differently, but maintain the phase coherence. We are therefore able to display the spatial, temporal and phase coherence change of the surface structure, and the spatial patterns are entirely describable within the framework of diffraction theory of condensed matter. The change in the structure is evident in the amplitude change of surface Si atoms. Unlike the result for bulk displacement (0.04 Å), the expansion of surface atoms is larger by an order of magnitude (0.2 Å). The cooling of the surface structure occurs on a time scale different from that of the bulk.

Structures in Phase Transitions

We also studied the structural changes involved in phase transitions when the temperature of the lattice is sufficiently high to cause large amplitude disorder. Initiating an ultrashort temperature jump of the amorphous structure with the infrared femtosecond pulse gives new diffraction ring patterns, which we followed as a function of time by referencing to the image of the ground state (Fig. 9). The instantaneous structural change is a phase transition to the liquid state. This is evident in the change of the radial distribution function, as we observe a depletion of the peak of the second coordination shell at ~4 Å, and the robustness of the first coordination shell at 2.35 Å. In amorphous silicon, atoms are topologically connected with fourfold coordination, but with the bond angles fluctuating substantially around 109.5°, the tetrahedron angle. In the liquid phase, these fluctuations are further enhanced, and the radial distribution function shows almost no second coordination shell. We note that the time scale reported here for the phase transition is different from that of the crystalline–to–liquid silicon, and that the restructuring occurs on a time scale much shorter than typically reported for equilibration (ns or longer).

5 Future Applications

While the first three generations of UED have been devoted to studies in the gas phase, the fourth-generation apparatus was specifically designed for diffraction studies of surfaces, crystals and macromolecules. With the electron pulse sequencing of the diffraction image referencing method, the six-orders-of-magnitude higher cross-section of electron scattering (compared to x-rays) and the femtosecond and sub-ångström temporal and spatial resolution, the

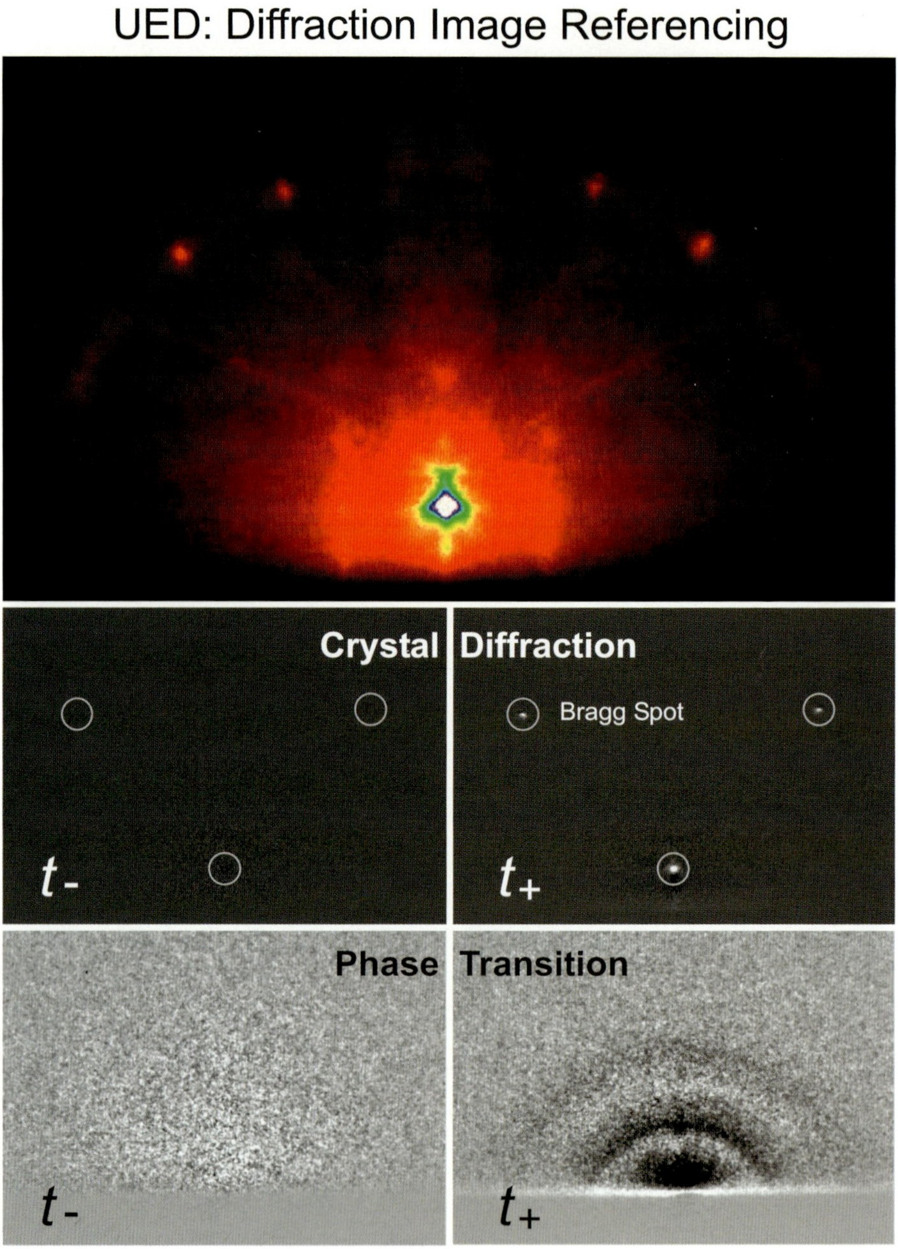

Fig. 9. UED of condensed phases — crystals and solid-to-liquid phase transitions. Shown are diffraction from the crystal (top), and diffraction image referencing of Bragg spots (middle) and of amorphous-to-liquid transition (bottom)

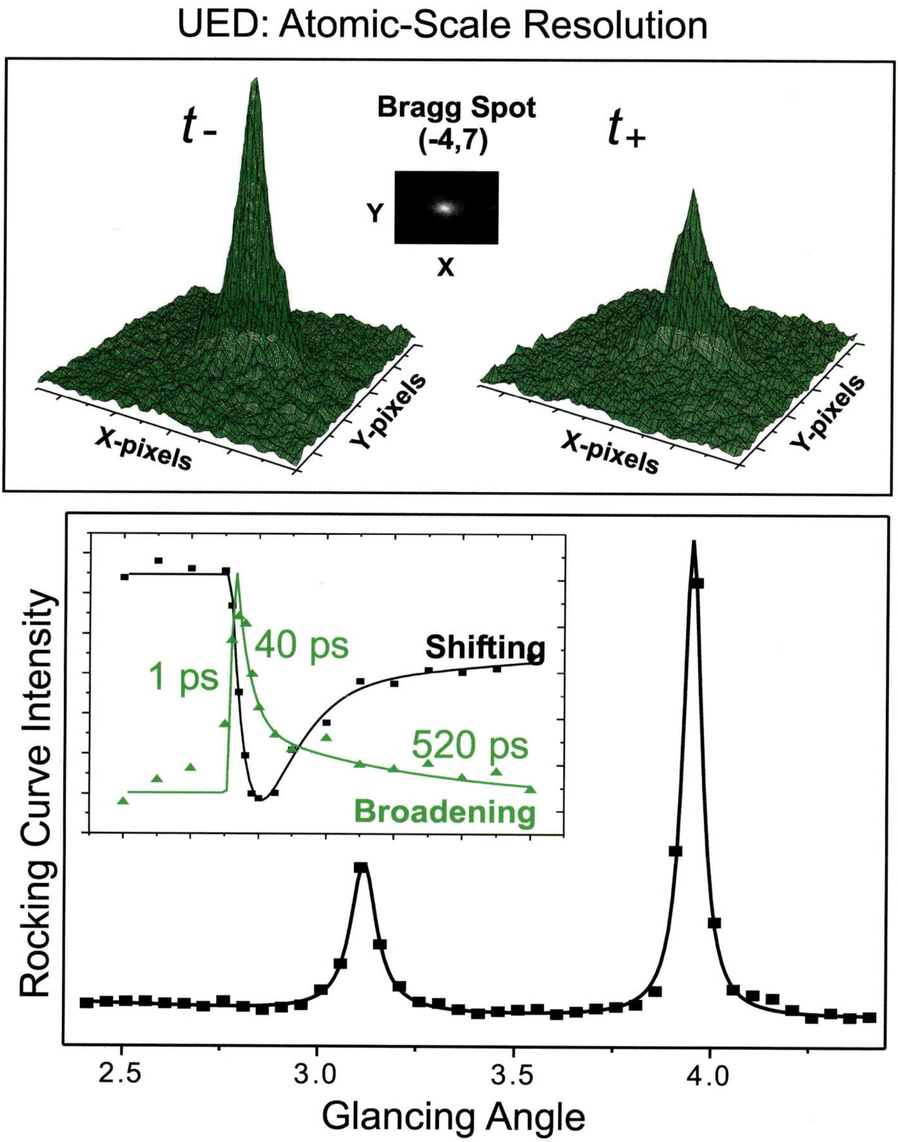

Fig. 10. Atomic–scale resolutions: spatial and temporal resolution of *one* Bragg spot, together with the shift in position and broadening as a function of time. The rocking curve is also shown.

technique is poised for many applications, including studies of proteins and other complex structures. UED has now emerged as a powerful method for mapping out, with atomic resolution, the ultrafast (femtosecond and picosecond) structural dynamics in chemistry, materials science, and potentially, in biology.

Acknowledgments

We gratefully acknowledge the generous support of the National Science Foundation for building the new generations of UED. Partial support was provided by the Air Force Office of Scientific Research and the Office of Naval Research.

References

1. Zewail, A.H.: Femtochemistry: Atomic-scale dynamics of the chemical bond using ultrafast lasers. In: Les Prix Nobel: The Nobel Prizes 1999, Frängsmyr, T. (ed). Almqvist & Wiksell, Stockholm (2000) and references therein
2. Zewail, A.H.: Femtosecond transition-state dynamics. Faraday Discuss., **91**, 207–237 (1991)
3. Williamson, J.C., Zewail, A.H.: Structural femtochemistry: Experimental methodology. Proc. Natl. Acad. Sci. USA, **88**, 5021–5025 (1991)
4. Williamson, J.C., Dantus, M., Kim, S.B., Zewail, A.H.: Ultrafast diffraction and molecular structure. Chem. Phys. Lett., **196**, 529–534 (1992)
5. Williamson, J.C., Zewail, A.H.: Ultrafast electron diffraction: Velocity mismatch and temporal resolution in crossed-beam experiments. Chem. Phys. Lett., **209**, 10–16 (1993)
6. Williamson, J.C., Zewail, A.H.: Ultrafast electron diffraction. 4. Molecular structures and coherent dynamics. J. Phys. Chem., **98**, 2766–2781 (1994)
7. Dantus, M., Kim, S.B., Williamson, J.C., Zewail, A.H.: Ultrafast electron diffraction. 5. Experimental time resolution and applications. J. Phys. Chem., **98**, 2782–2796 (1994)
8. Williamson, J.C., Cao, J., Ihee, H., Frey, H., Zewail, A.H.: Clocking transient chemical changes by ultrafast electron diffraction. Nature, **386**, 159–162 (1997)
9. Ihee, H., Cao, J., Zewail, A.H.: Ultrafast electron diffraction: Structures in dissociation dynamics of $Fe(CO)_5$. Chem. Phys. Lett., **281**, 10–19 (1997)
10. Cao, J., Ihee, H., Zewail, A.H.: Ultrafast electron diffraction: Determination of radical structure with picosecond time resolution. Chem. Phys. Lett., **290**, 1–8 (1998)
11. Cao, J., Ihee, H., Zewail, A.H.: Ultrafast electron diffraction and direct observation of transient structures in a chemical reaction. Proc. Natl. Acad. Sci. USA, **96**, 338–342 (1999)
12. Ihee, H., Zewail, A.H., Goddard III, W.A.: Conformations and barriers of haloethyl radicals (CH_2XCH_2, X = F, Cl, Br, I): Ab initio studies. J. Phys. Chem. A, **103**, 6638–6649 (1999)

13. Ihee, H., Lobastov, V.A., Gomez, U.M., Goodson, B.M., Srinivasan, R., Ruan, C.-Y., Zewail, A.H.: Direct imaging of transient molecular structures with ultrafast diffraction. Science, **291**, 458–462 (2001)
14. Ruan, C.-Y., Vigliotti, F., Lobastov, V.A., Chen, S., Zewail, A.H.: Ultrafast electron crystallography and transient structures of molecules, surfaces and phase transitions. Nature, submitted (2003)
15. Ihee, H., Cao, J., Zewail, A.H.: Ultrafast electron diffraction of transient $[Fe(CO)_4]$: Determination of molecular structure and reaction pathway. Angew. Chem. Int. Ed., **40**, 1532–1536 (2001)
16. Ihee, H., Kua, J., Goddard III, W.A., Zewail, A.H.: CF_2XCF_2X and $CF_2XCF_2\bullet$ radicals (X = Cl, Br, I): Ab initio and DFT studies and comparison with experiments. J. Phys. Chem. A, **105**, 3623–3632 (2001)
17. Ruan, C.-Y., Lobastov, V. A., Srinivasan, R., Goodson, B.M., Ihee, H., Zewail, A.H.: Ultrafast diffraction and structural dynamics: The nature of complex molecules far from equilibrium. Proc. Natl. Acad. Sci. USA, **98**, 7117–7122 (2001)
18. Lobastov, V.A., Srinivasan, R., Goodson, B.M., Ruan, C.-Y., Feenstra, J.S., Zewail, A.H.: Ultrafast diffraction of transient molecular structures in radiationless transitions. J. Phys. Chem. A, **105**, 11159–11164 (2001)
19. Ihee, H., Feenstra, J.S., Cao, J., Zewail, A.H.: Ultrafast electron diffraction of transient cyclopentadienyl radical: A dynamic pseudorotary structure. Chem. Phys. Lett., **353**, 325–334 (2002)
20. Ihee, H., Goodson, B.M., Srinivasan, R., Lobastov, V.A., Zewail, A.H.: Ultrafast electron diffraction and structural dynamics: Transient intermediates in the elimination reaction of $C_2F_4I_2$. J. Phys. Chem. A, **106**, 4087–4103 (2002)
21. Goodson, B.M., Ruan, C.-Y., Lobastov, V.A., Srinivasan, R., Zewail, A.H.: Ultrafast electron diffraction: Complex landscapes of molecular structures in thermal and light-mediated reactions. Chem. Phys. Lett., **374**, 417–424 (2003)
22. Srinivasan, R., Lobastov, V.A., Ruan, C.-Y., Zewail, A.H.: Ultrafast electron diffraction (UED) – A new development for the 4D determination of transient molecular structures. Helv. Chim. Acta, **86**, 1763–1838 (2003) and references therein

Some Medical and Biological Applications of Ultrafast Lasers

S. Svanberg

Department of Physics, Lund Institute of Technology P.O. Box 118, S-221 00 Lund, Sweden;
Lund Laser Centre, Lund University P.O. Box 118, S-221 00 Lund, Sweden

Summary. Ultrafast lasers are finding many applications in medicine and biology. Pulses from such systems can be used directly or after non-linear modification. Direct utilisation includes time-resolved fluorescence spectroscopy and imaging. Such studies can be performed for tissue diagnosis in the field of malignant diseases as well as in the cardiovascular sector. Two-photon induced fluorescence provides a basis for high-quality optical microscopy, and optical coherence tomography with ultra-short and thus broadband pulses allow a radar-like depth-ranging into tissue. Photon propagation in scattering media also has important applications in optical mammography, dosimetry for photodynamic therapy and species concentration assessment. Intense continua of electromagnetic radiation of very brief duration are formed in the interaction of focused ultra-short and intense laser pulses with matter. Focusing into water, or better, a photonic band-gap fibre, leads to the generation of a light continuum through self-phase modulation. The propagation of white light through tissue was studied addressing questions related to tissue chromophore concentration measurements employing their absorptive imprint in the light. Even free gas enclosures in solids and liquids give rise to an absorptive signature, which, however, is typically 10^4 times narrower than spectral structures in the matrix material. Such sharp features are best detected by wavelength modulation of a single-mode laser, and many new applications of gas-in-scattering-media absorption spectroscopy (GASMAS) are developing. When terawatt laser pulses are focused onto a solid target with high nuclear charge Z, intense X-ray radiation of few ps duration and with energies exceeding hundreds of keV is emitted. Biomedical applications of this radiation are described, including differential absorption and gated-viewing imaging.

1 Introduction

We will start our discussion of biological and medical applications of short-pulse optical radiation with some background regarding the interaction between electromagnetic radiation and biological tissue. Such interactions are discussed, e.g., in [1]. The interaction is governed by absorption and scattering processes. Absorption occurs mainly through the interaction with the major tissue constituents. Water absorbs all radiation below about 190 nm and is transparent up till about 1400 nm. Haem absorbs through the visible region with a maximum at about 400 nm (the Soret band) till about 600 nm, where the absorption falls off drastically, rendering blood its characteristic red colour. Aminoacids and melanin absorb through the UV and visible regime with an absorptivity falling of towards the red and near IR-regions. Practically, this means that the region from about 600 - 1400 nm is particularly transparent to radiation. Here scattering is instead the dominant interaction leading to the well-known phenomena of diffuse transmission of red light through tissue. While red light readily penetrates through a finger no shadow of the bones can be seen. When high-energy levels of CW radiation (more than 10W) in focused beams are used, strong tissue heating occurs, leading to coagulation or carbonisation of the tissue. Thermal treatment is the most wide-spread laser modality used, e.g. in surgery and eye treatments. Intense pulses in the UV region, where the penetration depth into tissue is very low, lead to ablation where cell layer by cell layer can be lifted off with high precision, e.g. in refractive eye surgery. Ultra-short laser pulses also lead to ablative interaction due to strong non-linear effects, and femto-second lasers are finding many applications for precision surgery.

A further laser-matter interaction mechanism of non-thermal nature is the photodynamic action. Photodynamic therapy (PDT) provides a method to selectively eradicate malignant cells while leaving the normal cells intact [2–4]. Here a particular chemical agent, a sensitiser, is administered to the body and localises selectively to malignant cells. Using light matching an electronic transition of the agent, it is excited. The excitation energy is transferred to oxygen molecules, which normally are in their ground triplet state. The resulting singlet oxygen is very toxic to tissue and induces necrosis. There is a certain parallel to photosynthesis in the environmental field. There the combination of a particular chemical agent, chlorophyll, and light brings about the build up of organic material out of carbon dioxide and water in a cold photochemical procedure. PDT is from a certain aspect the opposite process; the agent in combination with light instead breaks down organic material into its components, also in a cold process.

PDT was first introduced at the turn of the last century. However, with the development of pure and efficient sensitisers combined with laser sources at the relevant wavelengths, the field has expanded a lot, first through the work of T. Dougherty, demonstrating the treatment efficacy of haematoporphyrin derivative (Photofrin), absorbing at 630 nm. Actually, the sensitiser absorption must fall within the range 600 nm to 1400 nm to allow penetration into tissue. In practice it must also be below 1 μm in order to ensure enough energy to allow the energy transfer to oxygen. While most sensitisers such as mesotetrahydroxy phenyl chlorine (Foscan; 652 nm), benzoporphyrin (Verteporfin; 690 nm) and lutetium texaphyrin (Lutrin; 732 nm) must be administered intravenously, the protoporphyrin IX (PpIX) precursor δ-amino levulinic acid (ALA) has the advantage of additional possibilities of topical or oral administration. ALA is a natural constituent in the body, utilised as the starting material in the haem cycle. Strongly fluorescing and photodynamically active PpIX is produced at higher concentrations in skin tumours after applying an ALA-containing cream over the tumour a few hours before irradiation with light at 635 nm. The very efficient ALA-PDT was introduced by Z. Malik and J.C. Kennedy [5, 6] and has been used widely also in our group [7–9].

Sensitising agents frequently exhibit a strong and characteristic fluorescence signal that can be utilised for localising and demarcating tumours. Porphyrins are best excited at the peak of their Soret band around 400 nm. If a shorter excitation wavelength is used (optimally around 340 nm), autofluorescence from endogenous chromophores can be used for tumour diagnostics, since the signatures do frequently differ from normal tissue. Fibre-optic fluorosensors with a compact pulsed nitrogen laser emitting at 337 nm have been much utilised. The laser light is transmitted to the tissue through the fibre, which also picks up and conducts the fluorescence signal to the entrance slit of a gated and intensified detector, displaying the full LIF spectrum for each laser shot. Instrument constructed in this way are described, e.g. in [10, 11]. A simplified variety employs a violet diode laser at 397 nm to induce fluorescence, which is recorded by a compact integrated spectrometer with array detector read out [12].

Most medical diagnostic devices provide image information. It is also possible to build multi-spectral fluorescence imaging systems. Then a larger tissue area is illuminated and the fluorescence is simultaneously recorded in several fluorescence bands using special beam-splitting optics [13–15]. Tumour tissue is characterized by a strong intensity increase at 635 nm due to the sensitizer, at the same time as the tissue autofluorescence level around 470 nm is reduced. Three detection bands are used. By subtracting signals recorded at 635 nm by the signal at about 600 nm a background-free peak intensity is obtained, and this intensity is then divided by the blue intensity, resulting in a dimensionless quantity featuring enhanced contrast and immunity to detection geometry etc.

Fluorescence imaging can also be performed using a push-broom sensor, where multi-spectral imaging is performed line-wise in a similar way as in a satellite imaging spectrometer. Instead of using the platform movement for the line advancement, it is achieved by tilting a scanning mirror. This technique employing the strong mercury line at 365 nm for excitation has been applied to gynaecology [16]. By ALA instillation into the bladder, it is possible to detect dysplasia and carcinoma *in situ* using plain visual inspection, as performed in a commercial system by Storz.

2 Time-Resolved Fluorescence Spectroscopy and Imaging

An excited state is characterised not only by its energy but also by its lifetime. In the measurements just discussed only the spectral distribution is utilised. However, the decay time of the fluorescence, excited by a short laser pulse, can also be employed. Tissue consists of numerous constituents which contribute to the spectral shape of tissue and also to the decay characteristics. The excited state lifetimes in a number of the major constituents have been determined using picosecond laser excitation followed by time-resolved single-photon detection [17–19]. Frequently, several overlapping excited states with different lifetimes are excited, and several decay components with different time constants in the range of hundreds of picoseconds to 10 ns are observed.

The different temporal characteristics of excited states have also been used in fluorescence imaging experiments by Cubeddu et al. A short-pulse nitrogen laser is used to illuminate a larger tissue area, and the fluorescence is observed through a filter with a time-gated intensified CCD detector. By recording the images with different delays it is possible to generate a lifetime image. In tissue prepared with ALA, a tumour region is then characterised by longer lifetimes than the surrounding normal areas due to the longer lifetime of porphyrins [20].

The same type of lifetime measurements are performed on a microscopic level, frequently utilising two-photon-induced fluorescence. Such instruments have a very high optical resolution and are little vulnerable to scattering from out-of-focus regions. Commercial equipment employing also the temporal characteristics is available.

3 Ultra-Short Laser Pulse Propagation in Scattering Media

Multiple light scattering in weakly absorbing media is a common phenomenon observed in clouds, fog and in water. While some light is transmitted (there is still daylight on a cloudy day) image contrast disappears (the sun disc is not observed through the cloud). Milk is also a strongly scattering medium, and the presence of, e.g. a cherry, held by the stem inside a glass of milk, cannot be assessed in normal transillumination. If the image blurring due to scattering somehow could be defeated, red light tissue transillumination would be attractive for optical mammography without the use of ionising radiation. Also it would be very interesting to localise haematoma after trauma to the skull. Preceding such imaging applications, primary studies concerned the study of brain oxygenation [21]. First imaging recordings with scattering rejection through living tissue were reported in 1990 [22]. The basic idea behind imaging through turbid media using time-resolved spectroscopy is to restrict the detection to the photons which travel through the medium without being scattered. This corresponds to the first emerging photons following short-pulse injection. The non-scattered "ballistic" photons are recorded as the transmitter - detector line-of-sight is scanned across the sample. If an object of different absorbing and scattering properties is hidden inside the scattering medium the ballistic signal is changed.

Optical mammography using time-resolved detection of transmitted light has been demonstrated. In one approach a pulsed diode laser system was used in transillumination of breast specimen placed between two glass plates. The diode laser operated at 815 nm and produced 30 ps pulses at a rate of 10 MHz. The single-photon time-correlated detection technique was used. A ductal carcinoma could be revealed in time-gated transillumination, while the image corresponding to the time integrated signal did not contain any diagnostic value [23]. Many techniques for time-gated imaging have been developed, including the use of optical Kerr gating, stimulated Raman amplification and second-harmonic cross correlation. Other techniques, which focus on the distinct phase properties of the non-scattered light, are heterodyne detection and light-in-flight holography. Different aspects of light in scattering media such as tissue are covered in [24]. Time-resolved photon propagation studies have also been performed in green leaves from plants [25].

Related to time-resolved propagation studies in scattering media is the field of optical coherence tomography (OCT). This technique was first developed for probing the human fundus. An experiment by Fujimoto *et al.* in 1986 can be seen as a precursor to this technique [26]. Here short pulses were propagated into the optical system of an eye and reflexes at the passages through the optical elements were recorded similarly as in laser radar sounding. The arrival time of the echoes were detected by frequency doubling of the radiation if the backscattered light and a delayed reference pulse arrived simultaneously in a non-linear crystal, as is frequently utilised in autocorrelators. It was subsequently realised that an improved result could be obtained by using broad-band CW light in a Michelson interferometer and observe the intensity increase when the interferometer arms were equally long and all frequency components interfere constructively [27]. The broader the wavelength distribution, the better the spatial resolution. This has reintroduced the use of pulsed lasers in OCT, since a short femto-second laser pulse covers a spectral bandwidth of 100 nm or more. By carefully compensating for group-velocity dispersion a spatial resolution of 1 μm can be achieved [28].

4 Species Concentration Measurements in Turbid Media

Frequently it is of great interest to be able to measure the concentration of a substance in the presence of heavy scattering. The scattering results in the optical path length being undefined and the normal Beer-Lambertian law does not pertain to the measurements. It is then important to be able to "orthogonalise" the effects of absorption and scattering to be able to access both. This can be made by solving the transport equation in the turbid medium, or by relying on Monte-Carlo simulations. It is clearly necessary to measure at many different wavelengths, since absorption and scattering obviously respond to wavelength shifts differently. A number of discrete wavelengths can be employed, but even better "all" wavelengths, e.g. white light. When high-power laser pulses are focused into a condensed medium such as water, a light continuum is generated through self-phase modulation. The origin of the phenomenon is the dependence of the index of refraction on the light intensity, leading to frequency chirps when the pulse intensity is rising and falling. Following the injection of such pulses into strongly scattering media, such as tissue, the time dispersion of the photons can be studied. The spectral contents and the time duration of the radiation can be measured with a streak camera coupled to the exit slit of a spectrometer. A two-dimensional representation of time versus colour is obtained as demonstrated in [29].

In the white-light propagation studies, a chirped-pulse amplification (CPA) laser system based on Ti:Sa is conveniently used. The CPA technique allows to achieve extremely high optical intensities in compact laser systems. Recently, the introduction of photonic bandgap fibres allows a very efficient white-light generation, and a mode-locked femtosecond lasers can be used directly without the need for pulse amplification.

This type of measurements can be used for studying absorption spectra of sensitisers *in situ* in living tissue [30]. A rat tumour model was used, where a human adenocarcinoma, inoculated on the hind leg muscle produced tumours of about 15 mm diameter. Optical fibres were used to inject the light in the tissue and to collect scattered light emerging again at a distance of about 8 mm. The rats were injected with disulphonated aluminium phtalocyanine at a concentration of 2.5 or 5 mg/kg body weight and measurements were performed both on the tumour and normal tissue. From the experimental recordings the absorption and scattering coefficients can be extracted. This allows the plotting of the absorption profile of the sensitizer *in vivo* and an assessment of the tumour/normal tissue uptake ratio. Similar techniques can be used to extract tissue absorption and scattering properties related to optical mammography [31]. A further application is the assessment of the concentration of certain near-IR-absorbing molecules in scattering powders. Pharmaceutical preparations are of particular interest [32].

5 Gas in Scattering Media Absorption Spectroscopy (GASMAS)

The spectral absorptive features of solids and liquids are broad and seldom sharper than 10 nm due to the interactions between the participating atoms. Thus the spectroscopic equipment used for the study of such samples seldom has a very high resolution. However, in many samples there are structures which are 10,000 times sharper - enclosures of free gas. Enclosures mean that the materials are then also scattering due to the changes in refraction index. Thus there is again no straight-forward Beer-Lambertian law to analyse the results. Instead, the techniques wellknown in tissue optics as just discussed can be used. We have recently introduced a new spectroscopic method of considerable potential, the Gas in Scattering Media Absorption Spectroscopy (GASMAS) [33,34]. Here a light from a single-mode diode laser is conducted by a fibre to the sample, which is placed in proximity to the photocathode of a photomultiplier tube, in order to allow a maximum detectivity of the diffusely transmitted light, which might be very faint. By wavelength- or frequency-modulation techniques it is possible to sensitively detect the sharp absorptive imprint due to the gas. Our first measurement have concerned normal atmospheric oxygen molecules, which were studied in matrices as diverse as polystyrene foam, wood, and fruits. Concentration assessment can be achieved by also performing time-resolved measurements with a pulsed laser at the same wavelength using a set-up very similar to an optical mammograph. Then the time-dispersion curve will give information on the "history" of different photons travelling short or long distances through the sample. The mean travel time through the sample can be evaluated to be used for concentration evaluation.

A particularly interesting application is to study gas diffusion through porous material. Then the sample is first stored in a nitrogen atmosphere for few hours, e.g. by placing it in a plastic bag, which is briefly flushed from a nitrogen tank and sealed with a tape. After several hours the sample is taken out and placed in the GASMAS apparatus, and the successive reinvasion of the normal air into the sample can be followed by the oxygen signal [35]. E.g., it was found in studies of 10 mm thick samples of Norwegian Spruce and Balsa wood, that the latter exhibits a much longer time constant, probably related to well sealed individual cell compartments. Clearly, diffusion can be characterised by several time constants. The influence of surface coatings such as paints or plastic films can be readily studied. Internal pressure in a sample can also be monitored non-intrusively through the pressure-broadening of the absorption lines [33].

The monitoring of concentration ratios of two gases absorbing at close-lying lines eliminates the need for a detailed understanding of the photon transport through the sample and could give new insights in field as diverse as plant physiology and catalyst research.

Additional information on applications of laser spectroscopy to biology and medicine can be found, e.g. in [36–39].

6 Laser-Produced X-Rays

Medical application of X-rays has developed during 100 years and now constitute one of the most important diagnostic modalities. X-ray generation is accomplished through acceleration of electrons into a material with a high nuclear charge Z. With the development of short-pulse lasers with output powers in the terawatt range it has become possible to generate hard X-rays by focussing the radiation to a small spot on the surface of a high-Z material [40]. While this generation scheme is obviously more complicated and costly it can present some advantages which are mostly related to the small source size and the potential to reduce the X-ray dose by viewing techniques in an analogue way as discussed for optical mammography in Sect. 3. Early high-resolution X-ray imaging was reported in [41,42]. While it is easy to understand why scattered photons need to be eliminated in the optical regime due to the obvious and massive light scattering, it is not so evident for X-rays which are known to produce sharp images of ribs etc. However, in whole-body X-ray radiography without any special precautions actually 90% of the photons reaching the plate are Compton scattered and contribute only to a diffuse background. This background is frequently reduced using anti-scattering (straight-vision) grid plates between the patient and the recording plate, to reduce the photons having the wrong angle. The price is obviously that the patient has to be exposed to a higher dose of ionising radiation. The potential of medical X-ray gated viewing was first pointed out in [43]. We have performed detailed studies of gated X-ray detection using a 10 Hz high-energy laser system in combination with an X-ray streak camera [44], and also demonstrated gated tomographic X-ray imaging [45].

Contrast agents are frequently used in radiography. By utilising the sharp K emission lines from the laser-produced plasma and employing two different target materials so that the K-lines bridge the K absorption line of a contrast agent (at 33 keV for iodine, at 50 keV for gadolinium; both approved contrast agents). By dividing two images taken with different target materials only the contrast agent survives the division, and demonstrations of this kind are presented in [46]. Presently, medical image plates, which are much more sensitive than photographic plates and also re-usable, have been employed. The future for laser-produced X-rays in medicine will depend on the possibility to develop compact and efficient laser systems and gateable image plates. Recent experiments with a 1 kHz relatively compact system gave encouraging results [47].

A very important aspect in the potential use of laser-produced X-rays is the demonstrate that these widely spaced and extremely powerful X-ray spikes do not present an additional detrimental ionising radiation effect for the same integrated Gray dose. A first experiment using V79 Chinese hamster cells showed that this does not seem to be the case [48].

7 Discussion

Ultrafast laser provide many new possibilities for diagnostics and analysis in biology and medicine. Sometimes a moderate temporal resolution is sufficient, such as in time-resolved fluorescence imaging, while other applications, such as white-light or X-ray generation, benefit from the fastest possible pulses. Certain applications, such as two-photon-induced fluorescence microscopy and OCT, as well as certain aspects of precision medical surgery, have already reached a commercial stage. Very compact, rugged and reliable diode-laser pumped femto-second lasers are becoming readily available and will strongly influence the wide-spread use of the technology. Also large CPA-based terawatt systems have the potential for important applications in medicine. Induction of nuclear reactions leading to the production of short-lived radioactive isotopes for single-photon emission computerised tomography (SPECT) and positron emission tomography (PET) is an attractive possibility as is laser-driven acceleration of protons for tumour therapy [49].

Acknowledgement. The author is very grateful for a very fruitful collaboration with a large number of colleagues and students in the field of medical and biological laser applications. Present Lund collaborators include Ch. Abrahamsson, S. Andersson-Engels, A. Sjögren, M. Sjöholm, G. Somesfalean, K. Svanberg, and C.-G. Wahlström. Lund previous collaborators include R. Berg, K. Herrlin, O. Jarlman, J. Johansson, M. Grätz, C. Olsson, and C. Tillman. International collaborators also greatly contributing to the research reported include R. Cubeddu et al., Milano; E. Förster et al., Jena and J. Alnis, Riga. Financial support from the Swedish Science Research Council, The Swedish Strategic Research Foundation, the Knut and Alice Wallenberg Foundation and the European Community is gratefully acknowledged, as is a fruitful collaboration with several companies, including AstraZeneca, Johnson & Johnson, Science & Technology International and SpectraCure AB.

References

1. A.J. Welch, M van Gemert: *Optical-Thermal Response of laser-Irradiated Tissue* (Plenum, New York 1995)
2. S. L. Marcus: "Photodynamic therapy of human cancer: clinical status, potential and needs", in *Future Directions and Applications in Photodynamic Therapy,* C. J. Gomer (ed.), Proc. SPIE **IS-6,** (Soc. Photo-Opt. Instrum. Eng., 1990), pp. 5-56.
3. L. I. Grossweiner: *The Science of Phototherapy* (CRC Press, Boca Raton, FL, 1994)
4. T. J. Dougherty, C. J. Gomer, B. W. Henderson, G. Jori, D. Kessel, M. Korbelik, J. Moan, Q. Peng: Photodynamic therapy, J. Natl. Cancer Inst. **90,** 889 (1998)
5. J.C. Kennedy, R.H. Pottier, D.C. Pross: Photodynamic therapy with endogenous protoporphyrin IX: Basic principles and present clinical experience, J. Photochem. Photobiol. B **6,** 143 (1990)
6. J.C. Kennedy, R.H. Pottier: Endogenous protoporphyrin IX, a clinically useful photosensitizer for photodynamic therapy, J. Photochem. Photobiol. B **14,** 275 (1992)
7. K. Svanberg, T. Andersson, D. Killander, I. Wang, U. Stenram, S. Andersson-Engels, R. Berg, J. Johansson, S. Svanberg: Photodynamic therapy of non-melanoma malignant tumours of the skin using topical δ-amino levulinic acid sensitization and laser irradiation, Br. J. Dermatol. **130,** 743 (1994)

8. I. Wang, B. Bauer, S. Andersson-Engels, S. Svanberg, K. Svanberg: Photodynamic therapy utilising topical δ-aminolevulinic acid in non-melanoma skin malignancies of the eyelid and the periocular skin, Acta Ophthalmol. Scand. **77,** 182 (1999)
9. I. Wang, N. Bendsoe, C. af Klinteberg, A. M. K. Enejder, S. Andersson-Engels, S. Svanberg, K. Svanberg: Photodynamic therapy versus cryosurgery of basal cell carcinomas; results of a phase III randomized clinical trial, Br. J. Dermatol. **144,** 832 (2000)
10. S. Andersson-Engels, Å. Elner, J. Johansson, S.-E. Karlsson, L. G. Salford, L.-G. Strömblad, K. Svanberg, S. Svanberg: Clinical recordings of laser-induced fluorescence spectra for evaluation of tumour demarcation feasibility in selected clinical specialities, Lasers in Med. Sci. **6**, 415 (1991).
11. C. af Klinteberg, M. Andreasson, O. Sandström, S. Andersson-Engels, S. Svanberg: Compact medical fluorosensor for minimally invasive tissue characterization, Submitted to Review of Scientific Instruments
12. U. Gustafsson, S. Pålsson, S. Svanberg: Compact fiber-optic fluorosensor using a Continuous-wave Violet Diode Laser, Rev. Sci. Instr. **71**, 3004 (2000)
13. S. Andersson-Engels, J. Johansson, S. Svanberg: Medical diagnostic system based on simultaneous multi-spectral fluorescence imaging, Appl. Opt. **33**, 8022 (1994)
14. K. Svanberg, I. Wang, S. Colleen, I. Idvall, C. Ingvar, R. Rydell, D. Jocham, H. Diddens, S. Bown, G. Gregory, S. Montán, S. Andersson-Engels, S. Svanberg: Clinical multi-colour fluorescence imaging of malignant tumours - Initial experience, Acta Radiologica **39**, 2 (1998)
15. S. Andersson-Engels, G. Canti, R. Cubeddu, C. Eker, C. af Klinteberg, A. Pifferi, K. Svanberg, P. Taroni, G. Valentini, I. Wang: Preliminary evaluation of two fluorescence imaging methods for the detection and delineation of basal cell carcinomas of the skin, Lasers in Surg. Med. **26**, 76 (2000)
16. U. Gustafsson, E. McLaughlin, E. Jacobson, J. Håkansson, P. Troy, M.J. DeWeert, S. Pålsson, M. Soto Thompson, S. Svanberg, A. Vaitkuviene, K. Svanberg: Fluorescence and reflectance monitoring of human cervical tissue in vivo – A case study, SPIE **4959-**12 (2003)
17. S. Andersson-Engels, J. Johansson, U. Stenram, K. Svanberg, S. Svanberg: Time-resolved laser-induced fluorescence spectroscopy for enhanced demarcation of human atherosclerotic plaques: Photochem. Photobiol. **4**, 363 (1990)
18. S. Andersson-Engels, J. Johansson, S. Svanberg: The use of time-resolved fluorescence for diagnosis of atherosclerotic plaque and malignant tumours, Spectrochim. Acta **46A**, 1203 (1990)
19. S. Andersson-Engels, A. Gustafson, J. Johansson, U. Stenram, K. Svanberg, S. Svanberg: An investigation of possible fluorophores in human atherosclerotic plaque, Lasers in the Life Sciences **5**, 1 (1992)
20. S. Andersson-Engels, G. Canti, R. Cubeddu, C. Eker, C. af Klinteberg, A. Pifferi, K. Svanberg, P. Taroni, G. Valentini, I. Wang: Preliminary evaluation of two fluorescence imaging methods for the detection and Delineation of Basal Cell Carcinomas of the Skin, Lasers in Surg. Med. **26**, 76 (2000)
21. B. Chance, J.S. Leigh, H. Miyake, D.S. Smith, S. Nioka, R. Greenfeld, M. Finander, K. Kaufmann, W. Levy, M. Young, P. Cohen, H. Yoshioka, R. Boretsky: Proc. Natl. Acad. Sci. USA **85**, 4971 (1988)
22. S. Andersson-Engels, R. Berg, O. Jarlman, S. Svanberg: Time-resolved transillumination for medical diagnostics, Optics Letters **15**, 1179 (1990)

23. R. Berg, O. Jarlman, S. Svanberg: Medical transillumination imaging using short-pulse diode lasers, Appl. Opt. **32**, 574 (1993)
24. G. Müller *et al.*, Eds.: *Medical Optical Tomography, Functional Imaging and Monitoring (SPIE Institute Ser. Vol. 11)* (SPIE, Bellingham1993)
25. J. Johansson, R. Berg, A. Pifferi, S. Svanberg, L.O. Björn: Time-resolved studies of light propagation in *Crassula* and *Phaseolus* leaves, Photochem. Photobiol. **69**, 242 (1999)
26. J.G. Fujimoto *et al.*: Optics Letters **11**, 150 (1986)
27. J.G. Fujimoto *et al.*: Optics Letters **18,** 1864 (1993)
28. W. Drexler *et al.:* Arch. Ophtalmol. **121**, 695 (2003)
29. S. Andersson-Engels, R. Berg, A. Persson, S. Svanberg: Multispectral tissue characterization using time-resolved detection of diffusely scattered white light, Opt. Letters **18**, 1697 (1993)
30. C. af Klinteberg, A. Pifferi, S. Andersson-Engels, R. Cubeddu., S. Svanberg: *In vivo* absorption spectrum of disulphonated aluminium phtalocyanine (AlS_2Pc) in rats using femtosecond white light, to appear.
31. C. af Klinteberg, R. Berg, C. Lindquist, S. Andersson-Engels, S. Svanberg: SPIE **2626**, 149 (1995)
32. J. Johansson, S. Folestad, M. Josefson, A. Sparén, C. Abrahamsson, S. Andersson-Engels, S. Svanberg: Time-resolved NIR/VIS spectroscopy for analysis of solids: Pharmaceutical tablets, Appl. Spectrosc. **56**, 725 (2002)
33. M. Sjöholm, G. Somesfalean, J. Alnis, S. Andersson-Engels, S. Svanberg: Analysis of gas dispersed in scattering solids and liquids, Opt. Lett. **26**, 16 (2001)
34. G. Somesfalean, M. Sjöholm, J. Alnis, C. af Klinteberg, S. Andersson-Engels, S. Svanberg: Concentration measurement of gas imbedded in scattering media employing time and spatially resolved techniques, Appl. Optics **41**, 3538 (2002)
35. J. Alnis, B. Anderson, M. Sjöholm, G. Somesfalean, S. Svanberg: Analysis of gas in wood materials, Appl. Phys. B (2003), in press.
36. S. Svanberg: New developments in laser medicine, Physica Scripta **T72**, 69 (1997)
37. S. Svanberg: Some applications of ultrashort laser pulses in biology and medicine, Meas. Sci. Technology **12**, 1777 (2001)
38. S. Svanberg: Tissue diagnostics using lasers, Chap. 6 in *Lasers in Medicine,* ed. R. Waynant (CRC Press, Boca Raton 2002), p. 135-169
39. S. Andersson-Engels, K. Svanberg, S. Svanberg: Fluorescence imaging, Chapter in J. Fujimoto, Ed., *Lasers in Medicine*, to appear
40. J.D. Kmetec, C.L. Gordon III, J.J. Macklin, B.E. Lemoff, G.S. Brown, S.E. Harris: Phys. Rev. Lett. **68**, 1527 (1992)
41. K. Herrlin, G. Svahn, C. Olsson, H. Petterson, C. Tillman, A. Persson, C.-G. Wahlström, S. Svanberg: The generation of X-rays for medical imaging by high-power lasers - Present and future applications - Preliminary results, Radiology **189**, 65 (1993)
42. C. Tillman, A. Persson, C.-G. Wahlström, S. Svanberg, K. Herrlin: Imaging using hard X-rays from a laser-produced plasma, Appl. Phys. B**61**, 333 (1995)
43. C.L. Gordon III, G.Y. Yin, B.E. Lemoff, P.M. Bell, C.P.J. Barty: Opt. Lett. **20** 1056 (1995)
44. M. Grätz, A. Pifferi, C.-G. Wahlström, S. Svanberg: Time-gated imaging in radiology: Theoretical and experimental studies, IEEE J. Sel. Top. Quant. Electr. **2**, 1041 (1996)

45. M. Grätz, L. Kiernan, K. Herrlin, C.-G. Wahlström, S. Svanberg, K. Herrlin: Time-gated X-ray tomography, Appl. Phys. Lett. **73**, 2899 (1998)
46. C. Tillman, I. Mercer, S. Svanberg, K. Herrlin: Elemental biological imaging by differential absorption using a laser-produced X-ray source, J. Opt. Soc. Am. **13**, 209 (1996)
47. A. Sjögren, M. Harbst, C.-G. Wahlström, S. Svanberg, C. Olsson: High repetition rate, hard X-ray radiation from a laser produced plasma; Photon yield and applications considerations, Rev. Scient. Instr. **74**, 2300 (2003)
48. C. Tillman, G. Grafström, A-C. Jonsson, B-A. Jönsson, I. Mercer, S. Mattsson, S-E. Strand, S. Svanberg: Survival of mammalian cells exposed to ultrahigh dose-rates from a laser-produced plasma X-ray source, Radiology **213**, 860 (1999)
49. K. Ledingham, P.A. Norreys: Contemp. Phys. **40**, 367 (1999)

Adaptive Control of Two-Photon Excitation of Green Fluorescent Protein with Shaped Femtosecond Pulses

Hiroyuki Kawano[1], Yasuo Nabekawa[1], Akira Suda[1], Yu Oishi[1], Hideaki Mizuno[2], Atsushi Miyawaki[2], and Katsumi Midorikawa[1]

[1] Laser Technology Laboratory, RIKEN, 2-1 Hirosawa, Wako-shi, Saitama 351-0198, Japan kawano@riken.go.jp

[2] Laboratory for Cell Function and Dynamics, RIKEN, 2-1 Hirosawa, Wako-shi, Saitama 351-0198, Japan

1 Introduction

For many years, it has been believed that a Fourier-transform-limited (FTL) laser pulse is the most effective light source for the generation of nonlinear phenomena, since the FTL pulse has the shortest pulse duration, that is, the highest intensity, that can be limited by the spectral width due to the principle of uncertainty. Recently, many reports have been published on the adaptive control of nonlinear phenomena with shaped femtosecond excitation laser pulses [1,2]. Their reports have shown that the modification of the spectral and temporal phases of excitation pulses can increase or decrease the probabilities and efficiencies of such nonlinear phenomena. This method has been widely applied to studies on the active control of molecular motions or chemical reactions [3,4]. Considering further novel biological applications, we focus on the two-photon excited fluorescence (TPEF) of the green fluorescent protein (GFP) from the jellyfish *Aequorea victoria*. GFP is spontaneously fluorescent and is relatively *nontoxic* compared with other organic dyes used as optical markers. Therefore, it has been widely used as a "tag" material for the fluorescence observation of living cells [5]. Two-photon excitation microscopy (TPEM) is a powerful tool for biological real-time observation due to its various advantages, such as a clear contrast, good S/N ratio, and high spatial resolution [7]. From a practical point of view, however, there is a serious problem with TPEM, which is the photobleaching of a dye. The intensity of a fluorescence signal decreases significantly during observation. One of the reasons for this is that the chromophore structure is degraded by intense excitation laser pulses that are required for efficient two-photon excitation. In this study, therefore, we attempted to determine the optimal phase for maximizing the fluorescence efficiency of a GFP variant with excitation laser pulses of minimal intensity. We considered that GFP can be an ideal dye for the investigation of the dependence of photobleaching on the phase of excitation pulses, because the GFP chromophore is located at the center of β-can, that is, it is far from the site of reaction with activated molecular oxygen included in the solution, which can also cause photobleaching. The suppression of the photobleaching of a GFP variant in two-photon excitation is demonstrated.

2 Experimental

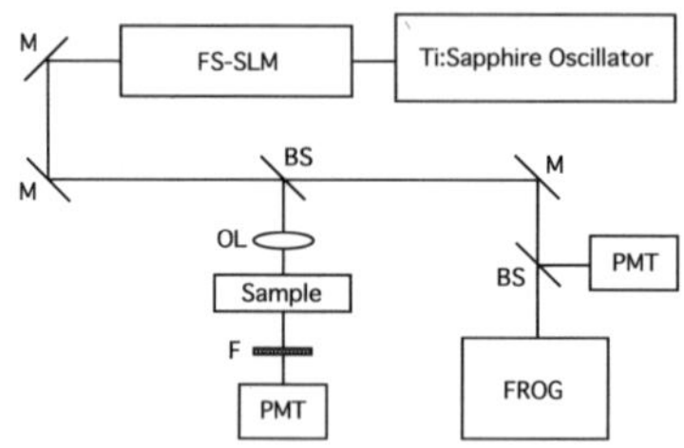

Fig. 1. Experimental setup. FS-SLM, fused-silica spatial light modulator; BS, beam splitter; OL, objective lens; F, color filter; M, mirror; PMT, photomultiplier tube; FROG, frequency-resolved optical gating system.

The experimental setup is illustrated in Fig. 1. Excitation laser pulses were obtained from a laboratory-built Ti:sapphire oscillator (28 fs, 815 nm at the center, 200 mW, 80 MHz). The spectral phase of the laser pulse was modulated with a fused-silica spatial light modulator (FS-SLM) originally developed in our laboratory [8]. The shaped pulse was divided into two pulses with a 50%-transmission beam splitter (BS, CVI). One of the resulting pulses was focused with an objective ($\times$ 20, Sigma Koki) into a container filled with a solution of a GFP variant, namely, an enhanced green fluorescent protein (EGFP, Clontech). The protein concentration was 2 mg/ml. When the photobleaching rate of the sample was measured, GFP molecules were immobilized on the inner surface of the container by a chemical treatment. Fluorescence (the peak wavelength was 510 nm) intensity was measured with a photomultiplier tube (H6780, Hamamatsu). The other pulses were introduced into a reference arm on which we obtained the second harmonic and FROG signals using a 30-μm-thick β-BaB_2O_4 (BBO) crystal with the same type of objective lens as above ($\times$20, Sigma Koki). According to a calculation based on a reference [9], this BBO is sufficiently thin to produce phase matching within the entire spectral bandwidth of our 28-fs laser pulses. The SH spectrum for a FROG trace was measured with a multichannel spectrometer (Ocean Optics Inc.). SH intensity for a feedback was detected with the same type of PMT as above (H6780, Hamamatsu). We placed additional dispersive materials on the beam line of the reference arm in order to balance the dispersion that the excitation pulses suffer on both arms. We used the simulated annealing (SA) method as a feedback algorithm.

3 Results and discussion

3.1 Increase in Fluorescence/SH Signal Intensity Ratio

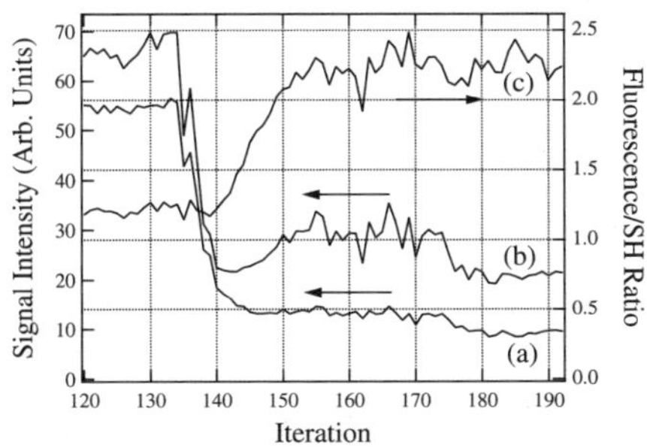

Fig. 2. Dependence of fluorescence and SH signals on feedback iterations with the ratio of fluorescence signal/SH signal as a feedback factor; (a) fluorescence intensity, (b) SH intensity, and (c) fluorescence/SH ratio. Feed back factor was changed at Iteration 134 from fluorescence intensity to fluorescence/SH ratio.

The result is shown in Fig. 2. Initially, we set the excitation pulses to be FTL in the first 134 iterations, and then started the optimization for increasing the fluorescence/SH signal ratio. The fluorescence and SH signals gradually decreased in intensity, and then the fluorescence/SH signal ratio increased, reaching its maximum at the iteration 154. The ratio increased by 100% compared with that for the FTL excitation pulses, indicating the validity of this method in reducing excitation intensity while preserving fluorescence signal intensity. The BBO crystal used here had a sufficiently broad acceptance bandwidth, as described in Section 2. The SH signal is, therefore, dependent on the square of the excitation intensity, regardless of the phase of the excitation pulse. In addition, the effect of linear chirps (positive chirps) of excitation pulses on fluorescence/SH signal ratio was investigated by inserting fused silica plates of various thicknesses into a beam axis. A small increase in the ratio was observed but it was only 1.5 at the maximum. It was found that extended pulses cannot sufficiently increase fluorescence efficiency and that randomly shaped excitation pulses are necessary for such a marked improvement.

3.2 Phase Measurements of Excitation Pulses

We measured the temporal phase of the excitation pulses. Figure 3 shows the FROG traces and reconstructed temporal profiles of the excitation pulses at iterations 134 and 192 in Fig. 2. At iteration 134 with a fluorescence/SH signal ratio of 1.2, the temporal phase of the excitation pulses was almost flat during

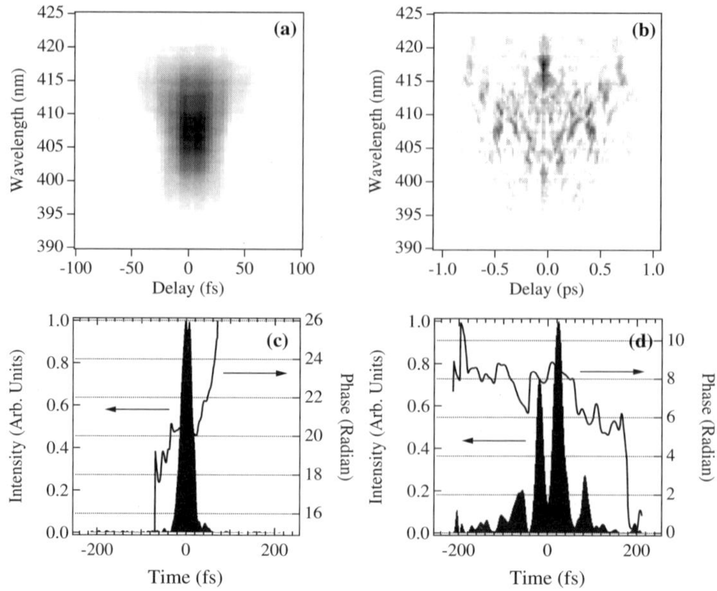

Fig. 3. FROG traces measured at iteration (a) 134 and (b) 192, respectively. The reconstructed temporal profiles are shown at the iteration (c) 134 and (d) 192, respectively. Solid lines are temporal phases.

the pulse duration and the reconstructed pulse width was approximately 28 fs, which was comparable with the FTL pulse duration. We measured the temporal phase of the excitation pulses at which the shaped pulses produced a fluorescence/SH signal ratio of 2.4 at iteration 192. The temporal profile was considerably complex and the phase was distributed among more than 3 radians during the pulse duration. Unfortunately, we cannot provide a clear explanation at present why this phase structure is effective in increasing fluorescence efficiency. Throughout our measurements, there was always a tendency for the temporal phase to increase towards the peak intensity and then decrease, when a 100% increase in fluorescence/SH signal ratio was achieved.

3.3 Decrease in GFP Photobleaching Ratio

We evaluated photobleaching of EGFP immobilized on a glass plate by measuring the temporal decay of the fluorescence signal during exposure to various excitation pulses (Fig. 4). Without any phase optimization, that is, when the excitation pulses were FTL with full energy, the fluorescence signal immediately decayed as shown in Fig. 4(c). On the contrary, with a shaped pulse in the case of (a), the photobleaching rate was significantly reduced despite the pulse having the same energy as the FTL pulse. One of the reasons for this

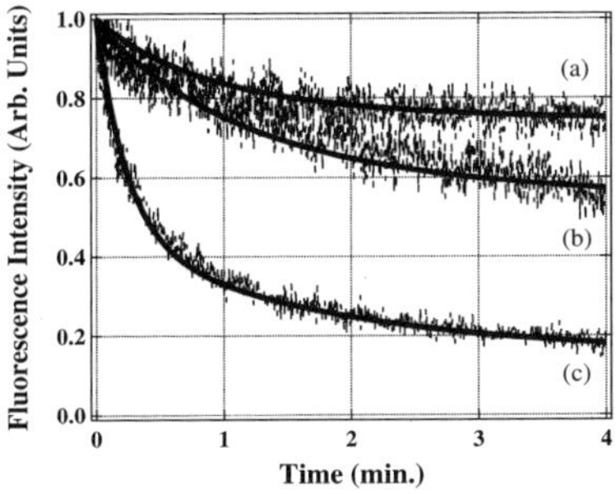

Fig. 4. Dependence of fluorescence signals on exposure time of excitation pulses. The sample was illuminated with (a) shaped pulses, (b) FTL pulses which provide the same fluorescence intensity as in the case (a) at time zero, and (c) FTL pulses which have the same pulse energy as in the case (a). Solid curves are double-exponential curve fittings.

reduced photobleaching rate is a lower peak intensity of shaped pulses than that of FTL pulses. For comparison, we produced FTL pulses that would initially have the same fluorescence intensity as that of the shaped excitation pulses mentioned above by inserting a ND filter to reduce the pulse energy and by operating the FS-SLM to compensate for the dispersion. Most researchers carrying out TPEM studies utilize relatively weak FTL pulses to prevent photobleaching of dyes. When such pulses were irradiated onto the sample, the fluorescence/SH signal ratio decayed faster than that for the shaped-pulse exposure as shown in Fig. 4(b). The decay curves were fitted using a double-exponential model. According to these results, when the shaped pulses (a) irradiated the sample, 25% of the fluorescence intensity was lost during the first 4.0 minutes. However, when the relatively weak FTL pulses (b) were employed for the excitation, it took only 1.0 minute to lose 25% of fluorescence intensity. This result indicates that one can extend the observation time of the fluorescence transmitted from the sample by a factor of approximately 4.

4 Conclusions

We demonstrated the adaptive control of two-photon-excited fluorescence from GFP using shaped femtosecond laser pulses. It was possible to achieve a 100% increase in the fluorescence/SH signal ratio, that is, we improved the fluorescence per pump intensity by this method. In our experiment, the use of phase-optimized shaped pulses reduced the photobleaching of a GFP variant by a factor of 4. This simple technique will be useful in overcoming severe difficulties, such as weak fluorescence signals and dye photobleaching caused by excitation pulses with high intensities. We demonstrated similar effects for other GFP variants. This method has a strong potential for immediate application to various important biological and medical studies.

References

1. Dudovich, N., Dayan, B., Gallagher Faeder, S. M., Silberberg, Y.: Transform-limited pulses are not optimal for resonant multiphoton transitions. Phys. Rev. Lett., **86**, 47–50 (2001)
2. Brixner, T., Damrauer, N. H., Niklaus, P., Gerber. G.: Photoselective adaptive femtosecond quantum control in the liquid phase. Nature, **414**, 57–60 (2001)
3. Weihacht, T. C., Bartels, R., Backus, S., Bucksbaum, P. H., Pearson, B., Geremia, J. M., Rabitz, H., Kapteyn, H. C., Murnane, M. M.: Coherent learning control of vibrational motion in room temperature molecular gases. Chem. Phys. Lett., **344**, 333–338 (2001)
4. Lee, S. H., Jung. K. H., Sung, J. H., Hong, K. H., Nam, C. H.: Adaptive quantum control of DCM fluorescence in the liquid phase. J. Chem. Phys., **117**, 9858–9861(2002)
5. Tsien, R. Y.: The green fluorescent protein. Annu. Rev. Biochemistry, **67**, 509–544 (1998)
6. Bardeen. C. J., Yakovlev, V. V., Squier, J. A., Wilson, K. R.: Quantum control of population transfer in green fluorescent protein by using chirped femtosecond pulses. J. Am. Chem. Soc. **120**, 13023–13027 (1998)
7. Denk, W.: Two-photon scanning photochemical microscopy: Mapping ligand-gated ion channel distributions. Proc. Natl. Acad. Sci. USA, **91**, 6629–6633 (1994)
8. Suda, A., Oishi, Y., Nagasaka, K., Wang, P., Midorikawa, K.: A spatial light modulator based on fused-silica plates for adaptive feedback control of intense femtosecond laser pulses. Opt. Express, **9**, 2–6 (2001)
9. Dmitriev, V. G., Gurzadyan, G. G., Nikogosyan, D. N.: Handbook of nonlinear optical crystals, 3rd edition. Springer, Berlin Heidelberg (1999)
10. Ando, R., Hama, H., Yamamoto-Hino, M., Mizuno, M., Miyawaki, A.: An optical marker based on the UV-induced green-to-red photoconversion of a fluorescent protein. Proc. Natl. Acad. Sci. USA, **99**, 12651–12656 (2002)
11. Heim, R., Tsien, R. Y.: Engineering green fluorescent protein for improved brightness, longer wavelengths and fluorescence energy transfer. Curr. Biol., **6**, 178–182 (1996)
12. Patterson, G. H., Knobel, S. M., Sharif, W. D., Kain, S. R., Piston, D. W.: Use of the green fluorescent protein and its mutants in quantitative fluorescence microscopy. Biophys. J., **73**, 2782–2790 (1997)

Highly Efficient Second-Harmonic Nanosource by Femtosecond Pulse Irradiation of a Metal Tip

M. Zavelani-Rossi[1], M. Labardi[2], D. Polli[1], G. Cerullo[1], S. De Silvestri[1], M. Allegrini[2], and O. Svelto[1]

[1] National Laboratory for Ultrafast and Ultraintense Optical Science - INFM, Istituto di Fotonica e Nanotecnologie IFN-CNR, Dipartimento di Fisica, Politecnico di Milano, Piazza Leonardo da Vinci 32, 20133 Milano, Italy
zavelani@mail.polimi.it

[2] INFM, Dipartimento di Fisica, Università di Pisa, via Buonarroti 2, 56127 Pisa, Italy

In the field of nano-optics there is a strong demand of confined radiation sources for local optical spectroscopy of single molecules or quantum structures, as well as for background-free Scanning Near-field Optical Microscopy (SNOM) and optical nanowriting applications towards the 10-nm resolution range.

Aperture SNOM [1], employing a metal-coated fiber tip, is limited to a resolution of the order of 50-100 nm by losses that strongly increase with decreasing fiber diameter, and by the skin depth of the metal coating. Scattering (or apertureless) SNOM [2] exploits the strong field enhancement at the end of a sharp tip to provide local illumination of the sample and can achieve a spatial resolution of the order of 5-10 nm [3–5]. However, it is characterized by a very poor signal to background ratio since the global elastic scattering of the illumination light overwhelms the local tip scattering signal. This problem can be overcome by using the second harmonic (SH) generated by the tip as a local illumination source [6,7]. SH generation at a metal tip results from a combination of field enhancement due to the electrostatic "lightning rod" effect induced by the tip sharpness and of the possible excitation of surface plasmon resonances. It is possible in this way to achieve a strongly confined light source at the tip largely suppressing the background at the fundamental wavelength. To obtain significant SH generation, ultrashort light pulses, with high energy, are required.

In this work, we combine advances in scattering SNOM with ultrashort high energy laser pulses to generate an highly localized SH at the apex of a pyramidal metal tip for atomic force microscopy (AFM) with unprecedented efficiency.

The laser source is a Kerr-lens mode-locked Ti:Sapphire oscillator, pumped by a cw argon laser, that generates 30 fs pulses at 800 nm (central wavelength). To obtain laser pulses with high energy we extended the cavity length of the "standard" oscillator, thus reducing the pulse repetition rate [8–10].

Figure 1 shows a schematic of the long-cavity laser. The cavity length is

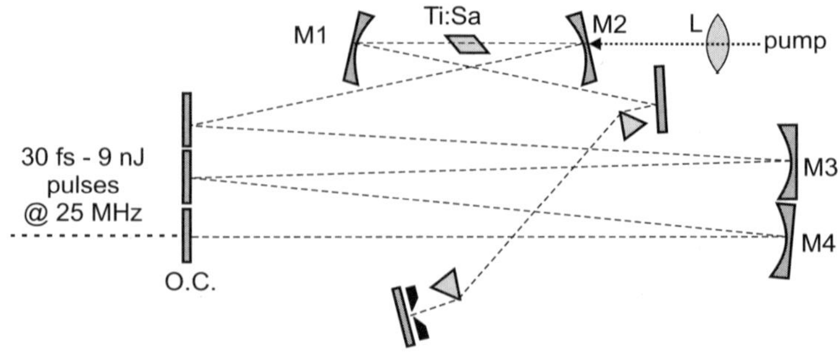

Fig. 1. Laser scheme. L: 10-cm focal-length pump beam focusing lens. M1 and M2: 10-cm radius of curvature focusing mirrors. Ti:Sa: 5-cm-long crystal, Brewster cut. OC: 5%-transmission output coupler. M3 and M4: 2-m telescope mirrors. The other mirrors are flat and are used to fold the beam, all mirrors are chirped and the prisms are made of fused silica.

stretched by simply introducing a 1:1 folded intracavity telescope in one arm, which has a unitary $ABCD$ matrix so that the optical stability limits of the cavity remain unchanged and our purpose is easily accomplished. A hard aperture is introduced in front of one end mirror to suppress multiple pulsing instabilities that may arise. The slit acts as a saturable absorber [11] and stabilizes the ML regime without reducing the output energy. The long cavity oscillator thus produces 9-nJ pulses at a repetition rate of 25 MHz. The reduced repetition rate increases the pulse energy for a given average power, thus finally resulting in a significant boosting of the SH generation for a safe level of average power incident on the metal tip.

PtIr-coated as well as Au-coated AFM tips with apical radius of about 20 nm are illuminated by the Ti:Sapphire laser. The experimental setup is sketched in figure 2. The light is focused on the tip by an aspheric lens with 11 mm focal length and 0.5 NA or by a 63× 0.85 N.A. microscope objective, with an average power density below 1 MW/cm^2, limited by thermal tip damage. The SH is collected by the same focusing optics to ensure optimal collection of the far-field emission, that is concentrated along the plane perpendicular to the tip dipole. A dichroic beamsplitter directs such light towards an UV-extended photomultiplier tube through a set of shortpass and interference filters to cut off the fundamental component. Previous experiments [12] used a tip oriented along the propagation direction of the excitation beam and achieved excitation with the longitudinal component near the focus of a strongly converging Gaussian beam. This geometry is however unfavorable for excitation of the tip dipole oscillation. We have therefore chosen to orient the tip perpendicularly to the beam propagation direction, applying a

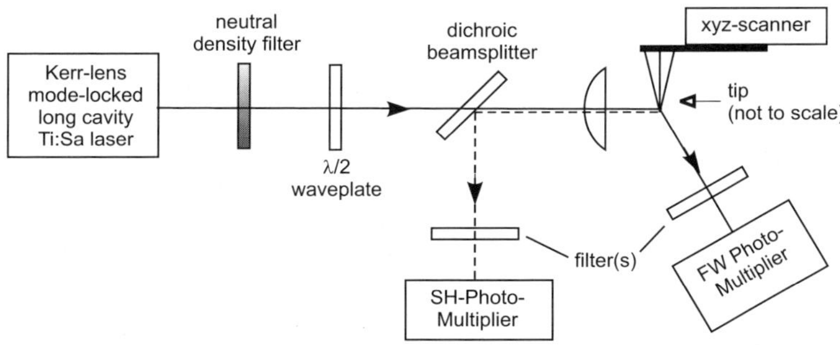

Fig. 2. Second Harmonic nanosource experimental setup. Continuous line: 800-nm Fundamental Wave (FW) beam; dashed line: SH beam.

more convenient geometry that has led to the best contrast achieved so far in elastic scattering-SNOM operation [5].

The collected SH light intensity amounts to about 3 pW with 13 mW input power, corresponding to 6×10^6 counts/s, which is at least a 3-order of magnitude improvement with respect to previous experiments [6, 7, 12, 13]. This high efficiency is made possible by the non-standard laser source as well as by the optimized SH generation and collection geometry. The nonlinearity of the scattering process is demonstrated by input/output characteristics. Figure 3 shows the quadratic dependence of the SH light collected with our setup on the incident average power, for different excitation beam polarizations. The efficiency clearly depends on the polarization but the quadratic

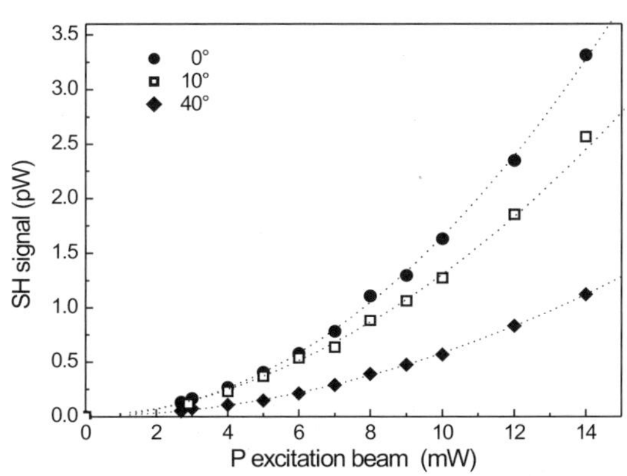

Fig. 3. Second Harmonic signal from the tip apex versus excitation signal, for different laser beam polarization (0° means parallel to the tip axes), and quadratic fits (dotted lines).

behavior remains unchanged in all cases. The best efficiency is achieved by orienting the laser beam polarization parallel to the tip axis. These data are consistent with the assumption of a nonlinear dipole axis parallel to the tip.

The efficiency of SH generation at the metal tip is extremely sensitive to the illumination spot location, consistently with the assumption of local generation at the tip apex. To check this issue the metal tip has been fixed to a $40 \times 40 \times 10\ \mu m^3$ range piezoscanner. Several scans of the tip location could be performed in the focal plane of the laser, at various incidence polarization directions and foci, due to the high signal level that allowed a scan rate typical for probe microscopy imaging with a good noise figure and drift. Such scans provide conclusive evidence that the SH generation is localized close to the tip apex. As an example, figure 4 shows the tip-induced SH image of the focal field (a), together with the corresponding background elastic scattering (b), for laser beam polarization parallel to the tip axis. The pyramidal shape of

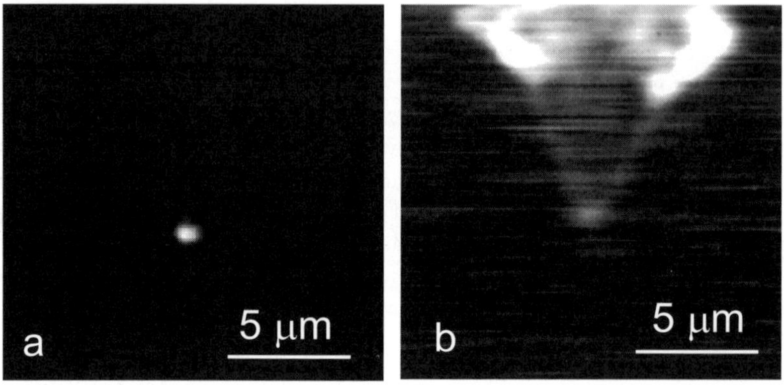

Fig. 4. Tip induced SH image (a) and corresponding FW elastic scattering (b) of the tip apex in the focal spot of a Gaussian beam, for laser beam polarization parallel to the tip axis. Scan size $15 \times 15\ \mu m^2$.

the tip can be easily recognized in the fundamental wavelength image where there is no significant difference between the scattering of the borders and that of the apex. On the contrary in the SH image a clear isolated nanosource appears which is due to the effect of field enhancement at the very apex of the tip. The image detail is here limited the spot size of the laser beam (~ 700 nm), further indicating the potentialities of this set-up.

Future studies will include the characterization of the produced near field, in addition to the far-field study carried out up to now, by means of scattering-SNOM techniques [2–5]. Other applications foreseen are towards improved-resolution background-free scattering SNOM, optical ultra-high density data storage on photosensitive polymers, and time resolved local spectroscopy of nanostructured functional materials for optoelectronics.

References

1. Pohl, D.W., Denk, W., Lanz, M.: Optical stethoscopy: Image recording with resolution $\lambda/20$. Appl. Phys. Lett. **44**, 651–653 (1984)
2. Zenhausern, F., O'Boyle, M.P., Wickramasinghe, H.K.: Apertureless near-field optical microscope. Appl. Phys. Lett. **65**, 1623–1625 (1994)
3. Hamann, H.F., Gallagher, A. Nesbitt, D.J.: Enhanced sensitivity near-field scanning optical microscopy at high spatial resolution. Appl. Phys. Lett. **73**, 1469–1471 (1998)
4. Labardi, M., Patanè, S., Allegrini, M.: Artifact-free near-field optical imaging by apertureless microscopy . Appl. Phys. Lett. **77**, 621–623 (2000)
5. Hillenbrand, R. and Keilmann, F.: Complex Optical Constants on a Subwavelength Scale . Phys. Rev. Lett. **85**, 3029–3032 (2000)
6. Sánchez, E.J., Novotny, L., Sunney Xie, X.: Near-field fluorescence microscopy based on two-photon excitation with metal tip. Phys. Rev. Lett. **82**, 4014–4017 (1999)
7. Zayats, A.V., Kalkbrenner, T. Sandoghdar, V. Mlynek, J.: Second-harmonic generation from individual surface defects under local excitation. Phys. Rev. B **61**, 4545–4548 (2000)
8. Ramaswamy, M., Ulman, M., Paye, J., Fujimoto, J.G.: Cavity-dumped femtosecond Kerr-lens mode-locked Ti:Al_2O_3 laser. Opt. Lett. **18**, 1822–1824 (1993)
9. Libertum, A.R., Shelton, R., Kapteyn, H.C., Murnane, M.M.: A 36 nJ-15 MHz extended-cavity Ti:sapphire oscillator. In: Conference on Lasers and Electro-Optics 1999, paper CThR3. OSA Technical Digest Series, Washington, D.C. (1999)
10. Cho, S.H., Kärtner, F.X., Morgner, U., Ippen, E.P., Fujimoto, J.G., Cunningham, J.E., Knox, W.H.: Generation of 90-nJ pulses with a 4-MHz repetition-rate Kerr-Lens mode-locked Ti:Al_2O_3 laser operating with net positive and negative intracavity dispersion. Opt. Lett. **26**, 560–562 (2001)
11. Magni, V., Cerullo, C., De Silvestri, S.: Closed form gaussian beam analysis of resonators containing a Kerr medium for femtosecond lasers. Opt. Commun. **101**, 365–370 (1993)
12. Bouhelier, A., Beversluis, M., Hartschuh, A., Novotny, L.: Near-field second-harmonic generation induced by local field enhacement. Phys. Rev. Lett. **90**, 013903-1–4 (2003)
13. Takashi, S. and Zayats, A.V.: Near-field second-harmonic generation at a metal tip apex. Appl. Phys. Lett. **80**, 3479–3481 (2002)

Femtosecond Electron Optics: Towards Atomically Resolved Transition States

Bradley J. Siwick, Jason R. Dwyer, Robert E. Jordan, Christoph T. Hebeisen, and R. J. Dwayne Miller[1]

Departments of Chemistry and Physics, University of Toronto, Toronto, Ontario, M5S 3H6 Canada dmiller@lphys.chem.utoronto.ca

1 Introduction

Time-resolved electron diffraction harbours great potential for following transition state processes with atomic resolution. To observe the largest amplitude motions typically involved in barrier crossings, the time resolution (including the velocity mismatch between the light and electron pulses) must approach 100 femtoseconds. By using thin solid-state samples, it is possible to avoid the problems associated with the velocity mismatch in these experiments. This approach also has the benefit of ensuring that the pump and probe volumes are well matched (i.e. little ground state contamination of the diffraction patterns) while still providing a strong coherent signal due to the large scattering cross-section of electrons.

2 Propagation Dynamics of Femtosecond Electron Packets

Femtosecond electron pulses of sufficient number density to determine structures with a modest number of pulses provide the means to overcome limitations imposed by the irreversibility of transition state crossings and limited sample size. This requirement, however, places severe constraints on the electron pulse propagation. Space-charge effects (Coulomb repulsion) and the initial kinetic energy distribution of the photoelectrons act to broaden the electron pulse as it propagates. The dependence of these effects on the number of electrons, propagation time, initial pulse duration and beam area are important considerations when designing photo-activated electron sources for these investigations. To gain a deeper understanding of the underlying physics, the propagation dynamics of femtosecond electron packets were investigated with an N-body numerical simulation and a mean-field model [1]. The mean-field model makes use of an approximate form for the on-axis potential of the electron pulse to obtain the pulse length equation:

$$\frac{d^2 l}{dt^2} = \frac{Ne^2}{m\varepsilon_0 \pi r^2}\left(1 - \frac{l}{\sqrt{l^2 + 4r^2}}\right) \tag{1}$$

where l is the pulse length, t the propagation time, N the number of electrons, r the beam radius, ε_0 the permittivity of free space, e and m the electron charge and rest mass, respectively. This approach gave access to microscopic details of the electron propagation, including the pulse-broadening dynamics (Fig. 1a). It was found that space-charge effects broaden the electron pulse to many times its original length and generate many eV of kinetic energy "bandwidth" in only a few nanoseconds of propagation. A number of novel features were also observed in the N-body numerical simulations, including the emergence of a linear chirp on the electron wavepacket during propagation that could be exploited for pulse compression (Fig. 1b). These results were used to design a photo-activated electron gun capable of generating sub-500fs electron pulses containing number densities sufficient for structure determinations with fewer than 100 shots.

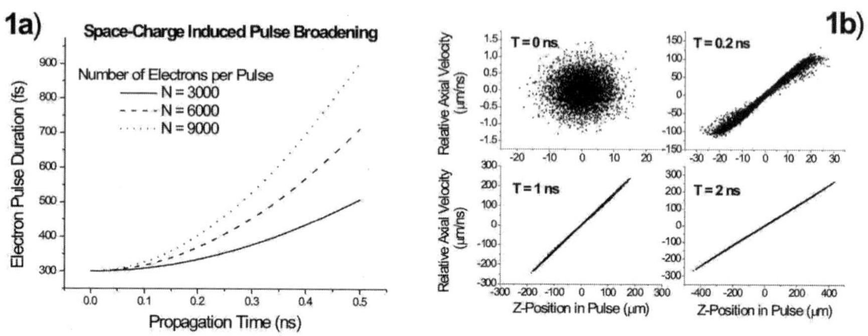

Fig. 1. a) Space-charge effects act to broaden the electron pulse as it propagates. By appropriately choosing the electron pulse parameters (beam radius, r=75 μm in (a)) and maintaining a short propagation distance, femtosecond electron pulses are obtained. b) Axial velocity (v_z) vs. axial position (z) for all electrons in the pulse at four times (T) during its propagation ($N = 10000$, initial pulse duration, $\tau_0 = 150$ fs, r=75 μm). The width of the velocity distribution becomes larger as the pulse lengthens. The spatial distribution of velocities also evolves. As electrons redistribute themselves inside the packet a linear velocity "chirp" develops on the electron pulse.

3 All-Optical Electron Pulse Duration Measurement

The pulse durations now available are too short to be easily characterized with conventional streak camera technology and further advances will only exaggerate this discrepancy in adequate diagnostics. New approaches, akin to nonlinear optical methods for laser pulse characterization, are needed. Here we propose a method to cross-correlate a femtosecond electron pulse with an intense laser pulse.

In the non-relativistic case the ponderomotive force acting on an electron in a laser field takes the form [2]:

$$\boldsymbol{F}_{\text{pond}} = -\nabla U_{\text{pond}} \tag{2}$$

$$U_{\text{pond}} = \frac{e^2\lambda^2 I}{2\pi m c^3} \tag{3}$$

where I is the intensity, λ is the wavelength, c is the speed of light, and e and m are the electron charge and rest mass, respectively.

The effect of the ponderomotive potential on an electron pulse was simulated using Monte-Carlo techniques (non-interacting electrons). The results are shown in Fig. 2. If a Gaussian femtosecond electron pulse ($\sim$ 30keV) is intersected at 90° with a laser pulse (1mJ, 400fs, 800nm, focused to 30μm), the pattern the electron beam forms on a detector 0.5 m away is no longer Gaussian (Fig. 2, inset). The height of the saddle point formed in the "far-field" beam pattern is sensitive to the delay between the laser and electron pulses and to the temporal duration of the two pulses. Inclusion of electron-electron interactions will likely reduce the contrast from the ideal case presented here, however, it is anticipated that this cross-correlation method will still be useful for the characterization of femtosecond electron pulses; particularly for pulses below 500fs.

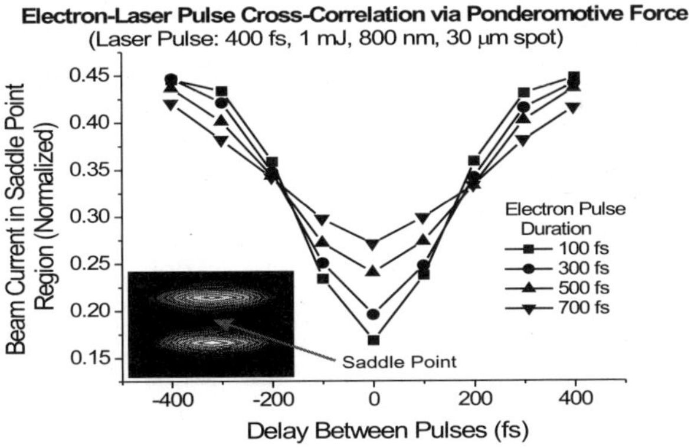

Fig. 2. An electron-laser pulse cross-correlation utilizing the ponderomotive interaction. By measuring the electron beam current through a slit (30μm wide in the above simulation, and centered on the unperturbed beam maximum) 50cm after the point of interaction with the laser pulse, one has a sensitive measure of the pulse overlap. This can also be used to determine T=0 in a time-resolved electron diffraction experiment.

4 Femtosecond Electron Diffraction Studies of Strongly Driven Phase Transitions

Time-resolved diffraction studies with this new instrument have focused on the strongly driven solid-liquid phase transition of aluminum as a model problem. There was evidence from optical studies to suggest that femtosecond laser excitation of Al near an interband absorption at 1.55eV could trigger a sub-picosecond order-disorder structural phase transition when that excitation exceeded a threshold fluence ($\sim 35\text{mJ/cm}^2$) [3]. In this work, the authors observed that the dielectric constant of the laser excited material approached that of the liquid within 500fs, and inferred that the lattice must disorder on this time-scale as well. Optical probe techniques, however, provide only a very indirect measure of the material's time-evolving structure. To determine the timescale and true nature of the structural transformation requires a direct measurement of the change in atomic configuration of the sort provided by time-resolved diffraction experiments.

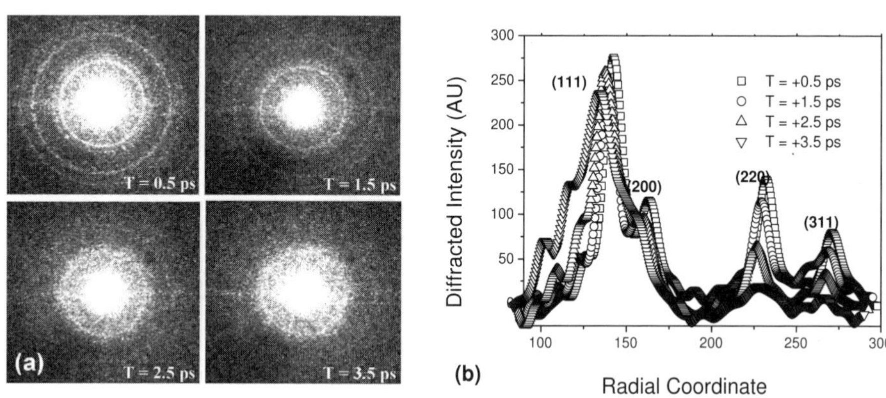

Fig. 3. a) Changes in the diffraction pattern of polycrystalline Al as a function of the time delay ($T \pm 250$fs) between pump (70mJ/cm^2 at 775nm) and probe pulses. These diffraction patterns were acquired with approximately 150 shots (6000 electrons/pulse) and temporal resolution of approximately 600fs. b) Radial average of the diffraction patterns in (a). The loss of long-range order is evident from the disappearance of the (220) and (311) rings and the broadening of the (111) and (200) rings to form a single peak.

The femtosecond electron diffractometer consists primarily of the 30keV femtosecond electron source, sample positioning system and microchannel plate-phosphor screen detector contained in a high vacuum chamber and an external, lens coupled CCD camera to image the diffraction patterns. As configured for these experiments, the detector collected scattering vectors up to a magnitude of $s = 2\sin(\theta)/\lambda = 1.35\text{Å}^{-1}$. This range of diffraction angles encompasses the first 10 rings of the powder diffraction pattern of Al. With this diffractometer we were able to obtain good signal-to-noise powder diffraction pattern of the Al samples (20 ± 2nm thick) with fewer than 150 shots, using sub-600fs electron pulses (6000 electrons/pulse).

Conceptually, the time-resolved electron diffraction method is similar to an ultrafast optical pump-probe experiment; the new ingredient is that the probing is done by electron diffraction. Dynamics are initiated with an ultrafast light pulse and then, at various delay times, the Al sample is probed with an electron pulse. By recording diffraction patterns as a function of delay time we are able to follow the real-space atomic configuration as it evolves in time during the phase transition. Electron diffraction studies with several picoseconds temporal resolution have been performed previously [4], but to our knowledge, these are the first diffraction experiments to be undertaken with sub-picosecond temporal resolution.

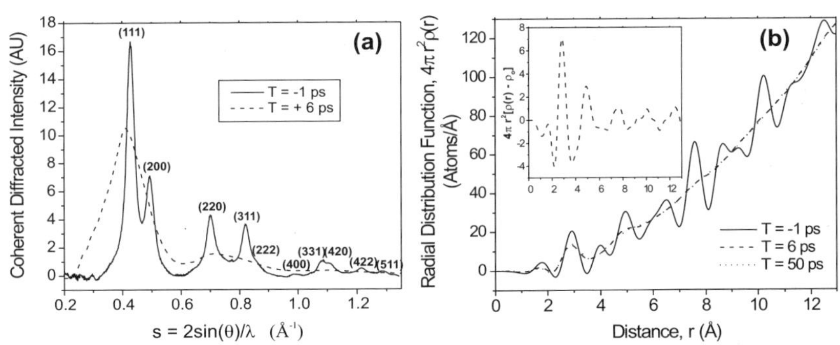

Fig. 4. a) Comparison of the diffraction pattern for polycrystalline Al, showing a well-defined ring structure, with the diffraction pattern of the disordered liquid-like state 6ps after excitation (70mJ/cm^2). b) The radial distribution function at -1, 6 and 50ps computed from the diffraction patterns. Long-range correlations in atomic position are lost through the phase-transition and the detailed structure of the FCC crystalline state is replaced by a simple shell-type liquid structure. The departure of the radial distribution function from the average density is also shown for the 6ps data (inset).

The phase transition dynamics in Al were studied under several laser fluence conditions. Under high fluence (70mJ/cm^2), long-range order is lost in less than 4ps: a liquid-like diffraction pattern emerges on this time-scale (Fig. 3). The changes to the diffraction pattern of Al during the phase transition are unambiguous and dramatic as the presence of discrete rings is a clear signature of the existence of long-range order. The disordered state (liquid) contains only short-range atomic correlations (short-range order), which also produces a characteristic diffraction pattern (Fig. 4a). The breakdown of long-range correlations in atomic position can be seen explicitly by computing the radial distribution function directly from the reduced diffracted intensity [5]. This function describes the atomic density as a function of the radial distance, r, from an atomic origin. The results of this analysis are shown in Fig. 4b. Before the phase-transition (-1ps), the diffraction spectrum consists of a set of discrete lines, each of which can be referred to by the Miller index of the particular set of lattice planes that give rise to that reflection. The peaks in the radial distribution function in this case correspond to distances between pairs of atoms in the face-centered cubic (FCC) structure of Al and are accurate to within ± 0.05 Å. The ultrafast transition from the solid to liquid state is clear from the radial distribution function obtained at 6ps after the laser excitation. The simple shell structure of liquids is apparent in the distribution (Fig. 4b, inset), which has lost the higher frequency oscillations associated with the FCC structure of the crystalline state as well as the associated long-range correlations. There are subtle changes on longer time-scales as well, in particular the reduction of the liquid correlation peak at $r \approx 5$ Åbetween 6 and 50 ps. This peak also shifts to a slightly larger distance which suggests that the liquid structure has not fully equilibrated by 6ps and that the disordered structure observed still contains correlations inherited from the crystalline state.

This work gives an atomic-level description of the ultrafast laser induced melting of aluminum in unprecedented detail. The time-scale for this transition suggests that this process is best understood as a thermally assisted phase transition, and that the electronically driven disordering suggested in [3] is unlikely. These results also illustrate the promise of atomically resolving a wide variety of transition state processes with this technique.

References

1. Siwick, B.J., Dwyer J.R., Jordan R.E., Miller, R.J.D.: Ultrafast electron optics: Propagation dynamics of femtosecond electron packets. J. Appl. Phys., **92**, 1643–1648 (2002)
2. Letokhov, V.S., Zewail, A.H., Kobayashi, T., Kohler, B.: Femtosecond laser control of electron beams for ultrafast diffraction. Adv. Chem. Phys., **101**, 185-192 (1997)
3. Guo, C., Rodriguez, G., Lobad, A., Taylor, A.J.: Structural phase transition of aluminum induced by electronic excitation. Phys. Rev. Lett., **84**, 4493–4496 (2000)
4. Ihee, H., Goodson, B.M., Srinivasan, R., Lobastov, V.A., Zewail, A.H.: Ultrafast electron diffraction and structural dynamics: Transient intermediates in the elimination reaction of $C_2F_4I_2$. , J. Phys. Chem. A **106**, 4087–4103 (2002)
5. Cockayne, D. J. H., McKenzie, D. R.: Electron-diffraction analysis of polycrystalline and amorphous thin-films. Acta Crystallogr. A, **44**, 870-878 (1988)

Application of High-Harmonic Radiation for EUV Interferometry and Spectroscopy

M. Wieland[1]*, Ch. Spielmann[2]**, U. Kleineberg[3], U. Heinzmann[3], and T. Wilhein[1]

[1] RheinAhrCampus, University of Applied Sciences, Südallee 2, 53424 Remagen, Germany
[2] University of Würzburg, Physics Department EP1, Am Hubland, 97074 Würzburg, Germany
[3] University of Bielefeld, Faculty of Physics, Universitätsstrasse 25, 33615 Bielefeld, Germany

Summary. We present the setup of a high-harmonic-beamline providing radiation in the 100 eV regime by nonlinear-frequency conversion of a 1 kHz, table-top 7 fs-laser system (high-harmonic generation). The beamline consists of a target chamber, a diagnostic sequence and a Mo/Si-multilayer-monochromator selecting a single harmonic out of the broad harmonic spectrum. First experiments from the fields of interferometry and absorption spectroscopy are reported.

The development of high-power fs-laser systems has established high-harmonic generation (HHG) as table-top laboratory sources in the VUV and EUV region in recent years [1–4]. Moreover, high-harmonic radiation down to the water-window region has been reported, as well [5,6]. The generation of harmonic radiation is temporally limited to a fraction of the driving laser pulse leading to either bursts or even isolated EUV pulses with durations of several 100 as [7–11]. These ultra-short electromagnetical pulses enable a new field of pump-probe experiments with temporal resolutions below 1fs. High-harmonic radiation is emitted in a well collimated beam of low divergence and offers the illumination of experimental setups with a high degree of spatial coherence [12,13]. Thus, HHG seems to be a promising approach for applications with strong requirements on the spatial coherence like, e.g. interferometry or the generation of high intensities by focusing the high-harmonic beam with x-ray optical elements. Typical conversion efficiencies from laser into high-harmonic radiation at 100 eV in the order of 10^{-8} demand for high repetitive laser systems to provide sufficient flux for a potential application.

Figure 1 shows a schematic of the setup of the beamline. The HHG takes place in the target chamber in a Ni capillary filled with Ne gas at backing pressures around 150 mbar. The following vacuum section contains several filter elements to suppress the visible and low harmonic radiation and

* *Corresponding author:* `wieland@rheinahrcampus.de`
** *former address:* Vienna University of Technology, Photonics Institute, Gusshausstrasse 27, 1040 Vienna, Austria

a simple monitor system consisting of a phosphor-coated CCD. The monitor allows for controlling the EUV-beam with respect to its position and intensity. The radiation is filtered by combinations of thin Si and Zr filters providing a bandpass between 70 . . . 100 eV photon energy. For selecting a single harmonic a multilayer-monochromator is used. It is equipped with two Mo/Si-multilayer mirrors in Z-configuration. The mirrors consist of 50 bilayers Mo/Si and reach a peak reflectivity at 96 eV of approximately 65%. Figure 2 shows the transmitted spectrum trough the monochromator measured with a transmission grating spectrograph [14], the bandwidth of the transmitted radiation is $\Delta E/E = 70$ at 97 eV. Mirror mounts on rotary and linear drives allow for fine tuning of the selected wavelength. The photon flux behind the monochromator within a single harmonic was measured with a calibrated back-illuminated, thinned CCD to approximately 3×10^6 photons per second.

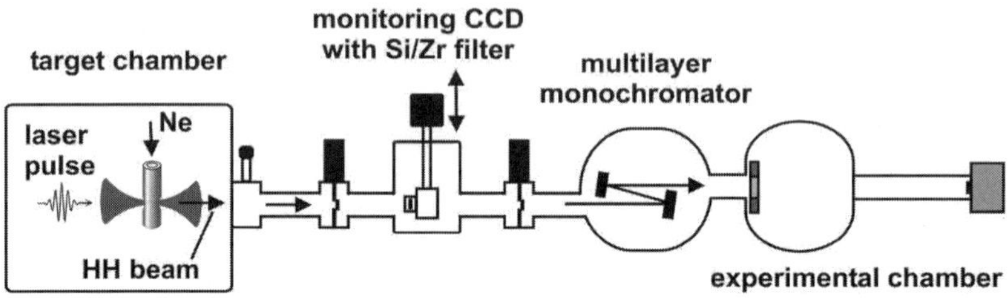

Fig. 1. Schematic of the "high-harmonic" beamline: target chamber, diagnostics chamber, multilayer-monochromator and experimental chamber.

In the following we describe two experiments performed at the "high-harmonic" beamline: i) the setup and demonstration of a common path zone plate interferometer and ii) absorption spectroscopy measurements at the Al-L-edge at 73 eV.

i) A common path zone plate interferometer consists of two zone plates arranged successively onto the optical axis of the system (see Fig. 3) [15]. Based on diffraction, a zone plate splits an incident plane wave into numerous spherical waves (diffraction orders, DO) and the 0th DO passing the zone plate undisturbed. The described setup operates with the +1 diffraction order of each zone plate, a detailed description can be found in [16]. The first zone plate acts as beam-splitter, the second one recombines the beams and creates the interference pattern. If the distance of the two zone plate matches the difference of their focal length, two spherical waves with origin in the back focal plane are created. The superposition of the two waves leads to the formation of the interference pattern in, e.g. the detector plane. Examples for obtained interferograms are shown in Figure 4. The different images correspond to different alignments of the two zone plates with respect

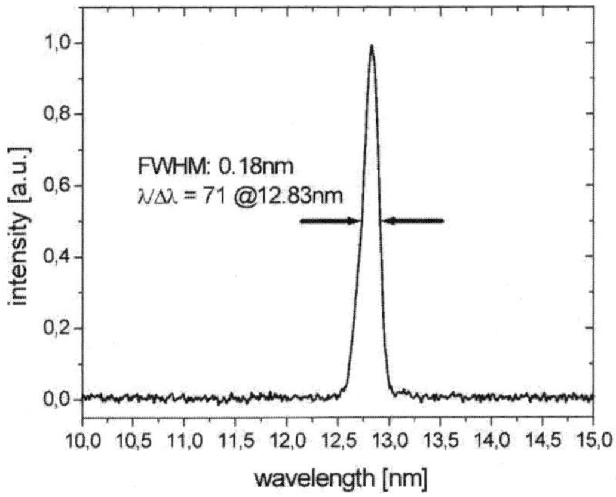

Fig. 2. Spectrum measured behind the multilayer-monochromator. A single harmonic is selected at 12.8 nm wavelength. The measured FWHM is $\lambda/\Delta\lambda \approx 70$, the transmission efficiency of the monochromator approx. 36%.

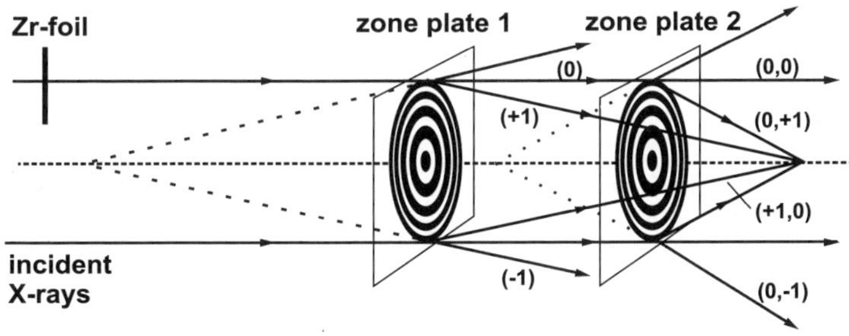

Fig. 3. Schematic of the zone plate interferometer. The diffraction orders and possible combinations for creating the interference pattern are labeled in brackets. For measuring optical properties a sample (Zr-foil) can be introduced into the beam path close to the front focal plane of zone plate 1.

to each other. Images a)-c) were taken with slight lateral shift of the zone plates while keeping the axial separation equal to the difference of their focal lengths. In this case linear fringes can be observed. Image d) shows an interference pattern occurring if the distance between the two zone plates along the optical axis is increased. While moving the zone plate on the optical axis, the linear fringes begin to bend and finally form a zone plate pattern.

The interferometer is suited for the measurement of optical constants in the 100 eV regime. With a single image it is possible to determine the complex index of refraction $n = 1 - \delta + i\beta$ completely. For this purpose, a sample

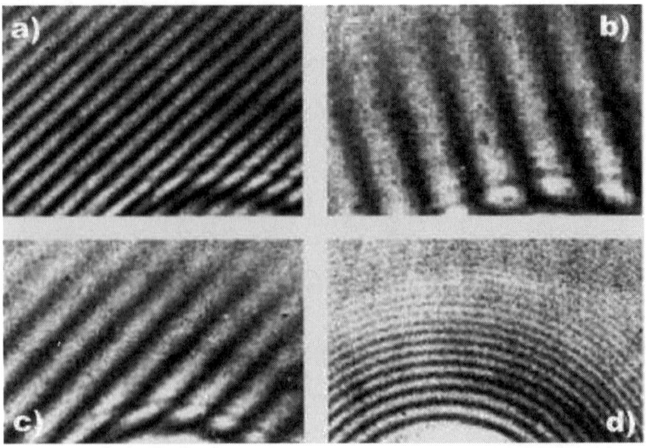

Fig. 4. Interferograms obtained with the current setup: a)-c) lateral shift of the zone plates, d) axial misalignment of zone plate 1.

is introduced into the beam path (see Fig. 3) and imaged simultaneously by the zone plates. The magnified images and the interference pattern are superposed in the detector plane and a shift of the interference fringes occurs at the location of the sample edge. By evaluating the shift of the interference fringes and the absorption of the sample, δ and β can be deduced. Figure 5 shows an interferogram obtained with a 200 nm Zr foil into the beam path, the measured phase shift introduced by the sample is 1.4π. This is in agreement with the expected value of 1.2π for bulk Zr of 200 nm thickness [17]. The transmission is measured to $T = 0.43$ ($T_{theorie} = 0.52$). Differences of experimental and theoretical values result from inaccurate sample thickness and a possible oxide layer on the surface of both foil sides. So far, the experimental setup is limited to the static case, i.e. no pump-probe feature is available. By completing the beamline with such a feature, it probably will be possible to follow changes of the complex index of refraction on a fs-timescale.

Moreover, the determination of spatial and temporal coherence properties of the illuminating radiation is possible. Depending on the particular interferometer setup, the requirements on spatial and temporal coherence can differ [15]. An accurate measurement of either spatial or temporal coherence is complicated by the entanglement of both coherence properties. Consequently, for an unambiguous determination of, e.g. the degree of spatial coherence, it has to be checked carefully, that the requirements on the temporal coherence are fulfilled completely, and vice versa.

So far, the interferometer is operated using a single harmonic, thus the pulse duration of the EUV-pulse is determined by the reflected bandwidth of the multilayer-mirrors. First tests indicate that the interferometer works with broadband radiation, i.e. a bandpass of several harmonics (70...100),

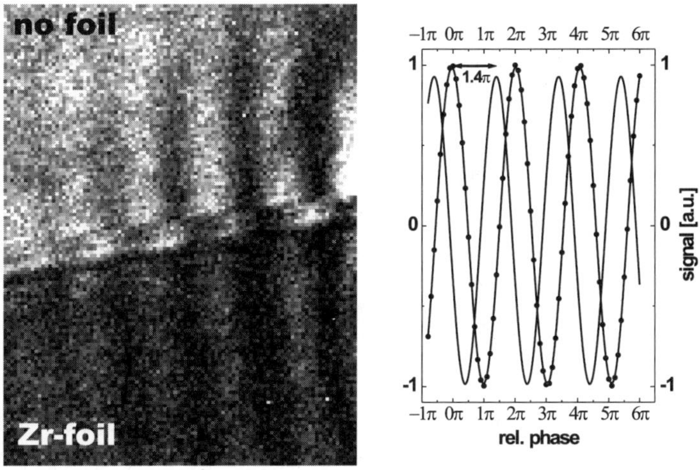

Fig. 5. Interferogram of a 200 nm Zr foil (left image part) mounted close to the front focal plane of the first zone plate. Thus, the object is imaged simultaneously with a magnification of approx. 30. At the foil edge the introduced phase shift is clearly visible. On the right hand side the phase shift of the fringes is shown (solid curve: foil; dotted curve: no foil). The experimental data was processed by Fourier-analysis and fit of a sine function. The determined phase shift is 1.4π.

as well. Due to the broadened spectrum, the pulse duration is expected to be substantially shorter. Assuming phase-locking of individual harmonics and bandwidth-limited radiation, a lower limit for the pulse duration can be directly determined by evaluating the interference pattern corresponding to an autocorrelation of the EUV-pulses.

ii) NEXAFS measurements at the Al-L-edge were performed using the "white" harmonic beam from the beamline without the multilayer-monochromator. In particular, the Al-L-absorption edge of an Al and an Al_2O_3 sample was compared and the chemical shift introduced by the oxygen in the Al_2O_3 sample could be observed. The samples (Al: 50 nm, Al_2O_3: 30 nm) were illuminated with the high-harmonic beam and absorption spectra were recorded with the transmission-grating spectrograph. The spectrograph consists of a Ni grating with 10000 l/mm and the back-illuminated CCD as detector. The spectral resolution of the used setup was better than $\lambda/\Delta\lambda = 300$ which was sufficient to observe the shift of the absorption edge. Figure 6 shows the experimental data, the wavelength axis was calibrated using the L-edge data for bulk Al (Al-$L_2p_{1/2}$: 72.95 eV = 17.00 nm) taken from [17]. For each absorption spectrum (exposure time 20 s) a reference spectrum without sample was recorded to obtain the optical density by taking the negative logarithm of the transmission at each wavelength. The L-edge of the oxide sample is measured to 77.10 eV which corresponds to a shift of 4.15 eV. Direction and

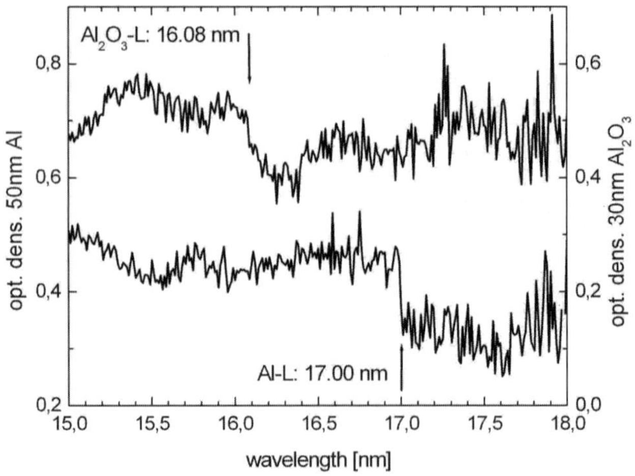

Fig. 6. Absorption spectra of the Al- and Al_2O_3-sample. The edge of the oxide sample is shifted by 4.15 eV towards higher energies.

amount of the shift agree with data reported in the literature [18], although the exact values differ. This discrepancy is probably found in the different sample configuration, in [18] single-crystal sapphire was investigated whereas we used a polycristalline sample produced by thermal evaporation.

Commonly, NEXAFS experiments are carried out at synchrotron sources delivering orders of magnitude higher fluxes in the discussed spectral range. However, the structure of the electron bunches limits the available temporal resolution in pump-probe experiments to some ps, normally requiring a single bunch operation of the storage ring. In addition, an elaborate synchronization to an external pump-laser is required. The laser-driven high-harmonic source inherently offers a synchronization of pump- and probe-beam and a significantly higher temporal resolution. The presented setup demonstrates absorption spectroscopy using high-harmonic radiation in the 70...100 eV-regime with a possible extension to time-resolved measurements.

Within this energy range, the $Si_{1/2}$-L-edge is located at 99.8 eV. As an important material in technological applications, Si is part of numerous (spectroscopic) investigations, in particular its electronic structure is of great interest. Some phenomena, like e.g. the visible light-induced shift of absorption edges in the EUV have been observed but not clearly understood due to a limited time-resolution of some ps available using laser-plasma EUV-sources for spectroscopy [19]. The HH-beamline offers the possibility to perform similar experiments with a higher temporal resolution. Recent developments of driving laser systems and HH-sources will improve the spectral characteristics and enable the generation of quasi-continuum radiation in the EUV which will be a benefit for absorption spectroscopy. Further, the use of tunable, high-efficiency monochromator systems for selecting a small bandpass out of the harmonic spectrum is the basis for spectromicroscopy as element-selective technique.

In conclusion, the presented experiments show the potential of high-harmonic sources in the EUV spectral range not only for fundamental investigations but also for applications like spectroscopy or interferometry basically relying on the special properties of HH-radiation. Techniques based on imaging optical elements like interferometry or microscopy still suffer from low photon fluxes present at HH sources and a continuous improvement of sources and optical setups is necessary to achieve high spatial resolution and short exposure times in pump-probe experiments for additional time resolution. Recent developments using a collimating multilayer-monochromator as condenser and a Fresnel zone plate as high-resolution objective indicate the feasibility of HH sources for high resolution imaging in the 100...200 nm regime.

Acknowledgement. The authors would like to thank F. Krausz, R. Kienberger, M. Hentschel and M. Uiberacker (Vienna University of Technology, Photonics Institute) for their support and help with the laser system and the Institute for X-ray physics, University of Göttingen, for providing X-ray optical elements. This work was supported by the DFG under Contract No. Wi1451/1-1, CS acknowledges support from the DFG grant SP 687/1-1 and the Austrian Science Fund grant F016.

References

1. A. L'Huillier and P. Balcou. High order harmonic generation in rare gases with a 1-ps 1053-nm laser. *Phys. Rev. Lett.*, 70(6):774, 1993.
2. J. J. Macklin, J. D. Kmetec, and C. L. Gordon. High-order harmonic generation using intense femtosecond pulses. *Phys. Rev. Lett.*, 70:766, 1993.
3. J. Zhou, J. Peatross, M. M. Murnane, and H. C. Kapteyn. Enhanced high-harmonic generation using 25 fs laser pulses. *Phys. Rev. Lett.*, 76(5):752, 1996.
4. M. Schnürer, Z. Cheng, M. Hentschel, G. Tempea, P. Kálmán, T. Brabec, and F. Krausz. Absorption limited generation of coherent ultrashort soft x-ray pulses. *Phys. Rev. Lett.*, 83(4):722, 1999.
5. Z. Chang, A. Rundquist, H. Wang, M. M. Murnane, and H. C. Kapteyn. Generation of coherent soft x-rays at 2.7nm using high harmonics. *Phys. Rev. Lett.*, 79(16):2967, 1997.
6. Ch. Spielmann, N. H. Burnett, S. Sartania, R. Koppitsch, M. Schnürer, C. Kan, M. Lenzner, P. Wobrauschek, and F. Krausz. Generation of coherent soft x-rays in the water window using 5-femtosecond laser pulses. *Science*, 278:661, 1997.
7. P. Antoine, A. L'Huillier, and M. Lewenstein. Attosecond pulse trains using high-order harmonics. *Phys. Rev. Lett.*, 77(7):1234, 1996.

8. I. P. Christov, M. M. Murnane, and H. C. Kapteyn. High-harmonic generation of attosecond pulses in the 'single-cycle' regime. *Phys. Rev. Lett.*, 78(16):1251, 1997.
9. N. A. Papadogiannis, B. Witzel, C. Kalpouzos, and D. Charalambidis. Observation of attosecond light localization in higher order harmonic generation. *Phys. Rev. Lett.*, 83(21):4289, 1999.
10. M. Hentschel, R. Kienberger, Ch. Spielmann, G. A. Reider, N. Milosevic, T. Brabec, P. Corkum, U. Heinzmann, M. Drescher, and F. Krausz. Attosecond metrology. *Nature*, 414:551, 2001.
11. P. M. Paul, E. S. Toma, P. Breger, G. Mullot, F. Augé, Ph. Balcou, H. G. Muller, and P. Agostini. Observation of a train of attosecond pulses from high harmonic generation. *Science*, 292:1689, 2001.
12. T. Ditmire, E. T. Gumbrell, R. A. Smith, J. W. G. Tisch, D. D. Meyerhofer, and M. H. R. Hutchinson. Spatial coherence measurement of soft x-ray radiation produced by high order harmonic generation. *Phys. Rev. Lett.*, 77(23):4756, 1996.
13. R. A. Bartels, A. Paul, H. Green, H. C. Kapteyn, M. M. Murnane, S. Backus, I. P. Christov, Y. Liu, D. Attwood, and C. Jacobsen. Generation of spatially coherent light at extreme ultraviolet wavelengths. *Science*, 297:376, 2002.
14. T. Wilhein, S. Rehbein, D. Hambach, M. Berglund, L. Rymell, and H. M. Hertz. A slit grating spectrograph for quantitative soft x-ray spectroscopy. *Rev. Sci. Instr.*, 70(3):1694, 1999.
15. T. Wilhein, B. Kaulich, and J. Susini. Two zone plate interference contrast microscopy at 4 kev photon energy. *Opt. Comm.*, 193:19, 2001.
16. M. Wieland, T. Wilhein, Ch. Spielmann, and U. Kleineberg. Zone plate interferometry at 13 nm wavelength. *Appl. Phys. B, accepted*, 2003.
17. http://www crxo.lbl.gov/optical_constants/. Center of X-Ray Optics, Berkeley. 2003.
18. J. M. Chen, J. K. Simons, K. H. Tan, and R. A. Rosenberg. Correlation between interatomic distances and the x-ray-absorption near-edge structure of single-crystal sapphire. *Phys. Rev. B*, 48(14):10047, 1993.
19. H. Nakano, Y. Goto, P. Lu, T. Nishikawa, and N. Uesugi. Time-resolved soft x-ray absorption spectroscopy of silicon using femtosecond laser plasma x rays. *Appl. Phys. Lett.*, 75(16):2350, 1999.

Time-Resolved Inner Shell Spectroscopy with Sub-fs EUV Pulses

M. Drescher[1], M. Hentschel[2], R. Kienberger[2], M. Uiberacker[2], V. Yakovlev[2], A. Scrinzi[2], Th. Westerwalbesloh[1], U. Kleineberg[1], U. Heinzmann[1], and F. Krausz[2]

[1] Fakultät für Physik, Universität Bielefeld, Universitätsstrasse 25, D-33615 Bielefeld, Germany
[2] Institut für Photonik, Technische Universität Wien, Gusshausstrasse 27, A-1040 Wien, Austria drescher@physik.uni-bielefeld.de

Summary. Ultrashort high harmonics are applied for a time-domain tracking of decaying atomic inner-shell vacancies. The novel EUV-pump / laser-probe scheme utilizes the delay-dependent momentum exchange between photoemitted electrons and an intense laser field. The relaxation of 3d core holes in krypton was studied in a benchmark experiment demonstrating the capabilities of the technique.

1 Introduction

If we are interested in following in the time-domain how the electronic structure in the interior of the atomic shell evolves after an excitation event not only a sufficiently high excitation energy is necessary but also a method enabling extremely fast stroboscopic snapshots of the emerging electron waves. Electronic detectors of sufficient velocity are not available and even the fastest streak-cameras are limited to a temporal resolution of a few $100fs$. Up to now information about the inner-shell decay dynamics was therefore deduced from measurements of the spectral linewidth $\Gamma = \hbar/\tau_h$, assuming an exponential decay with time-constant τ_h and a corresponding Lorentzian line shape [1].

On the other hand, extremely high temporal resolution is promised by the pump-probe technique, where a pump pulse excites the system under study which then develops for a well defined time until its state is sampled by a second delayed probe pulse. Due to the rapid progress in the development of ultrashort laser sources this universal method has proven its utility in the field of femtochemistry [2]. Here, the typical molecular photoreactions like vibrational excitation or dissociation, governed by the coupling between electronic and nuclear movement, occur on a femtosecond to picosecond time scale. A transfer of this technique to time-resolved studies of processes in the interior of the atomic shell requires sub-fs pulses of ionizing radiation in the extreme ultraviolet (EUV) or X-ray range, not provided by established light sources like X-ray tubes, synchrotrons or laser-driven plasmas. High harmonics of few-cycle laser pulses are capable of delivering the required photon energies together with an ultrashort - attosecond - time-structure, making them a unique source for EUV-pump / laser-probe experiments on atoms. The capability of these novel tools is demonstrated by a first example of a time-based tracking of an atomic Auger decay - the decay of a 3d core-hole in krypton.

2 Ultrafast excitation and sampling

For quite some time high harmonics of femtosecond laser pulses [3] generated as odd multiples $h\nu_{EUV} = n * h\nu_L$ of the laser frequency νL were considered as potential candidates for sub-fs light bursts. The resulting high orders $-n > 300$ have been reached [4] - correspond to wavelengths in the EUV and soft X-ray range. Utilization of fundamental laser pulses with a duration of a few tens of femtoseconds yields trains of radiation bursts [5,6] with a period of half a laser cycle ($T_L \sim 1.3fs$ at $\lambda_L \sim 750\ nm$). Accordingly, several schemes were proposed [7–9] for the selection of a more desirable isolated pulse. Recently, the generation of single EUV pulses with attosecond duration could be demonstrated [10] by compressing the fundamental laser pulses to a few - typically 3 - optical cycles and by selecting the high harmonic radiation at the spectral cut-off. Owing to the pronounced nonlinear character of tunnel-ionization only the highest field strengths in the laser pulse contribute to the cut-off harmonics.

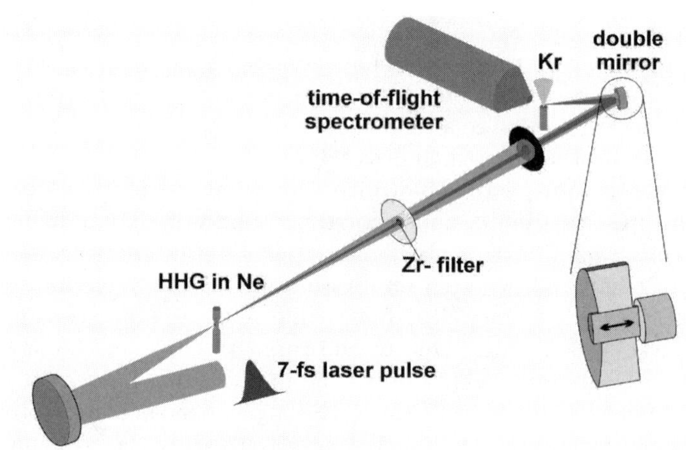

Fig. 1. Experimental set-up for time-resolving EUV-pump/laser-probe electron spectroscopy.

For sufficiently short pulses with a considerable contrast between subsequent field extrema radiation should consequently emerge only from a temporally well confined range around the strongest field amplitude thus forming an isolated attosecond EUV burst. The apparative backbone for a realization of this scheme is a Ti:sapphire CPA laser system [11] delivering intense near-infrared light pulses of only 3 oscillation periods (approx. $7fs$) duration (Fig. 1) at a repetition rate of $1kHz$. In the laser focus located within a neon-gas filled tube intensities exceeding $10^{14}W/cm^2$ lead to the generation of high harmonics with a cut-off at about 100 eV photon energy. Laser- and EUV beam are guided to interact once again in a second gas target using a dedicated double-reflector assembled from two concentric mirrors. The inner spherical Mo/Si multilayer mirror [12] filters a 3 eV to 10 eV wide spectral band in the 90 eV to 100 eV range from the high harmonic spectrum while the outer ring-shaped mirror of the same focal length refocusses the laser beam. In this way both beams are spatially superimposed in the second gas medium. The temporal overlap can be varied by a piezo-electrical translation of the inner mirror with nanometer precision and reproducibility thus enabling pump-probe delays with attosecond resolution.

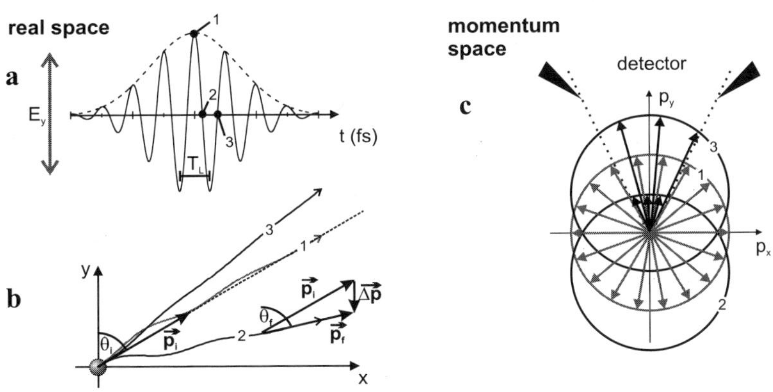

Fig. 2. The classical trajectories (b) of electrons liberated by the EUV pump pulse sensitively depend on the phase and magnitude of the probing laser pulse (a) at the instant of their emission. The momentum exchange (c) with the light-field leads to a bipolar energy shift oscillating with a $\frac{\pi}{2}$ shift to the laser field.

The key to taking snapshots of the electron emission with unprecedented temporal resolution is the electrons' interaction with an intense laser field, acting as an immaterial electron 'shutter' [13]. Let us consider the photoionization of an atom in the presence of a visible laser field. After excitation with an ultrashort EUV pulse with sufficient photon energy the atom is ionized and a photoelectron leaves the ion with a defined momentum $\mathbf{p_i}$. Once liberated, the electron experiences the laser field and is accelerated in the direction of the light polarization (Fig. 2). The classical analysis of the electron trajectories reveals, that the final momentum $\boldsymbol{p_f} = \boldsymbol{p_i} + \boldsymbol{\Delta p}$ when the laser pulse has faded away strongly depends on the phase and amplitude of the laser field at the instance of electron injection, i.e. the ionization event. The transferred momentum $\mathbf{\Delta p}$ vanishes when the ionization takes place in the field maximum and is strongest for the zero field transitions a quarter of a light period earlier or later (Fig. 2). This extreme sensitivity of the electron momentum on the light phase builds the footing of temporal measurements with a resolution of a fraction of the laser period (approx. $T_L/20 < 200\ as$ for the current laser period of $T_L \sim 2.6 fs$). To this end the corresponding modulation of kinetic electron energies is detected with a time-of-flight spectrometer (Fig. 1) with a varying delay between the EUV pump and the laser probe-pulse [13].

So far we assumed that the electron wave packet essentially mimics the intensity envelope of the incoming EUV pulse and has a duration τ_e of a fraction of the laser oscillation cycle, $\tau_e \ll T_L$. In this limit, the above classical description provides a reliable framework for the analysis of recent experiments characterizing the temporal properties of sub-fs EUV pulses. A duration of approx. $0.5 fs$ could be determined [10, 14] and the isolated nature of the EUV pulses verified. However, an extension of this sampling technique to secondary electron emission - e.g. Auger decay following the creation of a core-hole - must account for a possible temporal extension of the electron wave packet over one or more laser cycles, $\tau_e > T_L$. The analysis reveals [15], that in this regime different portions of the electron wave packet interact coherently with the laser field and lead to a modulation of the electron energy in the form of discrete side bands. The steady shift observed for $\tau_e \ll T_L$ is therefore substituted by a discrete sideband structure. This latter case applies for the time-resolved studies presented in the following.

3 Experimental Results

One of the most comprehensively studied systems (in the energy domain) involving an inner-shell relaxation is the MNN Auger decay in krypton following excitation of the $3d$ shell above the ionization threshold at $93.8eV$. For a first time-based measurement of the $Kr - 3d$ core-relaxation time τ_h by directly sampling the Auger wave packet krypton atoms were excited with a EUV pump-pulse in a $3\ eV$ wide band centered at $97eV$, corresponding to a (Fourier-limited) duration of $0.9 fs$. A typical electron spectrum after integration over 300,000 laser shots is shown in the upper right corner of (Fig. 3). Combining 20 of such spectra taken at different delays Δt between the EUV and the laser pulse a map of the temporal evolution of the electron emission is obtained, displayed as a surface plot in (Fig. 3). According to the above discussion for an electron emission extending over a few femtoseconds, the interaction with the laser field should lead to the emergence of sidebands separated by multiples of $h\nu_L \sim 1.6\ eV$ from each Auger line. Since the sideband area represents a convolution of the Auger emission rate and the laser envelope the latter must be precisely known for an extraction of the

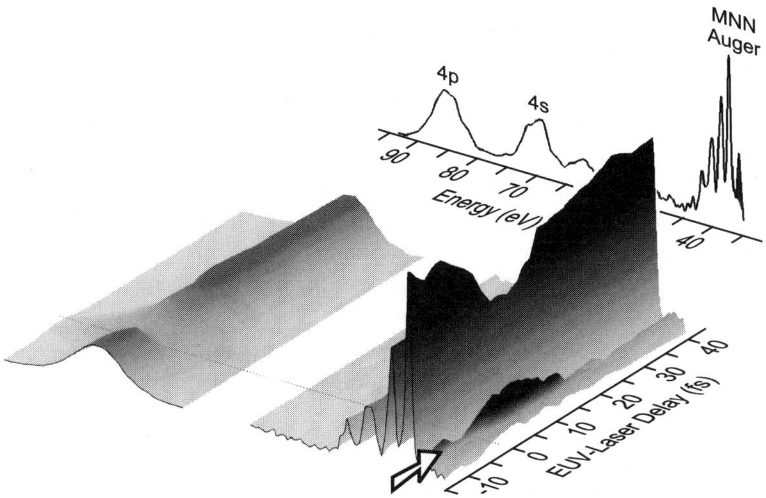

Fig. 3. Evolution of krypton electron spectra following 3d core excitation. The surface plot shows the evolution of the 4p photo line and the $M_{4,5}N_1N_{2,3}$ Auger lines with the delay between the 97 eV EUV- and the 1.6 eV laser pulse.

desired value of τ_h. This calibrating information is simultaneously embedded in the acquired electron spectra in the delay-dependent broadening of the $4p$ and $4s$ photo lines (Fig. 3). Since the direct ionization process can be considered as instantaneous on the relevant time scale the spectral broadening of the $4p$ photo line directly maps the evolution of the laser field envelope. With this information as an input a fit procedure with τ_h as the only varied parameter is capable of extracting the temporal evolution of the Auger emission rate. For the -1st sideband of a $3d^{-1}(^1P_1)$ Auger line (indicated by an arrow in (Fig. 3) this analysis yields $\tau_h = 7.9(+1.0/-0.9)fs$ for the lifetime of $M(3d_{5/2})$ vacancies in krypton. This value corresponds to a natural line width of $84 \pm 10meV$, well in accordance with the measured value of $88 \pm 4meV$ from [16].

4 Conclusion

The exponential decay of inner-shell vacancies - hitherto inferred from energy-domain measurements - has now been investigated directly in the time-domain. The utilized EUV-pump / laser-probe technique makes use of a momentum transfer between the laser field and electrons at the instant of their emission and does not therefore depend on particular transitions in the probing step. Accordingly, direct photoionization as well as indirect photoemission phenomena like Auger-decay are becoming accessible to time-resolved studies. Of particular interest will be such phenomena, that do not become manifest in time-integrated electron spectra, e.g. interferences of competing quantum-paths in an electronic transition. Advances in the light source development will lead to an extension of the experimental capabilities. With higher photon energies [4], deeper atomic shells - with generally faster decay processes - can be excited. The introduced ultrafast electron sampling principle is capable of resolving dynamics on a sub-laser-cycle time-scale. Very recently, fundamental laser pulses with a locked phase between the laser carrier and envelope have become available [17] and provide sufficient phase stability and reproducibility to improve the temporal resolution to a few hundred attoseconds.

This work was supported by the Fonds zur Förderung der wissenschaftlichen Forschung (Austria), the Deutsche Forschungsgemeinschaft (Germany) and by the European Atto Network.

References

1. V. Schmidt, "Electron Spectroscopy of Atoms using Synchrotron Radiation"; Cambridge University Press, Cambridge **54**, (1997)
2. A. Zewail, "Femtochemistry: atomic-scale dynamics of the chemical bond", J. Phys. Chem. A **104**, (2000), 5660
3. T. Brabec, F. Krausz, Rev. Mod. Phys. **72**, (2000), 545
4. M. Schnürer, Ch. Spielmann, P. Wobrauschek, C. Streli, N.H. Burnett, C. Kan, K. Ferenc, R. Koppitsch, Z. Cheng, T. Brabec, F. Krausz, Phys. Rev. Lett. **80**, (1998), 3236
5. N. A.Papadogiannis, B. Witzel, C. Kalpouzos, D. Charalambidis, Phys. Rev. Lett. **83** (1999), 4289
6. P.M. Paul, E.S. Toma, P. Breger, G. Mullot, F. Auge, P. Balcou, H.G. Muller, P. Agostini, Science **292**, (2001), 1689
7. V. T. Platonenko, V.V. Strelkov, Quant. Electron. **28** (1998), 749
8. I.P. Christov, M.M. Murnane, H.C. Kapteyn, Opt. Commun. **148**, (1998), 75
9. F.L. Kien, N. H. Shon, K. Hakuta, Phys. Rev. A **64**, (2001), 051803
10. M. Hentschel, R. Kienberger, Ch. Spielmann, G.A. Reider, N. Milosevic, T. Brabec, P.B. Corkum, U. Heinzmann, M. Drescher, F. Krausz, Nature **414**, (2001), 509
11. S. Sartania, Z. Cheng, M. Lenzner, G. Tempea, Ch. Spielmann, F. Krausz, Opt. Lett. **22**, (1997), 1562
12. Y.C. Lim, T. Westerwalbesloh, A. Aschentrup, O. Wehmeyer, G. Haindl, U. Kleineberg, U. Heinzmann, Appl. Phys. A **72**, (2001), 121
13. M. Drescher, M. Hentschel, R. Kienberger, G. Tempea, Ch. Spielmann, G.A. Reider, P.B. Corkum, F. Krausz, Science **291**, (2001), 1923
14. R. Kienberger, M. Hentschel, M. Uiberacker, Ch. Spielmann, M. Kitzler, A. Scrinzi, M. Wieland, Th. Westerwalbesloh, U. Kleineberg, U. Heinzmann, M. Drescher, F. Krausz, Science **297**, (2002), 1144
15. M. Drescher, M. Hentschel, R. Kienberger, M. Uiberacker, Th. Westerwalbesloh, U. Kleineberg, U. Heinzmann, F. Krausz, Nature **419**, (2002), 803
16. M. Jurvansuu, A. Kivimäki, S. Aksela, Phys. Rev. A **64**, (2001), 012502

17. A. Baltuska, Th. Udem, M. Uiberacker, M. Hentschel, E. Goulielmakis, Ch. Gohle, R. Holzwarth, V. Yakovlev, A. Scrinzi, T.W. Hänsch, F. Krausz, Nature **421**, (2003), 611

A Quantum Theory of Time-Resolved Auger Measurement

Olga Smirnova, Vladislav S. Yakovlev, and Armin Scrinzi

Photonics Institute, Vienna University of Technology
Gusshausstrasse 27/387, A-1040 Vienna, Austria/EU, `scrinzi@tuwien.ac.at`

Summary. We present a quantum mechanical model of the XUV-attosecond pump and laser probe measurement of an Auger decay [Drescher *et al.*, Nature, **419**, 803 (2002)] and investigate effects of quantum coherence. We demonstrate the enhancement of high- and low-energy tails of the Auger spectrum due to coherence between photoionization and Auger decay.

In a recent XUV pump – laser probe experiment the time-resolved image of an Auger decay [3] was obtained. Although the basic physical ideas of the measurements can be phrased in terms of classical mechanics [2], the observed electron spectra are determined by quantum interference phenomena. Here we unite core-hole formation, Auger decay, and Auger electron streaking by the laser in a consistent quantum theory of time-resolved Auger measurement. The primary purpose is to clarify the role which quantum coherence plays for the final electron spectra. A set of basic equations is derived that is solved numerically. With a few additional approximations we obtain an analytical solution, which allows for a clear physical interpretation. Finally, comparison to the *ad hoc* description of [3] shows that the Auger electron spectra are broadened due to coherence between photoionization and Auger decay.

1 Essential states model

A scheme of the pump-probe Auger measurement is shown in Fig. 1. An XUV pulse with frequency Ω excites a core electron with energy E_0 to a continuum energy near $W_x = E_0 + \Omega$. The vacancy is filled from the level E_a by Auger decay. The Auger electron is ejected from the level E_b to the final energy near $W_A = E_a + E_b - E_0$. A laser probe pulse modifies the Auger electron momentum and spectra are recorded as a function of the time delay Δt between XUV and laser pulse.

Our model is based on the essential states participating in the process: the ground state $|g\rangle$, the core hole states $|h\rangle$ with one electron in the continuum and the final states $|f\rangle$ with two continuum electrons. Disregarding for the present discussion correlation effects, the three states can be written as products of single electron orbitals in the form

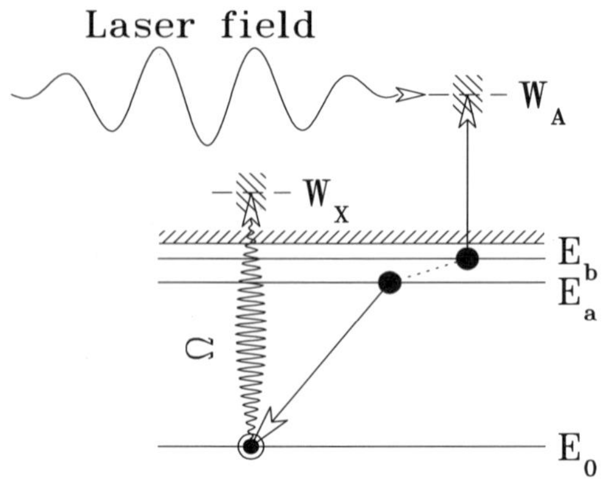

Fig. 1. Scheme of the Auger pump-probe measurement

$$|g\rangle := |0,t\rangle|a,t\rangle|b,t\rangle, \quad |h\rangle := |\mathbf{k}_{\mathrm{x}},t\rangle|a,t\rangle|b,t\rangle, \quad |f\rangle := |\mathbf{k}_{\mathrm{x}},t\rangle|0,t\rangle|\mathbf{k}_l,t\rangle. \quad (1)$$

The bound states $|0,t\rangle$, $|a,t\rangle$, and $|b,t\rangle$ with time dependence $\exp(-itE_\alpha)$, $\alpha = 0, a, b$ denote single-electron orbitals corresponding to the energies E_0, E_a and E_b in Fig. 1. The continuum states $|\mathbf{k}_{\mathrm{x}},t\rangle$ and $|\mathbf{k}_l,t\rangle$ are Volkov solutions for the free motion of an electron in the XUV and laser fields, respectively,

$$|\mathbf{k}_\eta,t\rangle = e^{-i\Phi_\eta(\mathbf{k}_\eta,t)}e^{i[\mathbf{k}_\eta-\mathbf{A}_\eta(t)]\cdot\mathbf{r}}, \quad \eta = l, \mathrm{x} \quad (2)$$

where $\mathbf{A}_\eta$ and the Volkov phases Φ_η are given by

$$\mathbf{A}_\eta(t) = -\int_{-\infty}^{t} \mathcal{E}_\eta(t')dt' \quad \text{and} \quad \Phi(\mathbf{k}_\eta,t) = \frac{1}{2}\int_{-\infty}^{t}[\mathbf{k}_\eta - \mathbf{A}_\eta(t')]^2dt' \quad (3)$$

and $\mathcal{E}_\eta$ denote the respective field strengths. Double occupation of the single-electron states is excluded by the construction of our essential states, if we assume that the XUV photoelectron and Auger electron spectra do not overlap $\mathbf{k}_{\mathrm{x}} \neq \mathbf{k}_l$.) We expand the full wave function in the essential states with expansion coefficients c_g, c_h and c_f and the initial condition $c_g(t=-\infty) = 1$ and $c_h(t=-\infty) = c_f(t=-\infty) = 0$. One obtains the equations for c_h and c_f

$$i\dot{c}_h(\mathbf{k}_{\mathrm{x}},t) = D(\mathbf{k}_{\mathrm{x}},t) + \int dk_{\mathrm{l}}^{(3)} M^*(\mathbf{k}_l,t)c_f(\mathbf{k}_{\mathrm{x}},\mathbf{k}_{\mathrm{x}},t) \quad (4)$$

$$i\dot{c}_f(\mathbf{k}_{\mathrm{x}},\mathbf{k}_l,t) = M(\mathbf{k}_l,t)c_h(\mathbf{k}_{\mathrm{x}},t). \quad (5)$$

The XUV dipole transition appears in

$$D(\mathbf{k}_{\mathrm{x}},t) = \mathcal{E}_0(t)d(\mathbf{k}_{\mathrm{x}} - \mathbf{A}_{\mathrm{x}})e^{i\Phi_{\mathrm{x}}(\mathbf{k}_{\mathrm{x}},t)-it(E_0+\Omega)}, \quad (6)$$

where $d(\mathbf{k})$ is the dipole matrix element of XUV-induced transition and $\mathcal{E}_0$ denotes the XUV pulse envelope. The Auger transition matrix element is

$$M(\mathbf{k}_l, t) = \langle 0, t| \langle \mathbf{k}_l, t| \frac{1}{|\mathbf{r}_2 - \mathbf{r}_3|} |b, t\rangle |a, t\rangle. \tag{7}$$

These equations include a few more approximations. Most importantly, we neglect continuum-continuum transitions between Volkov states. This excludes "post-collision interactions" [1,4] of photo- and Auger electron, which is outside the scope of the present work. Integrating (4) and substituting into (5) gives the integro-differential equation for c_h

$$i\dot{c}_h(\mathbf{k}_\mathrm{x}, t) = D(\mathbf{k}_\mathrm{x}, t) + \int_{-\infty}^{t} dt' \int dk_l^{(3)} M^*(\mathbf{k}_l, t) M(\mathbf{k}_l, t') c_h(\mathbf{k}_\mathrm{x}, t'). \tag{8}$$

2 Numerical solution

Eq. (8) was solved numerically for the parameters studied in Ref. [3]: XUV photon energy $\Omega = 90\,eV$, electron levels $E_0 = -75\,eV, E_a = E_b = -15\,eV$, and laser photon energy of $\omega_l = 1.55\,eV$ (λ= 800 nm). The $|0\rangle$-, $|a\rangle$-, and $|b\rangle$-states represent the $3d$-, $4s$- and $4p$-orbitals of krypton, respectively. For the questions investigated here, the exact shape of the electron orbitals is of little importance and therefore we chose hydrogenic wave functions with the corresponding principal and orbital quantum numbers. The radial coordinates of the orbitals were scaled to adjust them to the approximate spatial extension of the Kr orbitals and to produce the correct asymptotic behavior for the respective energies. The $4s \to 3d$ transition singles out the quadrupole term of the electron-electron interaction and justifies the factorization

$$M(\mathbf{k}_l, t) \approx Q_{a0}\, q(\mathbf{k}_l - \mathbf{A}_l(t)) e^{it(E_0 - E_a - E_b) + i\Phi_l(\mathbf{k}_l, t)}, \tag{9}$$

where $q(\mathbf{k})$ is the quadrupole matrix element for the transition from $|b\rangle$ into the continuum. The strength Q_{a0} of the quadrupole transition $|a\rangle \to |0\rangle$ was chosen as a parameter to control the Auger decay rate Γ.

3 Analytical solution

Closer inspection of (8) reveals that c_h is a smooth function of time and that the main contribution to the integral comes from $t' \approx t$. Replacing in Eq. (8) $c_h(\mathbf{k}_\mathrm{x}, t')$ with $c_h(\mathbf{k}_\mathrm{x}, t)$, leads to an ordinary differential equation with simple exponential decay for c_h. In absence of the laser field and using the rotating wave approximation for the XUV field, one finds the closed form solution

$$c_h(\mathbf{k}_\mathrm{x}, t) = \frac{d(\mathbf{k}_\mathrm{x})}{2\pi} \int_{-\infty}^{\infty} \frac{\tilde{\mathcal{E}}_0(\eta) \exp[i(\mathbf{k}_\mathrm{x}^2/2 - W_\mathrm{x} - \eta)t]}{\mathbf{k}_\mathrm{x}^2/2 - W_\mathrm{x} - \Delta - \eta - i\Gamma/2} d\eta, \tag{10}$$

where $\tilde{\mathcal{E}}_0$ is the Fourier transform of the XUV envelope $\mathcal{E}_0$. The Auger energy shift Δ and width Γ can be calculated in second order perturbation theory and by Fermi's golden rule, respectively. The final two-electron amplitude in absence of the laser field is obtained by substituting (10) into (5)

$$c_f(\mathbf{k}_\mathrm{x}, \mathbf{k}_l, \infty) = Q_{a0}\, d(\mathbf{k}_\mathrm{x}) q(\mathbf{k}_l) s(\mathbf{k}_\mathrm{x}^2, \mathbf{k}_l^2), \tag{11}$$

where we have defined the two-electron line shape

$$s(\mathbf{k}_\mathrm{x}^2, \mathbf{k}_l^2) := \frac{\tilde{\mathcal{E}}_0(\mathbf{k}_\mathrm{x}^2/2 + \mathbf{k}_l^2/2 - \Omega + E_a + E_b)}{\mathbf{k}_l^2/2 - W_A + i\Gamma/2}. \tag{12}$$

In general, dipole- and Auger matrix elements are slowly varying functions of $|\mathbf{k}_\mathrm{x}|$ and $|\mathbf{k}_l|$, respectively. The shape of the spectrum is determined by the Lorentzian line shape of the Auger decay and by the spectral envelope of the XUV pulse. The width of the spectrum is proportional to Γ.

A closed form solution for laser-streaked electron spectra can be obtained when one neglects the impact of the laser field on c_h by using expression (10) for c_h also when the laser is present. For a cw laser field $\mathbf{A}_l(t) = \mathbf{A}_0 \sin\omega(t + \Delta t)$ one then obtains

$$c_f(\mathbf{k}_\mathrm{x}, \mathbf{k}_l, \infty) = Q_{a0}\, d(\mathbf{k}_\mathrm{x}) q(\mathbf{k}_l) \sum_n e^{in(\omega_l \Delta t + \pi)} J_n(\mathbf{k}_l \cdot \mathbf{A}_0) s(\mathbf{k}_\mathrm{x}^2, \mathbf{k}_l^2 + n\omega). \tag{13}$$

4 Discussion

Figure 2 shows results using the integro-differential equation (8) for c_h and spectra according to the analytical solution (13). For the comparison, the matrix elements d and q were assumed to be independent of $\mathbf{A}_l$ in both calculations. As expected, there is exellent agreement, which demonstrates the validity of the key approximations made in deriving Eq. (13) from Eq. (8).

To complete the comparison, we have calculated spectra obtained by the *ad hoc* formula employed in Ref. [3]. That formula essentially amounts to setting the hole state amplitude to $c_h(\mathbf{k}_\mathrm{x}, t) \approx \sqrt{\rho(t)}$, where ρ is the population of the hole state obtained from the rate equations for XUV ionization and Auger decay. The hole state keeps a record about how it was formed in the $\mathbf{k}_\mathrm{x}$-dependent phase of $c_h \sim \exp(it\mathbf{k}_\mathrm{x}^2/2)$ (cf. Eq. 10), which ensures the coherence of hole state formation and decay. Due to this coherence the *sum* of the electron kinetic energies $\mathbf{k}_\mathrm{x}^2/2 + \mathbf{k}_l^2/2$ enters the argument of $\tilde{\mathcal{E}}_0$ in Eq. 12: the final state amplitude c_f (11) does not factorize into the product of single-electron amplitudes, indicating the presence of quantum correlations between the XUV- and Auger electrons. It is important to note that the phase coherence is the only source of electron correlations: no Coulomb or exchange interaction is assumed in the present model. A graphical representation of $c_f(\mathbf{k}_\mathrm{x}^2, \mathbf{k}_l^2)$ is given in Figure 3. The amplitude $c_f(\mathbf{k}_\mathrm{x}^2, \mathbf{k}_l^2)$ has its support on a 45°- slanted area of the $(\mathbf{k}_\mathrm{x}^2, \mathbf{k}_l^2)$-plane reflecting the correlations between

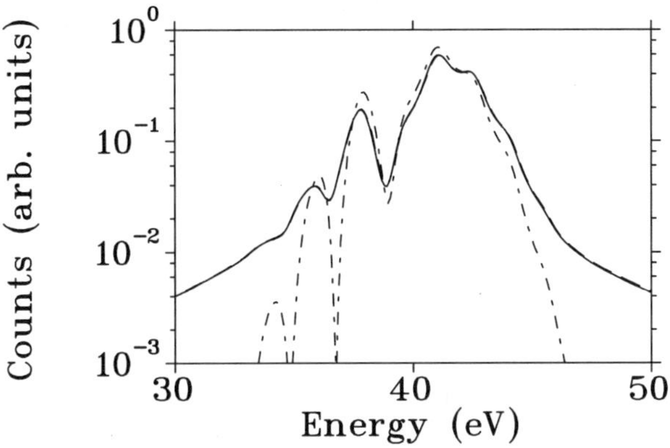

Fig. 2. Auger electron spectrum for a 500 as pulse and 500 as Auger life time. The full model (solid line) coincides with the spectrum using Eq. (13) within the resolution of the plot. The *ad hoc* formula used in Ref. [3] (dot-dashed) produces a narrower spectrum. The curves are normalized to have equal integral.

electron energies. If there were no correlation, the final amplitude would represent a rectangle in this diagram (Figure 3(c)). The degree of correlation is controlled by the parameter $\Gamma\tau_x$ and increases for $\Gamma\tau_x > 1$. Figure 3(a) corresponds to less correlated state ($\Gamma\tau_x << 1$), where the shape of the central (shaded) area can be approximated by a rectangle. On the contrary, for a highly correlated state ($\Gamma\tau_x >> 1$) the shape of the central area can not be approximated by a rectangle any more (Figure 3(b)).

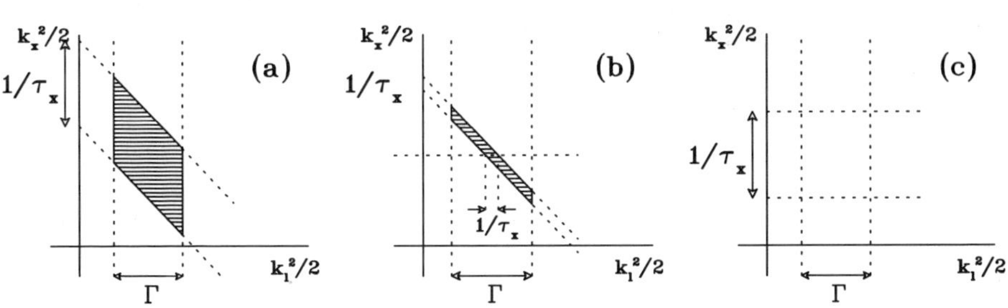

Fig. 3. Symbolic representation of the support of $c_f(\mathbf{k}_x^2, \mathbf{k}_l^2)$ (Eq.11) in the plane of independent electron energies $\mathbf{k}_x$ and $\mathbf{k}_l$. The degree of correlation is controlled by parameter $\Gamma\tau_x$, with the correlations increasing for $\Gamma\tau_x > 1$. (a) less correlated state $\Gamma\tau_x < 1$, (b) highly correlated state. An uncorrelated amplitude would have rectangular support (c).

The absence of $\mathbf{k}_x$-dependence in *ad hoc* formula for ρ removes all coherence between XUV ionization and Auger decay reducing the dynamics to a *two-step* process. The most conspicuous consequence is that the *ad hoc* formula underestimates the width of Auger spectra in the limit $\Gamma\tau_x > 1$

when the correlations are crucial. Omitting the $\mathbf{k}_x$-dependence of the phase of c_h amounts to the assumption that hole state has always the same central energy, irrespective of the photoelectron energy.

The difference between the quantum and the *ad hoc* model becomes important in the limit $\Gamma\tau_x >> 1$. In that limit the Auger spectrum at fixed $\mathbf{k}_x^2/2$ may be much narrower $\sim 1/\tau_x$ than the spectrum integrated over all $\mathbf{k}_x$ with width $\sim \Gamma$. In that case the *ad hoc* formula does not allow for correct reconstruction of temporal dynamics of Auger process from the Auger spectrogram.

The electron correlations may be exploited for refining the streaking measurement. When the photoelectron energy $\mathbf{k}_x^2/2$ is fixed, the range of possible Auger energies is reduced to to within the spectral width of the XUV pulse. In traditional spectroscopy, a similar effect is referred to as Raman narrowing [1]. In case of very fast Auger decay, where the large width Γ does not allow effective streaking of the whole Auger line, parts of the line can be selected in this way.

We are indebted to F. Krausz, M. Ivanov and J. Burgdörfer for numerous illuminating discussions. This work was supported by the Austrian Science Fund special research programs F011 and F016, and by project M718-N08.

References

1. G. B. Armen, H. Aksela, T. Aberg, S. Aksela. The resonant auger effect. *J. Phys. B*, 60:4667, 1999.
2. E. Constant, V. D. Taranukhin, A. Stolow, P. B. Corkum. Methods for the measurement of the duration of high-harmonic pulses. *Physical Review A*, 56:3870–3878, 1997.
3. M. Drescher, M. Hentschel, R. Kienberger, M. Uiberacker, V. Yakovlev, A. Scrinzi, Th. Westerwalbesloh, U. Kleineberg, U. Heinzmann, F. Krausz. Time-resolved atomic inner-shell spectroscopy. *Nature*, 419:803, 2002.
4. H. P. Saha. Near-threshold k-shell photoionization of argon. *Physical Review A*, 42:6507–6511, 1990.

Laser Induced Nuclear Reactions

Friederike Ewald[1], Heinrich Schwoerer[1], Joseph Magill[2], Jean Galy[2], Roland Schenkel[2], and Roland Sauerbrey[1]

[1] Institut für Optik und Quantenelektronik, Friedrich-Schiller-Universität Jena, Max-Wien-Platz 1, 07743 Jena, Germany
ewald@ioq.uni-jena.de

[2] European Commission, Joint Research Centre, Institut für Transurane, Postfach 2340, 76126 Karlsruhe, Germany

1 Introduction

In the last decade the intensities of light fields which can be produced in a laser focus increased by four orders of magnitude from 10^{16} to 10^{20} W/cm^2. Intensities exceeding 10^{18} W/cm^2 allow for the production of relativistic laser plasmas, that is the quiver energy of plasma electrons reaches the electron rest mass. These plasmas are sources of a whole spectrum of energetic particles, such as highly relativistic electrons, hard bremsstrahlung [13], protons with energies up to a few hundred MeV [7,14], neutrons [4,11,13] and deuterons [4, 17]. These particles can be used to induce nuclear reactions like photo-fission (γ,f) [3,6,8,12], neutron generation by (γ,n)- (p,n)- or (d,n)-reactions, neutron capture or fusion [4,17].

Up to now, nuclear reactions have mainly been used for measuring the temperatures of laser produced bremsstrahlung photons and electrons [8,15]. Nevertheless, various applications of laser induced nuclear reactions such as production of radioisotopes for medical applications increasingly come to the fore [5,9]. A very interesting application of laser generated bremsstrahlung is the investigation of photo-induced reaction cross-sections for (γ,n) and (γ,f)-reactions of long-lived actinides and fission products, which are both hazardous components of nuclear waste. Recently we have shown that – up to now unknown – (γ,n)-reaction cross sections of long lived fission products can be determined from measurements with a laser as γ-source. This topic is of particular interest in the context of transmutation of nuclear waste.

Considering photo induced fission (γ,f) of ^{232}Th and ^{238}U [12] as well as the (γ,n)-reaction of ^{181}Ta we show how nuclear reactions are used for the characterization of the bremsstrahlung spectrum from laser generated relativistic plasmas. With help of these results we determine the (γ,n)-cross section of ^{129}I.

2 The Jena High-Intensity Laser System

The nuclear activation experiments described herein were realised with the multi Tera Watt Ti:sapphire laser system in Jena (Germany). This tabletop laser reaches focused intensities of up to 10^{20} W/cm^2. Generating such high intensities necessitates a sophisticated laser technology as well as reliable laser diagnostics.

The laser is based on chirped pulse amplification. A fs-Ti:sapphire oscillator, delivering 45 fs, seeds the successive amplifier stages, as shown in fig. 1. The pulses are streched by a factor to 150 ps and amplified by a regenerative

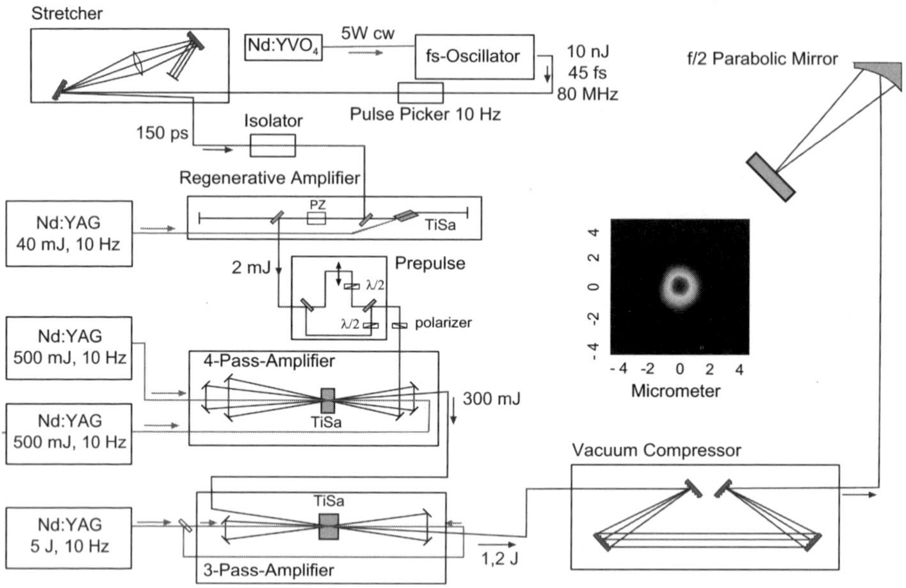

Fig. 1. Schematic illustration of the Jena multi Tera Watt Lasersystem. Additionally the spacial profile of the focused beam is shown.

as well as by two multi-pass amplifiers to a pulse energy of 1.3 J. Recompression to a final pulse duration of less than 80 fs FWHM is accomplished by a four grating vacuum compressor. The laser is run at a wavelength of $\lambda = 795$ nm with a repetition rate of 10 Hz. The whole system is a tabletop system, as small as 10 m $\times$ 3 m.

The laser pulses are focused with a f/2 45° off-axis parabolic mirror. For reliable determination of the focused laser intensity, the pulse duration τ, the pulse energy $E_{1/e}$ and the focus area $A_{1/e}$ have to be measured accurately, since the intensity is given by $I_{1/e} = E_{1/e}/(\tau A_{1/e})$. The index 1/e indicates that we use the mean value of the energy contained in the 1/e-area of the focal spot, resulting in a mean intensity in this area. The focal 1/e-area is directly obtained by imaging the focus by means of a microscope objective onto a CCD-camera. The spacial beam profile in the focal spot is characterized over 5 orders of magnitude using filters for beam attenuation (fig. 1). The measured 1/e-diameter is 2.5 μm which corresponds to an area of 5 μm^2. The diffraction limited focal spot diameter of the corresponding gaussian beam would be 1.4 μm. The absolute energy content in this area is determined from the energy transmitted through a 50 μm pinhole in a full power shot at the focus position as well as from the spacial intensity distribution of the focal spot. In total, the energy content in the focal 1/e-area is determined to 250 mJ, where the main losses are due to the reflectivity of the compressor gratings and to some small diffraction losses of the focusing optics.

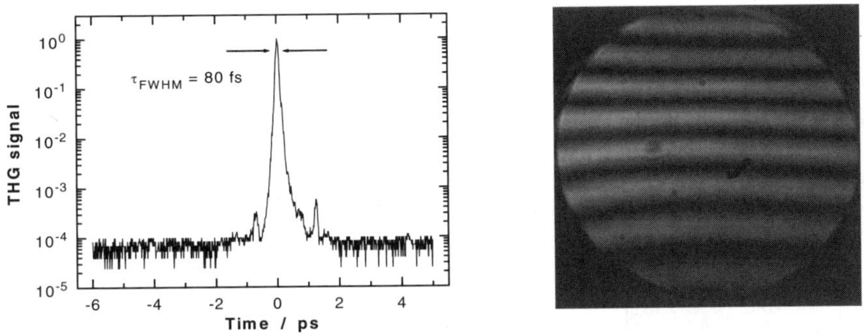

Fig. 2. Lefthand side: Third-order autocorrelation of the amplified laser pulse. The FWHM in this case is 80 fs. Righthand side: Interferogram of the non-tilted pulsefront. See text for explanation.

The temporal pulse structure is measured with a multi-shot third order autocorrelator which is positioned in the vacuum beam line close to the experiment. The dynamics of this measurement is about four orders of magnitude as can be seen in figure 2. Only one small prepulse at 600 fs can be identified within this dynamic range. The FWHM is measured to be less than 80 fs.

Since the pulse length has to be as short as possible at the focal spot, the pulse front has to be perpendicular to the propagation direction of the light pulse. This tilt of the pulse with regard to its propagation direction is measured interferometrically and corrected by adjusting the compressor gratings [10]. Figure 2 shows the interferogram of the non-tilted pulse front. A tilt would appear in the form of fringes which do not expand over the whole pulse diameter. The influence of a residual pulse front tilt on the effective pulse duration in the focal spot is determined to be less then a factor of 1.2.

Therewith the mean 1/e-intensity generated in the focal spot is $I_{1/e} = 6 \times 10^{19}\,\mathrm{W/cm^2}$ with an accuracy of the measurement of 50%.

3 Laser Induced Nuclear Reactions

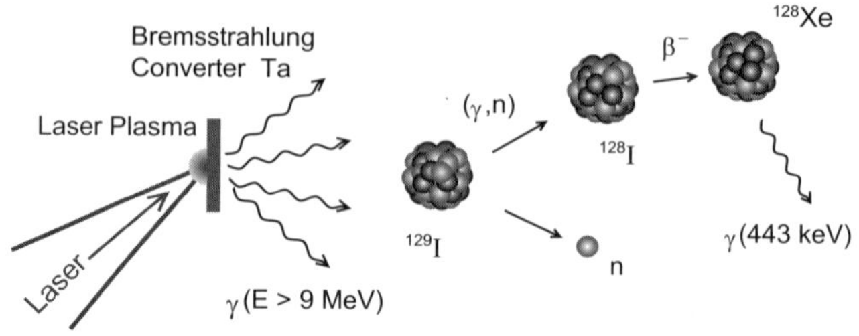

Fig. 3. Sketch of the experimental setup with the (γ,n)-reaction in ^{129}I as example.

The setup of the experiment is shown schematically in figure 3. The laser parameters are the same as described in 2. The intense laser pulse generates a plasma on the surface of a tantalum target. Simultaneously this light pulse accelerates plasma electrons up to energies which exceed the rest mass energy by one to two orders of magnitude. These suprathermal plasma electrons penetrate the solid phase of the target, where they generate highly energetic bremsstrahlung radiation. The energy distribution of the suprathermal electron fraction is commonly assumed to be a Boltzmann distribution with a temperature of about 2 MeV which is determined by the laser intensity of $5 \times 10^{19}\,\mathrm{W/cm^2}$ [16]. The resulting bremsstrahlung photons obey the same distribution function with a slightly lower temperature of $T_\gamma \approx 1.2\,\mathrm{MeV}$ [2]. The (γ,n)-reaction in Ta takes place inside the Ta converter itself. To induce nuclear reactions in the actinide samples ^{232}Th and ^{238}U, these samples were placed in this intense bremsstrahlung field, directly behind the tantalum sheet. About 10^4 laser shots were applied for each sample to accumulate photons and to enhance the signal to noise ratio of the measurement.

Nuclear reactions are detected via the radioactive decay of the reaction products. For (γ,f)-reactions these nuclei are fission products from ^{232}Th or ^{238}U, such as ^{134}I, ^{92}Sr or ^{138}Cs. The (γ,n)-reaction in ^{181}Ta is identified by the decay of ^{180}Ta. The characteristic γ- or x-ray radiation which is emitted during the decay process is measured by a Ge-semiconductor detector.

The number of reactions induced by one laser shot is deduced from the activity of the appropriate reaction product, the detector efficiency and the intensity of the observed line. In the case of fission the fission yield, which is the number of nuclei of a certain isotope produced by the fission reaction has additionally to be considered. Evaluation of the activation data gives numbers of 100, 130 and 160 nuclear reactions for ^{232}Th, ^{238}U and ^{181}Ta respectively, which could be induced with one single laser shot.

Since photo-induced nuclear reactions exhibit discrete energy thresholds below which the reaction cross section is equal to zero, activation experiments are suited for the determination of bremsstrahlung spectra from laser-matter interactions [3, 6, 8, 15]. From the measured number of reactions, the photon-temperature T_γ and the known reaction cross section $\sigma(E)$ the Boltzmann energy distribution of the bremsstrahlung photons is given.

From these values the number of photons with energies above for example 10 MeV can be calculated by considering the target characteristics and the cross section $\sigma(E)$. These energy dependent cross sections of the observed nuclear reactions are Lorentzian like [1] with maximum cross sections of a few tens to a few hundreds of mbarn at the Giant Resonance energy. The reaction thresholds for photofission of ^{232}Th and ^{238}U lie at 6 MeV and 5.8 MeV, respectively. The (γ,n)-reaction threshold of ^{181}Ta lies somewhat higher at 7.58 MeV. This leads to numbers of γ-photons with energies exceeding 10 MeV of 8×10^4 for ^{232}Th, 2×10^5 for ^{238}U and 2×10^4 for the ^{181}Ta measurement. The discrepancies between these numbers is due to slightly different experimental conditions under which the three experiments were performed. These photons in the energy range of about 10 MeV and higher are just the photons for which the cross sections of nuclear reactions like (γ,f) and (γ,n) reach the giant resonance maximum.

In the following we consider an experiment in which the photon distribution is known from TLD spectrometers [2] and from (γ,n)-activation of Ta as described above. In this case an unknown cross section of a photo induced reaction can be determined from the measured number of induced reactions. Recently we could demonstrate this procedure by means of the (γ,n)-reaction in ^{129}I.

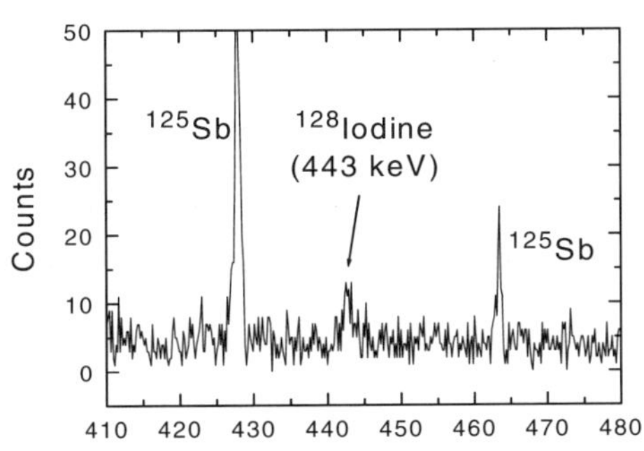

Fig. 4. γ-spectrum of the irradiated ^{129}I sample. The 443 keV-line which belongs to ^{128}I is the signature for the (γ,n)-reaction in ^{129}I. The ^{125}Sb-lines come from a residual contamination of the sample with this isotope.

The ^{129}I-sample was placed into the bremsstrahlung field behind the Ta sheet. By the (γ,n)-reaction the isotope ^{128}I is produced which decays within

25 min mainly into the stable isotope ^{128}Xe. Thereby a 443 keV-photon is emitted. Figure 4 shows a γ-spectrum which was accumulated during half an hour after the irradiation. In between two background lines from the sample, the 443 keV-line from iodine is visible. This signal leads to a number of 2 (γ,n)-reactions which have been induced per laser shot. Assuming that the shape of the energy dependent (γ,n)-cross section of ^{129}I is not very different from ^{127}I and knowing the threshold energy $E_{\mathrm{th}} = 8.8\,\mathrm{MeV}$ from giant dipole resonance models [1], the giant resonance maximum of the cross section has been determined to be $\sigma(E_{\max}) = (220^{+230}_{-100})$ mbarn.

This result demonstrates a first step towards a new application of high intensity laser plasmas: we are investigating the possibilities of obtainig nuclear data, such as different reaction cross sections, using high intensity laser plasmas. Since these plasmas are sources of a broad spectrum of emitted particles in the MeV range, different types of photo- and particle-induced reactions can be investigated with the same device. Furthermore the tabletop laser has the advantage of being a very compact particle source compared to conventional particle accelerators.

References

1. *Handbook on photonuclear data for applications – cross sections and spectra*, IAEA-TECDOC (2000).
2. R. Behrens, H. Schwoerer, S. Düsterer, P. Ambrosi, G. Pretzler, S. Karsch, and R. Sauerbrey, *A TLD-based few-channel spectrometer for simultaneous detection of electrons an photons from relativistic laser-produced plasmas*, Review of Scientific Instruments , accepted **74** (2003), no. 2, 961–968.
3. T. E. Cowan, A. W. Hunt, T. W. Phillips, S. C. Wilks, M. D. Perry, C. Brown, W. Fountain, S. Hatchett, J. Johnson, M. H. Key, T. Parnall, D. M. Pennington, R. A. Snavely, and Y. Takahashi, *Photonuclear fission from high energy electrons from ultraintensive laser-solid interractions*, Physical Review Letters **84** (2000), 903.
4. S. Karsch, S. Düsterer, H. Schwoerer, F. Ewald, D. Habs, M. Hegelich, G. Pretzler, A. Pukhov, K. Witte, and R. Sauerbrey, *High-intensity laser induced ion acceleration from heavy-water droplets*, Physical Review Letters, in press . (.), .
5. K. W. D. Ledingham, P. McKenna, and R. P. Singhal, *Applications for nuclear phenomena generated by ultra-intense lasers*, Science **300** (2003), 1107 – 1111.
6. K. W. D. Ledingham, I. Spencer, T. McCanny, R. P. Singhal, M. I. K. Santala, E. Clark, I. Watts, F. N. Beg, M. Zepf, K. Krushelnick, M. Tatarakis, A. E. Dangor, P. A. Norreys, R. Allot, D. Neely, R. J. Clark, A. C. Machacek, J. S. Wark, A. J. Cresswell, D. C. W. Sanderson, and J. Magill, *Photonuclear physics when a multiterawatt laser pulse interacts with solid targets*, Physical Review Letters **84** (2000), 899.
7. A. J. Mackinnon, Y. Sentoku, P. K. Patel, D. W. Price, S. Hatchett, M. H. Key, C. Andersen, R. Snavely, and R. R. Freeman, *Enhancement of proton acceleration by hot-electron recirculation in thin foils irradiated by ultraintense laser pulses*, Phys. Rev. Lett. **88** (2002), no. 21, 215006.

8. G. Malka, M. M. Aleonard, J. F. Chemin, G. Claverie, M. R. Harston, J. N. Scheurer, V. Tikhonchuk, S. Fritzler, V. Malka, P. Balcou, G. Grillon, S. Moustaizis, L. Notebaert, E. Lefebvre, and N. Cochet, *Relativistic electron generation in interactions of a 30 TW laser pulse with a thin foil target*, Physical Review E **66** (2002), 066402.
9. K. Nemoto, A. Maksimchuk, S. Banerjee, K. Flippo, G. Mourou, D. Umstadter, and V. Yu. bychenkov, *Laser-triggered ion acceleration and table top isotope production*, Appl. Phys. Lett. **78** (2001), no. 5, 595–597.
10. G. Pretzler, A. Kasper, and K. J. Witte, *Angular chirp and tilted light pulses in cpa lasers*, Appl. Phys. B (1999), DOI: 10.1007/s003409900076.
11. G. Pretzler, A. Saemann, A. Pukhov, D. Rudolph, T. Schätz, U. Schramm, P. Thirolf, D. Habs, K. Eidmann, G. D. Tsakiris, J. Meyer-ter Vehn, and K. J. Witte, *Neutron production by 200 mJ ultrashort laser pulses*, Phys. Rev. E **58** (1998), 1165–1168.
12. H. Schwoerer, F. Ewald, R. Sauerbrey, J. Galy, J. Magill, V. Rondinella, R. Schenkel, and T. Butz, *Fission of actinides using a tabletop laser*, Europhs. Lett. **61** (2003), 47.
13. H. Schwoerer, P. Gibbon, S. Düsterer, R. Behrens, C. Ziener, C. Reich, and R. Sauerbrey, *MeV x rays and photoneutrons from femtosecond laser-produced plasmas*, Physical Review Letters **86** (2001), 2317–2320.
14. R. A. Snavely, M. H. Key, S. P. Hatchett, T. E. Cowan, M. Roth, T. W. Phillips, M. A. Stoyer, E. A. Henry, T. C. Sangster, M. S. Singh, S. C. Wilks, A. MacKinnon, A. Offenberger, D. M. Pennington, K. Yasuike, A. B. Langdon, B. F. Lasnski, J. Johnson, M. D. Perry, and E. M. Campbell, *Intense high-energy proton beams from petawatt-laser irradiation of solids*, Physical Review Letters **85** (2000), no. 14, 2945 – 2348.
15. I. Spencer, K. W. D. Ledingham, R. P. Singhal, T. McCanny, P. McKenna, E. L. Clark, K. Krushelnick, M. Zeph, F. N. Beg, M. Tatarakis, A. E. Dangor, R. D. Edwards, M. A. Sinclair, P. A. Norreys, R. J. Clarke, and R. M. Allot, *A nearly real-time high temperature laser-plasma diagnostic using photonuclear reactions in tantalum*, Review of Scientific Instruments **73** (2002), 3801 – 3805.
16. S. C. Wilks, W. L. Kruer, M. Tabak, and A. B. Langdon, *Absorption of ultra-intense laser pulses*, Physical Review Letters **69** (1992), 1383.
17. J. Zweiback, R. A. Smith, T. E. Cowan, G. Hays, K. B. Wharton, V. P. Yanovsky, and T. Ditmire, *Nuclear fusion driven by coulomb explosions of large deuterium clusters*, Phys. Rev. Lett **84** (2000), no. 12, 2634–2637.

Springer Series in

OPTICAL SCIENCES

86 **Progress in Nano-Electro-Optics I**
Basics and Theory of Near-Field Optics
By M. Ohtsu (Ed.), 2003, 118 figs., XIV, 161 pages

87 **Optical Imaging and Microscopy**
Techniques and Advanced Systems
By P. Török, F.-J. Kao (Eds.), 2003, 260 figs., XVII, 395 pages

88 **Optical Interference Coatings**
By N. Kaiser, H.K. Pulker (Eds.), 2003, 203 figs., 50 tabs., XVI, 504 pages

89 **Progress in Nano-Electro-Optics II**
Novel Devices and Atom Manipulation
By M. Ohtsu (Ed.), 2003, 115 figs., XIII, 188 pages

90/1 **Raman Amplifiers for Telecommunications 1**
Physical Principles
By M.N. Islam (Ed.), 2004, 488 figs., XXVIII, 328 pages

90/2 **Raman Amplifiers for Telecommunications 2**
Sub-Systems and Systems
By M.N. Islam (Ed.), 2004, 278 figs., XXVIII, 420 pages

91 **Optical Super Resolution**
By Z. Zalevsky, D. Mendlovic, 2004, 164 figs., XVIII, 232 pages

92 **UV-Visible Reflection Spectroscopy of Liquids**
By J.A. Räty, K.-E. Peiponen, T. Asakura, 2004, 131 figs., XII, 219 pages

93 **Fundamentals of Semiconductor Lasers**
By T. Numai, 2004, 166 figs., XII, 264 pages

94 **Photonic Crystals**
Physics, Fabrication and Applications
By K. Inoue, K. Ohtaka (Eds.), 2004, 214 figs., IX, 331 pages

95 **Ultrafast Optics IV**
Selected Contributions to the 4th International Conference
on Ultrafast Optics, Vienna, Austria
By F. Krausz, G. Korn, P. Corkum, I.A. Walmsley (Eds.), 2004, 281 figs., XIV, 506 pages

96 **Progress in Nano-Electro Optics III**
Industrial Applications and Dynamics of the Nano-Optical System
By M. Ohtsu (Ed.), 2004, 155 figs., XIV, 226 pages

97 **Microoptics**
From Technology to Applications
By J. Jahns, K.-H. Brenner, 2004, 303 figs., XI, 335 pages